Electromagnetic Waves *and* Transmission Lines

R.S. RAO
Professor
Department of Electronics and Communication Engineering
Sree Vidyanikethan Engineering College
Tirupati, Andhra Pradesh

PHI Learning Private Limited
New Delhi-110001
2012

₹425.00

ELECTROMAGNETIC WAVES AND TRANSMISSION LINES
R.S. Rao

ISBN-978-81-203-4515-7

Published by Asoke K. Ghosh, PHI Learning Private Limited, M-97, Connaught Circus, New Delhi-110001 and Printed by Raj Press, New Delhi-110012.

To

The Lotus Feet of

Bhagwan Gautama, The Buddha

The Symbol of

Sympathetic Joy and

Compassionate Love

Contents

UNIT II ELECTROSTATICS

UNIT III STEADY MAGNETIC FIELDS

UNIT IV ELECTRODYNAMICS

UNIT V WAVE THEORY

Preface

Electromagnetics is one of the oldest and fully developed subjects. Still it is considered, by the students at large, the most difficult subject. Keeping this view in mind, the present book is written in an easy-to-understand manner, without missing the essence of the subject matter.

This book is an outgrowth of my long teaching experience at various institutes. It is intended as a textbook for the undergraduate students of electronics and electrical engineering, and electronics and communication engineering for a course on electromagnetic waves and transmission lines.

A unique feature of this book is that it covers all the three major areas of the subject, namely, fields, waves and lines, giving equitable importance and integrated approach. The topics included and discussed in the book are selected in such a way that they serve as a true text to the basic communication engineering subjects such as antennas, microwave engineering and so on. Another distinguishing feature of the book is the prolific use of curvilinear coordinate systems, which helps students in two different ways. First, the procedures and computations in the solutions of several problems become easy and second, the command over these coordinate systems helps in dealing with antenna theory that uses mostly and primarily spherical coordinate system and microwave engineering which requires cylindrical and spherical coordinate systems for analysis.

The book is organized in 13 chapters. Chapter 1 discusses vector calculus which is the language of field theory. Chapters 2 to 5 deal with electrostatics, an area which has plenty of applications. Steady magnetic fields are discussed in Chapter 6. Maxwell's equations that describe the time-varying fields are thoroughly dealt with in Chapter 7. The wave theory which is the basis of radiation and, consequently, the wireless communications is given due importance in the next three chapters (Chapters 8 to 10). Finally, the last three chapters (Chapters 11 to 13) focus on transmission line theory which is a foundation course in microwave engineering.

The book is supplemented by the inclusion of a large number of solved and unsolved problems, which will enhance the analytical skill of the students. The problems are framed to test the conceptual understanding of the students.

In the present times, it has become a common feature in almost all academic examinations of the Indian universities to test the students with objective type questions along with essay type questions and problems. This book contains innumerable number of objective type questions in different types, such as matching type, fill-in-the-blank type and multiple-choice type. These objective type questions have been framed in such a way that, besides being useful to the students, they will be equally helpful to the job seekers appearing in UPSC, DRDO and such other competitive examinations.

Although care has been taken to provide an error-free book, some errors are likely to be there. Positive suggestions from the students and teachers towards the improvement of the book will be gratefully acknowledged.

I hope the book will be highly useful to the students and practising engineers at large.

R.S. Rao

Acknowledgements

First and foremost, I would like to express my sincere thanks and gratefulness to our beloved chairman Dr. M. Mohan Babu and Professor T. Gopala Rao, Special Officer, Sree Vidyanikethan Educational Trust (SVET) for giving me an opportunity and necessary facilities in the campus to complete this work.

I am also thankful to Dr. Krishnamachari, Principal, Sree Vidyanikethan Engineering College (SVEC) and Dr. C. Subhas, HOD, Electronics and Communication Engineering, SVEC for their constant support and encouragement during the progress of the work.

I would also like to thank my colleagues Mr. P.V. Raman and Dr. Anitha, Professors, Department of Electronics and Communication Engineering; Mr. Girish and Dr. Polaiah, Professors, Department of Electronics and Instrumentation Engineering; and Dr. Deva Raj, Professor, Department of Electrical and Electronics Engineering for their suggestions and advice for improvements during the preparation of the manuscript.

On this occasion, I would also like to remember the kindness of Mr. Umamaheswara Rao, Administrative Officer, SVEC and Mr. Tulasi Naidu, Administrative Officer, Sree Vidyanikethan International School (SVIS) in providing the accommodation and necessary conducive atmosphere for the completion of the work successfully.

It is also my primary duty to remember and thank Smt. Sujatha, DTP operator, Department of Electronics and Communication Engineering for her constant cooperation and help in typing the manuscript and in the preparation of diagrams.

I would like to thank my students of past, present and future for their love and respect towards me and also for making me feel grow younger day by day.

I would also like to express my heartfelt deep sense of gratefulness to the editorial and production teams of PHI Learning, particularly to Smt. Shivani Garg and Sri Mayur Joseph for their constant encouragement and suggestions during the process of producing the book.

I am failing my duty if I do not thank Alekha Chandra Jena for his marvellous editing work.

Last but not least, I would also like to put on record the help rendered by Sri Balasubramaniam and Sri Chaitanya, both PHI Marketing Executives, for making this project successful and joyful.

R.S. Rao

UNIT I

VECTOR CALCULUS

1

Vector Calculus

Failure is success if we learn from it.
—Malcoln Forbes

Electromagnetic field theory has a language of its own: vector calculus. The tool of vector calculus is a must whether to understand or to analyse the electromagnetic phenomenon. The electric and magnetic field intensities and flux densities, which we are going to deal with, are all vector quantities.

We will come across many situations in the coming chapters where we need to use several vector identities for the solution of field problems. The relations that we need essentially are given as triple products, product rules and second derivatives without proofs.

To analyse a phenomenon that takes place in space, we need to have mathematical tools that can deal with three dimensions and if it is a time varying one, then one more, that is time, is also required. The del operators, Laplacian operators, surface integrals and volume integrals are some of such tools useful in field theory. All these operators along with some more are introduced and discussed.

There are relations connecting line, surface and volume integrals known as fundamental theorems of vector calculus. These theorems are so widely used that usually they are considered as an integral part of field theory. These are covered in detail.

Several problems in field theory exhibit some kind of symmetry and their solutions become simple and easy when dealt with respect to a reference coordinate system like spherical or cylindrical system. The fundamentals pertaining to these two coordinate systems are discussed in the present chapter under curvilinear coordinate systems.

However, here we concentrate only on the aspects of vector calculus or coordinate systems that are necessarily required in the study of the electromagnetic phenomenon. Derivations or proofs of the laws and formulae are avoided here and the interested reader may refer to a standard textbook for engineering mathematics for an exhaustive treatment of the topics.

1.1 VECTOR ALGEBRA

1.1.1 Definitions

Vector is a quantity having both magnitude and direction. Displacement, velocity and acceleration are known examples of vectors. Field intensity and flux density are vectors about which we are going to discuss soon. On the other hand, we have quantities like mass, charge and density that have magnitude but no direction. We denote vector quantities with **boldface** letters and their magnitude by ordinary *italic* letters. We use ordinary letters for scalars. Now we consider some of the important features of vectors which we use repeatedly in coming chapters.

Unit vector is a vector with a magnitude of unity and a unit vector along a vector, let us say **A**, is given by

$$\hat{\mathbf{a}} = \frac{\mathbf{A}}{A} \tag{1.1}$$

Vectors have directions and two vectors may subtend an angle. The angle between two vectors **A** and **B** can be found from

$$\alpha = \cos^{-1}\frac{\mathbf{A}\cdot\mathbf{B}}{AB} \tag{1.2}$$

It is always possible to have a plane with directions of two different vectors. The unit normal to the plane or to the two vectors is given by

$$\hat{\mathbf{n}} = \frac{\mathbf{A}\times\mathbf{B}}{|\mathbf{A}\times\mathbf{B}|} = \frac{\mathbf{A}\times\mathbf{B}}{AB\sin\alpha} \tag{1.3}$$

1.1.2 Multiplication of Vectors

Two vectors can get multiplied in two different ways: one is scalar or dot multiplication and the other is vector or cross multiplication. The first type results in a scalar quantity, whereas the second one results in vector quantity.

The dot product of two vectors is, by definition, equal to the product of the magnitudes of the vectors multiplied by the cosine of the angle between them. The dot product of two vectors **A** and **B** with an angle of α in between is given by

$$\mathbf{A}\cdot\mathbf{B} = AB\cos\alpha \tag{1.4}$$

The cross product, on the other hand, is a vector whose magnitude is equal to the product of the magnitudes of the vectors multiplied by the sine of the angle between them with direction normal to both the vectors. In finding the direction, the right-hand rule can be applied. The cross product of two vectors **A** and **B** with an angle of α in between is given by

$$\mathbf{A}\times\mathbf{B} = AB\sin\alpha\,\hat{\mathbf{n}} \tag{1.5}$$

Here $\hat{\mathbf{n}}$ is the normal to both the vectors.

Three vectors can be multiplied in two different ways and the resultant products, usually called triple products, are given by

$$\mathbf{A}\cdot(\mathbf{B}\times\mathbf{C})=\mathbf{B}\cdot(\mathbf{C}\times\mathbf{A})=\mathbf{C}\cdot(\mathbf{A}\times\mathbf{B}) \tag{1.6}$$

The above product among vectors is called scalar triple product and it can be assigned a geometrical interpretation: its absolute value gives the volume of the parallelepiped having the three vectors as three of its contiguous edges

$$\mathbf{A}\times(\mathbf{B}\times\mathbf{C})=\mathbf{B}(\mathbf{C}\cdot\mathbf{A})-\mathbf{C}(\mathbf{A}\cdot\mathbf{B}) \tag{1.7}$$

The above product is called triple cross product. Note that $\mathbf{A}\times(\mathbf{B}\times\mathbf{C})$ is not, in general, equal to $(\mathbf{A}\times\mathbf{B})\times\mathbf{C}$.

1.1.3 Direction Cosines

The component of a vector in a given direction is the projection of that vector along that direction. If A_x is x-component of vector **A**, then it is equal to $A\cos\alpha$, where α is the angle between the vector and x-direction. But from the definition of dot product, $A\cos\alpha=\mathbf{A}\cdot\hat{\mathbf{x}}$. Therefore, we can say that the component of a vector in a given direction is equal to the dot product of the vector and unit vector in that direction.

If α, β and γ are the angles made by a vector with the coordinate axes, then $\cos\alpha$, $\cos\beta$ and $\cos\gamma$ are known as the direction cosines of the vector.

1.2 DIFFERENTIAL CALCULUS

1.2.1 Differential Operators

Del and Laplacian are important three-dimensional operators widely used in field theory. Apart from these two, we also introduce some more like d'Alembertian and Helmholtz.

Operator del: Its symbol is '∇'. By definition, it is

$$\nabla=\frac{\partial}{\partial x}\hat{\mathbf{x}}+\frac{\partial}{\partial y}\hat{\mathbf{y}}+\frac{\partial}{\partial z}\hat{\mathbf{z}} \tag{1.8a}$$

It is a three-dimensional, partial differential vector operator defined in Cartesian system only. But it can be mapped into other coordinate systems. Its units are $(\text{mt})^{-1}$.

This operator can be applied over a scalar function to find its gradient, and over a vector function to find either its divergence or curl.

When applied to a position vector '$r\hat{\mathbf{r}}$' joining origin or (x_1, y_1, z_1) with (x, y, z), then

1. $$\nabla r=\hat{\mathbf{r}} \tag{1.8b}$$

2. $$\nabla\left(\frac{1}{r}\right)=-\frac{\hat{\mathbf{r}}}{r^2} \tag{1.8c}$$

3. $$\nabla\cdot(r^n\hat{\mathbf{r}}) = (n+2)r^n, \quad n \neq -2 \tag{1.8d}$$

$$= 4\pi\delta^3(r), \quad n=-2 \tag{1.8e}$$

4. $$\nabla\times(r^n\hat{\mathbf{r}}) = 0 \tag{1.8f}$$

The position vector is so called because it defines the position of the point in space relative to the origin.

Standard configuration with source and observation points is depicted in Figure 1.1.

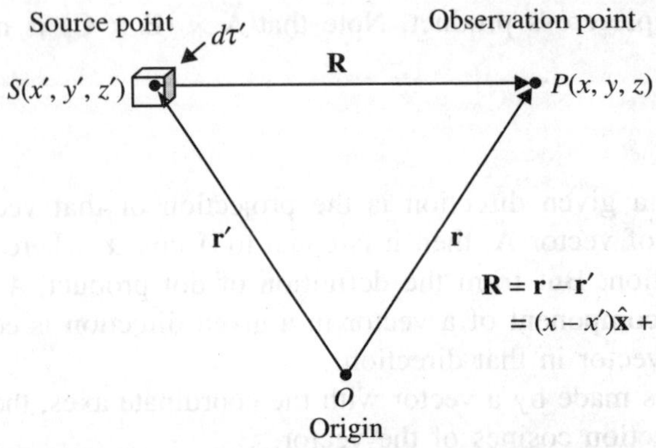

FIGURE 1.1 Standard configuration with source and observation points.

Operator Laplacian

$$\nabla^2 = \frac{\partial^2}{\partial x^2} + \frac{\partial^2}{\partial y^2} + \frac{\partial^2}{\partial z^2} \tag{1.9a}$$

- It is a three-dimensional, second-order, partial differential scalar operator defined in Cartesian system only. But it can be mapped into other coordinate systems. Its units are $(\text{mt})^{-2}$.
- This operator can be applied over a scalar function as well as a vector function.

When V is a scalar function, its Laplacian is defined as

$$\nabla^2 V = \nabla\cdot(\nabla V) = \frac{\partial^2 V}{\partial x^2} + \frac{\partial^2 V}{\partial y^2} + \frac{\partial^2 V}{\partial z^2} \tag{1.9b}$$

If **E** is a vector function, its Laplacian is defined as

$$\nabla^2\mathbf{E} = \left(\frac{\partial^2}{\partial x^2} + \frac{\partial^2}{\partial y^2} + \frac{\partial^2}{\partial z^2}\right)\mathbf{E} \tag{1.9c}$$

$$= \hat{\mathbf{x}}\nabla^2 E_x + \hat{\mathbf{y}}\nabla^2 E_y + \hat{\mathbf{z}}\nabla^2 E_z$$

Thus, in Cartesian coordinate system, the Laplacian of a vector is a vector whose components are equal to the Laplacian of the vector components.

It can be shown that

$$\nabla^2 \mathbf{E} = \nabla(\nabla \cdot \mathbf{E}) - \nabla \times (\nabla \times \mathbf{E}) \tag{1.9d}$$

Two operators connected to del and worth mentioning here are d'Alembertian and Helmholtz operators. The operator d'Alembertian is given by

$$\Box^2 = \nabla^2 - \mu_0 \varepsilon_0 \frac{\partial^2}{\partial t^2} \tag{1.10a}$$

and Helmholtz operator by

$$(\nabla \cdot \nabla + k^2) \tag{1.10b}$$

Both of them are three-dimensional partial differential operators. The first is normally applied upon scalar functions.

1.2.2 Gradient, Divergence and Curl

The del operator can be applied over various types of functions giving different results. When applied over a scalar function, the del operator gives gradient of the function. The del operator can be applied over a vector function in two different ways: one results in divergence and the other the curl of the vector function. We discuss each of them one by one.

Definition of gradient: It is a result of the application of del operator over a scalar function giving a vector whose magnitude is equal to the maximum rate of the change of the scalar function per unit distance and whose direction is along the direction of its maximum increase.

If f is a scalar function, then the gradient of f is indicated as ∇f. According to a theorem on partial derivatives,

$$df = \nabla f \cdot d\mathbf{l} \tag{1.11}$$

The above relation is used in the coming chapters.

Definition of curl: The curl of a vector field **A** is defined as the maximum possible value of circulation of **A** per unit area in the limit of the area shrinking to a point. Mathematically

$$\nabla \times \mathbf{A} = \lim_{\Delta \mathbf{a} \to 0} \left[\frac{\oint_C \mathbf{A} \cdot d\mathbf{l}}{\Delta \mathbf{a}} \right]_{\max n} \hat{\mathbf{n}} \tag{1.12}$$

The direction of the curl is the direction of the normal vector to the area in the limit that the area shrinks to the point and in the right-hand sense.

The curl of a vector field describes the rotational property or the circulation of that field. In fact, it is a measure of how much the vector curls around the point in question. If the curl of a vector field is non-zero, it does not necessarily mean that the field lines are curling. Likewise, if the curl is zero, it does not mean that the field lines are not curling. In fact, the curl of the vector field is a measure of the ability of the vector field to cause rotator motion of a device such as the paddle wheel, if it is a mechanical force field. For uniform fields circulation is zero.

If a vector function **F** is curl-less, i.e. $\nabla \times \mathbf{F} = 0$, it can be shown that in the case of curl-less fields all the following relations hold good and are also equivalent. If $\nabla \times \mathbf{F} = 0$ then it can be expressed as the gradient of a scalar V, i.e. $\nabla \times \mathbf{F} = 0 \leftrightarrow \mathbf{F} = \nabla V$.

1. $\mathbf{F} = \nabla V$
   From differential calculus, a curl-less vector function can always be expressed as gradient of some scalar function, i.e. $\nabla \times \mathbf{F} = 0 \leftrightarrow \mathbf{F} = \nabla V$
2. $\int_a^b \mathbf{F} \cdot d\mathbf{l}$ is independent of path for any given end points.
3. $\oint \mathbf{F} \cdot d\mathbf{l} = 0$ for any closed loop.
   The above two are from the gradient theorem.

Definition of divergence: The divergence of a vector field **A** is defined as the outward flux of **A** per unit volume in the limit that the volume shrinks to the point.

$$\nabla \cdot \mathbf{A} = \lim_{\Delta\tau \to 0} \left[\frac{\oint_s \mathbf{A} \cdot d\mathbf{a}}{\Delta\tau} \right] \tag{1.13}$$

The divergence of a vector field is a measure of how much the vector spreads out or diverges from the point in question.

If the divergence of a vector field is non-zero, it does not necessarily mean that the field lines are diverging. An example of such a vector function is

$$\mathbf{F} = e^{-x^2}\,\hat{\mathbf{x}}$$

It has non-zero divergence of $-2e^{-x^2}$ but its field lines do not diverge. In addition, if the divergence of a vector field is zero, it does not necessarily mean that the field lines are not diverging. As an example of such kind of behaviour is the vector function

$$\mathbf{F} = \frac{\hat{\mathbf{r}}}{r}$$

If divergence of a vector is zero, its field lines spread out. However, in general, the field lines of vector functions with non-zero divergence exhibit spreading out and as an example, we can consider the vector function

$$\mathbf{F} = r\hat{\mathbf{r}}$$

In fact, instead of considering the divergence simply in terms of spreading out or not of the flux lines, it is required to view it as a measure of the effort or ability of the vector field to cause expansion or contraction of a device such as the imaginary balloon, if it is a mechanical force field. Different vector fields are shown in Figure 1.2.

Note that the divergence is a differential operation that can work only on vector functions resulting in scalar functions. The divergence-less fields, i.e. $\nabla \cdot \mathbf{F} = 0$, are called *solenoidal fields*. From differential calculus, if a vector **F** is divergence-less i.e. $\nabla \cdot \mathbf{F} = 0$, then it can be expressed as the curl of another vector **A**.

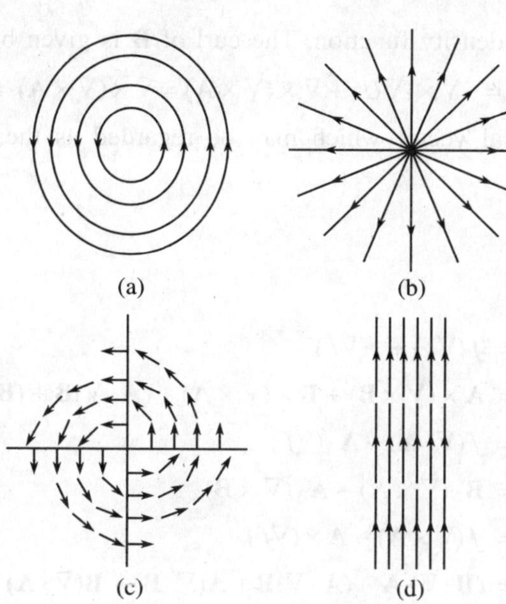

FIGURE 1.2 Vector fields: (a) curl but no divergence, (b) only divergent but no curl, (c) divergence plus curl and (d) divergence-less and curl-less.

Therefore, the following relations hold good for any divergence-less field:

1. $\nabla \cdot \mathbf{F} = 0$ everywhere.
2. $\mathbf{F}$ is curl of some vector function $\mathbf{A}$, so $\mathbf{F} = \nabla \times \mathbf{A}$.
3. $\int \mathbf{F} \cdot d\mathbf{a}$ is independent of surface for any given boundary line.
4. $\oint \mathbf{F} \cdot d\mathbf{a}$ for any closed surface.

However, note that here the vector $\mathbf{A}$ is not unique, since the gradient of any scalar function can be added to $\mathbf{A}$ without affecting the curl, since the curl of a gradient is always zero.

1.2.3 Helmholtz Theorem

The significance of this theorem lies in the fact, that it actually specifies the defining quantities of a vector field, i.e. divergence and curl. It states that a vector field is uniquely defined (within an additive constant) by specifying its divergence and its curl. It can be mathematically expressed as

$$\mathbf{B} = -\nabla U + \nabla \times \mathbf{A} \tag{1.14}$$

where $\mathbf{B}$ and $\mathbf{A}$ are vector fields and U is a scalar field. It implies that any vector field can be decomposed into two; one is gradient of a scalar function and the other is curl of a vector function. When $\mathbf{B}$ has this form, we can show that it can be defined by its divergence and curl. The divergence of $\mathbf{B}$ is given by

$$\nabla \cdot \mathbf{B} = -\nabla \cdot (\nabla U) + \nabla \cdot (\nabla \times \mathbf{A}) = -\nabla \cdot (\nabla U) = \rho \tag{1.15}$$

here ρ is a non-zero scalar density function. The curl of **B** is given by

$$\nabla \times \mathbf{B} = -\nabla \times (\nabla U) + \nabla \times (\nabla \times \mathbf{A}) = \nabla \times (\nabla \times \mathbf{A}) = \mathbf{J} \tag{1.16}$$

here **J** is a non-zero general vector which may be regarded as the strength of the vector source.

1.2.4 Useful Identities

Product rules:

$$\nabla(fg) = f(\nabla g) + g(\nabla f) \tag{1.17a}$$

$$\nabla(\mathbf{A} \cdot \mathbf{B}) = \mathbf{A} \times (\nabla \times \mathbf{B}) + \mathbf{B} \times (\nabla \times \mathbf{A}) + (\mathbf{A} \cdot \nabla)\mathbf{B} + (\mathbf{B} \cdot \nabla)\mathbf{A} \tag{1.17b}$$

$$\nabla \cdot (f\mathbf{A}) = f(\nabla \cdot \mathbf{A}) + \mathbf{A} \cdot (\nabla f) \tag{1.17c}$$

$$\nabla \cdot (\mathbf{A} \times \mathbf{B}) = \mathbf{B} \cdot (\nabla \times \mathbf{A}) - \mathbf{A} \cdot (\nabla \times \mathbf{B}) \tag{1.17d}$$

$$\nabla \times (f\mathbf{A}) = f(\nabla \times \mathbf{A}) - \mathbf{A} \times (\nabla f) \tag{1.17e}$$

$$\nabla \times (\mathbf{A} \times \mathbf{B}) = (\mathbf{B} \cdot \nabla)\,\mathbf{A} - (\mathbf{A} \cdot \nabla)\mathbf{B} + \mathbf{A}(\nabla \cdot \mathbf{B}) - \mathbf{B}(\nabla \cdot \mathbf{A}) \tag{1.17f}$$

Second derivatives:

$$\nabla \cdot (\nabla \times \mathbf{A}) = 0 \tag{1.18a}$$

A divergence-less vector can always be expressed as curl of an arbitrary vector function.

$$\nabla \times (\nabla \times \mathbf{A}) = \nabla(\nabla \cdot \mathbf{A}) - \nabla^2 \mathbf{A} \tag{1.18b}$$

$$\nabla \cdot (\nabla f) = \nabla^2 f \tag{1.18c}$$

$$\nabla \times (\nabla f) = 0 \tag{1.18d}$$

A curl-less vector can always be expressed as gradient of an arbitrary scalar function.

1.3 INTEGRAL CALCULUS

1.3.1 Fundamental Theorems of Calculus

As is well known, the volume is always enclosed by a closed surface and the surface is always enclosed by a closed path. The path, a directed curve, is a vector direction being indicated with an arrow over the curve. The surface is also, by definition, a vector and is always surrounded by a closed path. The direction of the surface at any point over it can be found by wrapping the point with the right hand. The direction pointed by the right-hand thumb when wrapped by the fingers around the point gives the direction of the surface at that point.

- In vector calculus, length and differential length, surface and differential surface are vector quantities.
- The volume and differential volumes are scalar quantities.

Line integral: Ordinarily, the value of a line integral depends critically on the particular path taken from the initial position to the end position, but there is an important special class of vector functions for which the line integral is independent of the path and is determined entirely by end points. The vector functions which can be expressed as gradient of some scalar functions have this special property, i.e. their line integrals are path independent.

Surface integral: The value of a surface integral, in general, depends on the particular surface chosen of integration, but there is an important special class of vector functions for which it is independent of the surface and is determined entirely by the boundary line. The vector functions which can be expressed as curl of some vector functions have this special property, i.e. their surface integrals are integration surface independent and they depend only on the boundary line.

Gradient Theorem

In word form, the gradient theorem can be stated as "the integral of the tangential component of gradient of a scalar function along a path from a to b is the difference of the function values at b and a, i.e. $f(b)$ and $f(a)$". Mathematically,

$$\int_a^b (\nabla f)\cdot d\mathbf{l} = f(b) - f(a) \tag{1.19}$$

Corollary I: $\int_a^b (\nabla f)\cdot d\mathbf{l}$ is independent of path taken from a to b.

Corollary II: $\oint (\nabla f)\cdot d\mathbf{l} = 0$ since the beginning and end points are identical and hence $f(b) - f(a) = 0$.

Divergence Theorem or Gauss Theorem

A vector function over a closed surface can be related to the same vector throughout *any* volume enclosed by the surface through a theorem known as the *Divergence theorem.* It connects surface integral with volume integral. In word form, it can be stated as "the integral of the normal component of a vector over a closed surface is equal to the integral of the divergence of the same vector through any volume enclosed by that surface." Mathematically,

$$\int \nabla\cdot\mathbf{F}\,d\tau = \oint \mathbf{F}\cdot d\mathbf{a} \tag{1.20}$$

Curl Theorem or Stokes Theorem

A vector function along a closed path can be related to the same vector function over *any* surface enclosed by the path through *curl theorem.* It connects line integral with surface integral. In word form it can be stated as "the integral of the tangential component of a vector around a closed path is equal to the integral of the normal component of the curl of the same vector through any surface enclosed by the path." Mathematically,

$$\int (\nabla \times \mathbf{F}) \cdot d\mathbf{a} = \oint \mathbf{F} \cdot d\mathbf{l} \tag{1.21}$$

Corollary I: $\int (\nabla \times \mathbf{F}) \cdot d\mathbf{a}$ depends only on the boundary line, not on the particular surface used.

Corollary II: $\oint (\nabla \times \mathbf{F}) \cdot d\mathbf{a} = 0$ for any closed surface since any boundary line, like the mouth of a balloon, shrinks down to a point, and hence the right-hand side of the equation becomes zero.

Another useful relation is

$$\int (\nabla \times \mathbf{F})\, d\tau = -\oint \mathbf{F} \times d\mathbf{a} \tag{1.22}$$

Significance

- These theorems are useful in converting the Maxwell's equations from point form to integral form and vice versa.
- They relate a surface integral to its corresponding volume integral and also a line integral to its corresponding surface integral.

Gradient theorem:	$\int_a^b (\nabla f) \cdot d\mathbf{l} = f(b) - f(a)$
Divergence theorem:	$\int \nabla \cdot \mathbf{F} d\tau = \oint \mathbf{F} \cdot d\mathbf{a}$
Curl theorem:	$\int (\nabla \times \mathbf{F}) \cdot d\mathbf{a} = \oint \mathbf{F} \cdot d\mathbf{l}$
	$\int (\nabla \times \mathbf{F}) d\tau = -\oint \mathbf{F} \times d\mathbf{a}$

1.3.2 Integration by Parts

According to the product rule for derivatives,

$$\frac{d}{dx}(fg) = f\frac{dg}{dx} + g\frac{df}{dx} \tag{1.23}$$

Integrating both sides

$$\int_a^b \frac{d}{dx}(fg)\, dx = fg\Big|_a^b = \int_a^b f\frac{dg}{dx}\, dx + \int_a^b g\frac{df}{dx}\, dx$$

or

$$\int_a^b f\frac{dg}{dx}\, dx = fg\Big|_a^b - \int_a^b g\frac{df}{dx}\, dx \tag{1.24}$$

Its significance lies in the fact that during the integration of one function and the derivative of another function, it is possible to transfer the derivative from one to another making the integration possibly easier.

1.4 COORDINATE SYSTEMS

In electromagnetics, the physical quantities we deal with are, in general, functions of space and time. A three-dimensional coordinate system allows us to uniquely specify the location of a point in space or the direction of a vector quantity. The coordinate systems can be orthogonal (Figure 1.3) or non-orthogonal.

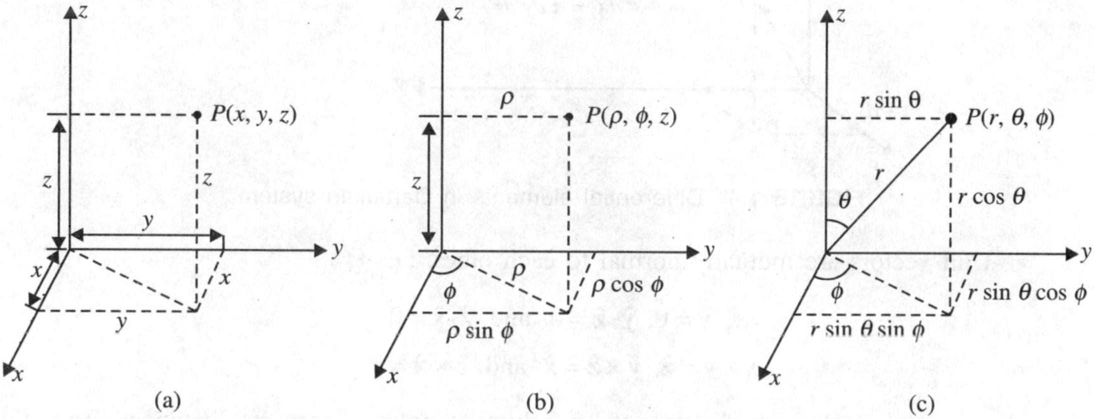

FIGURE 1.3 Orthogonal coordinate systems: (a) Cartesian, (b) cylindrical and (c) spherical.

An *orthogonal coordinate* system is one whose coordinates are mutually perpendicular, whereas in a non-orthogonal system not all the three coordinates are mutually perpendicular. Non-orthogonal systems are specialized and seldom used in solving the practical problems. Many orthogonal coordinate systems have been devised, but the most standard and commonly used are

- Cartesian or rectangular coordinate system
- Cylindrical coordinate system
- Spherical coordinate system

The need for more than one coordinate system arises from the fact that the solution of a given practical problem can be greatly facilitated by the proper choice of a coordinate system that best fits the geometry of the problem.

1.4.1 Features of Coordinate Systems

Out of the several systems available, the Cartesian coordinate system is one that is simple as well as widely used. Differential elements in the Cartesian coordinate system are shown in Figure 1.4. The salient features of this system are listed as follows:

- The coordinate variables are x, y, z.
- All the coordinates are uniform.
- The ranges of the coordinates are

$$x \to -\infty \text{ to } \infty,\ y \to -\infty \text{ to } \infty,\ z \to -\infty \text{ to } \infty$$

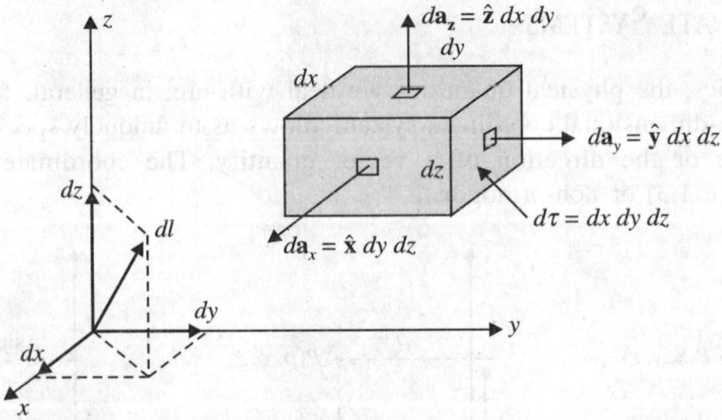

FIGURE 1.4 Differential elements in Cartesian system.

- Unit vectors are mutually normal to each other, i.e.

$$\hat{\mathbf{x}}\cdot\hat{\mathbf{y}}=0,\ \hat{\mathbf{y}}\cdot\hat{\mathbf{z}}=0 \ \text{ and } \ \hat{\mathbf{z}}\cdot\hat{\mathbf{x}}=0$$

$$\hat{\mathbf{x}}\times\hat{\mathbf{y}}=\hat{\mathbf{z}},\ \hat{\mathbf{y}}\times\hat{\mathbf{z}}=\hat{\mathbf{x}} \ \text{ and } \ \hat{\mathbf{z}}\times\hat{\mathbf{x}}=\hat{\mathbf{y}}$$

- x, y and z, the coordinates of an arbitrary point P, are its distances from the yz-plane, xz-plane and xy-plane.
- The constant x, constant y and constant z surfaces are plane surfaces parallel to the yz-plane, xz-plane and xy-plane, respectively.

Apart from the Cartesian system, we use cylindrical coordinate system, in the solution of field theory problems that involve cylindrical symmetry. Differential elements in the cylindrical system is shown in Figure 1.5. The salient features of this system are listed as follows:

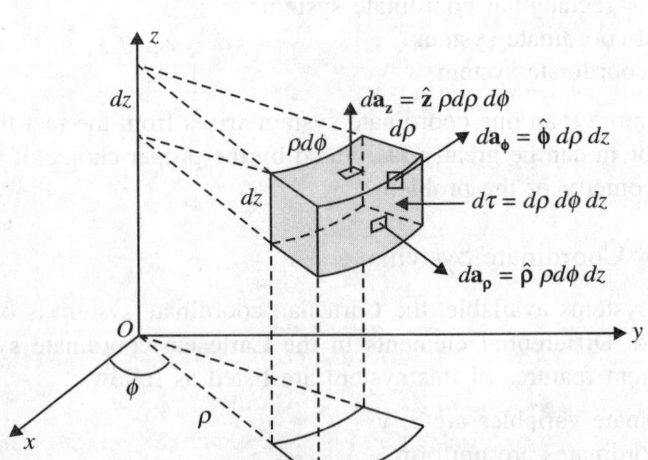

FIGURE 1.5 Differential elements in cylindrical system.

- The coordinate variables are ρ, ϕ, z.
- Only one coordinate z is uniform.
- The ranges of the coordinates are $\rho \to 0$ to ∞, $\phi \to 0$ to 2π, $z \to -\infty$ to ∞.
- Unit vectors: The coordinates are mutually perpendicular to each other and hence the unit vectors satisfy the following relations:

$$\hat{\boldsymbol{\rho}} \cdot \hat{\boldsymbol{\phi}} = 0,\ \hat{\boldsymbol{\phi}} \cdot \hat{\mathbf{z}} = 0 \text{ and } \hat{\mathbf{z}} \cdot \hat{\boldsymbol{\rho}} = 0$$

$$\hat{\boldsymbol{\rho}} \times \hat{\boldsymbol{\phi}} = \hat{\mathbf{z}},\ \hat{\boldsymbol{\phi}} \times \hat{\mathbf{z}} = \hat{\boldsymbol{\rho}} \text{ and } \hat{\mathbf{z}} \times \hat{\boldsymbol{\rho}} = \hat{\boldsymbol{\phi}}$$

- $\rho \to$ is called cylindrical radial coordinate and represents radial distance from z-axis.
- $\phi \to$ is called azimuth or longitude angle. It is the angle between the positive x-axis and the line formed from the origin to the P projected on the xy-plane. It is always measured in counterclockwise direction.
- $z \to$ is same as defined in the Cartesian system, sometimes called height coordinate. It represents the height of the point above the xy-plane.
- The constant ρ surface is a cylindrical curved surface coaxial with the z-axis.
- Thc constant ϕ surface is a plane vertical surface normal to the xy-plane.
- The constant z surface is a plane horizontal surface parallel to the xy-plane.

There exist relations between Cartesian and cylindrical systems which are given hereunder.

Cartesian to cylindrical transformation:

$$\rho = \sqrt[+]{x^2 + y^2} \qquad A_\rho = A_x \cos\phi + A_y \sin\phi$$

$$\phi = \tan^{-1}\left(\frac{y}{x}\right) \qquad A_\phi = -A_x \sin\phi + A_y \cos\phi$$

$$z = z \qquad A_z = A_z$$

Cylindrical to Cartesian transformation:

$$x = \rho\cos\phi \qquad A_x = A_\rho \cos\phi - A_\phi \sin\phi$$

$$y = \rho\sin\phi \qquad A_y = A_\rho \sin\phi + A_\phi \cos\phi$$

$$z = z \qquad A_z = A_z$$

The spherical coordinate system is another curvilinear one apart from the cylindrical system that is required to be used in the solution of several problems in field theory. Differential elements in the sperical system are shown in Figure 1.6. The significant features of the system are given as follows:

- The coordinate variables are r, θ, ϕ.
- No coordinate is uniform.
- The ranges are $r \to 0$ to ∞, $\theta \to 0$ to π, $\phi \to 0$ to 2π.

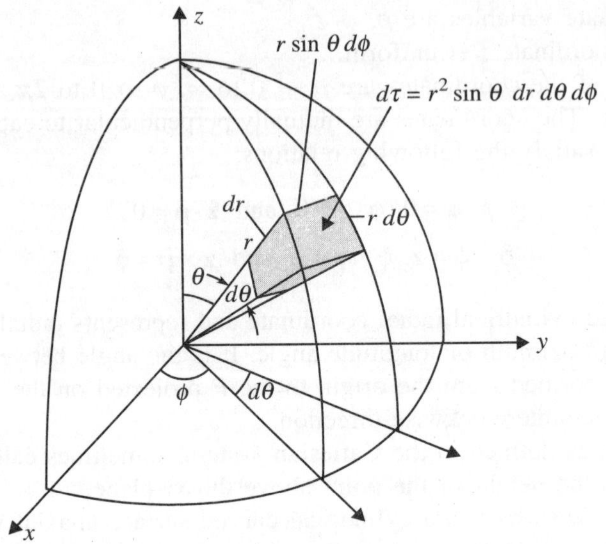

FIGURE 1.6 Differential elements in spherical system.

- Unit vectors: The coordinates are mutually perpendicular satisfying the following relations:

$$\hat{\mathbf{r}} \cdot \hat{\boldsymbol{\theta}} = 0,\ \hat{\boldsymbol{\theta}} \cdot \hat{\boldsymbol{\phi}} = 0 \ \text{ and } \ \hat{\boldsymbol{\phi}} \cdot \hat{\mathbf{r}} = 0$$

$$\hat{\mathbf{r}} \times \hat{\boldsymbol{\theta}} = \hat{\boldsymbol{\phi}},\ \hat{\boldsymbol{\theta}} \times \hat{\boldsymbol{\phi}} = \hat{\mathbf{r}} \ \text{ and } \ \hat{\boldsymbol{\phi}} \times \hat{\mathbf{r}} = \hat{\boldsymbol{\theta}}$$

- $r \rightarrow$ is called range coordinate and represents the distance from the origin to a given point P.
- $\theta \rightarrow$ is called zenith or polar angle or colatitude. It is the angle between the positive z-axis and the line formed between the origin and the point P.
- $\phi \rightarrow$ is called azimuth angle or longitude angle. It is the angle between the positive x-axis and the line formed from the origin to the point P projected on the xy-plane. It is always measured in counterclockwise direction.
- Coordinates θ and ϕ lose significance when $r = 0$ and ϕ loses significance when $\sin\theta = 0$.
- The constant r surface is a spherical curved surface concentric with the origin.
- The constant θ surface is a conical curved surface surrounding the z-axis with the apex at origin.
- The constant ϕ surface is a plane vertical surface normal to the xy-plane.

The interrelations between spherical and other coordinate systems are given hereunder various transformations (Figure 1.7).

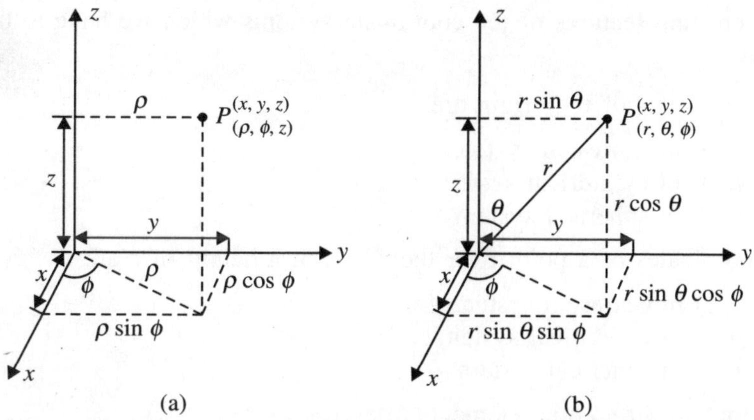

FIGURE 1.7 Interrelations between Cartesian and (a) cylindrical and (b) spherical systems.

Cartesian to spherical transformation:

$$r = \sqrt[+]{x^2 + y^2 + z^2} \qquad A_r = A_x \sin\theta\cos\phi + A_y \sin\theta\sin\phi + A_z\cos\theta$$

$$\theta = \tan^{-1}\left(\frac{\sqrt[+]{x^2+y^2}}{z}\right) \qquad A_\theta = A_x\cos\theta\cos\phi + A_y\cos\theta\sin\phi - A_z\sin\theta$$

$$\phi = \tan^{-1}\left(\frac{y}{x}\right) \qquad A_\phi = -A_x\sin\phi + A_y\cos\phi$$

Spherical to Cartesian transformation:

$$x = r\sin\theta\cos\phi \qquad A_x = A_r\sin\theta\cos\phi + A_\theta\cos\theta\cos\phi - A_\phi\sin\phi$$

$$y = r\sin\theta\sin\phi \qquad A_y = A_r\sin\theta\sin\phi + A_\theta\cos\theta\sin\phi - A_\phi\cos\phi$$

$$z = r\cos\theta \qquad A_z = A_r\cos\theta - A_\theta\sin\theta$$

Cylindrical to spherical transformation:

$$r = \sqrt[+]{\rho^2 + z^2} \qquad A_r = A_\rho\sin\theta + A_z\cos\theta$$

$$\theta = \tan^{-1}\left(\frac{\rho}{z}\right) \qquad A_\theta = A_\rho\cos\theta - A_z\sin\theta$$

$$\phi = \phi \qquad A_\phi = A_\phi$$

Spherical to cylindrical transformation:

$$\rho = r\sin\theta \qquad A_\rho = A_r\sin\theta + A_\theta\cos\theta$$

$$\phi = \phi \qquad A_\phi = A_\phi$$

$$z = r\cos\theta \qquad A_z = A_r\cos\theta - A_\theta\sin\theta$$

Some of the interesting features of the coordinate systems which we have to take note of are given as follows:

1. The coordinates of the origin are
   - (0, 0, 0) in Cartesian system
   - (0, ϕ, 0) in cylindrical system
   - (0, θ, ϕ) in spherical system
2. The coordinates of a point over the z-axis at a height of z are
   - (0, 0, z) in Cartesian system
   - (0, ϕ, z) in cylindrical system
   - (z, 0, ϕ) in spherical system
3. Directions of differential elements often used:
   - The direction of spherical surface is radial (sph) and its differential area is $\mathbf{da} = r^2 \sin\theta\, d\theta\, d\phi \hat{\mathbf{r}}$.
   - The direction of cylindrical surface is radial (cyl) and its differential area is $\mathbf{da} = \rho d\phi\, dz \hat{\boldsymbol{\rho}}$. It is for the curved surface and for top and bottom faces, the direction is $\hat{\mathbf{z}}$ and differential area is $\mathbf{da} = \rho d\rho\, d\phi(\pm\hat{\mathbf{z}})$.
   - When the line charge is over the x-axis, its differential length is $dx\hat{\mathbf{x}}$, on the y-axis, it is $dy\hat{\mathbf{y}}$ and on the z-axis, it is $dz\hat{\mathbf{z}}$.
4. Vector and distance between two points:
   Let us suppose there are two points whose coordinates are
   - (x_1, y_1, z_1) and (x_2, y_2, z_2) in Cartesian system
   - (ρ_1, ϕ_1, z_1) and (ρ_2, ϕ_2, z_2) in cylindrical system
   - (r_1, θ_1, ϕ_1) and (r_2, θ_2, ϕ_2) in spherical system

The vector and the distance between the points, in the Cartesian system, can be found to be

$$\mathbf{D} = (x_2 - x_1)\hat{\mathbf{x}} + (y_2 - y_1)\hat{\mathbf{y}} + (z_2 - z_1)\hat{\mathbf{z}}$$

$$d = |\mathbf{D}| = \sqrt{(x_2 - x_1)^2 + (y_2 - y_1)^2 + (z_2 - z_1)^2}$$

In other coordinate systems, first the coordinates of the points need to be converted into Cartesian system and then the above formulae are to be applied.

The vector and the distance between them, in the cylindrical system, can be found to be

$$\mathbf{D} = (\rho_2 \cos\phi_2 - \rho_1 \cos\phi_1)\hat{\mathbf{x}} + (\rho_2 \sin\phi_2 - \rho_1 \sin\phi_1)\hat{\mathbf{y}} + (z_2 - z_1)\hat{\mathbf{z}}$$

$$d = \sqrt{\rho_2^2 + \rho_1^2 - 2\rho_1\rho_2 \cos(\phi_2 - \phi_1) + (z_2 - z_1)^2}$$

The vector and the distance between them, in spherical system, can be found to be

$$\mathbf{D} = (r_2 \sin\theta_2 \cos\phi_2 - r_1 \sin\theta_1 \cos\phi_1)\hat{\mathbf{x}} + (r_2 \sin\theta_2 \sin\phi_2 - r_1 \sin\theta_1 \sin\phi_1)\hat{\mathbf{y}}$$
$$+ (r_2 \cos\theta_2 - r_1 \cos\theta_1)\hat{\mathbf{z}}$$

$$d = \sqrt{r_2^2 + r_1^2 - 2r_1 r_2 [\cos\theta_1 \cos\theta_2 + \sin\theta_1 \sin\theta_2 \cos(\phi_2 - \phi_1)]}$$

The unit vectors can then be evaluated by dividing thus obtained vectors by their magnitudes.

Proof:

- Vector and distance from point $P_1(\rho_1, \phi_1, z_1)$ to point $P_2(\rho_2, \phi_2, z_2)$
  As the coordinates are in cylindrical system, we have to first convert them from cylindrical to Cartesian system.
  For point P_1, $x_1 = \rho_1 \cos \phi_1$, $y_1 = \rho_1 \sin \phi_1$ and $z_1 = z_1$
  For point P_2, $x_2 = \rho_2 \cos \phi_2$, $y_2 = \rho_2 \sin \phi_2$ and $z_2 = z_2$
  Now we have the coordinates of the points in Cartesian system as $(\rho_1 \cos \phi_1, \rho_1 \sin \phi_1, z_1)$ and $(\rho_2 \cos \phi_2, \rho_2 \sin \phi_2, z_2)$.
  The vector joining point P_1 to point P_2 then becomes
  $$(x_2 - x_1)\hat{\mathbf{x}} + (y_2 - y_1)\hat{\mathbf{y}} + (z_2 - z_1)\hat{\mathbf{z}}$$
  $$= (\rho_2 \cos\phi_2 - \rho_1 \cos\phi_1)\hat{\mathbf{x}} + (\rho_2 \sin\phi_2 - \rho_1 \sin\phi_1)\hat{\mathbf{y}} + (z_2 - z_1)\hat{\mathbf{z}}$$
  The magnitude of this vector then becomes the distance between the points which is
  $$[(\rho_2 \cos\phi_2 - \rho_1 \cos\phi_1)^2 + (\rho_2 \sin\phi_2 - \rho_1 \sin\phi_1)^2 + (z_2 - z_1)^2]^{1/2}$$
  On expansion and simplification, we get
  $$d = \sqrt{\rho_2^2 + \rho_1^2 - 2\rho_1\rho_2 \cos(\phi_2 - \phi_1) + (z_2 - z_1)^2}$$
  A very common case we come across is the one in which one of the points is over the horizontal plane $(\rho, \phi, 0)$ and another over the z-axis $(0, \phi, z)$. In this case, the vector joining the first point to the second point becomes
  $$(\rho_2 \cos\phi_2 - \rho_1 \cos\phi_1)\hat{\mathbf{x}} + (\rho_2 \sin\phi_2 - \rho_1 \sin\phi_1)\hat{\mathbf{y}} + (z_2 - z_1)\hat{\mathbf{z}}$$
  $$= (-\rho \cos\phi)\hat{\mathbf{x}} + (-\rho \sin\phi)\hat{\mathbf{y}} + z\hat{\mathbf{z}} = -\rho\hat{\mathbf{r}} + z\hat{\mathbf{z}}$$

- Vector and distance from point $P_1(r_1, \theta_1, \phi_1)$ to point $P_2(r_2, \theta_2, \phi_2)$
  As the coordinates are in spherical system, we have to first convert them into Cartesian system.
  For point P_1, $x_1 = r_1 \sin\theta_1 \cos\phi_1$, $y_1 = r_1 \sin\theta_1 \sin\phi_1$ and $z_1 = r_1 \cos\theta_1$
  For point P_2, $x_2 = r_2 \sin\theta_2 \cos\phi_2$, $y_2 = r_2 \sin\theta_2 \sin\phi_2$ and $z_2 = r_2 \cos\theta_2$
  Now we have the coordinates of the points in Cartesian system as $(r_1 \sin\theta_1 \cos\phi_1, r_1 \sin\theta_1 \sin\phi_1, r_1 \cos\theta_1)$ and $(r_2 \sin\theta_2 \cos\phi_2, r_2 \sin\theta_2 \sin\phi_2, r_2 \cos\theta_2)$.
  The vector joining point P_1 to point P_2 then becomes
  $$(x_2 - x_1)\hat{\mathbf{x}} + (y_2 - y_1)\hat{\mathbf{y}} + (z_2 - z_1)\hat{\mathbf{z}}$$
  $$= (r_2 \sin\theta_2 \cos\phi_2 - r_1 \sin\theta_1 \cos\phi_1)\hat{\mathbf{x}} + (r_2 \sin\theta_2 \sin\phi_2 - r_1 \sin\theta_1 \sin\phi_1)\hat{\mathbf{y}}$$
  $$+ (r_2 \cos\theta_2 - r_1 \cos\theta_1)\hat{\mathbf{z}}$$

The magnitude of this vector then becomes the distance between the points which is

$$[(r_2 \sin\theta_2 \cos\phi_2 - r_1 \sin\theta_1 \cos\phi_1)^2 + (r_2 \sin\theta_2 \sin\phi_2 - r_1 \sin\theta_1 \sin\phi_1)^2 + (r_2 \cos\theta_2 - r_1 \cos\theta_1)^2]^{1/2}$$

On expansion and simplification, we get

$$d = \sqrt{r_2^2 + r_1^2 - 2r_1 r_2 [\cos\theta_1 \cos\theta_2 + \sin\theta_1 \sin\theta_2 \cos(\phi_2 - \phi_1)]}$$

1.4.2 Symmetry

Symmetry is an important aspect pertaining to a configuration and the knowledge of this aspect is useful, in field theory, to find an appropriate coordinate system in the solution procedure. Various types of symmetries are described and discussed here.

- *Plane symmetry:* If a quantity depends only on one Cartesian coordinate while independent of the other two coordinates, then the quantity is said to have plane symmetry.

  The analytical relation for the quantity contains only one of the three coordinates, the other two being missing.
- *Cylindrical symmetry:* If the quantity depends only on one cylindrical coordinate ρ while maintaining independence of the other two coordinates, ϕ and z, then the quantity is said to have cylindrical symmetry.

  In the analytical relation of the quantity, only one coordinate ρ appears, while the other two being missing.
- *Spherical symmetry:* If the quantity depends only on one spherical coordinate r while maintaining independence of the other two coordinates, θ and ϕ, then the quantity is said to have spherical symmetry.

  In the analytical relation of the quantity, only one coordinate r appears, while the other two being missing.

1.4.3 Differential Length, Surface and Volume

Differential length is the vector joining two points whose coordinates differ by differential amounts, i.e.

(x, y, z) to $(x + dx, y + dy, z + dz)$ in Cartesian system,
(ρ, ϕ, z) to $(\rho + d\rho, \phi + d\phi, z + dz)$ in cylindrical system and
(r, θ, ϕ) to $(r + dr, \theta + d\theta, \phi + d\phi)$ in spherical system

Differential areas are the areas of the faces formed with different components of the differential length. The differential length is always indicated by $d\mathbf{l}$, whereas differential areas are indicated by $d\mathbf{a}$ or $d\mathbf{s}$.

Differential volume is the volume of the region which is formed with the different components of the differential area. Differential volumes are indicated by dv or $d\tau$. Differential lengths and surfaces are vector quantities, whereas the differential volume is a scalar quantity.

The direction of differential length is the direction of the tangent drawn to the length at the point of its location and the direction of the differential area is by definition the direction of the normal to the surface at the point of its location. The unit normal to a surface is given by

$$\hat{\mathbf{n}} = \frac{d\mathbf{l}_1 \times d\mathbf{l}_2}{|d\mathbf{l}_1 \times d\mathbf{l}_2|}$$

where $d\mathbf{l}_1$ and $d\mathbf{l}_2$ are two differential length vectors at the point under consideration and tangential to the two curves on the surface.

Below are listed all the three differential elements in all the three coordinate systems. The differential areas are given even component-wise.

Cartesian coordinate system

$$d\mathbf{l} = dx\hat{\mathbf{x}} + dy\hat{\mathbf{y}} + dz\hat{\mathbf{z}}$$
$$d\mathbf{a}_x = dydz\hat{\mathbf{x}}$$
$$d\mathbf{a}_y = dxdz\hat{\mathbf{y}}$$
$$d\mathbf{a}_z = dxdy\hat{\mathbf{z}}$$
$$d\tau = dxdydz$$

Cylindrical coordinate system

$$d\mathbf{l} = d\rho\hat{\boldsymbol{\rho}} + \rho d\phi\hat{\boldsymbol{\phi}} + dz\hat{\mathbf{z}}$$
$$d\mathbf{a}_\rho = \rho d\phi dz\hat{\boldsymbol{\rho}}$$
$$d\mathbf{a}_\phi = d\rho dz\hat{\boldsymbol{\phi}}$$
$$d\mathbf{a}_z = \rho d\rho d\phi\hat{\mathbf{z}}$$
$$d\tau = \rho d\rho d\phi dz$$

Spherical coordinate system

$$d\mathbf{l} = dr\hat{\mathbf{r}} + rd\theta\hat{\boldsymbol{\theta}} + r\sin\theta d\phi\hat{\boldsymbol{\phi}}$$
$$d\mathbf{a}_r = r^2\sin\theta d\theta d\phi\hat{\mathbf{r}}$$
$$d\mathbf{a}_\theta = r\sin\theta dr d\phi\hat{\boldsymbol{\theta}}$$
$$d\mathbf{a}_\phi = rdrd\theta\hat{\boldsymbol{\phi}}$$
$$d\tau = r^2\sin\theta dr\, d\theta\, d\phi$$

1.4.4 Mapping of Del Operator

The del operator is defined with respect to rectangular coordinate system only. But in field theory, we come across problems that require the application of del over functions that exist in other coordinate systems like cylindrical or spherical. To overcome this difficulty, the del operator is mapped into other coordinate systems and then applied over to the functions. The mapping procedure (not proof) is an easy one that starts with the expression for differential length and it is given here.

There exists a simple and easy method by which one expresses the gradient, divergence and curl in different orthogonal coordinate systems. Let us suppose the differential length in the coordinate system under consideration be

$$d\mathbf{l} = h_1 du\hat{\mathbf{u}} + h_2 dv\hat{\mathbf{v}} + h_3 dw\hat{\mathbf{w}} \tag{1.25}$$

then the gradient, curl, divergence and Laplacian can be found using the following expressions:

Gradient:

$$\nabla V = \frac{1}{h_1 h_2 h_3}\left(h_2 h_3 \frac{\partial V}{\partial u}\hat{\mathbf{u}} + h_1 h_3 \frac{\partial V}{\partial v}\hat{\mathbf{v}} + h_1 h_2 \frac{\partial V}{\partial w}\hat{\mathbf{w}}\right) \tag{1.26}$$

Curl:

$$\nabla \times \mathbf{D} = \frac{1}{h_1 h_2 h_3}\begin{vmatrix} h_1\hat{\mathbf{u}} & h_2\hat{\mathbf{v}} & h_3\hat{\mathbf{w}} \\ \dfrac{\partial}{\partial u} & \dfrac{\partial}{\partial v} & \dfrac{\partial}{\partial w} \\ h_1 D_u & h_2 D_v & h_3 D_w \end{vmatrix} \tag{1.27}$$

Divergence:

$$\nabla \cdot \mathbf{D} = \frac{1}{h_1 h_2 h_3}\left\{\left[\frac{\partial}{\partial u}(h_2 h_3 D_u)\right] + \left[\frac{\partial}{\partial v}(h_3 h_1 D_v)\right] + \left[\frac{\partial}{\partial w}(h_1 h_2 D_w)\right]\right\} \tag{1.28}$$

Laplacian:

$$\nabla^2 V = \frac{1}{h_1 h_2 h_3}\left\{\left[\frac{\partial}{\partial u}\left(\frac{h_2 h_3}{h_1}\frac{\partial V}{\partial u}\right)\right] + \left[\frac{\partial}{\partial v}\left(\frac{h_3 h_1}{h_2}\frac{\partial V}{\partial v}\right)\right] + \left[\frac{\partial}{\partial w}\left(\frac{h_1 h_2}{h_3}\frac{\partial V}{\partial w}\right)\right]\right\} \tag{1.29}$$

Now we apply the above templates to various orthogonal coordinate systems and try to find out actual expressions. In the case of Cartesian system, differential length is given by

$$d\mathbf{l} = dx\hat{\mathbf{x}} + dy\hat{\mathbf{y}} + dz\hat{\mathbf{z}} \tag{1.30a}$$

By comparing with Eq. (1.25), we can have

$$u = x,\ v = y \text{ and } w = z$$

$$h_1 = 1,\ h_2 = 1 \text{ and } h_3 = 1$$

Now, we get

$$\nabla V = \frac{\partial V}{\partial x}\hat{\mathbf{x}} + \frac{\partial V}{\partial y}\hat{\mathbf{y}} + \frac{\partial V}{\partial z}\hat{\mathbf{z}} \tag{1.30b}$$

$$\nabla \times \mathbf{D} = \begin{vmatrix} \hat{\mathbf{x}} & \hat{\mathbf{y}} & \hat{\mathbf{z}} \\ \dfrac{\partial}{\partial x} & \dfrac{\partial}{\partial y} & \dfrac{\partial}{\partial z} \\ D_u & D_v & D_w \end{vmatrix} \tag{1.30c}$$

$$\nabla \cdot \mathbf{D} = \frac{\partial D_x}{\partial x} + \frac{\partial D_y}{\partial y} + \frac{\partial D_z}{\partial z} \tag{1.30d}$$

$$\nabla^2 V = \frac{\partial^2 V}{\partial x^2} + \frac{\partial^2 V}{\partial y^2} + \frac{\partial^2 V}{\partial z^2} \tag{1.30e}$$

In the case of cylindrical system, differential length is given by

$$d\mathbf{l} = d\rho\hat{\boldsymbol{\rho}} + \rho d\phi\hat{\boldsymbol{\phi}} + dz\hat{\mathbf{z}} \tag{1.31a}$$

By comparing with Eq. (1.25), we can have

$$u = \rho,\ v = \phi \text{ and } w = z$$
$$h_1 = 1,\ h_2 = \rho \text{ and } h_3 = 1$$

Now, we get

$$\begin{aligned}\nabla V &= \frac{1}{\rho}\left(\frac{\partial V}{\partial \rho}\hat{\boldsymbol{\rho}} + \frac{\partial V}{\partial \phi}\hat{\boldsymbol{\phi}} + \rho\frac{\partial V}{\partial z}\hat{\mathbf{z}}\right)\\ &= \frac{\partial V}{\partial \rho}\hat{\boldsymbol{\rho}} + \frac{1}{\rho}\frac{\partial V}{\partial \phi}\hat{\boldsymbol{\phi}} + \frac{\partial V}{\partial z}\hat{\mathbf{z}}\end{aligned} \tag{1.31b}$$

$$\nabla \times \mathbf{D} = \frac{1}{\rho}\begin{vmatrix} \hat{\boldsymbol{\rho}} & \rho\hat{\boldsymbol{\phi}} & \hat{\mathbf{z}} \\ \dfrac{\partial}{\partial \rho} & \dfrac{\partial}{\partial \phi} & \dfrac{\partial}{\partial z} \\ D_\rho & \rho D_\phi & D_z \end{vmatrix} \tag{1.31c}$$

$$\nabla \cdot \mathbf{D} = \frac{1}{\rho}\frac{\partial}{\partial \rho}(\rho D_\rho) + \frac{1}{\rho}\frac{\partial D_\phi}{\partial \phi} + \frac{1}{\rho}\frac{\partial}{\partial z}(\rho D_z) \tag{1.31d}$$

$$\nabla^2 V = \frac{1}{\rho}\frac{\partial}{\partial \rho}\left(\rho\frac{\partial V}{\partial \rho}\right) + \frac{1}{\rho^2}\left(\frac{\partial^2 V}{\partial \phi^2}\right) + \frac{\partial^2 V}{\partial z^2} \tag{1.31e}$$

In the case of spherical system, differential length is given by

$$d\mathbf{l} = dr\hat{\mathbf{r}} + rd\theta\hat{\boldsymbol{\theta}} + r\sin\theta d\phi\hat{\boldsymbol{\phi}} \tag{1.32a}$$

By comparing with Eq. (1.25), we can have

$$u = r,\ v = \theta \text{ and } w = \phi$$
$$h_1 = 1,\ h_2 = r \text{ and } h_3 = r\sin\theta$$

Now, we get

$$\begin{aligned}\nabla V &= \frac{1}{r^2\sin\theta}\left(r^2\sin\theta\frac{\partial V}{\partial r}\hat{\mathbf{r}} + r\sin\theta\frac{\partial V}{\partial \theta}\hat{\boldsymbol{\theta}} + r\frac{\partial V}{\partial \phi}\hat{\boldsymbol{\phi}}\right)\\ &= \frac{\partial V}{\partial r}\hat{\mathbf{r}} + \frac{1}{r}\frac{\partial V}{\partial \theta}\hat{\boldsymbol{\theta}} + \frac{1}{\sin\theta}\frac{\partial V}{\partial \phi}\hat{\boldsymbol{\phi}}\end{aligned} \tag{1.32b}$$

$$\nabla \times \mathbf{D} = \frac{1}{r^2 \sin\theta} \begin{vmatrix} \hat{\mathbf{r}} & r\hat{\boldsymbol{\theta}} & r\sin\theta\hat{\boldsymbol{\phi}} \\ \dfrac{\partial}{\partial r} & \dfrac{\partial}{\partial \theta} & \dfrac{\partial}{\partial \phi} \\ D_r & rD_\theta & r\sin\theta D_\phi \end{vmatrix} \tag{1.32c}$$

$$\nabla \cdot \mathbf{D} = \frac{1}{r^2 \sin\theta}\left\{\left[\frac{\partial}{\partial r}(r^2 \sin\theta D_r)\right] + \left[\frac{\partial}{\partial \theta}(r\sin\theta D_\theta)\right] + \left[\frac{\partial}{\partial \phi}(rD_\phi)\right]\right\} \tag{1.32d}$$

$$\nabla^2 V = \frac{1}{r^2}\frac{\partial}{\partial r}\left(r^2 \frac{\partial V}{\partial r}\right) + \frac{1}{r^2 \sin\theta}\frac{\partial}{\partial \theta}\left(\sin\theta \frac{\partial V}{\partial \theta}\right) + \frac{1}{r^2 \sin^2\theta}\left(\frac{\partial^2 V}{\partial \phi^2}\right) \tag{1.32e}$$

1.5 DIRAC DELTA FUNCTION

The mathematical functions that represent electromagnetic fields in general are well behaved, i.e. they are continuous and their derivatives are also continuous. However, the presence of discrete sources like point charges, filamentary currents, etc. causes singularities, and to represent these discontinuities, a specialized function is required. The Dirac delta function is useful for this purpose.

Dirac delta or unit impulse is not an ordinary function; it is defined as

$$\delta(x) = 0 \text{ at } t \neq 0$$

and

$$\int_{-\infty}^{\infty} \delta(x)dx = 1 \tag{1.33}$$

A unit impulse can be considered as a unit area, zero width pulse. The pulse can be a rectangular, Gaussian, triangular, exponential or of many other forms.

Its important and useful properties are listed as follows:

- $\int_{-\infty}^{\infty} f(x)\,\delta(x - x_0)\,dx = f(x_0)$ → Sampling/sifting property of impulse function.
- $\int_{-\infty}^{\infty} f(x)\,\delta(x)\,dx = f(0)$
- $\int_{-\infty}^{\infty} f(x_0)\,\delta(x - x_0)\,dx_0 = f(x) * \delta(x) = f(x)$ → Replication property of impulse function.
- $\delta(-x) = \delta(x)$
- $\delta(ax) = \dfrac{1}{|a|}\,\delta(x)$

- $x\delta'(x) = -\delta(x)$
- $f(x)\ \delta(x) = f(0)\ \delta(x)$
- $\int_{-\infty}^{\infty} f(x)\ \delta'(x)\,dx = f'(0)$

The derivative of unit impulse is useful to represent charge dipoles and electric double layers.

The representation of a point charge or a current element requires the use of the three-dimensional impulse function. The three-dimensional impulse at the origin is indicated by $\delta(\mathbf{r})$, whereas at point $\mathbf{r} = \mathbf{r}_0$ it is designated by $\delta(\mathbf{r} - \mathbf{r}_0)$.

It has the following properties

$$\int_v \delta(\mathbf{r} - \mathbf{r}_0)d\tau = 1 \qquad \text{if } \mathbf{r}_0 \text{ is in } v \tag{1.34a}$$

$$= 0 \qquad \text{if } \mathbf{r}_0 \text{ is not in } v \tag{1.34b}$$

$$\int_v f(\mathbf{r})\ \delta(\mathbf{r} - \mathbf{r}_0)d\tau = f(\mathbf{r}_0) \qquad \text{if } \mathbf{r}_0 \text{ is in } v \tag{1.35a}$$

$$= 0 \qquad \text{if } \mathbf{r}_0 \text{ is not in } v \tag{1.35b}$$

In Cartesian coordinate system, the three-dimensional impulse can be expressed in terms of one-dimensional impulses as

$$\delta(\mathbf{r} - \mathbf{r}_0) = \delta(x - x_0)\ \delta(y - y_0)\ \delta(z - z_0) \tag{1.36}$$

In cylindrical coordinate system, the three-dimensional impulse can be expressed in terms of one-dimensional impulses as

$$\delta(\mathbf{r} - \mathbf{r}_0) = \frac{\delta(\rho - \rho_0)\ \delta(\phi - \phi_0)\ \delta(z - z_0)}{\rho_0} \tag{1.37}$$

In spherical coordinate system, the three-dimensional impulse can be expressed in terms of one-dimensional impulses as

$$\delta(\mathbf{r} - \mathbf{r}_0) = \frac{\delta(r - r_0)\ \delta(\theta - \theta_0)\ \delta(\phi - \phi_0)}{r_0^2 \sin\theta_0} \tag{1.38}$$

SOLVED PROBLEMS

1. A vector field is specified by $\mathbf{F} = z\hat{\mathbf{x}} + x\hat{\mathbf{y}} + 2y\hat{\mathbf{z}}$.

(a) Find the field at point (2, 3, 4).

(b) At this point, find the scalar and vector components of the field in the direction of the unit vector $\hat{\mathbf{n}} = \left(\frac{1}{\sqrt{26}}\right)(3\hat{\mathbf{x}} + \hat{\mathbf{y}} + 4\hat{\mathbf{z}})$.

Solution:

(a) The field at the point (2, 3, 4) is

$$\mathbf{F}\big|_{(2,3,4)} = 4\hat{\mathbf{x}} + 2\hat{\mathbf{y}} + 6\hat{\mathbf{z}}$$

(b) The scalar component along the unit vector can be obtained by computing the dot product.

$$\mathbf{F}\big|_{(2,3,4)} \cdot \hat{\mathbf{n}} = (4\hat{\mathbf{x}} + 2\hat{\mathbf{y}} + 6\hat{\mathbf{z}}) \cdot \left(\frac{1}{\sqrt{26}}\right)(3\hat{\mathbf{x}} + \hat{\mathbf{y}} + 4\hat{\mathbf{z}})$$

$$= \left(\frac{1}{\sqrt{26}}\right)(12 + 2 + 24) = \left(\frac{1}{\sqrt{26}}\right)38$$

The vector component along the unit vector can be obtained by multiplying the scalar component with the unit vector.

$$\left(\mathbf{F}\big|_{(2,3,4)} \cdot \hat{\mathbf{n}}\right)\hat{\mathbf{n}} = \left(\frac{1}{\sqrt{26}}\right)38 \times \left(\frac{1}{\sqrt{26}}\right)(3\hat{\mathbf{x}} + \hat{\mathbf{y}} + 4\hat{\mathbf{z}}) = \left(\frac{38}{26}\right)(3\hat{\mathbf{x}} + \hat{\mathbf{y}} + 4\hat{\mathbf{z}})$$

Vectors have directions and two vectors may subtend an angle. The angle between two vectors **A** and **B** can be found from

$$\alpha = \cos^{-1}\frac{\mathbf{A}\cdot\mathbf{B}}{AB}$$

$$= \cos^{-1}\frac{\mathbf{F}\big|_{(2,3,4)} \cdot \hat{\mathbf{n}}}{\left|\mathbf{F}\big|_{(2,3,4)}\right| |\hat{\mathbf{n}}|} = \cos^{-1}\frac{\left(\frac{1}{\sqrt{26}}\right)38}{\sqrt{56}}$$

$$= \cos^{-1}\frac{38}{38.157} = 5.2°$$

2. Consider the following three vectors: $\mathbf{D}_1 = -\sqrt{2}\hat{\mathbf{x}} + \sqrt{2}\hat{\mathbf{y}} + 3\hat{\mathbf{z}}$, $\mathbf{D}_2 = \sqrt{3}\hat{\boldsymbol{\rho}} + \hat{\boldsymbol{\phi}} + 3\hat{\mathbf{z}}$ at $(2, 5\pi/6, 3)$ and $\mathbf{D}_3 = -3\hat{\mathbf{r}} - \sqrt{3}\hat{\boldsymbol{\theta}} + \hat{\boldsymbol{\phi}}$ at $(4, 2\pi/3, \pi/6)$. Determine which of these three vectors are equal in magnitude and also determine which of them are equal magnitude-wise as well as direction-wise.

Solution: Since each set of unit vectors is an orthogonal vector system, the magnitude is simply the square root of the sum of the squares of the components.

$$|\mathbf{D}_1| = \sqrt{2+2+9} = \sqrt{13}$$

$$|\mathbf{D}_2| = \sqrt{3+1+9} = \sqrt{13}$$

$$|\mathbf{D}_3| = \sqrt{9+3+1} = \sqrt{13}$$

We can observe magnitude-wise all the three vectors are equal. To find out whether they are same as vectors, we need to convert $\mathbf{D}_2$ and $\mathbf{D}_3$ into Cartesian system and compare with $\mathbf{D}_1$. Now converting $\mathbf{D}_2$ into Cartesian system, we have

$$\mathbf{D}_2 = \sqrt{3}\hat{\boldsymbol{\rho}} + \hat{\boldsymbol{\phi}} + 3\hat{\mathbf{z}} \text{ at } \left(2, \frac{5\pi}{6}, 3\right)$$

$$= \sqrt{3}\left(\cos\frac{5\pi}{6\hat{\mathbf{x}}} + \sin\frac{5\pi}{6\hat{\mathbf{y}}}\right) + \left(-\sin\frac{5\pi}{6\hat{\mathbf{x}}} + \cos\frac{5\pi}{6\hat{\mathbf{y}}}\right) + 3\hat{\mathbf{z}}$$

$$= -2\hat{\mathbf{x}} + 3\hat{\mathbf{z}}$$

Similarly, $\mathbf{D}_3$ can also be converted into Cartesian systems giving

$$\mathbf{D}_3 = -3\hat{\mathbf{r}} - \sqrt{3}\hat{\boldsymbol{\theta}} + \hat{\boldsymbol{\phi}} \text{ at } \left(4, \frac{2\pi}{3}, \frac{\pi}{6}\right)$$

$$= -2\hat{\mathbf{x}} + 3\hat{\mathbf{z}}$$

Therefore, we can say that $\mathbf{D}_2$ is equal to $\mathbf{D}_3$ but they are not equal to $\mathbf{D}_1$.

3. Transform the vector $B = y\hat{\mathbf{x}} - x\hat{\mathbf{y}} + z\hat{\mathbf{z}}$ into cylindrical coordinates.

Solution: The new components are

$$B_\rho = \mathbf{B}\cdot\hat{\boldsymbol{\rho}} = y\hat{\mathbf{x}}\cdot\hat{\boldsymbol{\rho}} - x\hat{\mathbf{y}}\cdot\hat{\boldsymbol{\rho}}$$

$$= y\cos\phi - x\sin\phi = \rho\sin\phi\cos\phi - \rho\cos\phi\sin\phi = 0$$

$$B_\phi = \mathbf{B}\cdot\hat{\boldsymbol{\phi}} = y\hat{\mathbf{x}}\cdot\hat{\boldsymbol{\phi}} - x\hat{\mathbf{y}}\cdot\hat{\boldsymbol{\phi}}$$

$$= -y\sin\phi - x\cos\phi = \rho\sin^2\phi - \rho\cos^2\phi = -\rho$$

Thus,

$$\mathbf{B} = \rho\hat{\boldsymbol{\phi}} + z\hat{\mathbf{z}}$$

4. Check the fundamental theorem of gradients for the function $\mathbf{F} = xy\hat{\mathbf{x}} + 2yz\hat{\mathbf{y}} + 2zx\hat{\mathbf{z}}$, between the origin and the point (1, 1, 1) by three different routes.

(a) (0, 0, 0) → (1, 0, 0) → (1, 1, 0) → (1, 1, 1)

(b) (0, 0, 0) → (0, 0, 1) → (0, 1, 1) → (1, 1, 1)

(c) Parabolic path $z = x^2$ and $y = x$

Solution:

(a) $f(1, 1, 1) = 1 + 4 + 2 = 7$ and $f(0, 0, 0) = 0$

$$f(1, 1, 1) - f(0, 0, 0) = 7$$

$$\nabla f = (2x + 4y)\hat{\mathbf{x}} + (4x + 2z^3)\hat{\mathbf{y}} + (6yz^2)\hat{\mathbf{z}}$$

$$\nabla f \cdot d\mathbf{l} = (2x + 4y)dx + (4x + 2z^3)dy + (6yz^2)dz$$

Segment I: $y = z = dy = dz = 0$, x: $0 \to 1$

$$\int \nabla f \cdot d\mathbf{l} = 2\int_0^1 x\,dx = x^2\Big|_0^1 = 1$$

Segment II: $x = 1$, $z = 0$, $dz = dx = 0$, y: $0 \to 1$

$$\int \nabla f \cdot d\mathbf{l} = \int_0^1 (4x + 2z^3)\,dy = \int_0^1 4dy = 4$$

Segment III: $x = y = 1$, $dx = dy = 0$, z: $0 \to 1$

$$\int \nabla f \cdot d\mathbf{l} = \int_0^1 (6yz^2)dz = \int_0^1 6z^2 dz = 2$$

By adding the contributions from the three segments computed above results in 7. Now we try the second part of the problem.

(b) Segment I: $x = y = dx = dy = 0$, z: $0 \to 1$

$$\int \nabla f \cdot d\mathbf{l} = \int_0^1 (6yz^2)\,dz = 0$$

Segment II: $x = 0$, $z = 1$, $dx = dz = 0$, y: $0 \to 1$

$$\int \nabla f \cdot d\mathbf{l} = \int_0^1 (4x + 2z^3)dy = \int_0^1 2dy = 2$$

Segment III: $y = z = 1$, $dy = dz = 0$, x: $0 \to 1$

$$\int \nabla f \cdot d\mathbf{l} = \int_0^1 (2x + 4y)\,dx = \int_0^1 (2x + 4)\,dx = 5$$

By addition we get 7. Now we move to the third part of the problem.

(c) It is given that $z = x^2$ and $y = x$ resulting in $dz = 2xdx$ and $dy = dx$.

$$\begin{aligned}\nabla f \cdot d\mathbf{l} &= (2x + 4y)dx + (4x + 2z^3)dy + (6yz^2)dz \\ &= (2x + 4y)dx + (4x + 2z^3)dy + (6yz^2)dz \\ &= (2x + 4y)dx + (4x + 2z^3)dy + (6yz^2)dz \\ \int_{0,0,0}^{1,1,1} \nabla f \cdot d\mathbf{l} &= \int_0^1 (10x + 14x^6)dx = 5x^2 + 2x^7\Big|_0^1 = 5 + 2 = 7\end{aligned}$$

5. Check the divergence theorem for the function $\mathbf{F} = xy\hat{\mathbf{x}} + 2yz\hat{\mathbf{y}} + 3zx\hat{\mathbf{z}}$ by considering the volume of a cube with sides of length 2 located with its sides on positive axes and one corner over the origin.

Solution:

$$\begin{aligned}\nabla \cdot \mathbf{F} &= y + 2z + 3x \\ \int (\nabla \cdot \mathbf{F})\,d\tau &= \int (y + 2z + 3x)\,dxdydz \\ &= \int \left[(y + 2z)x + \frac{3}{2}x^2\right]_0^2 dydz = \int (2y + 4z + 6)dydz \\ &= \int \left[(4z + 6)y + y^2\right]_0^2 dz = \int (8z + 16)dz \\ &= (4z^2 + 16z)\Big|_0^2 = 48\end{aligned}$$

Now we evaluate the surface integral over various faces of the cube.

$$\mathbf{da} = dydz\hat{\mathbf{x}},\ x = 2,\ \mathbf{F}\cdot\mathbf{da} = \mathbf{F}\cdot dydz\hat{\mathbf{x}} = 2ydydz,\ \int \mathbf{F}\cdot\mathbf{da} = \int 2ydydz = 8$$

$$\mathbf{da} = -dydz\hat{\mathbf{x}},\ x = 0,\ \mathbf{F}\cdot\mathbf{da} = \mathbf{F}\cdot dydz\hat{\mathbf{x}} = 0,\ \int \mathbf{F}\cdot\mathbf{da} = 0$$

$$\mathbf{da} = dydz\hat{\mathbf{y}},\ y = 2,\ \mathbf{F}\cdot\mathbf{da} = \mathbf{F}\cdot dxdz\hat{\mathbf{y}} = 4zdxdz,\ \int \mathbf{F}\cdot\mathbf{da} = \int 4zdxdz = 16$$

$$\mathbf{da} = -dxdz\hat{\mathbf{y}},\ y = 0,\ \mathbf{F}\cdot\mathbf{da} = \mathbf{F}\cdot dxdz\hat{\mathbf{y}} = 0,\ \int \mathbf{F}\cdot\mathbf{da} = 0$$

$$\mathbf{da} = dxdy\hat{\mathbf{z}},\ z = 2,\ \mathbf{F}\cdot\mathbf{da} = \mathbf{F}\cdot dxdz\hat{\mathbf{y}} = 6xdxdy,\ \int \mathbf{F}\cdot\mathbf{da} = \int 6xdxdy = 24$$

$$\mathbf{da} = -dxdy\hat{\mathbf{z}},\ z = 0,\ \mathbf{F}\cdot\mathbf{da} = \mathbf{F}\cdot dxdz\hat{\mathbf{y}} = 0,\ \int \mathbf{F}\cdot\mathbf{da} = 0$$

$$\oint \mathbf{F}\cdot da = 8 + 16 + 24 = 48$$

6. Check the Stokes' theorem for the function $\mathbf{F} = xy\hat{\mathbf{x}} + 2yz\hat{\mathbf{y}} + 3zx\hat{\mathbf{z}}$ using a triangular area whose corners are located at (0, 0, 0), (0, 2, 0) and (0, 0, 2).

Solution: Stokes' theorem says

$$\int (\nabla\times\mathbf{F})\cdot da = \oint \mathbf{F}\cdot dl$$

We compute left side and right side of this equation and observe their equality. Note here that we assume the closed path in the integral in counterclockwise sense.

$$\nabla\times\mathbf{F} = -2y\hat{\mathbf{x}} - 3z\hat{\mathbf{y}} - x\hat{\mathbf{z}}$$

and

$$\mathbf{da} = dydz\hat{\mathbf{x}}$$

$$(\nabla\times\mathbf{F})\cdot\mathbf{da} = -2ydydz$$

$$\int (\nabla\times\mathbf{F})\cdot da = -2\int_0^2\int_0^{2-z} ydydz = -\int_0^2 (2-z)^2\,dz = -\frac{8}{3}$$

$$\mathbf{F}\cdot d\mathbf{l} = xydx + 2yzdy + 3zxdz$$

There are three segments in the closed path and we calculate the line integral over each of the segments. For the segment from (0, 0, 0) to (0, 2, 0), we have

$$x = z = 0 = dx = dz,\ y = 0 \to 2,\ \mathbf{F}\cdot d\mathbf{l} = 0,\ \int \mathbf{F}\cdot d\mathbf{l} = 0$$

For the segment from (0, 2, 0) to (0, 0, 2), we have

$$x = 0,\ z = 2 - y,\ dx = 0, dz = -dy,\ y = 2 \to 0,\ \mathbf{F}\cdot d\mathbf{l} = 2yzdy$$

$$\int \mathbf{F}\cdot d\mathbf{l} = \int_2^0 2y(2-y)\,dy = -\int_0^2 (4y - 2y^2)\,dy$$

$$= -\left(2y^2 - \frac{2}{3}y^3\right)\Bigg|_0^2 = -\frac{8}{3}$$

For the segment from (0, 0, 2) to (0, 0, 0), we have

$$x = y = 0 = dx = dy,\ z = 2 \to 0,\ \mathbf{F}\cdot d\mathbf{l} = 0,\ \int \mathbf{F}\cdot d\mathbf{l} = 0$$

Therefore,

$$\oint \mathbf{F}\cdot d\mathbf{l} = -\frac{8}{3}$$

7. Let **R** be the separation vector from a fixed point (x', y', z') to another point (x, y, z). Show that $\nabla R^2 = 2\mathbf{R}$ and $\nabla(1/R) = -\hat{\mathbf{R}}/R^2$. Also find the general formula for ∇R^n.

Solution:

$$\mathbf{R} = (x - x')\hat{\mathbf{x}} + (y - y')\hat{\mathbf{y}} + (z - z')\hat{\mathbf{z}}$$

$$R = \sqrt{(x - x')^2 + (y - y')^2 + (z - z')^2}$$

$$\nabla R^2 = \frac{\partial R^2}{\partial x}\hat{\mathbf{x}} + \frac{\partial R^2}{\partial y}\hat{\mathbf{y}} + \frac{\partial R^2}{\partial z}\hat{\mathbf{z}}$$

$$= 2(x - x')\hat{\mathbf{x}} + 2(y - y')\hat{\mathbf{y}} + 2(z - z')\hat{\mathbf{z}} = 2\mathbf{R}$$

$$\nabla\left(\frac{1}{R}\right) = \frac{\partial R^{-1}}{\partial x}\hat{\mathbf{x}} + \frac{\partial R^{-1}}{\partial y}\hat{\mathbf{y}} + \frac{\partial R^{-1}}{\partial z}\hat{\mathbf{z}}$$

$$= -\frac{1}{2}2(x - x')R^{-3/2}\,\hat{\mathbf{x}} - \frac{1}{2}2(y - y')R^{-3/2}\hat{\mathbf{y}} - \frac{1}{2}2(z - z')R^{-3/2}\hat{\mathbf{z}}$$

$$= -R^{-3/2}\left[(x - x')\hat{\mathbf{x}} + (y - y')\hat{\mathbf{y}} + (z - z')\hat{\mathbf{z}}\right]$$

$$= -R^{-3/2}\mathbf{R} = -\frac{\hat{\mathbf{R}}}{R^2}$$

The general formula can be arrived at by the following procedure:

$$\nabla R^n = \frac{\partial R^n}{\partial x}\hat{\mathbf{x}} + \frac{\partial R^n}{\partial y}\hat{\mathbf{y}} + \frac{\partial R^n}{\partial z}\hat{\mathbf{z}}$$

Consider the first term

$$\frac{\partial R^n}{\partial x} = nR^{n-1}\frac{\partial R}{\partial x} = nR^{n-1}\left(\frac{1}{2}\frac{1}{R}2R_x\right) = nR^{n-1}\left(\frac{R_x}{R}\right)$$

Similarly,

$$\frac{\partial R^n}{\partial y} = nR^{n-1}\left(\frac{R_y}{R}\right) \quad\text{and}\quad \frac{\partial R^n}{\partial z} = nR^{n-1}\left(\frac{R_z}{R}\right)$$

$$\nabla R^n = \frac{\partial R^n}{\partial x}\hat{\mathbf{x}} + \frac{\partial R^n}{\partial y}\hat{\mathbf{y}} + \frac{\partial R^n}{\partial z}\hat{\mathbf{z}}$$

$$= nR^{n-1}\left(\frac{R_x\hat{\mathbf{x}} + R_y\hat{\mathbf{y}} + R_z\hat{\mathbf{z}}}{R}\right) = nR^{n-1}\left(\frac{\mathbf{R}}{R}\right) = nR^{n-1}\hat{\mathbf{R}}$$

8. Evaluate the following set of integrals:

(a) $\int_{2}^{6} (3x^2 - 2x - 1)\, \delta(x-3)dx$

(b) $\int_{0}^{5} \cos x \delta(x-\pi)dx$

(c) $\int_{0}^{3} x^3 \delta(x+1)dx$

(d) $\int_{-\infty}^{\infty} \ln(x+3)\, \delta(x+2)dx$

Solution: We use the sampling property of the delta function in evaluating the above integrals.

(a) $\int_{2}^{6} (3x^2 - 2x - 1)\, \delta(x-3)dx = (3x^2 - 2x - 1)\Big|_{x=3} = 20$

(b) $\int_{0}^{5} \cos x \delta(x-\pi)dx = \cos x\Big|_{x=\pi} = -1$

(c) $\int_{0}^{3} x^3 \delta(x+1)dx$

The value of this integral is zero, because the delta located at $x = -1$ is not in the range of the integral.

(d) $\int_{-\infty}^{\infty} \ln(x+3)\, \delta(x+2)dx = \ln(x+3)\Big|_{x=-2} = \ln(-2+3) = 0$

9. Evaluate the following set of integrals:

(a) $\int_{-2}^{2} (2x+3)\, \delta(3x)\, dx$

(b) $\int_{0}^{2} (x^3 + 3x + 2)\delta(1-x)\, dx$

(c) $\int_{-1}^{1} 9x^2 \delta(3x+1)\, dx$

(d) $\int_{-\infty}^{a} \delta(x-b)\, dx$

Solution:

(a) $$\int_{-2}^{2}(2x+3)\,\delta(3x)\,dx = \int_{-2}^{2}(2x+3)\frac{1}{3}\delta(x)\,dx = \frac{1}{3}(2x+3)\Big|_{x=0} = 1$$

(b) $$\int_{0}^{2}(x^3+3x+2)\delta(1-x)\,dx = (x^3+3x+2)\Big|_{x=1} = 6$$

(c) $$\int_{-1}^{1}9x^2\delta(3x+1)\,dx = \int_{-1}^{1}9x^2\frac{1}{3}\delta\left(x+\frac{1}{3}\right)dx = \frac{1}{3}9x^2\Big|_{x=-1/3} = \frac{1}{3}$$

(d) $$\int_{-\infty}^{a}\delta(x-b)\,dx$$

Its value is 1 if $a > b$ and is 0 if $a < b$.

10. Show that

$$x\frac{d}{dx}\delta(x) = -\delta(x)$$

$$\frac{d}{dx}u(x) = \delta(x)$$

Solution:
$$x\frac{d}{dx}\delta(x) = -\delta(x)$$

Consider the integral

$$\int_{-\infty}^{\infty} f(x)\left[x\frac{d}{dx}\delta(x)\right]dx = xf(x)\delta(x)\Big|_{-\infty}^{\infty} - \int_{-\infty}^{\infty}\left\{\frac{d}{dx}[xf(x)]\,\delta(x)\right\}dx$$

$$\frac{d}{dx}u(x) = \delta(x)$$

The first term is zero as $\delta(x)$ is zero at infinity. And

$$\frac{d}{dx}xf(x) = f(x)\frac{d}{dx}x + x\frac{d}{dx}f(x) = f(x) + x\frac{d}{dx}f(x)$$

$$-\int_{-\infty}^{\infty}\left\{\left[f(x)+x\frac{d}{dx}f(x)\right]\delta(x)\right\}dx = 0 - f(0) = -\int_{-\infty}^{\infty}f(x)\delta(x)dx$$

Therefore, by comparing both sides, we get

$$x\frac{d}{dx}\delta(x) = -\delta(x)$$

Now we consider differential of unit step function.

$$\frac{d}{dx}u(x) = \delta(x)$$

Consider the integral

$$\int_{-\infty}^{\infty} f(x)\frac{du}{dx}dx = f(x)\,u(x)\Big|_{-\infty}^{\infty} - \int_{-\infty}^{\infty}\left[\frac{df}{dx}u(x)\right]dx$$

$$= f(\infty) - \int_{0}^{\infty}\frac{df}{dx}dx = f(\infty) - [f(\infty) - f(0)]$$

$$= f(0) = \int_{-\infty}^{\infty} f(x)\,\delta(x)dx$$

Comparing both sides, we get

$$\frac{d}{dx}u(x) = \delta(x)$$

11. Show that

(a) $\int_{v}(\nabla f)\,d\tau = \oint_{s} f\,d\mathbf{a}$

(b) $\int_{v}(\nabla\times\mathbf{F})d\tau = -\oint_{s}\mathbf{F}\times d\mathbf{a}$

(c) $\int_{v}[f\nabla^2 g + (\nabla f)\cdot(\nabla g)]\,d\tau = \oint_{s}(f\nabla g)\cdot d\mathbf{a}$

(d) $\int_{v}[f\nabla^2 g - g\nabla^2 f]\,d\tau = \oint_{s}(f\nabla g - g\nabla f)\cdot d\mathbf{a}$

(e) $\int_{s}(\nabla f)\times d\mathbf{a} = -\oint_{c} f\,d\mathbf{l}$

Solution:

(a) $\int_{v}(\nabla f)\,d\tau = \oint_{s} f\,d\mathbf{a}$

In the divergence theorem

$$\int\nabla\cdot\mathbf{F}\,d\tau = \oint\mathbf{F}\cdot d\mathbf{a}$$

Let $\mathbf{F} = \mathbf{c}f$, where $\mathbf{c}$ is a constant vector.

$$\int\nabla\cdot\mathbf{c}f\,d\tau = \mathbf{c}\cdot\oint f\,d\mathbf{a}$$

$$\mathbf{c}\cdot\int\nabla f\,d\tau = \mathbf{c}\cdot\oint f\,d\mathbf{a} \rightarrow \int\nabla f\,d\tau = \oint f\,d\mathbf{a}$$

(b) $\int_v (\nabla \times \mathbf{F})\, d\tau = -\oint_s \mathbf{F} \times d\mathbf{a}$

In the divergence theorem

$$\int \nabla \cdot \mathbf{F}\, d\tau = \oint \mathbf{F} \cdot d\mathbf{a}$$

Let **F** = **F** × **c** where **c** is a constant vector.

$$\int \nabla \cdot (\mathbf{F} \times \mathbf{c})\, d\tau = \oint (\mathbf{F} \times \mathbf{c}) \cdot d\mathbf{a}$$

$$\int [\mathbf{c} \cdot (\nabla \times \mathbf{F}) - \mathbf{F} \cdot (\nabla \times \mathbf{c})]\, d\tau = \oint (\mathbf{F} \times \mathbf{c}) \cdot d\mathbf{a}$$

$$\int \mathbf{c} \cdot (\nabla \times \mathbf{F})\, d\tau = -\oint \mathbf{c} \cdot (\mathbf{F} \times d\mathbf{a}) \Rightarrow \int_v (\nabla \times \mathbf{F})\, d\tau = -\oint_s \mathbf{F} \times d\mathbf{a}$$

(c) $\int_v [f\nabla^2 g + (\nabla f) \cdot (\nabla g)]\, d\tau = \oint_s (f\nabla g) \cdot d\mathbf{a}$

In the divergence theorem

$$\int \nabla \cdot \mathbf{F}\, d\tau = \oint \mathbf{F} \cdot d\mathbf{a}$$

If we let $\mathbf{F} = f\nabla g$.

$$\int \nabla \cdot f\nabla g\, d\tau = \oint f\nabla g \cdot d\mathbf{a}$$

$$\int [f\nabla \cdot \nabla g + \nabla f \cdot \nabla g]\, d\tau = \oint f\nabla g \cdot d\mathbf{a}$$

$$\int [f\nabla^2 g + \nabla f \cdot \nabla g]\, d\tau = \oint f\nabla g \cdot d\mathbf{a}$$

Hence proved.

(d) $\int_v [f\nabla^2 g - g\nabla^2 f]\, d\tau = \oint_s (f\nabla g - g\nabla f) \cdot d\mathbf{a}$

To prove, we use the result of the previous problem.

$$\int_v [f\nabla^2 g + \nabla f \cdot \nabla g]\, d\tau = \oint f\nabla g \cdot d\mathbf{a}$$

Interchange the scalar functions in the result

$$\int_v [g\nabla^2 f + fg \cdot \nabla f]\, d\tau = \oint g\nabla f \cdot d\mathbf{a}$$

Subtracting the above two equations from one another, we get

$$\int_v [f\nabla^2 g - g\nabla^2 f]\, d\tau = \oint_s (f\nabla g - g\nabla f) \cdot d\mathbf{a}$$

Hence proved.

(e) $\int_s (\nabla f) \times d\mathbf{a} = -\oint_c f\,d\mathbf{l}$

Using the Stokes' theorem

$$\int (\nabla \times \mathbf{F}) \cdot d\mathbf{a} = \oint \mathbf{F} \cdot d\mathbf{l}$$

Let $\mathbf{F} = \mathbf{c}f$ where $\mathbf{c}$ is a constant vector.

$$\int (\nabla \times \mathbf{c}f) \cdot d\mathbf{a} = \oint \mathbf{c}f \cdot d\mathbf{l}$$

$$\int [f(\nabla \times \mathbf{c}) - \mathbf{c} \times \nabla f] \cdot d\mathbf{a} = \oint \mathbf{c}f \cdot d\mathbf{l}$$

$$-\int \mathbf{c} \times \nabla f \cdot d\mathbf{a} = \oint \mathbf{c}f \cdot d\mathbf{l}$$

$$-\int \mathbf{c} \cdot (\nabla f \times d\mathbf{a}) = \oint \mathbf{c}f \cdot d\mathbf{l} \Rightarrow \int_s (\nabla f) \times d\mathbf{a} = -\oint_c f\,d\mathbf{l}$$

12. Find the divergence of the function

$$\mathbf{F} = \frac{\hat{\mathbf{r}}}{r}$$

What is the general formula for the divergence and curl of $r^n\hat{\mathbf{r}}$?

Solution:
$$\nabla \cdot \mathbf{F} = \frac{1}{r^2}\frac{\partial}{\partial r}\left(r^2 \frac{1}{r}\right) = \frac{1}{r^2}\frac{\partial}{\partial r}(r) = \frac{1}{r^2}$$

$$\nabla \cdot r^n\hat{\mathbf{r}} = \frac{1}{r^2}\frac{\partial}{\partial r}(r^2 r^n) = \frac{1}{r^2}\frac{\partial}{\partial r}(r^{n+2}) = \frac{1}{r^2}(n+2)r^{n+1} = (n+2)r^{n-1}$$

$$\nabla \times r^n\hat{\mathbf{r}} = \frac{1}{r^2 \sin\theta}\begin{vmatrix} \hat{\mathbf{r}} & r\hat{\boldsymbol{\theta}} & r\sin\theta\hat{\boldsymbol{\phi}} \\ \dfrac{\partial}{\partial r} & \dfrac{\partial}{\partial \theta} & \dfrac{\partial}{\partial \phi} \\ r^n & 0 & 0 \end{vmatrix} = 0$$

EXERCISE SECTION

Essay Type Questions

1. Show that the absolute value of a scalar triple product is equal to the volume of the parallelepiped having the three vectors as three of its contiguous edges.
2. Show that a divergence-less vector can be expressed as curl of an arbitrary vector function and curl-less vector can be expressed as gradient of an arbitrary scalar function.
3. State the gradient, divergence and curl theorems along with their implications.

4. Describe the coordinate systems: Cartesian, cylindrical and spherical mentioning names, ranges of coordinates along with coordinate surfaces.
5. Define gradient, divergence and curl operations giving physical meaning of each of these operations.

Numerical Problems

1. Derive the conversion relations between
   (i) Cartesian and cylindrical
   (ii) Cartesian and spherical
   (iii) Cylindrical and spherical
2. Derive expressions for differential length, area and volume in the case of
   (i) Cartesian
   (ii) Cylindrical
   (iii) Spherical
3. Derive expressions for distance between two arbitrary points in
   (i) Cylindrical
   (ii) Spherical
4. Derive expressions for gradient and divergence in the case of
   (i) Cartesian
   (ii) Cylindrical
   (iii) Spherical
5. Derive expressions for curl and Laplacian in the case of
   (i) Cartesian
   (ii) Cylindrical
   (iii) Spherical

Fill in the Blanks

1. A divergence-less vector can be expressed as __________.
2. A curl-less vector can be expressed as __________.
3. The circulation $\oint \mathbf{A} \cdot d\mathbf{l} = 0$ only when vector function **A** is __________.
4. The flux $\oint \mathbf{A} \cdot d\mathbf{a} = 0$ only when vector function **A** is __________.
5. Circulation of a vector **A**, by definition, is __________.
6. Flux of a vector **A**, by definition, is __________.
7. Curl of a vector field is __________.
8. Divergence of a vector field is __________.
9. The direction of a spherical surface is __________.

10. The differential area over the *xy*-plane is __________.
11. The differential area of a spherical surface is __________.
12. The differential area of the curved face of a cylinder is __________.
13. The coordinates of the origin in a spherical coordinate system are __________.
14. The coordinates of the point on the *z*-axis in a cylindrical coordinate system is __________.
15. The range of a polar angle θ is __________.

Answers

1. curl of another vector function
2. gradient of a scalar function
3. curl of another vector function
4. curl of another vector function
5. $\oint \mathbf{A} \cdot d\mathbf{l}$
6. $\oint \mathbf{A} \cdot d\mathbf{a}$
7. circulation per unit area
8. flux per unit volume
9. radial $\hat{\mathbf{r}}$
10. $dxdy\hat{\mathbf{z}}$
11. $r^2 \sin\theta \, d\theta \, d\phi \hat{\mathbf{r}}$
12. $\rho d\phi dz\hat{\boldsymbol{\rho}}$
13. $(0, \theta, \phi)$
14. $(0, \phi, z)$
15. 0 to π

Match the Following

Set I

1. x, y, z	(a) 0 to π
2. θ	(b) $-\infty$ to ∞
3. ϕ	(c) 0 to ∞
4. r, ρ	(d) 0 to 2π
5. r, ρ, θ, ϕ	(e) non-uniform

Set II

1. x	(a) $\tan^{-1}(\rho/z)$
2. y	(b) $r \sin\theta$
3. z	(c) $r \cos\theta$
4. ρ	(d) $r \sin\theta \sin\phi$
5. θ	(e) $r \sin\theta \cos\phi$

Set III

1. Divergence-less vector	(a) gradient of scalar
2. Curl-less vector	(b) curl of another vector

3. Circulation	(c) $\oint \mathbf{A} \cdot d\mathbf{a}$
4. Flux	(d) $\oint \mathbf{A} \cdot d\mathbf{l}$
5. Laplacian	(e) ∇^2

Set IV

1. Divergence	(a) $(0, \theta, \phi)$
2. Curl	(b) $\hat{\boldsymbol{\rho}}$ direction
3. Spherical surface	(c) $\hat{\mathbf{r}}$ direction
4. Cylindrical surface	(d) flux per unit volume
5. Origin	(e) circulation per unit area

Answers

Set I:	1. (b)	2. (a)	3. (d)	4. (c)	5. (e)
Set II:	1. (e)	2. (d)	3. (c)	4. (b)	5. (a)
Set III:	1. (b)	2. (a)	3. (d)	4. (c)	5. (e)
Set IV:	1. (e)	2. (d)	3. (c)	4. (b)	5. (a)

Multiple Choice Questions

1. A divergence-less vector can be expressed as
   (a) curl of another vector function (b) div of another vector function
   (c) gradient of a scalar function (d) hybrid
2. A curl-less vector can be expressed as
   (a) curl of another vector function (b) div of another vector function
   (c) gradient of a scalar function (d) hybrid
3. $\mathbf{A} \cdot (\mathbf{B} \times \mathbf{C})$ is equal to
   (a) $\mathbf{B} \cdot (\mathbf{C} \times \mathbf{A})$ (b) $\mathbf{C} \cdot (\mathbf{A} \times \mathbf{B})$
   (c) both (a) and (b) (d) none of these
4. $\nabla \cdot (\mathbf{A} \times \mathbf{B})$ is equal to
   (a) $\mathbf{A} \cdot (\nabla \times \mathbf{B}) + \mathbf{B} \cdot (\nabla \times \mathbf{A})$ (b) $-\mathbf{A} \cdot (\nabla \times \mathbf{B}) + \mathbf{B} \cdot (\nabla \times \mathbf{A})$
   (c) both (a) and (b) (d) none of these
5. The line integral becomes independent of path when the vector function is
   (a) gradient of a scalar function (b) curl of another vector function
   (c) both (a) and (b) (d) none of these
6. The surface integral depends only on the boundary line when the vector function is
   (a) gradient of a scalar function (b) curl of another vector function
   (c) both (a) and (b) (d) none of these

7. The circulation $\oint \mathbf{A} \cdot d\mathbf{l} = 0$ only when vector function **A** is
   (a) gradient of a scalar function
   (b) curl of another vector function
   (c) both (a) and (b)
   (d) none of these
8. The flux $\oint \mathbf{A} \cdot d\mathbf{a} = 0$ only when vector function **A** is
   (a) gradient of a scalar function
   (b) curl of another vector function
   (c) both (a) and (b)
   (d) none of these
9. The operator del ∇ is
   (a) vector operator
   (b) three-dimensional operator
   (c) partial differential operator
   (d) all of these
10. The operator Laplacian ∇^2 is
    (a) a scalar operator
    (b) a three-dimensional operator
    (c) a partial differential operator
    (d) all of these
11. Circulation of a vector **A**, by definition, is
    (a) $\oint \mathbf{A} \cdot d\mathbf{l}$
    (b) $\int \mathbf{A} \cdot d\mathbf{a}$
    (c) $\oint \mathbf{A} \cdot d\mathbf{a}$
    (d) none of these
12. Flux of a vector **A**, by definition, is
    (a) $\oint \mathbf{A} \cdot d\mathbf{l}$
    (b) $\int \mathbf{A} \cdot d\mathbf{a}$
    (c) $\int \mathbf{A} \cdot d\mathbf{l}$
    (d) none of these
13. Curl of a vector field is
    (a) circulation per unit area
    (b) circulation per unit volume
    (c) flux per unit area
    (d) flux per unit volume
14. Divergence of a vector field is
    (a) circulation per unit area
    (b) circulation per unit volume
    (c) flux per unit area
    (d) flux per unit volume
15. The direction of a spherical surface is
    (a) radial $\hat{\mathbf{r}}$
    (b) radial $\hat{\boldsymbol{\rho}}$
    (c) upwards $\hat{\mathbf{z}}$
    (d) downwards $\hat{\mathbf{z}}$
16. The directions of surfaces of a cylinder are
    (a) downwards $-\hat{\mathbf{z}}$
    (b) radial $\hat{\boldsymbol{\rho}}$
    (c) upwards $\hat{\mathbf{z}}$
    (d) all of these
17. The differential area over the *xy*-plane is
    (a) $dxdy$
    (b) $dxdy\hat{\mathbf{z}}$
    (c) $dz\hat{\mathbf{z}}$
    (d) none of these

18. The differential area of a spherical surface is
    (a) $r\sin\theta d\theta d\phi\hat{\mathbf{r}}$ (b) $r^2\sin\theta d\theta dz\hat{\mathbf{r}}$
    (c) $\sin\theta d\theta d\phi\hat{\mathbf{r}}$ (d) $r^2\sin\theta d\theta d\phi\hat{\mathbf{r}}$
19. The differential area of the curved face of a cylinder is
    (a) $d\phi dz\hat{\boldsymbol{\phi}}$ (b) $\rho d\phi dz\hat{\boldsymbol{\phi}}$
    (c) $\rho d\phi d\rho\hat{\boldsymbol{\phi}}$ (d) $\rho d\rho dz\hat{\boldsymbol{\phi}}$
20. The coordinates of the origin in a spherical coordinate system are
    (a) (0, 0, 0) (b) $(0, 0, \pi)$
    (c) $(0, \theta, \phi)$ (d) $(0, \pi, \pi)$
21. The coordinates of the point on the z-axis in a cylindrical coordinate system are
    (a) $(0, 0, z)$ (b) $(0, \phi, 0)$
    (c) $(0, \phi, z)$ (d) $(0, \pi, z)$
22. The range of polar angle θ is
    (a) 0 to 2π (b) 0 to π
    (c) 0 to $\pi/2$ (d) none of these

Answers

1. (a)	2. (c)	3. (c)	4. (b)	5. (a)
6. (b)	7. (b)	8. (b)	9. (d)	10. (d)
11. (a)	12. (b)	13. (a)	14. (d)	15. (a)
16. (d)	17. (b)	18. (d)	19. (b)	20. (c)
21. (c)	22. (b)			

UNIT II

ELECTROSTATICS

2

Electric Force and Field Intensity

The most important single ingredient in the formula of success is knowing how to get along with people.

—Theodore Roosevelt

In this chapter, charges, which are sources of fields, are introduced describing their properties and shapes or distributions. They can exert force, called electrostatic force, on each other even without coming into contact. A law which quantifies the electrostatic force, Coulomb's law, is described. This law is considered as one of the two fundamental laws based on which the entire electrostatics is built.

The regions that surround the charge distributions are special in more than one way and they are called electrostatic fields. The concept of field is introduced after a thorough discussion on electrostatic force.

2.1 Charge Distributions

The source of the electric field is electric charge and it exhibits three important and well-known properties. The first one is regarding polarity: in nature, we can find charges with two polarities only. The second one is conservation of electric charge which means that the charge can neither be created nor be destroyed. The third property is regarding quantization which means the charge is always available as integer multiples of charge on electron.

Charge, in nature, is available in only two varieties, i.e. positive and negative polarities. When it comes to their distribution, i.e. shape, it exists, in general in four types: point charges, line charges, surface charges and volume charges (Figure 2.1). Sometimes the

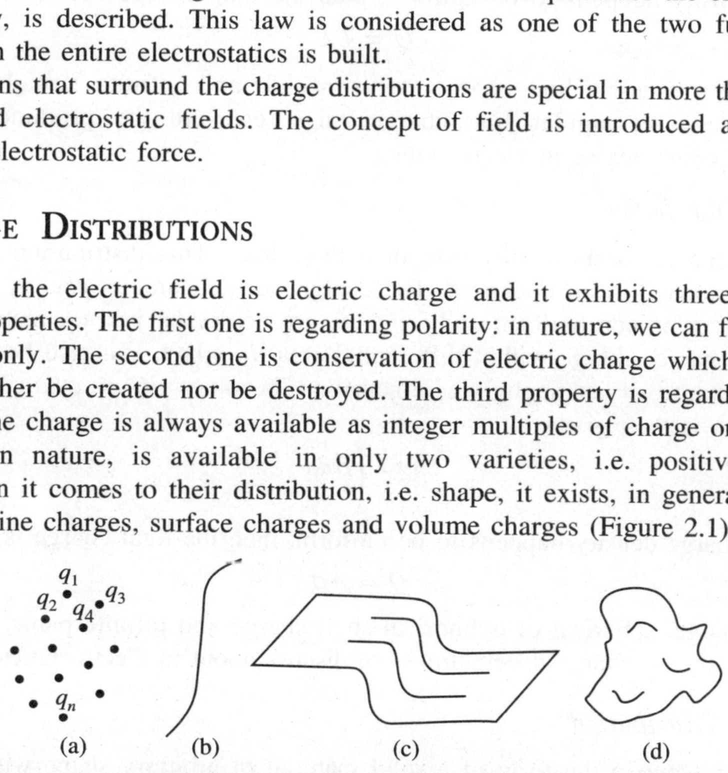

FIGURE 2.1 Types of charge distributions: (a) discrete, (b) line, (c) surface and (d) volume.

dimensions of the charge are very small and they are considered dimensionless in all the calculations. These are called point charges as they are similar to geometrical points dimension-wise. Apart from the discrete or point form, the charge also occurs in line form, thin sheet form and in a solid form having some volume; all the three forms are called continuous charge distributions.

Line Charge Distribution

When the shape of the charge is in the form of a thin line, then it is said to be line charge distribution. This distribution has only the length dimension with no area or volume. The charge per unit length is usually indicated by λ which may be uniform or non-uniform. Its units are Coulombs per metre or C/m.

The charge within a differential length dl called differential charge becomes $dQ = \lambda dl$. The dimensions of the differential charge are too small; we can consider it as a point charge in all the calculations. The total charge within a length L can be obtained from the differential charge using the relation

$$Q = \int_L \lambda dl \tag{2.1}$$

If the charge density happens to be uniform, then the total charge becomes

$$Q = L\lambda \tag{2.2}$$

Finite length straight line charge, semi-infinite and infinite straight line charges, circular, square or rectangular line charge rings, etc. are the prominent line charge distributions we come across in electrostatics.

Surface Charge Distribution

Sometimes the charge exists in the form of a thin sheet. This distribution has both length and width dimensions but no thickness. The charge per unit area is usually indicated by σ. It may be uniform or non-uniform and its units are coulombs per sq. metre (C/m^2). Then the differential charge, charge within differential area da which can also be considered as a point charge because of its negligible dimensions, becomes $dQ = \sigma da$. The total charge within an area A is then given by

$$Q = \int_A \sigma da \tag{2.3}$$

If the surface charge density happens to be uniform, then the total charge is given by

$$Q = A\sigma \tag{2.4}$$

Circular disc charge, spherical or cylindrical shell charge and infinite plane sheet charge are some of the prominent charge distributions we discuss about in electrostatics.

Volume Charge Distribution

Charge can also occur in the form of a solid material of arbitrary shape with some volume. This distribution can have all the three dimensions: length, width and thickness. The charge per unit volume is usually indicated by ρ. It may be uniform or non-uniform and its units are coulombs per cubic metre (C/m^3). Then the differential charge, the charge within differential volume $d\tau$ which can also be considered as a point charge because of its negligible dimensions, becomes $dQ = \rho d\tau$. The total charge within a volume V is then given by

$$Q = \int_V \rho d\tau \tag{2.5}$$

If the volume charge density happens to be uniform, then the total charge is given by

$$Q = V\rho \tag{2.6}$$

Solid spherical charge is one that belongs to this category.

Dirac Delta Representation of Charges

In the coming sections, we come across situations where we need to integrate point charges, line charges and surface charges over an enclosing volume. In such cases, it is required and useful to represent them with delta function.

A point charge Q located at $\mathbf{r} = \mathbf{r}_0$ is represented by a volume charge density given by

$$\rho(\mathbf{r}) = Q\delta(\mathbf{r} - \mathbf{r}_0) \tag{2.7}$$

And in terms of one-dimensional deltas

$$\rho(\mathbf{r}) = Q\delta(x - x_0)\ \delta(y - y_0)\ \delta(z - z_0) \tag{2.8}$$

For a point charge Q at the origin, the volume charge density can be represented as

$$\rho(\mathbf{r}) = Q\delta(x)\ \delta(y)\ \delta(z) \tag{2.9}$$

Line and surface charge distributions also can be represented with Dirac delta. The volume charge density of a line charge with density λ C/m lying along the x-axis is given by

$$\rho(\mathbf{r}) = \lambda\delta(y)\ \delta(z) \tag{2.10}$$

And for a charge sheet lying over the xy-plane with a surface charge density σ C/m^2 is represented with a volume charge density given by

$$\rho(\mathbf{r}) = \sigma\delta(z) \tag{2.11}$$

2.2 COULOMB'S LAW

Since time immemorial, it is known to the mankind regarding the existence of charge, its dual polarity, the force between charges and the dependence of force on medium properties. During the last decade of the sixteenth century, it was Dr. Gilbert who really did some truly experimentally work on static electricity. Later, Colonel Charles Augustine de Coulomb of France performed a series of experiments to determine the force, quantitatively, that exists between charges and published his results which later came to be known as Coulomb's law.

It states that the force on a stationary point charge due to another stationary point charge is:

1. proportional to the product of the point charges
2. inversely proportional to square of the distance between them
3. along the line joining the point charges

The electric force, in addition to the above-mentioned three points, is also characterized by the following three points:

1. It is repulsion if the charges are like and attraction in the case of unlike charges.
2. It is inversely proportional to the ε_r of the medium.
3. The proportionality constant is $1/4\pi$ in MKS system.

Mathematically, all the above points, describing the electric force can be put in the following form:

$$\mathbf{F} = \frac{1}{4\pi\varepsilon}\frac{Qq}{R^2}\hat{\mathbf{R}} \tag{2.12}$$

Here, **F** is the force on a point charge q due to another point charge Q separated by a distance R with unit vector $\hat{\mathbf{R}}$ pointing from Q to q.

The charge Q, which is the cause of the force in the present context, is called 'source charge' and its location is called 'source point'. Whereas the charge q on which force is being observed is called 'test charge' and it is required to be so small, for the validity of the above formula, that the existing field should no way get disturbed by its presence. The location of test charge is called 'observation point' or 'field point'. Both the charges must be point charges, i.e. their dimensions must be negligibly small. In addition, the source charge also experiences force due to the test charge and it is, according to Newton's third law, equal in magnitude but opposite in direction to the force due to source charge.

Point charges: For the validity of Coulomb's law, the charges need to be point charges. It does not necessarily mean that the absolute dimensions should be small or negligible. It only means that their sizes need to be negligibly small when compared to the distance between them. It implies that the absolute dimensions of the charges can be large but their relative sizes need to be negligible. As an example, we can consider the forces to be revolving electrons and nucleus of an atom and that between the protons of a nucleus. In the first case, the Coulomb's law gives exact results, whereas in the second case, as the separation distance between the protons is comparable to their dimensions, the law cannot be applied to estimate the force. Another examples is also worth considering where an earth-sized charge is separated from a moon-sized charge by a distance equal to the distance between the earth and the moon. In this case, we can apply the law to find the force, because even though. The absolute sizes of the charges are large, their relative sizes in comparison with their separation distance are negligibly small and hence the application of the law gives valid results.

Permittivity: In the above expressions, the parameter ε represents the nature of medium in which the charges are placed. It is called the permittivity of the medium and it is the product of two factors given by $\varepsilon = \varepsilon_0\varepsilon_r$. The first factor is called permittivity of free space and its value is $\varepsilon_0 = 10^{-9}/36\pi$ F/m. And the second factor, a dimensionless quantity, is known as relative permittivity or dielectric constant of the medium and for free space, its value is one.

If ε remains constant, i.e. the same throughout the medium, then the medium is called homogeneous, otherwise the medium is said to be inhomogeneous. For certain media, called isotropic, it is a scalar quantity. There exist some media for which the permittivity is direction dependent: in that case, the media are called anisotropic, about which we discuss further in the section on dielectrics. The relative permittivities for different materials are given in Table 2.1.

Table 2.1 Relative permittivity ε_r

Material	ε_r	Material	ε_r
Air	1	Sands (dry)	4
Alcohol (ethyl)	25	Silica (fused)	3.8
Bakelite	4.8	Snow	3.3
Glass	4–7	Sodium chloride	5.9
Ice	4.2	Soil (dry)	2.8
Mica (ruby)	5.4	Styrofoam	1.03
Nylon	4	Teflon	2.1
Paper	2–4	Water (distilled)	80
Plexiglass	2.6–3.5	Water (sea)	20
Polyethylene	2.25	Water (dehydrated)	1
Polystyrene	2.55	Wood (dry)	1.5–4
Porcelain (dry process)	6	Ground (wet)	5–30
Quartz (fused)	3.80	Ground (dry)	2–5
Rubber	2.5–4	Water (fresh)	80

The significance of this law extends far beyond to be a simple expression, merely giving electrostatic force. It is the law which is the starting point of electrostatics. In fact, it is one of the two laws, the other being Gauss law, on which the entire electrostatics is based. As such, its importance can never be undermined. However, there are two fundamental limitations as far as the applicability of this law is concerned. One is that it can be applied only to point charges. The second one is that both the charges must remain stationary.

The force relation given by Eq. (2.7) is said to be in coordinate-free form because it does not refer to any coordinate system. We will see soon that it is more useful if the relation is available with reference to a coordinate system, either in terms of the position vectors or the coordinates of the charges. Now we try to derive the relations.

Let us suppose a point charge q is located at the tip of position vector $\mathbf{r}_2$ and another charge Q is located at the tip of position vector $\mathbf{r}_1$. The vector joining the charge Q to charge q is, let us say, $\mathbf{R}$. It is related to the position vectors of the charges by

$$\mathbf{R} = \mathbf{r}_2 - \mathbf{r}_1 \tag{2.13}$$

And the distance between the locations of the charges is

$$R = |\mathbf{r}_2 - \mathbf{r}_1| \tag{2.14}$$

The unit vector along this vector is

$$\hat{\mathbf{R}} = \frac{\mathbf{R}}{|\mathbf{R}|} = \frac{(\mathbf{r}_2 - \mathbf{r}_1)}{|\mathbf{r}_2 - \mathbf{r}_1|} \tag{2.15}$$

Substituting these values in Eq. (2.12), we obtain

$$\mathbf{F} = \frac{1}{4\pi\varepsilon} \frac{Qq(\mathbf{r}_2 - \mathbf{r}_1)}{|\mathbf{r}_2 - \mathbf{r}_1|^3} \tag{2.16}$$

Therefore, if the locations of the charges are specified through position vectors as $Q(\mathbf{r}_1)$ and $q(\mathbf{r}_2)$, then the force on q due to Q is as given by Eq. (2.16), which is the desired relation or Coulomb's force law in terms of position vectors of the charges. Now we move to find out the force law in terms of the coordinates of the charges. Coulomb's law in different forms is shown in Figure 2.2.

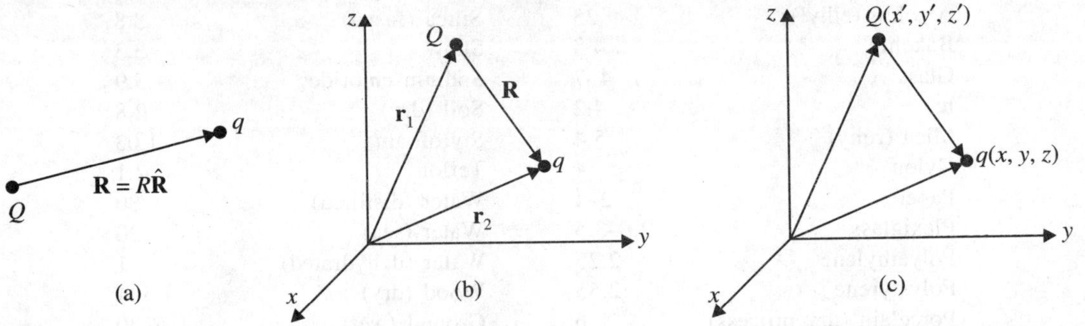

FIGURE 2.2 Coulomb's law in (a) coordinate-free form, (b) in terms of position vectors and (c) in Cartesian system.

Let us suppose that charge q is located at (x, y, z) and another point charge Q is located at (x', y', z'). It is common practice to denote coordinates of source charge with a prime and those of test charge without a prime.

The distance between the locations of the charges is

$$R = [(x - x')^2 + (y - y')^2 + (z - z')^2]^{1/2} \tag{2.17}$$

And the vector joining the charge Q to charge q is

$$\mathbf{R} = (x - x')\hat{\mathbf{x}} + (y - y')\hat{\mathbf{y}} + (z - z')\hat{\mathbf{z}} \tag{2.18}$$

And the unit vector along this vector is

$$\hat{\mathbf{R}} = \frac{\mathbf{R}}{|\mathbf{R}|} = \frac{\mathbf{R}}{R} = \frac{(x - x')\hat{\mathbf{x}} + (y - y')\hat{\mathbf{y}} + (z - z')\hat{\mathbf{z}}}{[(x - x')^2 + (y - y')^2 + (z - z')^2]^{1/2}} \tag{2.19}$$

Substituting these expressions, in the above Coulomb's formula for force, we get

$$\mathbf{F} = \frac{1}{4\pi\varepsilon} \frac{Qq\left[(x - x')\hat{\mathbf{x}} + (y - y')\hat{\mathbf{y}} + (z - z')\hat{\mathbf{z}}\right]}{\left[(x - x')^2 + (y - y')^2 + (z - z')^2\right]^{3/2}} \tag{2.20}$$

Therefore, if the locations of the charges are specified as $Q(x', y', z')$, $q(x, y, z)$ and $q(x_2, y_2, z_2)$, then the force on q due to Q can be found using Eq. (2.20). This is the Coulomb's force law in terms of coordinates of the charges.

Force due to several discrete charges: An important property of electric forces is the principle of linear superposition, which states that the total electric force on a point charge due to a system of point charges is equal to the vector sum of the electric forces on that point charge due to the individual charges. We can apply the linear property of the electric force to find the force due to several point charges as follows:

Let us suppose, the force on test q charge due to Q_1 is $\mathbf{F}_1$, due to Q_2 is $\mathbf{F}_2$,... and due to Q_n is $\mathbf{F}_n$ when these charges are acting alone. Then, by superposition, the total force on q due to the entire set of n discrete charges Q_1, Q_2, ..., Q_n becomes

$$\mathbf{F} = \mathbf{F}_1 + \mathbf{F}_2 + \cdots + \mathbf{F}_n = \frac{1}{4\pi\varepsilon}\left[\frac{Q_1 q}{R_1^2}\hat{\mathbf{R}}_1 + \frac{Q_1 q}{R_1^2}\hat{\mathbf{R}}_2 + \cdots + \frac{Q_n q}{R_n^2}\hat{\mathbf{R}}_n\right]$$

$$= \frac{1}{4\pi\varepsilon}\sum_{k=1}^{n}\frac{Q_k q}{R_k^2}\hat{\mathbf{R}}_k \tag{2.21}$$

2.3 ELECTRIC FIELD INTENSITY

In this section, we move into a new concept, called electric field, smoothly from a well-understood phenomenon of electric force.

The regions that exist in the vicinity of stationary charges exhibit some special properties and these regions are given a name *electrostatic fields*, i.e. electric fields due to stationary charges. One important property of these fields is that they exert force on charges that are placed within the regions of their influence. The amount of force the field exerts on charges depends on one property of the field called electric field intensity which is indicated by English capital letter **E**. It is a vector quantity and because of this reason, boldface letter is used for its representation. Its units are volts per metre (V/m). Formally it can be defined as the force exerted by the field on a unit positive charge.

An expression for the electric field intensity due to a point charge can be derived by combining Coulomb's law with the definition of the field intensity. According to the Coulomb's law, the force on a point charge q due to another point charge Q is given by

$$\mathbf{F} = \frac{1}{4\pi\varepsilon}\frac{Qq}{R^2}\hat{\mathbf{R}} \qquad \text{[Refer to Eq. (2.12)]}$$

Now, the force on unit positive charge, i.e. if q = +1, by definition, becomes the field intensity due to Q. Therefore, the electric field intensity **E** due to a point charge Q becomes

$$\mathbf{E} = \frac{\mathbf{F}}{q} = \frac{1}{4\pi\varepsilon}\frac{Q}{R^2}\hat{\mathbf{R}} \tag{2.22}$$

This is an important relation, which is considered a fundamental one, used in the development of electrostatics. Several points regarding the field due to a point charge can be observed from the above relation. They are given as follows:

1. The electric field intensity depends on the amount and polarity of charge and the distance from the charge.

2. If the charge is positive, the direction of the electric field intensity is from the point charge radially outward and if the charge is negative, it is directed radially towards the charge.
3. The electric field intensity varies inversely with the square of the distance from the charge. At equal distances from the charge, the magnitude of the electric field intensity is constant.

Now we try to find out the field intensity due to a set of discrete charges using linearity and consequent superposition property of the electric force (Figure 2.3). The force due to several discrete charges is

$$\mathbf{F} = \frac{1}{4\pi\varepsilon} \sum_{k=1}^{n} \frac{Q_k q}{R_k^2} \hat{\mathbf{R}}_k \quad \text{[Refer to Eq. (2.21)]}$$

The force then on a unit positive charge becomes

$$\mathbf{E} = \frac{\mathbf{F}}{q} = \frac{1}{4\pi\varepsilon} \sum_{k=1}^{n} \frac{Q_k}{R_k^2} \hat{\mathbf{R}}_k \tag{2.23}$$

which is the expression for field intensity due to several discrete charges. We can also note that the electric field intensity obeys the superposition principle.

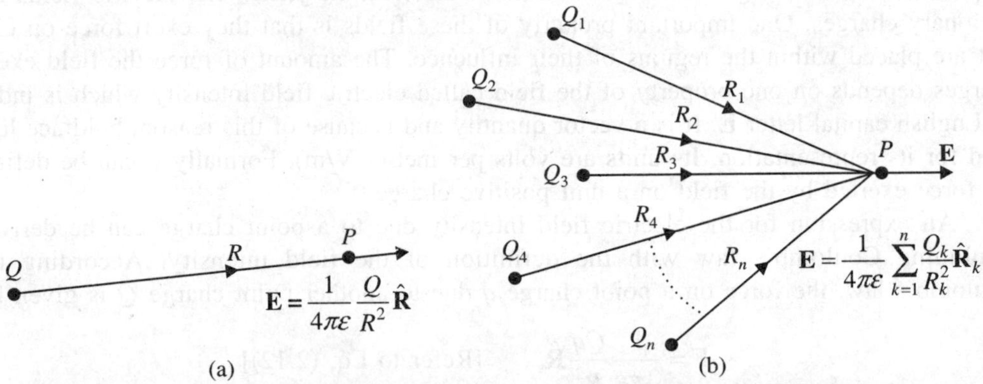

FIGURE 2.3 Electric field intensity due to discrete charge distributions.

Continuous charge distributions: To find out the field intensity due to continuous charge distributions, we follow a strategy: we divide the total charge distribution into pieces, each of which is small enough to be considered as differential charge and hence point charge. Each piece of the charge gives rise to its own field and according to the superposition principle, the field intensity due to the entire charge distribution, then, is the vector sum of the intensities due to the individual pieces. We have already expressions for the field intensity of point charge and using them we express the field due to charge within a piece and then it is integrated over the entire source distribution with appropriate limits to get the field intensity.

The charge dQ, called differential charge, that exists within a differential element, i.e. differential length or area or volume can be considered as a point charge because of its negligibly small dimension (Figure 2.4). If we call the field intensity due to differential charge as differential field $d\mathbf{E}$, then

$$d\mathbf{E} = \frac{1}{4\pi\varepsilon}\frac{dQ}{R^2}\hat{\mathbf{R}} \tag{2.24}$$

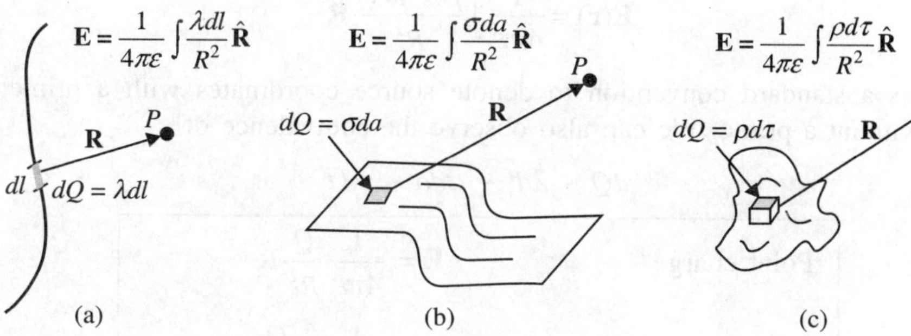

FIGURE 2.4 Electric field intensity of continuous charge distributions (a) line chargo, (b) surface charge and (c) volume charge.

Then the field intensity due to the entire continuous charge distribution becomes

$$\mathbf{E} = \int d\mathbf{E} = \frac{1}{4\pi\varepsilon}\int\frac{dQ}{R^2}\hat{\mathbf{R}} \tag{2.25}$$

Now for line charge distributions, the differential charge $dQ = \lambda dl$. Hence,

$$\mathbf{E} = \frac{1}{4\pi\varepsilon}\int\frac{dQ}{R^2}\hat{\mathbf{R}} = \frac{1}{4\pi\varepsilon}\int\frac{\lambda dl}{R^2}\hat{\mathbf{R}} \tag{2.26}$$

for line charge distribution.

Similarly, for surface charge distributions, the differential charge $dQ = \sigma da$. Hence,

$$\mathbf{E} = \frac{1}{4\pi\varepsilon}\int\frac{dQ}{R^2}\hat{\mathbf{R}} = \frac{1}{4\pi\varepsilon}\int\frac{\sigma da}{R^2}\hat{\mathbf{R}} \tag{2.27}$$

And for volume charge distributions, the differential charge $dQ = \rho d\tau$. Hence, the field expression is

$$\mathbf{E} = \frac{1}{4\pi\varepsilon}\int\frac{dQ}{R^2}\hat{\mathbf{R}} = \frac{1}{4\pi\varepsilon}\int\frac{\rho d\tau}{R^2}\hat{\mathbf{R}} \tag{2.28}$$

If the position vectors of the source point and field point are $\mathbf{r}'$ and $\mathbf{r}$, respectively, then the vector joining the source point with the field point becomes $\mathbf{R} = \mathbf{r} - \mathbf{r}'$, and the electric field intensity for line charge distribution becomes

$$\mathbf{E}(\mathbf{r}) = \frac{1}{4\pi\varepsilon}\int\frac{\lambda(\mathbf{r}')dl'}{R^2}\hat{\mathbf{R}} \tag{2.29}$$

Similarly, for surface charge distributions, it becomes

$$\mathbf{E}(\mathbf{r}) = \frac{1}{4\pi\varepsilon}\int \frac{\sigma(\mathbf{r}')da'}{R^2}\hat{\mathbf{R}} \tag{2.30}$$

And for volume charge distributions, the field expression is

$$\mathbf{E}(\mathbf{r}) = \frac{1}{4\pi\varepsilon}\int \frac{\rho(\mathbf{r}')d\tau'}{R^2}\hat{\mathbf{R}} \tag{2.31}$$

Note that it is a standard convention to denote source coordinates with a prime and field coordinates without a prime. We can also observe the equivalence of

$$dQ \sim \lambda dl \sim \sigma da \sim \rho d\tau \tag{2.32}$$

Point charge	$\mathbf{E} = \frac{1}{4\pi\varepsilon}\frac{Q}{R^2}\hat{\mathbf{R}}$
Discrete distribution	$\mathbf{E} = \frac{1}{4\pi\varepsilon}\sum_{k=1}^{n}\frac{Q_k}{R_k^2}\hat{\mathbf{R}}_k$
Line charge	$\mathbf{E}(\mathbf{r}) = \frac{1}{4\pi\varepsilon}\int \frac{\lambda(\mathbf{r}')dl'}{R^2}\hat{\mathbf{R}}$
Surface charge	$\mathbf{E}(\mathbf{r}) = \frac{1}{4\pi\varepsilon}\int \frac{\sigma(\mathbf{r}')da'}{R^2}\hat{\mathbf{R}}$
Volume charge	$\mathbf{E}(\mathbf{r}) = \frac{1}{4\pi\varepsilon}\int \frac{\rho(\mathbf{r}')d\tau'}{R^2}\hat{\mathbf{R}}$

2.3.1 Infinite Straight Line Charge

The infinite line charge exhibits cylindrical symmetry as line is a zero-radius cylinder and hence we use cylindrical coordinate system in solving the problem.

Place the line charge along the z-axis with the observation point P in the xy-plane. Note that the charge it extends is from $z = -\infty$ to $z = \infty$. The electric field intensity of an infinite straight line charge is shown in Figure 2.5.

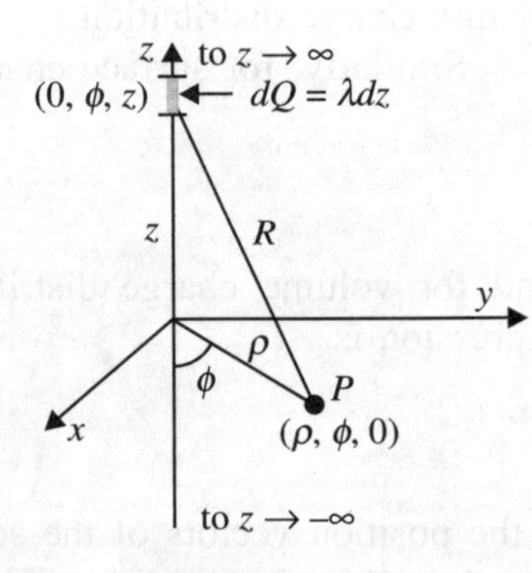

FIGURE 2.5 Electric field intensity of an infinite straight line charge.

The field intensity at $(\rho, \phi, 0)$ due to differential line charge at $(0, \phi, z)$ is given by

$$d\mathbf{E} = \frac{1}{4\pi\varepsilon_0}\frac{dQ}{R^2}\hat{\mathbf{R}} = \frac{1}{4\pi\varepsilon_0}\frac{\lambda dz}{(\rho^2+z^2)}\frac{(\rho\hat{\boldsymbol{\rho}} - z\hat{\mathbf{z}})}{(\rho^2+z^2)^{1/2}} \tag{2.33}$$

Since for every dQ at z, there is another charge dQ at $-z$ resulting in the cancellation of the z-component of the field intensity. Thus,

$$\mathbf{E} = \frac{1}{4\pi\varepsilon_0}\int_{-\infty}^{\infty}\frac{\lambda\rho dz\hat{\boldsymbol{\rho}}}{(\rho^2+z^2)^{3/2}} = \frac{2}{4\pi\varepsilon_0}\int_{0}^{\infty}\frac{\lambda\rho dz\hat{\boldsymbol{\rho}}}{(\rho^2+z^2)^{3/2}}$$

$$= \frac{2\lambda\rho\hat{\boldsymbol{\rho}}}{4\pi\varepsilon_0}\left[\frac{z}{\rho^2(\rho^2+z^2)^{3/2}}\right]_0^{\infty} = \frac{2\lambda}{4\pi\varepsilon_0\rho}\hat{\boldsymbol{\rho}} \tag{2.34}$$

In the evaluation of the integral above, we used the integral identity

$$\int\frac{dx}{(x^2+a^2)^{3/2}} = \frac{\frac{x}{a^2}}{\sqrt{(x^2+a^2)}} \tag{2.35}$$

We can observe that the field intensity due to the infinite line charge is

1. inversely proportional to the distance of the observation point from the line charge and
2. radial, i.e. always normal to the length of the line charge.

2.3.2 Infinite Plane Sheet Charge

The infinite plane sheet charge exhibits cylindrical symmetry as the infinite plane sheet is an infinite-radius zero-height solid cylinder and hence the solution becomes simple if we use cylindrical coordinate system in solving the problem. The electric field intensity of an infinite plane sheet charge is shown in Figure 2.6.

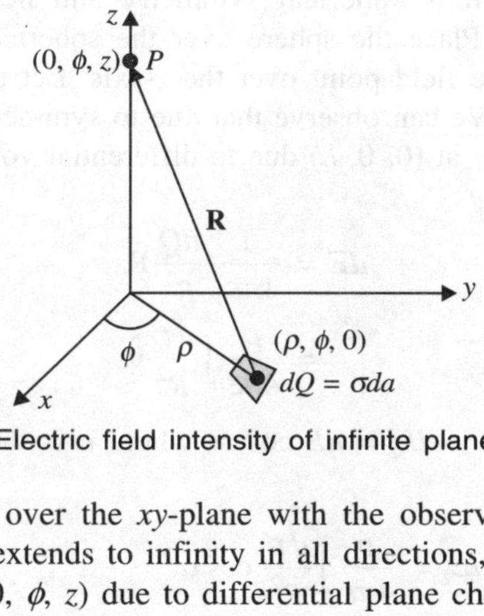

FIGURE 2.6 Electric field intensity of infinite plane sheet charge.

Place the sheet charge over the *xy*-plane with the observation point over the positive *z*-axis. Note that the charge extends to infinity in all directions, i.e. $\rho \to \infty$.

The field intensity at $(0, \phi, z)$ due to differential plane charge at $(\rho, \phi, 0)$ is given by

$$d\mathbf{E} = \frac{1}{4\pi\varepsilon_0}\frac{dQ}{R^2}\hat{\mathbf{R}} = \frac{1}{4\pi\varepsilon_0}\frac{\sigma\rho z\, d\rho\, d\phi}{(\rho^2+z^2)}\frac{(z\hat{\mathbf{z}}-\rho\hat{\boldsymbol{\rho}})}{(\rho^2+z^2)^{1/2}} \tag{2.36}$$

Since for every dQ over the plane charge, there is another charge dQ on the sheet diametrically opposite to it resulting in the cancellation of the ρ-component. Thus,

$$\mathbf{E} = \frac{1}{4\pi\varepsilon_0} \int_0^{2\pi}\int_0^{\infty} \frac{\sigma\rho z\, d\rho\, d\phi}{(\rho^2 + z^2)^{3/2}} \hat{\mathbf{z}}$$

$$= \frac{\sigma z \hat{\mathbf{z}}}{2\varepsilon_0} \left[\frac{-1}{(\rho^2 + z^2)^{1/2}} \right]_0^{\infty} = \frac{\sigma}{2\varepsilon_0} \hat{\mathbf{z}} \tag{2.37}$$

If the observation is below the sheet, then the expression for field becomes

$$\mathbf{E} = \frac{\sigma}{2\varepsilon_0}(-\hat{z}) \tag{2.38}$$

In the evaluation of the integral above, we used the identity

$$\int \frac{x}{(x^2 + a^2)^{3/2}} dx = -\frac{1}{\sqrt{(x^2 + a^2)}} \tag{2.39}$$

We observe that the field intensity due to the infinite sheet charge is

1. independent of the distance of the observation point from the sheet charge and
2. normal to the plane of the sheet charge.

2.3.3 Solid Spherical Charge

The charge distribution exhibits spherical symmetry and hence it is appropriate to use spherical coordinate system. Place the sphere over the spherical coordinate system with its centre over the origin and the field point over the z-axis. Let us suppose the coordinates of the field point are (0, 0, h). We can observe that due to symmetry, the field is entirely in the z-direction. The field intensity at (0, 0, h) due to differential volume charge $d\tau'$ at (r', θ', ϕ') is given by

$$d\mathbf{E} = \frac{1}{4\pi\varepsilon_0} \frac{dQ}{R^2} \hat{\mathbf{R}}$$

$$= \frac{\rho}{4\pi\varepsilon} \int \frac{d\tau'}{R^2} \hat{\mathbf{R}} \tag{2.40}$$

The z-component of the field is $dE_z = dE \cos\alpha$, where α is the angle between the vector **R** and the z-axis. Therefore,

$$dE_z = \frac{\rho}{4\pi\varepsilon} \int \frac{d\tau'}{R^2} \cos\alpha$$

or

$$dE_z = \frac{\rho}{4\pi\varepsilon} \int \frac{r'^2 \sin\theta\, dr'\, d\theta'\, d\phi'}{R^2} \cos\alpha \tag{2.41}$$

The electric field intensity of a solid spherical charge is shown in Figure 2.7.

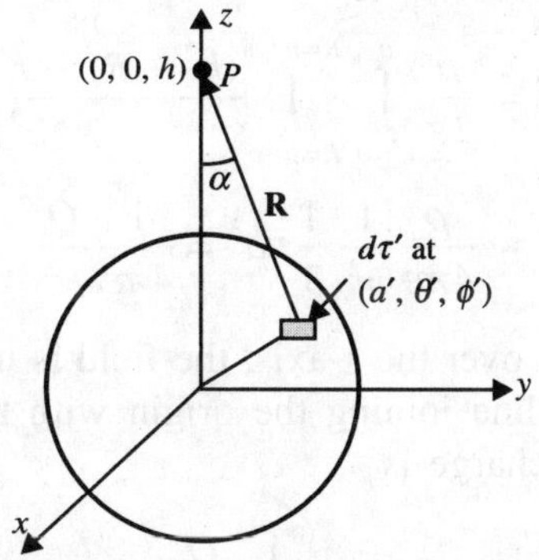

FIGURE 2.7 Electric field intensity of a solid spherical charge.

From the geometry, we can observe that

$$r'^2 = h^2 + R^2 - 2hR \cos \alpha$$

or

$$\cos \alpha = \frac{h^2 + R^2 - r'^2}{2hR}$$

Also,

$$R^2 = h^2 + r'^2 - 2hr' \cos \theta'$$

or

$$\cos \theta' = \frac{h^2 + r'^2 - R^2}{2hr'}$$

By differentiating, we get

$$\sin \theta' \, d\theta' = \frac{RdR}{hr'}$$

Substituting the above relations, we get

$$dE_z = \frac{\rho}{4\pi\varepsilon} \int \frac{r'^2 \sin \theta \, dr' d\theta' d\phi'}{R^2} \cos \alpha$$

$$= \frac{\rho}{4\pi\varepsilon} \int \frac{r'^2 dr' d\phi'}{R^2} \frac{RdR}{hr'} \frac{h^2 + R^2 - r'^2}{2hR}$$

$$= \frac{\rho}{4\pi\varepsilon} \int \frac{(h^2 + R^2 - r'^2)(r'dr'd\phi'dR)}{2h^2 R^2}$$

Integration with respect to ϕ' results in a multiplication factor of 4π

$$dE = dE_z = \frac{2\pi\rho}{4\pi\varepsilon} \int \frac{(h^2 + R^2 - r'^2)(r'dr'dR)}{2h^2 R^2}$$

To cover the entire sphere, the range of r' is 0 to a radius of the sphere, and that of the R is from $h - a$ to $h + a$. Thus,

$$E = \frac{\rho}{2\varepsilon} \int_{r'=0}^{a} \int_{R=h-a}^{R=h+a} \frac{(h^2 + R^2 - r'^2)(r'dr'dR)}{2h^2R^2}$$

$$= \frac{\rho}{4\pi\varepsilon} \frac{1}{h^2} \frac{4}{3}\pi a^3 = \frac{1}{4\pi\varepsilon} \frac{Q}{h^2}$$

Note that when the field point is over the z-axis, the field is upwards. In general, the direction of the field is radial, along the line joining the origin with the observation point. Therefore, the field due to solid spherical charge is

$$\mathbf{E} = \frac{1}{4\pi\varepsilon} \frac{Q}{r^2} \hat{\mathbf{r}} \tag{2.42}$$

It is the same as that of a point charge, magnitude-wise as well as direction-wise.

2.3.4 Spherical Shell Charge

The charge distribution exhibits spherical symmetry and hence it is appropriate to use spherical coordinate system. Place the sphere over the spherical coordinate system with its centre over the origin and the field point over the z-axis. Let us suppose the coordinates of the field point be (0, 0, h). We can observe that due to symmetry, the field is entirely in the z-direction.

The field intensity at (0, 0, h) due to differential surface charge da' at (a', θ', ϕ') is given by

$$d\mathbf{E} = \frac{1}{4\pi\varepsilon_0} \frac{dQ}{R^2} \hat{\mathbf{R}}$$

$$= \frac{\sigma}{4\pi\varepsilon} \int \frac{da'}{R^2} \hat{\mathbf{R}} \tag{2.43}$$

The z-component of the field is $dE_z = dE \cos \alpha$, where α is the angle between the vector $\mathbf{R}$ and the z-axis. Therefore,

$$dE_z = \frac{\sigma}{4\pi\varepsilon} \int \frac{da'}{R^2} \cos \alpha$$

or

$$dE_z = \frac{\sigma}{4\pi\varepsilon} \int \frac{a^2 \sin \theta \, d\theta' d\phi'}{R^2} \cos \alpha \tag{2.44}$$

The electric field intensity of a spherical shell charge is shown in Figure 2.8. From the geometry, we can observe that

$$a^2 = h^2 + R^2 - 2hR \cos \alpha$$

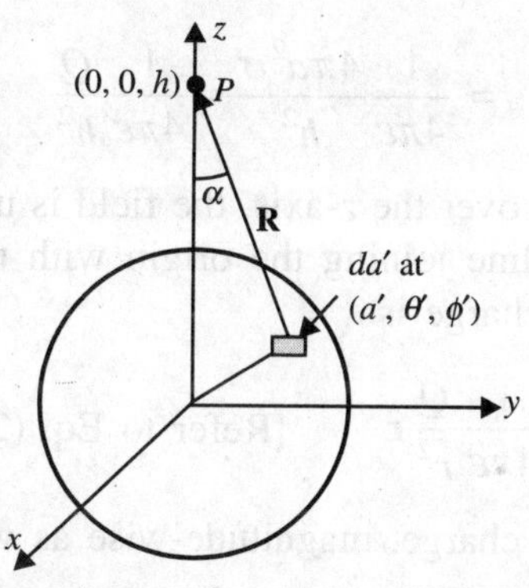

FIGURE 2.8 Electric field intensity of spherical shell charge.

or

$$\cos \alpha = \frac{h^2 + R^2 - a^2}{2hR}$$

Also,

$$R^2 = h^2 + a^2 - 2ha \cos \theta'$$

or

$$\cos \theta' = \frac{h^2 + a^2 - R^2}{2ha}$$

By differentiating, we get

$$\sin \theta' \, d\theta' = \frac{RdR}{ha}$$

Substituting the above relations results in

$$dE_z = \frac{\sigma}{4\pi\varepsilon} \int \frac{a^2 \sin \theta \, d\theta' d\phi'}{R^2} \cos \alpha$$

$$= \frac{\sigma}{4\pi\varepsilon} \int \frac{a^2 \, d\phi'}{R^2} \frac{RdR}{ha} \frac{h^2 + R^2 - a^2}{2hR}$$

$$= \frac{\sigma}{4\pi\varepsilon} \int \frac{(h^2 + R^2 - a^2)(ad\phi' dR)}{2h^2 R^2}$$

Integration with respect to ϕ' results in a multiplication factor of 4π

$$dE = dE_z = \frac{2\pi\rho}{4\pi\varepsilon} \int \frac{(h^2 + R^2 - a^2)(adR)}{2h^2 R^2}$$

To cover the entire sphere, the range of the R is from $h - a$ to $h + a$. Thus,

$$E = \frac{\sigma}{2\varepsilon} \int_{R=h-a}^{R=h+a} \frac{(h^2 + R^2 - a^2)(adR)}{2h^2 R^2}$$

$$= \frac{1}{4\pi\varepsilon} \frac{4\pi a^2 \sigma}{h^2} = \frac{1}{4\pi\varepsilon} \frac{Q}{h^2}$$

Note that when the field point is over the z-axis, the field is upwards. In general, the direction of the field is radial, along the line joining the origin with the observation point. Therefore, the field due to solid spherical charge is

$$\mathbf{E} = \frac{1}{4\pi\varepsilon} \frac{Q}{r^2} \hat{\mathbf{r}} \qquad \text{[Refer to Eq. (2.42)]}$$

It is the same as that of a point charge, magnitude-wise as well as direction-wise.

SOLVED PROBLEMS

1. Find the force on point charge q = 30 μC located at (3, 0, 0) m due to another point charge Q = 300 μC located at (1, 2, 1) m.

Solution: The distance between the locations of the charges is

$$R = [(3 - 1)^2 + (-2)^2 + (-1)^2]^{1/2}$$

And the vector joining the charge Q to charge q is

$$\mathbf{R} = (3-1)\hat{\mathbf{x}} + (-2)\hat{\mathbf{y}} + (-1)\hat{\mathbf{z}}$$

And the unit vector along this vector is

$$\mathbf{R} = \frac{\mathbf{R}}{|\mathbf{R}|} = \frac{(3-1)\hat{\mathbf{x}} + (-2)\hat{\mathbf{y}} + (-1)\hat{\mathbf{z}}}{\left[(3-1)^2 + (-2)^2 + (-1)^2\right]^{1/2}}$$

Substituting these expressions, in the above Coulomb's formula, we get

$$\mathbf{F} = \frac{1}{4\pi\varepsilon} \frac{300 \times 10^{-6} \times 30 \times 10^{-6} [(3-1)\hat{\mathbf{x}} + (-2)\hat{\mathbf{y}} + (-1)\hat{\mathbf{z}}]}{\left[(3-1)^2 + (-2)^2 + (-1)^2\right]^{1/2}}$$

$$= \frac{10^{-9}}{12\pi\varepsilon} (2\hat{\mathbf{x}} - 2\hat{\mathbf{y}} - \hat{\mathbf{z}}) \text{ N}$$

2. A charge q is located at (4, 0, –3) and another charge Q = 4 nC is located at (2, 0, 1). Find q such that the force on a unit positive charge located at (5, 0, 6) m has

   (a) no z-component and
   (b) no x-component.

Solution: The force on q due to Q is given by

$$\mathbf{F} = \frac{1}{4\pi\varepsilon} \frac{Qq[(x-x')\hat{\mathbf{x}} + (y-y')\hat{\mathbf{y}} + (z-z')\hat{\mathbf{z}}]}{\left[(x-x')^2 + (y-y')^2 + (z-z')^2\right]^{3/2}}$$

The force due to q at the observation point is

$$\mathbf{F}_q = \frac{1}{4\pi\varepsilon}\frac{q(\hat{\mathbf{x}} + 9\hat{\mathbf{z}})}{(1^2 + 9^2)^{3/2}}$$

The force due to Q at the observation point is

$$\mathbf{F}_Q = \frac{1}{4\pi\varepsilon}\frac{4\times 10^{-9}(3\hat{\mathbf{x}} + 5\hat{\mathbf{z}})}{(3^2 + 5^2)^{3/2}}$$

(a) The z-component of the total force is

$$F_z = \frac{1}{4\pi\varepsilon}\frac{9q}{(1^2 + 9^2)^{3/2}} + \frac{1}{4\pi\varepsilon}\frac{4\times 10^{-9}\times 5}{(3^2 + 5^2)^{3/2}}$$

Given that it is zero. Therefore,

$$\frac{9q}{(1^2 + 9^2)^{3/2}} + \frac{4\times 10^{-9}\times 5}{(3^2 + 5^2)^{3/2}} = 0$$

or

$$q = -\frac{4\times 10^{-9}\times 5\;(1^2 + 9^2)^{3/2}}{(3^2 + 5^2)^{3/2}\quad 9}$$

$$= -\frac{2\times 10^{-8}}{34^{3/2}}\cdot\frac{82^{3/2}}{9} = -8.32 \text{ nC}$$

(b) The x-component of the total force is

$$F_x = \frac{1}{4\pi\varepsilon}\frac{q}{(1^2 + 9^2)^{3/2}} + \frac{1}{4\pi\varepsilon}\frac{4\times 10^{-9}\times 3}{(3^2 + 5^2)^{3/2}}$$

It is said that it is zero. Therefore,

$$\frac{q}{(1^2 + 9^2)^{3/2}} + \frac{4\times 10^{-9}\times 3}{(3^2 + 5^2)^{3/2}} = 0$$

or

$$q = -\frac{4\times 10^{-9}\times 3\;(1^2 + 9^2)^{3/2}}{(3^2 + 5^2)^{3/2}\quad 1} = -44.9 \text{ nC}$$

3. Find the force on a point charge Q = 100 μC at (0, 0, 5) m due to a charge of 500π μC that is uniformly distributed over a circular disc of radius 5 m located over the xy-plane.

Solution: The scheme of the solution is: first divide the disc into smaller areas called differential areas, each of which can be approximated to a point charge. Apply the Coulomb's law to each of the differential areas to find the force and then add the forces vectorially due to all the areas to get the net force. Electric forces due to circular charge disc are shown in Figure 2.9.

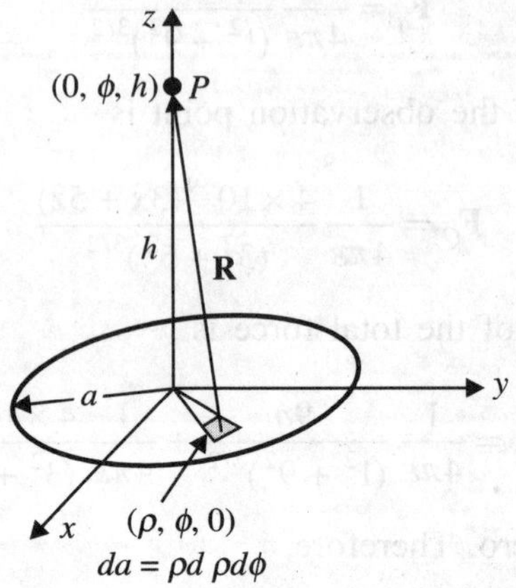

FIGURE 2.9 Electric forces due to circular charge disc.

The surface charge density over the disc is

$$\sigma = \frac{Q}{A} = \frac{500\pi \times 10^{-6}}{\pi \times 5^2} = 0.2 \times 10^{-4} \text{ C/m}^2$$

Consider a general differential area at (ρ, ϕ) and the charge over the differential area is

$$dQ = \sigma\rho d\rho d\phi$$

The distance vector from the differential area to the observation point is

$$\mathbf{R} = -\rho\hat{\boldsymbol{\rho}} + 5\hat{\mathbf{z}}$$

The force due to the differential area is

$$d\mathbf{F} = \frac{1}{4 \times \pi \times \left(\frac{10^{-9}}{36\pi}\right)} \frac{100 \times 10^{-6} \times \sigma\rho d\rho d\phi}{(\rho^2 + 25)} \frac{(-\rho\hat{\boldsymbol{\rho}} + 5\hat{\mathbf{z}})}{\sqrt{\rho^2 + 25}}$$

We can observe that the radial components of the force get cancelled because the differential charges which are diametrically opposite give rise to forces with radial components and are equal but opposite in direction. We are left with only the vertical component of the force. The total force we get by addition or integration of forces due to differential charges is

$$\mathbf{F} = \int d\mathbf{F} = \frac{100 \times 10^{-6} \times 5 \times 0.2 \times 10^{-4}}{4 \times \pi \times \left(\frac{10^{-9}}{36\pi}\right)} \int_0^{2\pi}\int_0^{5} \frac{\rho d\rho d\phi \hat{\mathbf{z}}}{(\rho^2 + 25)^{3/2}} = 33.12\,\hat{\mathbf{z}}\text{ N}$$

4. Find the electric field intensity due to straight line charge of length $2L$ at height h from its centre. It is carrying a uniform charge density of λ C/m. Extend the result to find the intensity of an infinite straight line charge.

Solution: Place the line charge along the z-axis with the observation point P in the xy-plane. Note that the charge extends from $z = -L$ to $z = L$. The electric field intensity is shown in Figure 2.10.

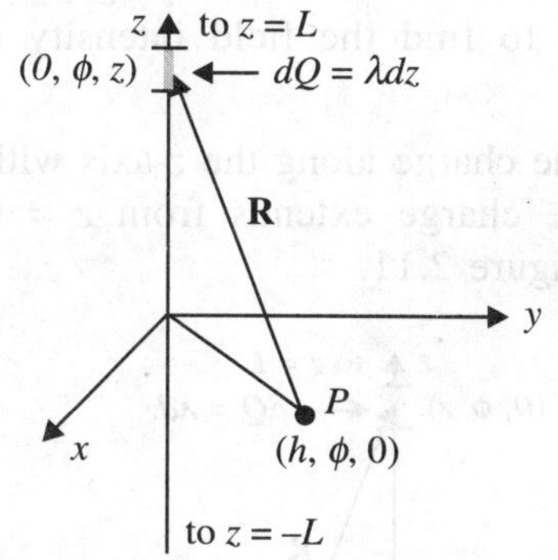

FIGURE 2.10 Electric field intensity due to finite length line charge above the middle point.

The field intensity at $(h, \phi, 0)$ due to differential line charge at $(0, \phi, z)$ is

$$d\mathbf{E} = \frac{1}{4\pi\varepsilon_0}\frac{dQ}{R^2}\hat{\mathbf{R}} = \frac{1}{4\pi\varepsilon_0}\frac{\lambda dz}{(h^2+z^2)}\frac{(h\hat{\boldsymbol{\rho}} - z\hat{\mathbf{z}})}{(h^2+z^2)^{1/2}}$$

Since for every dQ at z, there is another charge dQ at $-z$ resulting in the cancellation of the z-component. Thus,

$$\mathbf{E} = \frac{1}{4\pi\varepsilon_0}\int_{-L}^{L}\frac{\lambda h dz\hat{\boldsymbol{\rho}}}{(h^2+z^2)^{3/2}} = \frac{\lambda h\hat{\boldsymbol{\rho}}}{4\pi\varepsilon_0}\left[\frac{z}{h^2(h^2+z^2)^{1/2}}\right]_{-L}^{L}$$

$$= \frac{\lambda h\hat{\boldsymbol{\rho}}}{4\pi\varepsilon_0}\left[\frac{L}{h^2(h^2+L^2)^{1/2}} + \frac{L}{h^2(h^2+L^2)^{1/2}}\right] = \frac{\lambda\hat{\boldsymbol{\rho}}}{2\pi\varepsilon_0 h}\left[\frac{L}{(h^2+L^2)^{1/2}}\right]$$

In the evaluation of the integral above, we used an identity

$$\int\frac{dx}{(x^2+a^2)^{3/2}} = \frac{\dfrac{x}{a^2}}{\sqrt{(x^2+a^2)}}$$

The above result can be extended to find the field intensity of an infinite length line charge by tending the length to infinity, i.e. $L \to \infty$. Thus,

$$\lim_{L\to\infty}\frac{\lambda\hat{\boldsymbol{\rho}}}{2\pi\varepsilon_0 h}\left[\frac{L}{(h^2+L^2)^{1/2}}\right] = \frac{\lambda\hat{\boldsymbol{\rho}}}{2\pi\varepsilon_0 h}$$

Therefore, the field intensity due to infinite line charge is

$$\mathbf{E} = \frac{2\lambda}{4\pi\varepsilon_0 \rho}\hat{\boldsymbol{\rho}}$$

Observe the fact that this result is tallying with Eq. (2.34).

5. Find the electric field intensity due to straight line charge of length L at a height h from one of its ends. Assume that it is carrying a uniform charge density of λ C/m. Also extend the result to find the field intensity of a semi-infinite straight line charge.

Solution: Place the line charge along the z-axis with the observation point P in the xy-plane. Note that the charge extends from $z = 0$ to $z = L$. The electric field intensity is shown in Figure 2.11.

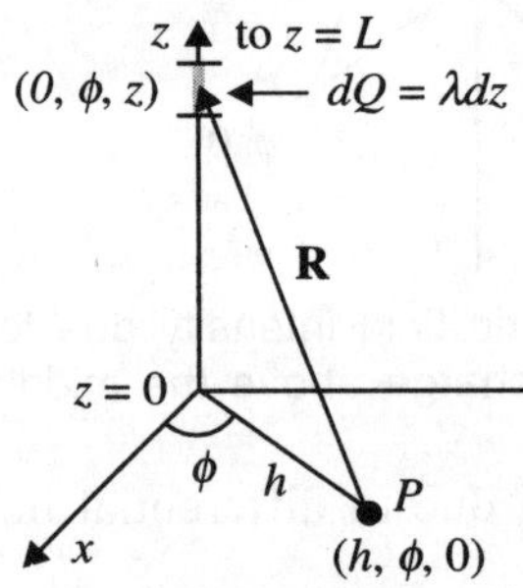

FIGURE 2.11 Electric field intensity due to finite length line charge above the end point.

The field intensity at $(h, \phi, 0)$ due to differential line charge at $(0, \phi, z)$ is

$$d\mathbf{E} = \frac{1}{4\pi\varepsilon_0}\frac{dQ}{R^2}\hat{\mathbf{R}} = \frac{1}{4\pi\varepsilon_0}\frac{\lambda dz}{(h^2+z^2)}\frac{(h\hat{\boldsymbol{\rho}} - z\hat{\mathbf{z}})}{(h^2+z^2)^{1/2}}$$

Now,

$$\mathbf{E} = \frac{\lambda}{4\pi\varepsilon_0}\int_0^L \frac{hdz\hat{\boldsymbol{\rho}}}{(h^2+z^2)^{3/2}} - \frac{\lambda}{4\pi\varepsilon_0}\int_0^L \frac{zdz\hat{\mathbf{z}}}{(h^2+z^2)^{3/2}}$$

$$= \frac{\lambda}{4\pi\varepsilon_0}\frac{z}{h^2}\frac{h\hat{\boldsymbol{\rho}}}{(h^2+z^2)^{1/2}}\Bigg|_0^L + \frac{\lambda}{4\pi\varepsilon_0}\frac{\hat{\mathbf{z}}}{(h^2+z^2)^{1/2}}\Bigg|_0^L$$

or

$$\mathbf{E} = \frac{\lambda}{4\pi\varepsilon_0}\left[\frac{L}{h}\frac{\hat{\boldsymbol{\rho}}}{(h^2+L^2)^{1/2}} + \frac{\hat{\mathbf{z}}}{(h^2+L^2)^{1/2}} - \frac{\hat{\mathbf{z}}}{h}\right]$$

Semi-infinite line charge is one which extends from origin to infinity. We can find the field intensity due to this charge distribution by tending the length to infinity, i.e. $L \to \infty$. Thus,

$$\lim_{L\to\infty} \frac{\lambda}{4\pi\varepsilon_0}\left[\frac{L}{h}\frac{\hat{\boldsymbol{\rho}}}{(h^2+L^2)^{1/2}} + \frac{\hat{\mathbf{z}}}{(h^2+L^2)^{1/2}} - \frac{\hat{\mathbf{z}}}{h}\right]$$

$$= \frac{\lambda}{4\pi\varepsilon_0}\left[\frac{1}{h}\hat{\boldsymbol{\rho}} - \frac{1}{h}\hat{\mathbf{z}}\right]$$

Therefore, the field intensity due to semi-infinite line charge becomes

$$\mathbf{E} = \frac{\lambda}{4\pi\varepsilon_0}\left[\frac{1}{h}\hat{\boldsymbol{\rho}} - \frac{1}{h}\hat{\mathbf{z}}\right]$$

6. Find the electric field intensity at an arbitrary point due to vertical straight line charge with its ends making angles of α_1 and α_2 with the horizontal. Assume that its uniform charge density is of λ C/m.

Solution: This is one problem whose solution has several applications. It is worth considering all the aspects seriously.

Arrange the finite length line along the z-axis as shown in Figure 2.12 such that the observation point falls over the xy-plane. Let θ_1, α_1 correspond to the lower end and θ_2, α_2 correspond to the upper end. Consider a differential length over the conductor at a height of z with angles θ, α.

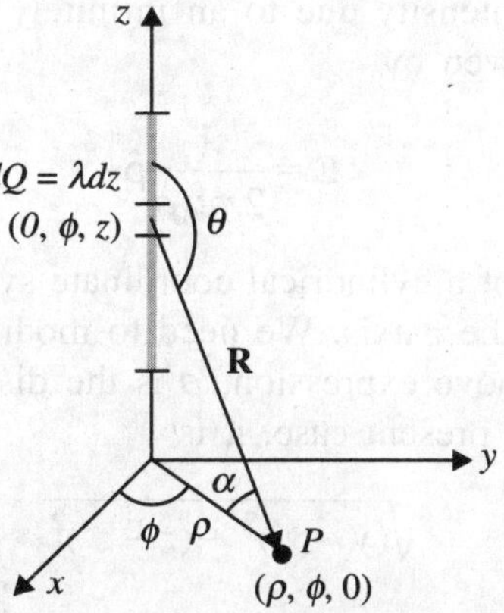

FIGURE 2.12 Electric field intensity due to finite length line charge at an arbitrary point.

From the geometry depicted in the figure, we get

$$z = \rho \tan \alpha \quad \Rightarrow \quad dz = \rho \sec^2 \alpha d\alpha$$

The distance between the differential length and observation point is $R = \rho \sec \alpha$. Substituting all these values in the expression, we get

$$d\mathbf{E} = \frac{1}{4\pi\varepsilon_0}\frac{dQ}{R^2}\hat{\mathbf{R}} = \frac{1}{4\pi\varepsilon_0}\frac{\lambda dz}{(\rho \sec \alpha)^2}\frac{(\rho\hat{\boldsymbol{\rho}} - z\hat{\mathbf{z}})}{\rho \sec \alpha}$$

$$= \frac{1}{4\pi\varepsilon_0} \frac{\lambda\rho \sec^2 \alpha d\alpha}{(\rho \sec \alpha)^2} \frac{(\rho\hat{\rho} - \rho \tan \alpha\hat{z})}{\rho \sec \alpha}$$

$$= \frac{\lambda}{4\pi\varepsilon_0\rho}(\cos \alpha\hat{\rho} - \sin \alpha\hat{z})d\alpha$$

Now,

$$E = \int d\mathbf{E} = \frac{\lambda}{4\pi\varepsilon_0\rho} \int_{\alpha_1}^{\alpha_2} (\cos \alpha\hat{\rho} - \sin \alpha\hat{z})\, d\alpha$$

$$= \frac{\lambda}{4\pi\varepsilon_0\rho}\left[(\sin \alpha_2 - \sin \alpha_1)\hat{\rho} + (\cos \alpha_2 - \cos \alpha_1)\hat{z}\right]$$

From the geometry shown above, we can observe that $\theta = 90 + \alpha$. And if θ_1, α_1 corresponds to the lower end and θ_2, α_2 corresponds to the upper end then $\alpha_1 = \theta_1 - 90^\circ$ and $\alpha_2 = \theta_2 - 90^\circ$. With these relations, it is possible to express the field intensity in terms of θ_1 and θ_2 also.

7. Find the electric field intensity due to an infinitely long uniform straight line charge λ C/m parallel to the x-axis and passing through points $y = y'$, $z = z'$ at a point P given by (x, y, z).

Solution: The field intensity due to an infinitely long uniform straight line charge of density λ C/m is given by

$$\mathbf{E} = \frac{\lambda}{2\pi\varepsilon_0\rho}\hat{\rho}$$

The above formula is of a cylindrical coordinate system. And it is derived by placing the line charge along the z-axis. We need to modify it so that it is applicable to the present case. In the above expression, ρ is the distance between the field point and the line charge. In the present case, it is

$$\sqrt{(y - y')^2 + (z - z')^2}$$

And $\hat{\rho}$ is the unit vector from the line charge towards the field point. In the present case, it is

$$\frac{(y - y')\hat{\mathbf{y}} + (z - z')\hat{\mathbf{z}}}{\sqrt{(y - y')^2 + (z - z')^2}}$$

Therefore, for the present case, the electric field intensity is given by

$$\mathbf{E} = \frac{\lambda}{2\pi\varepsilon_0} \frac{(y - y')\hat{\mathbf{y}} + (z - z')\hat{\mathbf{z}}}{\left[(y - y')^2 + (z - z')^2\right]^{3/2}}$$

8. Find the electric field intensity at a distance z above the centre of a circular loop of side a carrying a uniform line charge λ C/m.

Solution: Place the loop over the xy-plane with its centre over the origin. The observation point P now is over the z-axis at a height of z (Figure 2.13). Thus,

$$dQ = \lambda a d\phi$$

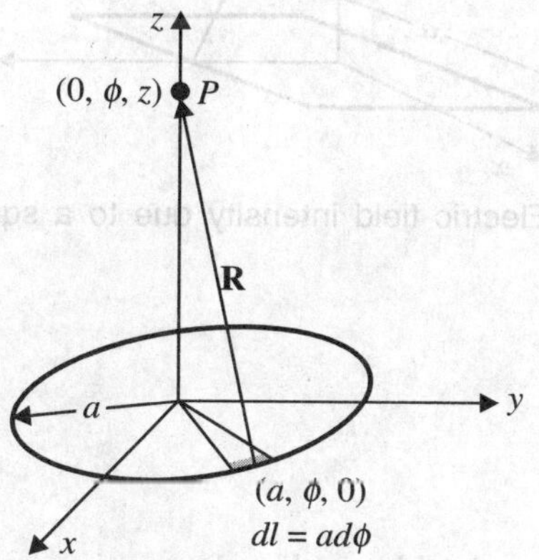

FIGURE 2.13 Field intensity due to a circular loop.

The field intensity at $(0, \phi, z)$ due to differential line charge at $(a, \phi, 0)$ is

$$d\mathbf{E} = \frac{1}{4\pi\varepsilon_0}\frac{dQ}{R^2}\hat{\mathbf{R}} = \frac{1}{4\pi\varepsilon_0}\frac{\lambda a d\phi}{(a^2+z^2)}\frac{(-a\hat{\boldsymbol{\rho}}+z\hat{\mathbf{z}})}{(a^2+z^2)^{1/2}}$$

The radial components of the field get cancelled due to symmetry of the source leaving behind only the vertical component. Thus,

$$\mathbf{E} = \frac{1}{4\pi\varepsilon_0}\int_0^{2\pi}\frac{\lambda a d\phi z\hat{\mathbf{z}}}{(a^2+z^2)^{3/2}} = \frac{1}{4\pi\varepsilon_0}\frac{2\pi\lambda a z\hat{\mathbf{z}}}{(a^2+z^2)^{3/2}}$$

$$= \frac{\lambda a z\hat{\mathbf{z}}}{2\varepsilon_0(a^2+z^2)^{3/2}}$$

9. Find the electric field intensity at a distance z above the centre of a square loop of side a carrying a uniform line charge of λ C/m.

Solution: Place the square over the xy-plane with its centre over the origin and sides parallel to the x-and y-axes. The observation point P now is over the z-axis at a height of z (Figure 2.14).

Using the result of the above problem, the field due to one side at the observation point P can be estimated as follows:

$$\mathbf{E}(P) = \frac{\lambda\hat{\boldsymbol{\rho}}}{2\pi\varepsilon_0 h}\left[\frac{L}{(h^2+L^2)^{1/2}}\right]$$

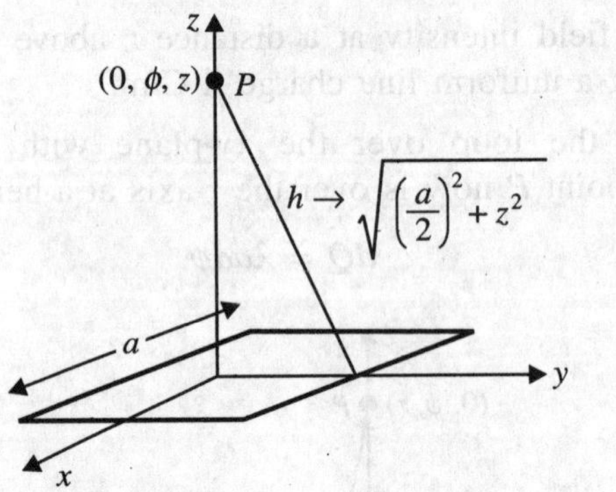

FIGURE 2.14 Electric field intensity due to a square line charge.

(a) $L \to \dfrac{a}{2}$

(b) $h \to \sqrt{\left(\dfrac{a}{2}\right)^2 + z^2}$

(c) The field due to each side at the observation point can be decomposed into z- and non-z-components and the non-z-components of opposite sides get cancelled. Only the z-components due to all the four sides exist and then get added. Thus,

$$E_z = 4E\cos\theta = \frac{4Ez}{\sqrt{\left(\dfrac{a}{2}\right)^2 + z^2}}$$

Incorporating the three changes, we get

$$\mathbf{E}(P) = \frac{\lambda\hat{\mathbf{z}}}{2\pi\varepsilon_0\sqrt{\left(\dfrac{a}{2}\right)^2 + z^2}}\left\{\frac{\dfrac{a}{2}}{\left[\left(\dfrac{a}{2}\right)^2 + z^2 + \left(\dfrac{a}{2}\right)^2\right]^{1/2}}\right\}4\cos\theta$$

$$= \frac{\lambda\hat{\mathbf{z}}}{4\pi\varepsilon_0\sqrt{\left(\dfrac{a}{2}\right)^2 + z^2}}\left\{\frac{a}{\left[2\left(\dfrac{a}{2}\right)^2 + z^2\right]^{1/2}}\right\}\frac{4z}{\sqrt{\left(\dfrac{a}{2}\right)^2 + z^2}}$$

$$= \frac{4\lambda az\hat{\mathbf{z}}}{4\pi\varepsilon_0\left[\left(\dfrac{a}{2}\right)^2 + z^2\right]\left[2\left(\dfrac{a}{2}\right)^2 + z^2\right]^{1/2}}$$

10. Find the field intensity due to a circular charged sheet of radius a carrying a uniform surface charge density of σ C/m^2 at a height of h from its centre.

Solution: As the circular charged sheet is a solid cylinder with zero-height and radius a, the solution becomes simple if we tackle the problem in cylindrical coordinate system.

Place the sheet charge over the xy-plane with the observation point P over the positive z-axis (Figure 2.15). Let us suppose the radius of the sheet be a and hence the charge extends to a in all directions, i.e. $\rho \to a$.

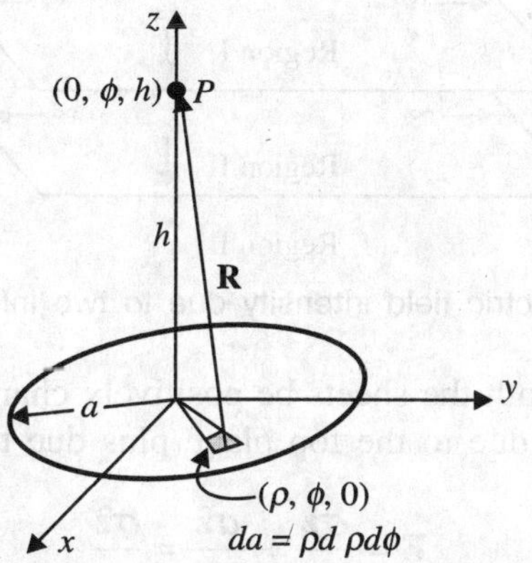

FIGURE 2.15 Electric field intensity due to circular charge disc.

The field intensity at $(0, \phi, h)$ due to differential plane charge at $(\rho, \phi, 0)$ is

$$d\mathbf{E} = \frac{1}{4\pi\varepsilon_0}\frac{dQ}{R^2}\hat{\mathbf{R}} = \frac{1}{4\pi\varepsilon_0}\frac{\sigma\rho d\rho\, d\phi}{(\rho^2+h^2)}\frac{(h\hat{\mathbf{z}}-\rho\hat{\boldsymbol{\rho}})}{(\rho^2+h^2)^{1/2}}$$

Since for every dQ over the plane charge, there is another charge dQ on the sheet diametrically opposite to it resulting in the cancellation of ρ components. Thus,

$$\mathbf{E} = \frac{1}{4\pi\varepsilon_0}\int_0^{2\pi}\int_0^{a}\frac{\sigma h\, \rho d\rho d\phi}{(\rho^2+h^2)^{3/2}}\hat{\mathbf{z}} = \frac{\sigma h\hat{\mathbf{z}}}{2\varepsilon_0}\left[\frac{-1}{(\rho^2+h^2)^{1/2}}\right]_0^a$$

$$= \frac{\sigma h}{2\varepsilon_0}\left[\frac{-1}{(a^2+h^2)^{1/2}}+\frac{1}{h}\right]\hat{\mathbf{z}}$$

If the observation is below the sheet then the expression for the field becomes

$$\mathbf{E} = \frac{\sigma h}{2\varepsilon_0}\left[\frac{-1}{(a^2+h^2)^{1/2}}+\frac{1}{h}\right](-\hat{\mathbf{z}})$$

We can observe that the circular disc becomes an infinite plane sheet if we extend the sheet to infinity $\rho \to a \to \infty$. The field intensity of the disc also becomes that of the infinite plane sheet.

11. Find the electric field intensity due to two parallel infinite plane sheets carrying a uniform surface charge density of σ C/m^2 above, below and in between the plates.

Solution: Let us suppose the planes be horizontal and for convenience let us designate the region above the top plate as 'I' and below the bottom plate as 'III' and the region in between them as 'II' (Figure 2.16). The field in any region is the sum of the fields due to the top plane plus due to the bottom plane, according to the superposition principle.

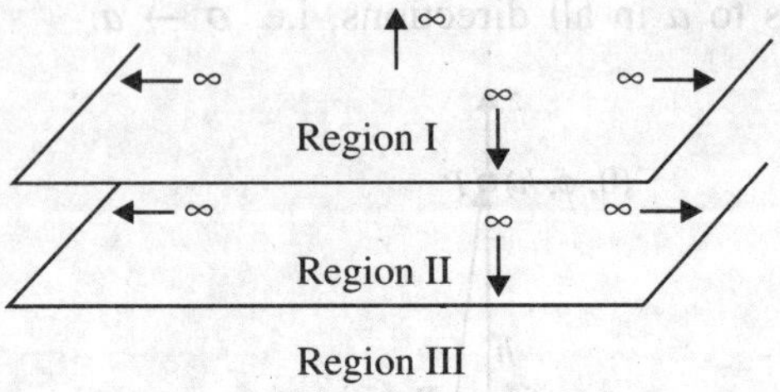

FIGURE 2.16 Electric field intensity due to two infinite parallel plates.

Now let us suppose both the sheets be positively charged. The field in the region I is the sum of the fields due to the top plane plus due to the bottom plane, which is

$$\mathbf{E} = \frac{\sigma\hat{\mathbf{z}}}{2\varepsilon_0} + \frac{\sigma\hat{\mathbf{z}}}{2\varepsilon_0} = \frac{\sigma\hat{\mathbf{z}}}{\varepsilon_0}$$

In the same manner, the field in the region III is

$$\mathbf{E} = \frac{\sigma}{2\varepsilon_0}(-\hat{\mathbf{z}}) + \frac{\sigma}{2\varepsilon_0}(-\hat{\mathbf{z}}) = \frac{\sigma}{\varepsilon_0}(-\hat{\mathbf{z}})$$

The field in the region II is

$$\mathbf{E} = \frac{\sigma}{2\varepsilon_0}(-\hat{\mathbf{z}}) + \frac{\sigma}{2\varepsilon_0}(\hat{\mathbf{z}}) = 0$$

The other situations like both sheets being negative or one negative and the other positive can be similarly analysed.

12. Find the electric field intensity at the origin due to a large plane $2x - 3y + 2z = 6$ m carrying a surface charge density $\sigma = 0.6$ nC/m^2.

Solution: The field intensity due to a large plane sheet charge is given by

$$\hat{\mathbf{E}} = \frac{\sigma}{2\varepsilon}\hat{\mathbf{n}} = \frac{0.6 \times 10^{-9}}{2\varepsilon}\hat{\mathbf{n}} \text{ V/m}$$

From the geometry, we know that the unit normal to a plane $Ax + By + Cz$ is given by

$$\hat{\mathbf{n}} = \pm\frac{A\hat{\mathbf{x}} + B\hat{\mathbf{y}} + C\hat{\mathbf{z}}}{\sqrt{A^2 + B^2 + C^2}}$$

For the present case, it is

$$\hat{\mathbf{n}} = \pm\frac{2\hat{\mathbf{x}} - 3\hat{\mathbf{y}} + 2\hat{\mathbf{z}}}{\sqrt{17}}$$

The unit vector towards the origin is given by the negative sign. Therefore, the field intensity due to the large plane sheet charge is given by

$$\hat{\mathbf{E}} = \frac{\sigma}{2\varepsilon}\hat{\mathbf{n}} = -\frac{0.6\times10^{-9}(2\hat{\mathbf{x}} - 3\hat{\mathbf{y}} + 2\hat{\mathbf{z}})}{2\sqrt{17}\varepsilon}\text{V/m}$$

EXERCISE SECTION

Essay Type Questions

1. What are the various charge distributions we come across in electrostatics? Write down integral expressions for total charge of these distributions.
2. Define line charge density, surface charge density and volume charge density giving their symbols and units.
3. State and explain the Coulomb's law of force and the terms point charge, source point and field point.
4. Define electric field intensity giving standard symbol and units. Derive the expression of field intensity due to a point charge from the fundamentals.
5. Derive the integral expressions of the electric field intensity in terms of the source charge distributions for line, surface and volume charge distributions using Coulomb's law and superposition principle.
6. Find the field intensity of a finite length straight line charge and then extend the result to find the field intensity of infinite straight line charge.
7. Derive the expression for the field intensity due to a circular sheet charge and then extend it to find the field due to an infinite plane sheet charge.
8. Find the field intensity at the centre of a square loop carrying a uniform line charge density.
9. Find the field intensity at a height from the centre of a circular loop carrying a uniform line charge density.

Numerical Problems

1. Three point charges 4 mC, 5 mC and 6 mC are located at (4, 0, 0), (0, 5, 0) and (0, 0, 6). Find the force on another charge of 10 mC located at (0, 0, 0).
2. Two charges Q and $2Q$ arc placed with a separation of 3 m. Find the location and value of the charge which can keep the two charge system in equilibrium.
3. Three equal point charges of 20 mC are placed at the corners of an equilatera' triangle of side 1 m. Find the field intensity at the centre of the triangle.

4. Find the electric field intensity at (1, 1, 2) due to a charge distribution consisting of a point charge of 10 mC at (2, 2, –4), a line charge of density 3 mC/m over the x-axis and a sheet charge of density 10 mC/m^2 over the z-axis.
5. A point charge of 300 μC is located at point (1, 2, 1) m. Find the force on another point charge of 30 μC at the point (3, 0, 0) m and also calculate the electric field at (3, 0, 0) m.
6. Point charges Q_1 and Q_2 are respectively located at (4, 0, –3) and (2, 0, 1). If Q_2 = 4 nC, find such that
   (a) the E at (5, 0, 6) has no z-component and
   (b) the force on test charge at (5, 0, 6) has no x-component.
7. A uniform line charge λ = 25 nC/m lies on the line x = –3 and z = 4 m in free space. Find the electric field intensity at a point (2, 5, 3).

Fill in the Blanks

1. Electrostatic fields are given by ________ charge distributions.
2. Group of point charges is an example of ________ charge distribution.
3. Line, surface and volume charges belong to ________ charge distributions.
4. Coulomb's law is similar to Newton's law of ________.
5. Coulomb's law gives ________ and ________ of the force between charges.
6. The magnitude of Coulomb's force is directly proportional to the ________ of the charges and ________ proportional to the square of the distance.
7. Direction of Coulomb's force is ________ the line joining the charges.
8. Free space permittivity is ________ F/m.
9. The electric field intensity is the force on unit positive charge and its units are ________.
10. The direction of field due to point charges is ________.
11. The field intensity due to point charge is inversely proportional to distance ________.
12. The direction of field due to infinite straight line charge is ________ to line.
13. The field intensity due to infinite straight line charge is ________ proportional to distance.
14. The direction of field due to infinite plane sheet of negative charge is ________ the sheet.
15. The field intensity due to infinite plane sheet charge is ________ of distance.

Answers

1. stationary
2. discrete
3. continuous
4. gravitation
5. magnitude, direction
6. product, inversely

7. along
8. $\dfrac{10^{-9}}{36\pi}$
9. volts/metre
10. radial
11. square
12. normal
13. inversely
14. into
15. independent

Match the Following

Set I

1. Like charges	(a) scalar ε
2. Unlike charges	(b) size of a point
3. Field intensity	(c) force per coulomb
4. Point charges	(d) attraction
5. Isotropic	(e) repulsion

Set II

1. Direction	(a) product of charges
2. Proportional	(b) along the joining line of charges
3. Inversely proportional	(c) permittivity
4. Media dependence	(d) square of the distancc
5. Proportionality constant	(e) $\dfrac{1}{4\pi}$

Set III

1. Point charge	(a) $\mathbf{E(r)} = \dfrac{1}{4\pi\varepsilon}\displaystyle\int \dfrac{\rho(\mathbf{r}')d\tau'}{R^2}\hat{\mathbf{R}}$
2. Discrete charge distribution	(b) $\mathbf{E(r)} = \dfrac{1}{4\pi\varepsilon}\displaystyle\int \dfrac{\sigma(\mathbf{r}')da'}{R^2}\hat{\mathbf{R}}$
3. Line charge	(c) $\mathbf{E(r)} = \dfrac{1}{4\pi\varepsilon}\displaystyle\int \dfrac{\lambda(\mathbf{r}')dl'}{R^2}\hat{\mathbf{R}}$
4. Surface charge	(d) $\mathbf{E} = \dfrac{1}{4\pi\varepsilon}\displaystyle\sum_{k=1}^{n} \dfrac{Q_k}{R_k^2}\hat{\mathbf{R}}_k$
5. Volume charge	(e) $\mathbf{E} = \dfrac{1}{4\pi\varepsilon}\dfrac{Q}{R^2}\hat{\mathbf{R}}$

Set IV

1. Point charge	(a) $\dfrac{10^{-9}}{36\pi}$
2. Free space permittivity	(b) radial
3. Line charge	(c) into
4. Negative plane sheet	(d) normal
5. Dielectric constant	(e) relative permittivity

Set V

1. Point charge	(a) zero to infinity
2. Line charge	(b) $-\infty$ to ∞
3. Sheet charge	(c) distance independence
4. Infinite line	(d) distance
5. Semi-infinite	(e) distance square

Answers

Set I:	1. (e)	2. (d)	3. (c)	4. (b)	5. (a)
Set II:	1. (b)	2. (a)	3. (d)	4. (c)	5. (e)
Set III:	1. (e)	2. (d)	3. (c)	4. (b)	5. (a)
Set IV:	1. (b)	2. (a)	3. (d)	4. (c)	5. (e)
Set V:	1. (e)	2. (d)	3. (c)	4. (b)	5. (a)

Multiple Choice Questions

1. Coulomb's law is about the electrostatic force between
   (a) point charges (b) line charges
   (c) sheet charges (d) volume charges
2. The electric force between non-point charges is governed by
   (a) Coulomb's law (b) superposition principle
   (c) both (a) and (b) (d) none of these
3. The electric field intensity **E** due to point charges is
   (a) inversely proportional to distance (b) directly proportional to distance square
   (c) independent of the distance (d) none of these
4. The electric field intensity **E** due to infinite line charges is
   (a) inversely proportional to distance (b) directly proportional to distance square
   (c) independent of the distance (d) none of these
5. The electric field intensity **E** due to infinite sheet charges is
   (a) inversely proportional to distance (b) directly proportional to distance square
   (c) independent of the distance (d) none of these

6. The electric field intensity **E** due to solid spherical charges is
   (a) inversely proportional to distance square
   (b) directly proportional to distance square
   (c) independent of the distance
   (d) none of these
7. The direction of field intensity **E** due to point charges
   (a) radial (spherical) (b) radial (cylindrical)
   (c) azimuthal (d) none of these
8. The direction of field intensity **E** due to infinite straight line charge is
   (a) normal to line charge (b) radial (cylindrical)
   (c) both of (a) and (b) (d) none of these
9. The direction of field intensity **E** due to infinite plane charge is
   (a) normal to the sheet (b) radial (cylindrical)
   (c) parallel to the sheet (d) none of these
10. The direction of field intensity **E** due to spherical charge is
   (a) radial (spherical) (b) radial (cylindrical)
   (c) azimuthal (d) none of these
11. The electric field intensity **E** inside spherical charge shell is
   (a) non-zero (b) variable
   (c) both (a) and (b) (d) zero

Answers

1. (a) 2. (c) 3. (d) 4. (a) 5. (c)
6. (a) 7. (a) 8. (c) 9. (a) 10. (a)
11. (d)

3

Electric Displacement and Gauss Law

I never see what has been done; I only see what remains to be done.

—**Buddha**

Flux is an important concept in electromagnetic field theory. It is a result of several experiments performed by Michael Faraday on static electricity. It is useful to denote the fields graphically.

Electric flux is usually called electric displacement. Its importance lies in the fact that the Gauss law, one of the Maxwell's equations, is stated in terms of electric displacement. In this chapter, the idea of flux or displacement is introduced and its importance as part of the Gauss law as well as a useful tool in graphical representation of the field is stressed.

The Gauss law is introduced along with its utility and applications. Solutions of several problems are given which illustrate the simplicity of the solution when the Gauss law is applied. Divergence and curl are two aspects of fields which are more than sufficient to describe vector fields completely. They are computed for electrostatic fields considering a general charge distribution.

3.1 Displacement Density D

During the second half of the 1830s, Michael Faraday performed several experiments on static electric fields and the impact of dielectrics on these fields. One of his experiments consists of two concentric conducting spherical shells separated by a thick dielectric material. He placed some amount of positive charge on the inner conducting shell before assembling. After assembling he measured the charge on the outer conducting shell just after a momentary earthing. It was found that a positive charge on the outer shell is equal in magnitude to the charge on the inner shell.

To explain the observation, he assumed the existence of a kind of 'displacement' from the inner to the outer spherical shell independent of the intermediate dielectric. This flow or flux is now called *displacement flux* or *electric flux,* with a symbol ψ.

Another observation of the experimental results is regarding the equality of the charges on the spherical shells. To explain this observation, he came to a conclusion that the electric flux and the charge on the inner sphere are directly proportional to each other. The proportionality factor happened to be one in SI system of units. As a consequence, the electric flux and charge on the inner sphere became equal, i.e. $Q = \psi$.

Even though the flux and charge are the same, quantity-wise, conceptually they are poles apart. Flux exists in the fields that surround the charge distributions. Literal meaning of flux is flow and it is represented, graphically, by sketching curves known as flux lines. When lines are drawn, the field appears as if there is flow or flux in the region. As the lines drawn are required to be proportional to charge, line density requires to be more for large charges resulting in smaller spacing between the lines. The net result is appearance of more flow or flux for large charges. In the case of smaller charges, the lines are relatively sparse and there appears to be little flow or flux in the field. The net result is the existence of a flux in proportion to the charge that causes the field.

The number of flux lines crossing a unit cross-sectional area is called displacement density **D**. It is a vector quantity with units of C/m^2 and it can be shown equal to $\varepsilon\mathbf{E}$, i.e. $\mathbf{D} = \varepsilon\mathbf{E}$. One important feature of the electric displacement density is its media independence. The total number of lines crossing an arbitrary surface area s is defined as the electric displacement or flux ψ through the surface and mathematically, it is given by

$$\psi = \int_s \mathbf{D} \cdot d\mathbf{a} \tag{3.1}$$

3.1.1 Electric Flux Lines

The concept of electric flux lines or field lines was first mooted by Faraday to represent the electric fields graphically. A vector field can be described by a set of lines known as direction lines, also known as stream lines and flux lines. Direction line is, by definition, a curve drawn in the field in such a way that the field is tangential to the curve at all points on the curve. The electric field lines exhibit certain important features. Two of the most important are: they originate on positive charge and terminate either on negative charge or at infinity and they never cross each other.

While sketching the lines, the following points should be observed scrupulously:

- While sketching the lines to represent the field, there is no restriction on the number of lines drawn, but the number should be sufficiently large so that the detailed variations of field are exhibited.
- Consistency should be maintained in the number of lines so that the number must be proportional to the charge.
- The lines must be placed fairly, i.e. the spacing must be as uniform as possible, as if they emanate from a point charge symmetrically in all directions.

The fact that the field intensity is always tangential to flux lines can be used to derive the differential equations that represent the flux lines (Figure 3.1). If the lines are confined to, let us say, the *xy*-plane, then

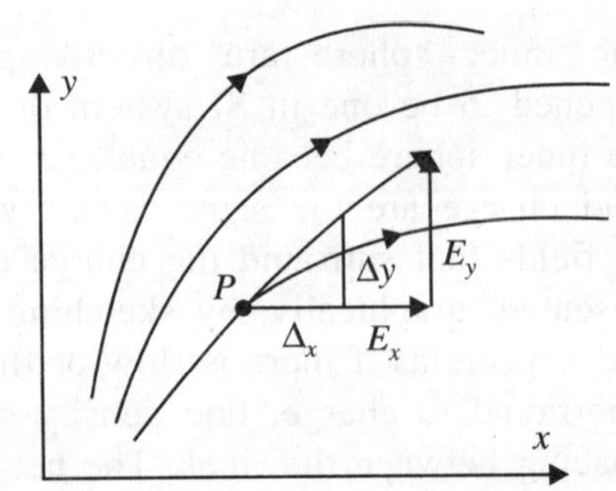

FIGURE 3.1 Derivation of equations of electric flux lines.

$$\frac{E_y}{E_x} = \frac{dy}{dx}$$

or

$$\frac{dx}{E_x} = \frac{dy}{E_y} \tag{3.2}$$

We can extend it to three dimensions as well as to other coordinate systems resulting in the following differential equations for the flux lines:

$$\frac{dx}{E_x} = \frac{dy}{E_y} = \frac{dz}{E_z} \quad \text{in Cartesian system} \tag{3.3}$$

$$\frac{d\rho}{E_\rho} = \frac{\rho d\phi}{E_\phi} = \frac{dz}{E_z} \quad \text{in cylindrical system} \tag{3.4}$$

$$\frac{dr}{E_r} = \frac{rd\theta}{E_\theta} = \frac{r \sin \theta d\phi}{E_\phi} \quad \text{in spherical system} \tag{3.5}$$

3.2 GAUSS LAW

Faraday's experimental work on dielectric separated concentric spherical shells led to an observation of far-reaching consequence that electric displacement through any imaginary closed surface lying in between the conducting shells is equal to the total charge enclosed by that imaginary surface irrespective of the shape or configuration of the enclosed charge distribution and the outer conducting shell. Further generalizations led to the enunciation of a famous law known as *Gauss law* in the name of one of the greatest mathematicians of the world. Gauss not only gave a law form to the experimental findings of Faraday on the concentric spheres but also provided a mathematical form to it. In recognition of his contribution to its development, the law thereafter is being called in his name as Gauss law.

According to this law, "the net electric displacement through any closed surface is equal to the net charge enclosed by that surface". Mathematically, it can be stated as

$$\oint_s \mathbf{D} \cdot d\mathbf{a} = Q_{\text{enc}} \tag{3.6}$$

This expression is said to be the Gauss law in integral form as it involves integration. It has another form known as differential form given by

$$\nabla \cdot \mathbf{D} = \rho \tag{3.7}$$

These two forms are related through the divergence theorem of vector calculus. One can be derived from the other as shown. Let us suppose that is a volume charge equal to Q_{enc} coulombs enclosed by a (closed) surface and ρ is the volume charge density. Then we know that these two quantities are related through

$$Q_{\text{enc}} = \oint_s \rho d\tau \tag{3.8}$$

The divergence theorem of calculus states that

$$\oint_s \mathbf{A} \cdot d\mathbf{a} = \int \nabla \cdot \mathbf{A} d\tau \tag{3.9}$$

where **A** is an arbitrary vector. Using this theorem, we get

$$\oint_s \mathbf{D} \cdot d\mathbf{a} = \int \nabla \cdot \mathbf{D} d\tau \tag{3.10}$$

With these two substitutions in the integral form of the Gauss law, we get

$$\oint_s \mathbf{D} \cdot d\mathbf{a} = \int \nabla \cdot \mathbf{D} d\tau = Q_{\text{enc}} = \oint_s \rho d\tau \tag{3.11}$$

Comparing both sides, we get

$$\nabla \cdot \mathbf{D} = \rho \qquad \text{[as given in Eq. (3.7)]}$$

Even though the two forms are equivalent, they differ from each other on two important aspects. One is that, in the integral form, it is valid for all types of charge distributions, i.e. discrete, line, surface and volume, but in the differential form as shown above it is valid only for volume charge distributions. The second aspect they differ with each other is regarding applications: the integral form is widely used in finding the field intensity of several charge distributions without much computational burden.

Applications: The Gauss law in the integral form has an extraordinary power of computing fields quickly and easily when the charge distribution exhibits high level symmetry, i.e. planar or cylindrical or spherical symmetry.

Depending on the symmetry exhibited by the charge distributions, closed surfaces known as *Gaussian surfaces* (surfaces over which integration is performed) are selected and most of the times, they are one of a pillbox, coaxial cylinder or a concentric sphere. The surface is selected in such a way that the integration involved in the Gauss law becomes least complicated and in fact it becomes easiest possible when the field is either normal or tangential to the closed surface and when normal, its value is constant over the surface.

It is also possible to apply the Gauss law to combinations of objects possessing symmetry, even when the arrangement as a whole is not symmetrical. The Gauss law can be applied to each individual objects separately; the resulting field intensities are added to find the net field intensity using the superposition principle.

Limitations: In the integral form it is useful to find the field intensity by mere observation in some cases. But it has certain limitations like it is useful to find the field intensity only when the charge distribution is highly symmetrical. Another limitation is that its application to find the field intensity requires the prior information regarding the direction of the field.

3.3 DIVERGENCE AND CURL

The most important characteristics of vector fields like electric and magnetic fields are divergence and curl.

The fields are regions surrounding the charge distributions. They are not just ordinary regions: they exhibit some special kind of properties. The fields are invisible and they make their presence felt only by affecting charges and currents. Their nature, in general, is difficult to describe but the Helmholtz theorem says that the divergence and curl of the fields together are able to give the total and complete information regarding them. As such, divergence and curl are important and critical aspects of vector fields in general. The static electric fields exhibit the following important properties:

(a) The electric field is conservative in nature because the line integral of $\mathbf{E} \cdot d\mathbf{l}$ around any closed path is zero. This nature of the field is also described as irrotational as its curl is zero. Mathematically,

$$\oint \mathbf{E} \cdot d\mathbf{l} = 0 \tag{3.12}$$

and

$$\nabla \times \mathbf{E} = 0 \tag{3.13}$$

(b) The divergence of electric displacement or flux is non-zero and also the net flux through any closed surface is non-zero. Mathematically,

$$\nabla \cdot \mathbf{D} = \rho \quad \text{and} \quad \oint \mathbf{D} \cdot d\mathbf{a} = Q \tag{3.14}$$

Derivation: Consider the electrostatic field $\mathbf{E}$ at a point P due to a volume charge distribution ρ. If the source and field points are $\mathbf{r}'$ and $\mathbf{r}$, respectively, then

$$\mathbf{E}(\mathbf{r}) = \frac{1}{4\pi\varepsilon} \int \frac{\rho(\mathbf{r}')d\tau'}{R^2} \hat{\mathbf{R}}$$

Here $\mathbf{E}$ is a function of $\mathbf{r}$ and ρ is a function of $\mathbf{r}'$ and the vector joining the source point with field point is $\mathbf{R} = \mathbf{r} - \mathbf{r}'$. Note that in Cartesian system, tip of $\mathbf{r}$ is (x, y, z), that of $\mathbf{r}'$ is (x', y', z'), differential volume of source is $d\tau' = dx'dy'dz'$ and $\mathbf{R} = (x - x')\hat{\mathbf{x}} + (y - y')\hat{\mathbf{y}} + (z - z')\hat{\mathbf{z}}$. Also note that in the above integral relation, the integration is over primed coordinates and divergence and curl are with respect to unprimed quantities.

3.3.1 Divergence

Consider the integral expression for field intensity due to volume charge distribution

$$\mathbf{E}(\mathbf{r}) = \frac{1}{4\pi\varepsilon}\int \frac{\rho(\mathbf{r}')d\tau'}{R^2}\hat{\mathbf{R}}$$

Now applying divergence on both sides of the equation, we have

$$\nabla\cdot\mathbf{E} = \frac{1}{4\pi\varepsilon}\int \nabla\cdot\frac{\hat{\mathbf{R}}}{R^2}\rho(\mathbf{r}')d\tau'$$

But we know that

$$\nabla\cdot\frac{\hat{\mathbf{R}}}{R^2} = 4\pi\delta^3\,(\mathbf{R}) = 4\pi\delta^3(\mathbf{r}-\mathbf{r}')$$

Therefore, we have

$$\nabla\cdot\mathbf{E}(\mathbf{r}) = \frac{1}{4\pi\varepsilon}\int 4\pi\delta^3\,(\mathbf{r}-\mathbf{r}')\,\rho(\mathbf{r}')\,d\tau'$$

This integral can be evaluated using the sampling/sifting property of impulse function

$$\int_{-\infty}^{\infty} f(t)\,\delta\,(t-t_0)\,dt = f(t_0)$$

Now we can have

$$\begin{aligned}\nabla\cdot\mathbf{E}(\mathbf{r}) &= \frac{1}{4\pi\varepsilon}\int 4\pi\delta^3(\mathbf{r}-\mathbf{r}')\,\rho(\mathbf{r}')\,d\tau' \\ &= \frac{1}{\varepsilon}\rho(\mathbf{r})\end{aligned}$$

It results in an expression for the divergence of electric field. In the case of field in free space $\varepsilon = \varepsilon_0$,

$$\nabla\cdot\mathbf{D} = \rho_v$$

3.3.2 Curl

If we apply curl on both sides, we get

$$\nabla\times\mathbf{E} = \frac{1}{4\pi\varepsilon}\int\left(\nabla\times\frac{\hat{\mathbf{R}}}{R^2}\right)\rho(\mathbf{r}')d\tau'$$

But

$$\nabla\times\frac{\hat{\mathbf{R}}}{R^2} = 0$$

Therefore,

$$\nabla\times\mathbf{E} = 0$$

The result can also be derived as follows:
We know that

$$\mathbf{E} = -\nabla V$$

Now applying closed line integral on both sides, we get

$$\oint \mathbf{E} \cdot d\mathbf{l} = -\oint \nabla V \cdot d\mathbf{l}$$

But according to the theorem of gradients,

$$\oint \nabla V \cdot d\mathbf{l} = 0$$

which results in

$$\oint \mathbf{E} \cdot d\mathbf{l} = 0$$

Therefore, $\nabla \times \mathbf{E} = 0$.

SOLVED PROBLEMS

1. Find the electric field intensity due to infinite straight line charge with a charge density of λ C/m at a distance of h.

Solution: Because of symmetry considerations, we can observe that the field due to infinite line charge is radial and also it is constant over any coaxial cylindrical surface.

Now consider a coaxial cylinder as Gaussian surface and apply Gauss law Figure 3.2. Thus,

$$\oint_s \mathbf{D} \cdot d\mathbf{a} = \oint_{s_1} \mathbf{D} \cdot d\mathbf{a} + \oint_{s_2} \mathbf{D} \cdot d\mathbf{a} + \oint_{s_3} \mathbf{D} \cdot d\mathbf{a} = Q_{\text{enc}}$$

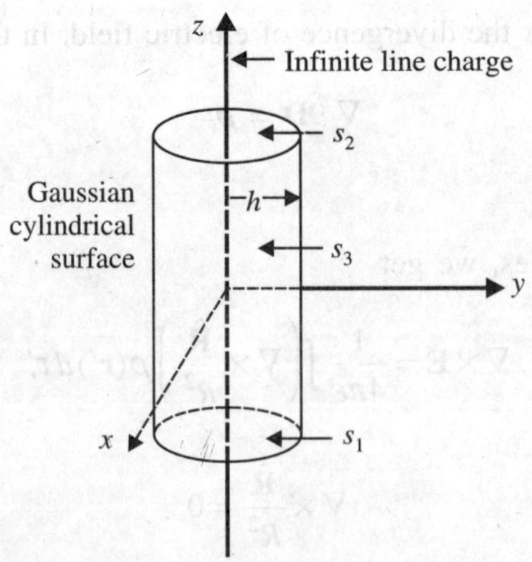

FIGURE 3.2 Electric field intensity by Gauss law infinite line charge.

As the field is normal to the bottom and top surfaces, the corresponding integrals become zero. The field is radial and hence its direction is the same as that of the curved face of the cylinder. Therefore, the dot can be replaced by ordinary multiplication. As the field intensity is constant over the curved surface, the displacement density can be brought out of the integral. Incorporating all these observations, we get

$$\oint_s \mathbf{D} \cdot d\mathbf{a} = 0 + 0 \int_{s_3} D da$$

$$= D \int_{s_3} da = D2\pi hL$$

Here L is the length of the Gaussian cylinder. The total charge enclosed by the cylinder is λL C. Thus,

$$D2\pi hL = \lambda L \text{ or } \varepsilon E2\pi h = \lambda$$

Now,

$$\mathbf{E} = \frac{1}{2\pi\varepsilon} \frac{\lambda}{h} \hat{\boldsymbol{\rho}}$$

Note that the direction is not given by Gauss law. We have to get it by applying the Coulomb's law.

2. Consider a long coaxial cable carrying a charge of Q C/m on its inner conductor. Find the electric field intensity using Gauss law in between the conductors.

Solution: From the symmetry of the charge distribution, we can observe that the field in between the conductors can be entirely radial and also it is constant over any coaxial cylindrical surface. The electric field intensity by the Gauss law is shown in Figure 3.3.

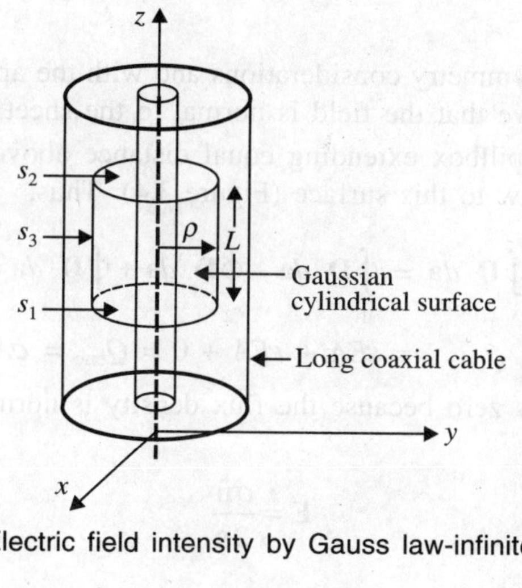

FIGURE 3.3 Electric field intensity by Gauss law-infinite coaxial cable.

Now consider a coaxial cylinder as Gaussian surface and apply the Gauss law. Thus,

$$\oint_s \mathbf{D}\cdot d\mathbf{a} = \oint_{s_1} \mathbf{D}\cdot d\mathbf{a} + \oint_{s_2} \mathbf{D}\cdot d\mathbf{a} + \oint_{s_3} \mathbf{D}\cdot d\mathbf{a} = Q_{\text{enc}}$$

As the field is normal to bottom and top surfaces, the corresponding integrals become zero. The field is radial and hence its direction is the same as that of the curved face of the cylinder. Therefore, the dot can be replaced by ordinary multiplication. As the field intensity is constant over the curved surface, the displacement density can be brought out of the integral. Incorporating all these observations, we get

$$\oint_s \mathbf{D}\cdot d\mathbf{a} = 0 + 0 \int_{s_3} D da$$

$$= D\int_{s_3} da = D2\pi\rho L$$

Here L is the length of the Gaussian cylinder. The total charge enclosed by the cylinder is QL C. Thus,

$$D2\pi\rho L = QL \text{ or } \varepsilon E 2\pi\rho = Q$$

Now,

$$\mathbf{E} = \frac{1}{2\pi\varepsilon}\frac{Q}{\rho}\hat{\boldsymbol{\rho}}$$

Note that the direction is not given by the Gauss law. We have to get it by applying Coulomb's law.

3. Consider a plane sheet charge of infinite extent carrying a uniform surface charge density of σ C/m^2. Find the electric field intensity at a height of h m from its surface.

Solution: From symmetry considerations and with the application of the Coulomb's law, we can observe that the field is normal to the sheet.

Draw a Gaussian pillbox extending equal distance above and below the plane and apply the Gauss law to this surface (Figure 3.4). Thus,

$$\oint_s \mathbf{D}\cdot d\mathbf{a} = \oint_{s_1} \mathbf{D}\cdot d\mathbf{a} + \oint_{s_2} \mathbf{D}\cdot d\mathbf{a} + \oint_{s_3} \mathbf{D}\cdot d\mathbf{a} = Q_{\text{enc}}$$

$$= \varepsilon EA + \varepsilon EA + 0 = Q_{\text{enc}} = \sigma A$$

The last integral is zero because the flux density is normal to the side curved face. Thus,

$$\mathbf{E} = \frac{\sigma\hat{\mathbf{n}}}{2\varepsilon}$$

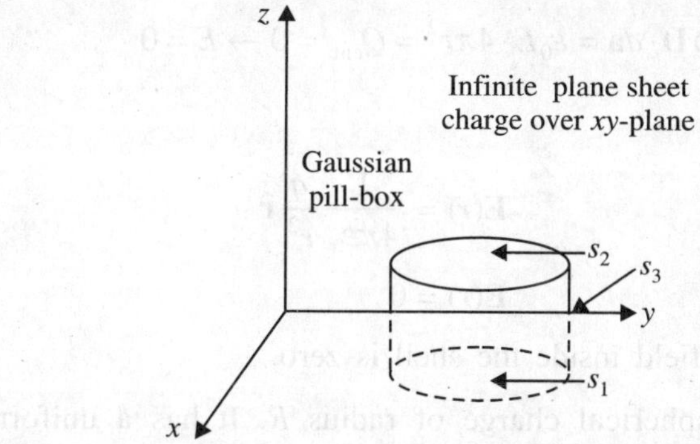

FIGURE 3.4 Electric field intensity by Gauss law—Infinite sheet charge.

4. The surface of a spherical conducting shell carries a uniform charge density σ C/m^2. Find the field intensity due to a spherical shell both outside and inside.

Solution: Let us suppose the spherical shell be with radius R and the total charge on the shell be q. The electric field intensity by the Gauss law is shown in Figure 3.5.

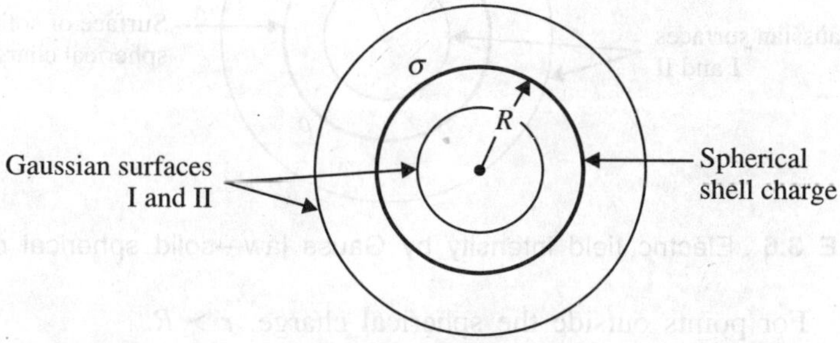

FIGURE 3.5 Electric field intensity by Gauss law—spherical sheet charge.

(a) For points outside the spherical shell, $r > R$.
Consider a Gaussian surface concentric with the spherical shell with radius $r > R$ and apply the Gauss law. Thus,

$$\oint_s \mathbf{D} \cdot d\mathbf{a} = \varepsilon_0 E \cdot 4\pi r^2 = Q_{\text{enc}} = 4\pi R^2 \sigma \to E = \frac{\sigma R^2}{\varepsilon_0 r^2}$$

In terms of total charge $4\pi R^2 \sigma = q$,

$$\mathbf{E}(r) = \frac{1}{4\pi\varepsilon_0} \frac{q}{r^2} \hat{\mathbf{r}}$$

(b) For points inside the spherical shell, $r < R$.
Consider a Gaussian surface concentric with the spherical shell with radius $r < R$ and apply Gauss law. Thus,

$$\oint_s \mathbf{D} \cdot d\mathbf{a} = \varepsilon_0 E \cdot 4\pi r^2 = Q_{\text{enc}} = 0 \rightarrow E = 0$$

Now,

$$\mathbf{E}(r) = \frac{1}{4\pi\varepsilon_0} \frac{q}{r^2} \hat{\mathbf{r}}$$

$$\mathbf{E}(r) = 0$$

Note that the field inside the shell is zero.

5. Consider a solid spherical charge of radius R. It has a uniform volume charge density ρ C/m^3. Find its field intensity both inside and outside.

Solution: Given that the solid spherical charge is with radius R and carrying uniform volume charge density ρ. Let us suppose the total charge on the sphere be q (see Figure 3.6).

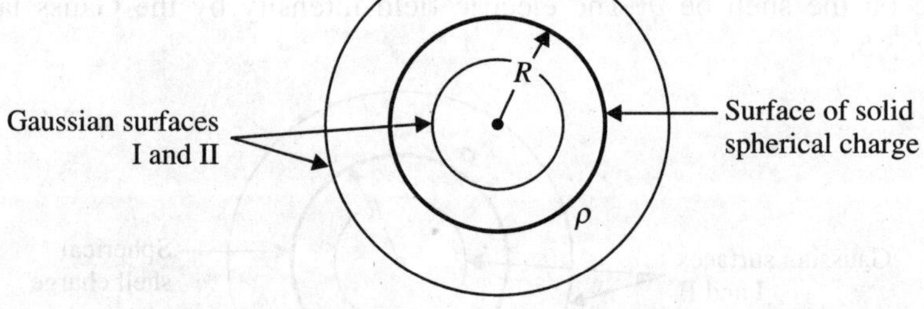

FIGURE 3.6 Electric field intensity by Gauss law—solid spherical charge.

Case I: For points outside the spherical charge, $r > R$.
Consider a Gaussian surface concentric with the spherical shell with radius $r > R$ and apply the Gauss law. Thus,

$$\oint_s \mathbf{D} \cdot d\mathbf{a} = \varepsilon_0 E \cdot 4\pi r^2 = Q_{\text{enc}} = \frac{4}{3}\pi R^3 \rho \rightarrow E = \frac{\rho R^3}{3\varepsilon_0 r^2}$$

In terms of total charge $\frac{4}{3}\pi R^3 \rho = q$,

$$\mathbf{E}(r) = \frac{1}{4\pi\varepsilon_0} \frac{q}{r^2} \hat{\mathbf{r}}$$

Case II: For points inside the spherical charge, $r < R$.
Consider a Gaussian surface concentric with the spherical shell with radius $r < R$ and apply the Gauss law. Thus,

$$\oint_c \mathbf{D} \cdot d\mathbf{a} = \varepsilon_0 E \cdot 4\pi r^2 = Q_{\text{enc}} = \frac{4}{3}\pi r^3 \rho \rightarrow E = \frac{\rho r}{3\varepsilon_0}$$

The charge density can be expressed in terms of total charge. As

$$\frac{4}{3}\pi R^3 \rho = q$$

the charge density becomes

$$\rho = q\frac{3}{4\pi R^3}$$

Substituting the above expression for charge density results in

$$\mathbf{E} = \frac{q}{4\pi\varepsilon_0}\frac{r}{R^3}\hat{\mathbf{r}}$$

EXERCISE SECTION

Essay Type Questions

1. Describe the evolution of the concept of electric displacement from the Faraday's law.
2. Describe the relation between intensity, displacement and displacement density. State the units of each of these quantities.
3. State and explain the Gauss law. Brief on the applications, utility and limitations.
4. Derive the differential form of the Gauss law from the integral form. How does this law appear for non-volume charge distributions in the two forms?
5. What is Gaussian surface? How many types are there? Give the criterion for the selection of Gaussian surface.
6. Consider a volume charge distribution and find out its divergence and curl.
7. Find the field intensity due to an infinite length straight line charge.
8. Applying the Gauss law, find out the field intensity due to an infinite plane sheet charge.
9. Find the field intensity due to a uniformly charged solid sphere both inside and outside.
10. Find the field intensity due to a uniformly charged spherical shell both inside and outside.

Numerical Problems

1. Determine the charge densities corresponding to the given displacement densities

(a) $\mathbf{D} = 4x^2y\hat{\mathbf{x}} + 3xy^2\hat{\mathbf{y}}$ C/m^2

(b) $\mathbf{D} = 2\rho\cos\phi\hat{\boldsymbol{\rho}} + 3\rho\sin\phi\hat{\boldsymbol{\phi}} + 5z\hat{\mathbf{z}}$ C/m^2

(c) $\mathbf{D} = 4\frac{\sin\theta}{r^2}\hat{\mathbf{r}} + \frac{\cos\theta}{r^3}\hat{\boldsymbol{\theta}}$ C/m^2

2. A point charge of 100 mC is placed at the centre of a cube. Find the total flux passing out of one face of the cube.

Fill in the Blanks

1. Electric flux is the result of the experiments performed by ________.
2. Flux lines represent the direction of the ________.
3. The electric flux lines ________ cross each other.
4. Units of displacement are ________ whereas those of displacement density are ________.
5. Displacement of charge is ________ of media whereas field intensity is ________ dependent.
6. Gauss law is a consolidation of results of experiments performed by ________.
7. Application of Gauss law requires integration over ________ surface.
8. Application of Gauss law requires the charge distribution to be highly ________.
9. Application of Gauss law requires ________ information regarding field.
10. Point form of Gauss law involves the ________ charge density.
11. For a spherical charge the Gaussian surface is ________.
12. For an infinite straight line charge the Gaussian surface is ________.
13. For an infinite plane sheet charge the Gaussian surface is ________.
14. Field inside a charged spherical shell is ________.
15. Field inside a charged solid sphere is ________.

Answers

1. Faraday
2. field
3. never
4. C, C/m^2
5. independent, media
6. Faraday
7. Gaussian
8. symmetrical
9. prior
10. volume
11. spherical
12. cylindrical
13. thin pillbox
14. zero
15. non-zero

Match the Following

Set I

1. Displacement	(a) C/m^2
2. Displacement density	(b) charge
3. Div of **D**	(c) charge enclosed

4. Closed surface integral of **D**	(d) volume charge density
5. Field inside charged spherical shell	(e) zero

Set II

1. Straight line charge	(a) integral form
2. Spherical charge distribution	(b) symmetry
3. Plane sheet charge distribution	(c) thin pillbox
4. Gauss law	(d) spherical surface
5. Field computation	(e) cylindrical surface

Set III

1. Quick computation of field	(a) limitation
2. Symmetry of charge distribution	(b) advantage
3. Direction of field	(c) scalar
4. Displacement	(d) normal or tangential to Gaussian surface
5. Displacement density	(e) vector

Set IV

1. Infinite straight line charge	(a) integral form
2. Solid spherical charge	(b) field independent of distance
3. Spherical charge shell	(c) field inside zero
4. Infinite plane charge	(d) field inside non-zero
5. Point charge	(e) inverse to distance

Answers

Set I:	1. (b)	2. (a)	3. (d)	4. (c)	5. (e)
Set II:	1. (e)	2. (d)	3. (c)	4. (b)	5. (a)
Set III:	1. (b)	2. (a)	3. (d)	4. (c)	5. (e)
Set IV:	1. (e)	2. (d)	3. (c)	4. (b)	5. (a)

Multiple Choice Questions

1. The integral form of Gauss law $\oint_c \mathbf{D} \cdot d\mathbf{a} = Q_{\text{enc}}$ is valid for
   (a) all types of charge distributions (b) volume charge distributions only
   (c) surface charge distributions only (d) point charge distributions only
2. The point form of Gauss law $\nabla \cdot \mathbf{D} = \rho$ is valid for
   (a) all types of charge distributions (b) volume charge distributions only
   (c) surface charge distributions only (d) point charge distributions only

3. The number of basic types of Gaussian surfaces is
   (a) two (b) three
   (c) four (d) zero
4. The limitation of the Gauss law in finding electric field intensity **E** is
   (a) prior knowledge of the field (b) highly symmetrical charge distributions
   (c) both (a) and (b) (d) none of these
5. The direction of field **E** on the Gaussian surface must be
   (a) normal (b) tangential
   (c) either (a) or (b) (d) none of these
6. When the field **E** is normal, its magnitude on the Gaussian surface
   (a) must be constant over that portion
   (b) can be variable over that portion
   (c) no restrictions on the magnitude
   (d) none of these
7. When the field **E** is tangential, its magnitude on the Gaussian surface
   (a) need not be constant over that portion
   (b) can be variable over that portion
   (c) both (a) and (b)
   (d) none of these

Answers

1. (a)	2. (b)	3. (a)	4. (c)	5. (a)
6. (c)	7. (c)			

4

Electric Potential and Energy

The most important single ingredient in the formula of success is knowing how to get along with people.

—Theodore Roosevelt

Electrostatic fields can also be described by potential functions. Even though field intensity and flux density are useful for this purpose, they have the disadvantage of being vectors which are usually considered difficult to manipulate when compared to scalar functions.

In this chapter, the concept and definition of electric scalar potential are given. Its variants, absolute and relative potentials, and their applicability to various types of charge distributions are discussed in detail.

The fields are reservoirs of energy and the source of this energy is explained. Then the expressions for the energy content of discrete as well as continuous charge distributions are derived.

4.1 Electric Scalar Potential V

The scalar potential V at a point P in an electric field $\mathbf{E}$ is defined as the external work done to bring a unit positive charge from infinity to the point P.

Mathematically, it can be stated as

$$V = -\int_{\infty}^{P} \mathbf{E} \cdot d\mathbf{l} \tag{4.1}$$

The reference point (lower limit on the integral sign) can be considered to be at infinity in ordinary cases except when the charge distribution itself extends to infinity like infinite straight line or infinite sheet charge. In such cases, if the reference point is infinity, the above expression does not give a finite value for the potential. To overcome this problem, in the case of infinite straight line charge, the reference is selected as a point which is at a finite distance from the line and in the case of infinite sheet charge, it can be either the origin or a point at a finite distance from the origin. Changing reference point amounts to adding a constant to the potential. One important property of potential function is that it always obeys the superposition principle.

Proof: Let us suppose a unit positive charge at a point P_1 in an electrostatic field **E** is moved to point P by the application of a force F_{ext}. Then the work done on this charge is given by,

$$W = \int_{P_1}^{P} \mathbf{F}_{ext} \cdot d\mathbf{l}$$

As the unit positive charge is in the field, it experiences a Coulomb's force which is given by

$$\mathbf{F}_c = Q\mathbf{E} = +1\mathbf{E} = \mathbf{E}$$

Observe that the force required to move the unit charge, i.e. $\mathbf{F}_{ext}$, is equal to $\mathbf{F}_c$ magnitude-wise but opposite in direction. Thus,

$$\mathbf{F}_{ext} = -\mathbf{F}_c = -\mathbf{E}$$

Therefore, the work by the external agency in terms of field intensity becomes

$$W = \int_{P_1}^{P} \mathbf{F}_{ext} \cdot d\mathbf{l} = -\int_{P_1}^{P} \mathbf{E} \cdot d\mathbf{l}$$

If we want the expression for external work done in moving the unit charge from infinity then P_1 should be replaced by infinity. And of course, this work is scalar potential at point P, by definition. Hence,

$$W = -\int_{\infty}^{P} \mathbf{E} \cdot d\mathbf{l} \equiv V$$

Note that this equation is an integral relation for scalar potential which can be used to compute the potential when the expression for the field intensity is available.

4.1.1 Differential Relation

A differential relation connecting electric potential and field intensity can also be derived. It is useful to compute the intensity when the potential function is available.

The basic definition of the scalar potential tells

$$V = -\int_{\infty}^{P} \mathbf{E} \cdot d\mathbf{l} \qquad \text{[Refer to Eq. (4.1)]}$$

Differentiating both sides, we get an expression for differential potential

$$dV = -\mathbf{E} \cdot d\mathbf{l} \tag{4.2}$$

But, from differential calculus, the differential potential can be expressed as

$$dV = \frac{\partial V}{\partial x}dx + \frac{\partial V}{\partial y}dy + \frac{\partial V}{\partial z}dz$$

$$= \left(\hat{\mathbf{x}}\frac{\partial V}{\partial x} + \hat{\mathbf{y}}\frac{\partial V}{\partial y} + \hat{\mathbf{z}}\frac{\partial V}{\partial z}\right) \cdot (\hat{\mathbf{x}}dx + \hat{\mathbf{y}}dy + \hat{\mathbf{z}}dz) = \nabla V \cdot d\mathbf{l} \quad (4.3)$$

Comparing both sides of the equation, we get results in the following identity:

$$\mathbf{E} = -\nabla V \quad (4.4)$$

It says that 'the electric field intensity is equal to the negative gradient of electric scalar potential'. As mentioned it is highly useful to deduce the field intensity when potential function is available.

Depending upon the reference used in the computation, the potential of a point can be one of the two types: absolute potential and relative potential. They are discussed here further.

4.1.2 Absolute Potential

It is the potential computed with respect to a reference point whose potential is zero. For most of the finite charge distributions, the zero potential point lies at infinity; therefore, we get absolute potential for them by setting the lower limit over the integral sign of the relation given above to infinity. In the case of charge distributions which extend up to infinity like infinite straight line charge or infinite plane sheet charge, zero potential point does not exist at infinity. For the first charge distribution, zero potential point exists nowhere and hence absolute potential cannot be computed. In the case of infinite plane sheet charge distribution, the zero potential point lies at the origin. Hence, absolute potential can be calculated for infinite sheet charge distributions by considering the reference at the origin.

4.1.3 Relative Potential

The potential that is computed with respect to a point where the potential is finite non-zero is called relative potential. Finite charge distributions in general can have both absolute and relative potentials, whereas certain infinite charge distributions like infinite line charge cannot have absolute potentials and we can calculate only the relative potentials in such cases. But infinite plane sheet charge distribution is a special type of distribution extending to infinity and still having absolute potential.

Mathematically the relative potential can be stated as

$$V_{ab} = -\int_b^a \mathbf{E} \cdot d\mathbf{l} \quad (4.5)$$

It is the expression for potential of a point a with respect to b. The electric potential of a single print charge and discrete charge distribution is shown in Figure 4.1.

4.1.4 Conservative Property

In an electrostatic field, the net work done in moving a charge in a closed path is always zero. It means the net amount of energy taken from the field is equal to the net amount of

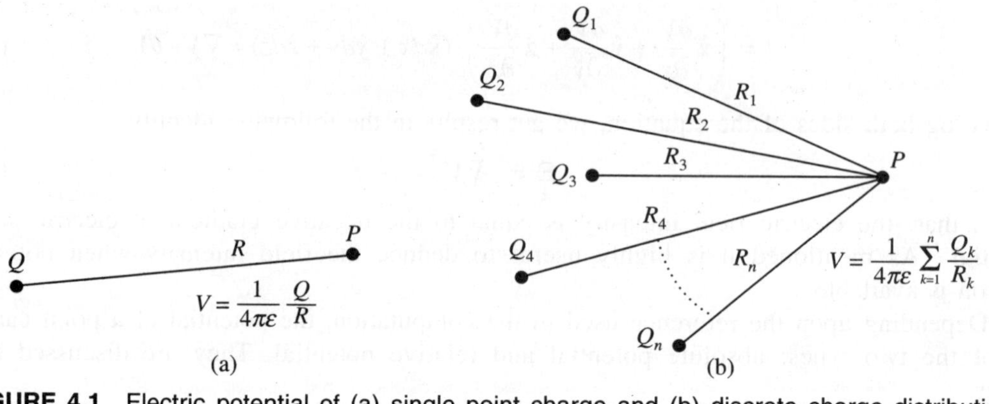

FIGURE 4.1 Electric potential of (a) single point charge and (b) discrete charge distribution.

energy given to the field by the particle while it is traversing the closed path. The energy of the field as well as of the particle is conserved in the rotation of the charge and hence this property is named as *conservative property*.

Proof: We consider the movement of a charged particle Q in a closed path c in an electrostatic field. Let us select any two distinct points 'a' and 'b' over the path. The work done in moving the particle from 'a' to 'b' is

$$W_{ab} = -Q\int_a^b \mathbf{E}\cdot d\mathbf{l}$$

And the work done in moving the particle from 'b' to 'a' is

$$W_{ba} = -Q\int_b^a \mathbf{E}\cdot d\mathbf{l}$$

The net work done while the particle traverses the closed path once from 'a' to 'b' and back to 'a' is the sum of works done while moving from 'a' to 'b' and 'b' to 'a'. Therefore,

$$W = W_{ab} + W_{ba} = -Q\int_a^b \mathbf{E}\cdot d\mathbf{l} - Q\int_b^a \mathbf{E}\cdot d\mathbf{l}$$

$$= -Q\int_a^b \mathbf{E}\cdot d\mathbf{l} + Q\int_a^b \mathbf{E}\cdot d\mathbf{l} = 0 \tag{4.6}$$

As the net work done in moving the charged particle over a closed path is zero, the electrostatic field is a conservative field. Note that only electrostatic fields are conservative; in due course of time we will see that time-varying electric fields are not conservative in general.

Curl of Electrostatic Fields

From the relation given by Eq. (4.6), a mathematical statement pertaining to the conservative nature of the electric field can be deduced. Now,

$$W = -Q\int_a^b \mathbf{E}\cdot d\mathbf{l} - Q\int_b^a \mathbf{E}\cdot d\mathbf{l} = 0$$

Thus,

$$-\int_a^b \mathbf{E}\cdot d\mathbf{l} - \int_b^a \mathbf{E}\cdot d\mathbf{l} = 0$$

$$\oint_c \mathbf{E}\cdot d\mathbf{l} = 0 \tag{4.7}$$

Equation (4.7) involves a closed line integral over a quantity $\mathbf{E}\cdot d\mathbf{l}$. In terms of this quantity, we can state the above relation as 'the line integral of $\mathbf{E}\cdot d\mathbf{l}$ over a closed path in electrostatic fields is always zero'. This relation can be put into the differential form invoking the Stokes' theorem of vector calculus. Thus,

$$\oint_c \mathbf{E}\cdot d\mathbf{l} = \int_s \nabla\times\mathbf{E}\cdot d\mathbf{a} \tag{4.8}$$

Here, the closed contour of the left-hand side and the surface of the right-hand side of the equation are related. The integral of the right-hand side can be evaluated on any surface bounded by the closed contour over which the left-hand side integral is evaluated. Now as the closed line integral is zero, i.e.

$$\oint_c \mathbf{E}\cdot d\mathbf{l} = 0$$

the surface integral must also be zero, i.e.

$$\int_s \nabla\times\mathbf{E}\cdot d\mathbf{a} = 0$$

As the surface over which the integration is performed is arbitrary, we can conclude that

$$\nabla\times\mathbf{E} = 0 \tag{4.9}$$

The above equation states that the curl of electrostatic fields is always zero or electrostatic fields are curl-less fields. In fact, it is one of the Maxwell's four famous equations.

- Potential and potential energy are completely different.
- Potential function contains all the information contained by the three components of electric field.
- Changing reference point amounts to adding a constant to the potential function.
- Potential difference is independent of the reference point.
- In the case of charge distributions of infinite extent, infinity is not a suitable reference.
- Potential obeys the superposition principle.
- Unit of potential is volt.

4.1.5 Equipotential Contours

Depending upon the charge distribution, different points surrounding the charge in space may have the same potentials. If all the points with the same potentials are joined, sometimes they form surfaces. Over these surfaces, at any point the potential has the same value. Such surfaces are called *equipotential surfaces* or *contours*. The shape of the contours depends upon the charge distribution. In the case of point charges or charges with spherical symmetry, the equipotential surfaces are concentric spherical shells. For long straight line charge distributions, the equipotential surfaces are coaxial circular cylinders. Flux lines and equipotential contours are shown in Figure 4.2(a–d).

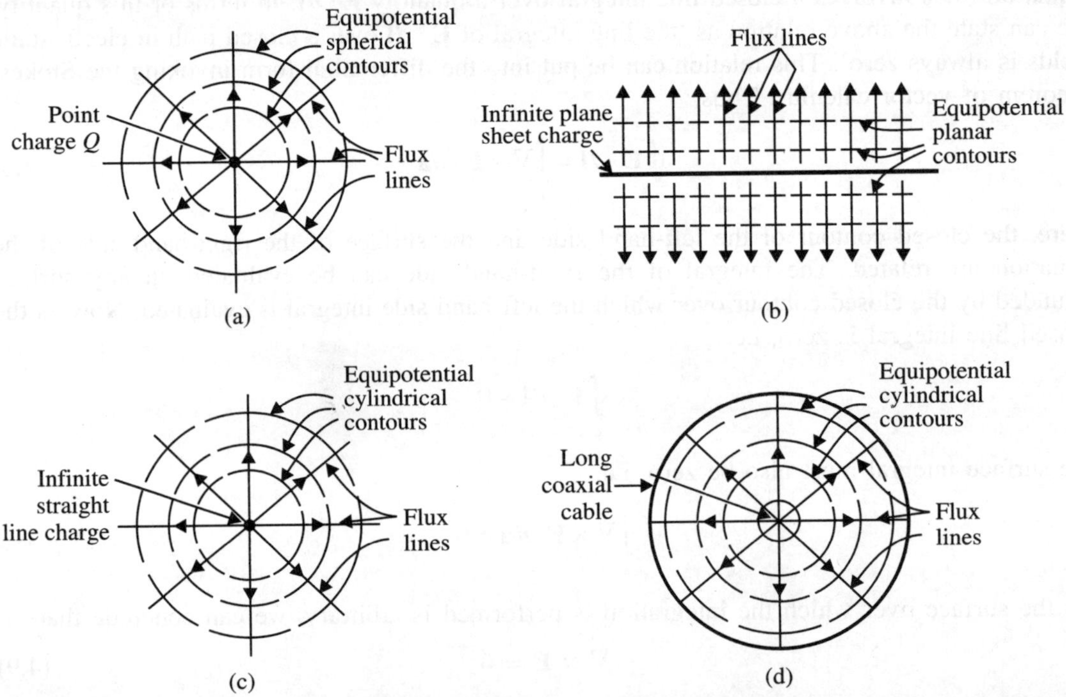

FIGURE 4.2 Flux lines and equipotential contours of (a) point charge, (b) large sheet charge (c) large straight line charge and (d) long coaxial cable.

Note the difference between flux lines of the field and equipotential surfaces. Even though both pertain to regions surrounding charge distributions, one represents the direction of the field, whereas the other represents the surfaces of equal potential. It can be shown that electric flux lines are always orthogonal to equipotential contours.

4.1.6 Potential Functions

With the zero potential reference point at infinity, the potential function for a point charge Q is given by

$$V = \frac{1}{4\pi\varepsilon}\int \frac{Q}{R} \tag{4.10}$$

where, R represents the distance between the source point and the field point. The potential functions for various continuous charge distributions are given as follow:

- For line charge distribution

$$V(\mathbf{r}) = \frac{1}{4\pi\varepsilon}\int \frac{\lambda(\mathbf{r}')dl'}{R} \tag{4.11}$$

- For surface charge distribution

$$V(\mathbf{r}) = \frac{1}{4\pi\varepsilon}\int \frac{\sigma(\mathbf{r}')da'}{R} \tag{4.12}$$

- For volume charge distribution

$$V(\mathbf{r}) = \frac{1}{4\pi\varepsilon}\int \frac{\rho(\mathbf{r}')d\tau'}{R} \tag{4.13}$$

In the above expressions, the position vectors of the source point and the field point are $\mathbf{r}'$ and $\mathbf{r}$ respectively and $\mathbf{R}$ is the vector joining the source point with the field point given by $\mathbf{R} = \mathbf{r} - \mathbf{r}'$. Electric potential of continuous charge distributions is shown in Figure 4.3(a–c).

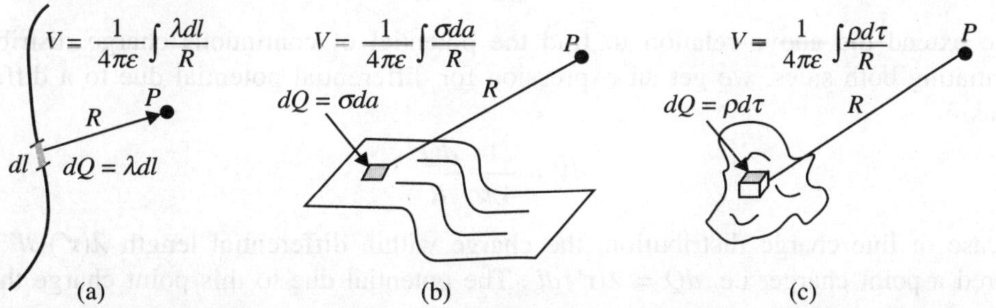

FIGURE 4.3 Electric potential of continuous charge distributions: (a) line charge, (b) sheet charge and (c) volume charge distributions.

Proof: To find the general expressions for potentials of various charge distributions, we make use of the basic defining relation and spherical coordinate system.

$$V = -\int_{\infty}^{P} \mathbf{E}\cdot d\mathbf{l}$$

We know that for a point charge Q located at the origin of the spherical coordinate system, the field intensity $\mathbf{E}$ is given by

$$\mathbf{E} = \frac{1}{4\pi\varepsilon}\frac{Q}{r^2}\hat{\mathbf{r}}$$

Here $\hat{\mathbf{r}}$ is the position vector of the observation point P. And the differential length in the spherical coordinate system is

$$d\mathbf{l} = dr\hat{\mathbf{r}} + rd\theta\hat{\boldsymbol{\theta}} + r\sin\theta d\phi\hat{\boldsymbol{\phi}}$$

Substituting the above two relations in the defining relation of potential, we get

$$V = -\frac{Q}{4\pi\varepsilon}\int_{\infty}^{P}\frac{1}{2}\hat{\mathbf{r}}\cdot(dr\hat{\mathbf{r}} + rd\theta\hat{\boldsymbol{\theta}} + r\sin\theta d\phi\hat{\boldsymbol{\phi}})$$

$$= -\frac{Q}{4\pi\varepsilon}\int_{\infty}^{P}\frac{dr}{r^2} = -\frac{Q}{4\pi\varepsilon}\left.\frac{-1}{r}\right|_{\infty}^{P} = \frac{Q}{4\pi\varepsilon r}$$

This gives the expression for the potential due to a point charge Q at a distance of r from the location of charge.

In the derivation, it is assumed that the source charge is over the origin. We can apply the above formula to find the potential even when it is located elsewhere. If the position vectors of the source point and the field point are $\mathbf{r}'$ and $\mathbf{r}$ respectively and if $\mathbf{R} = \mathbf{r} - \mathbf{r}'$ is the vector joining the source point with the field point, then the electric scalar potential for the point charge becomes

$$V(\mathbf{r}) = \frac{1}{4\pi\varepsilon}\int\frac{Q(\mathbf{r}')}{R}$$

Now we extend the above relation to find the potential of continuous charge distributions. Differentiating both sides, we get an expression for differential potential due to a differential charge as

$$dV = \frac{1}{4\pi\varepsilon}\frac{dQ}{R}$$

In the case of line charge distribution, the charge within differential length $\lambda(\mathbf{r}')\,dl'$ can be considered a point charge, i.e. $dQ = \lambda(\mathbf{r}')\,dl'$. The potential due to this point charge therefore becomes

$$dV = \frac{1}{4\pi\varepsilon}\frac{dQ}{R} = \frac{1}{4\pi\varepsilon}\frac{\lambda(\mathbf{r}')dl'}{R}$$

The potential due to the entire charge distribution then becomes

$$V = \int dV = \frac{1}{4\pi\varepsilon}\int\frac{\lambda(\mathbf{r}')dl'}{R}$$

Similarly, for other continuous charge distributions, the differential charge $dQ = \sigma da$ for surface charge distributions and $dQ = \rho d\tau$ for volume charge distributions result in

$$V(\mathbf{r}) = \frac{1}{4\pi\varepsilon}\int\frac{\sigma(\mathbf{r}')da'}{R}$$

$$V(\mathbf{r}) = \frac{1}{4\pi\varepsilon}\int \frac{\rho(\mathbf{r}')d\tau'}{R}$$

for surface and volume charge distributions respectively. Note that the above relations for scalar potential are arrived at assuming zero potential point at infinity. So they give absolute potentials of charge distributions. If relative potential is required, then we have to use the relation which gives potential in terms of field intensity with appropriate lower limit.

Point charge	$V = \frac{1}{4\pi\varepsilon}\frac{Q}{R}$
Discrete charge distribution	$V = \frac{1}{4\pi\varepsilon}\sum_{k=1}^{n}\frac{Q_k}{R_k}$
Line charge	$V(\mathbf{r}) = \frac{1}{4\pi\varepsilon}\int \frac{\lambda(\mathbf{r}')dl'}{R}$
Surface charge	$V(\mathbf{r}) = \frac{1}{4\pi\varepsilon}\int \frac{\sigma(\mathbf{r}')da'}{R}$
Volume charge	$V(\mathbf{r}) = \frac{1}{4\pi\varepsilon}\int \frac{\rho(\mathbf{r}')d\tau'}{R}$

4.2 ELECTRIC DIPOLE

The concept of electric dipole is useful in understanding the phenomenon of polarization that occurs in dielectrics in the presence of fields. It is an arrangement with two point charges of equal magnitude and opposite polarity, separated by a small distance. An important parameter pertaining to dipole is *dipole moment*: it is the product of charge and the separation distance.

A dipole gives rise its own field and when it is placed in an external field, it experiences torque and if the field happens to be non-uniform, it also experiences a net force. In this section, we try to estimate all the quantities required in the analysis of dielectrics.

4.2.1 Potential and Field Intensity

A dipole gives rise to a scalar potential given by

$$V = \frac{qd\cos\theta}{4\pi\varepsilon_0 r^2} \tag{4.14}$$

with a corresponding field strength of

$$\mathbf{E} = \frac{qd}{4\pi\varepsilon_0 r^3}(2\cos\theta\hat{\mathbf{r}} + \sin\theta\hat{\boldsymbol{\theta}}) \tag{4.15}$$

Proof: Let us suppose that the dipole consists of two point charges $+q$ and $-q$ separated by a distance d. Align the dipole along the z-axis and consider a field point P at a large distance r (Figure 4.4). Let us suppose that the positive and negative charges are at distances of r_1 and r_2 from that point. We know that the potential due to a point charge q at a distance of r is

$$V = \frac{q}{4\pi\varepsilon_0 r}$$

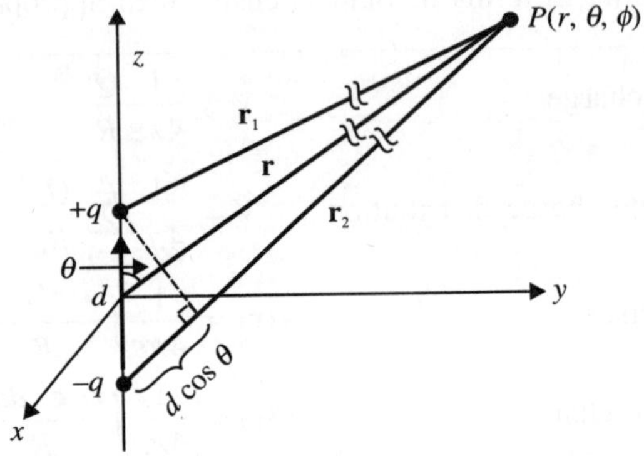

FIGURE 4.4 Electric potential of electric dipole.

The application in this relation for the two charges of the electric dipole results in

$$V = \frac{1}{4\pi\varepsilon_0}\left(\frac{q}{r_1} + \frac{-q}{r_2}\right) = \frac{q}{4\pi\varepsilon_0}\left(\frac{r_2 - r_1}{r_1 r_2}\right)$$

Since $d \ll r$, it can be approximated $r_2 - r_1 \simeq d \cos\theta$ and $r_1 r_2 = r^2$. Hence,

$$V = \frac{qd\cos\theta}{4\pi\varepsilon_0 r^2}$$

The numerator of the above equation can be written as the dot product of $q\mathbf{d}$, where $\mathbf{d}$ is the distance vector from the charge $-q$ to the charge $+q$ and the unit vector $\hat{\mathbf{r}}$ pointing from the centre of the dipole towards the observation point P.

$$qd\cos\theta = q\mathbf{d}\cdot\hat{\mathbf{r}} = \mathbf{p}\cdot\hat{\mathbf{r}}$$

where $\mathbf{p} = q\mathbf{d}$ is called *dipole moment* of the electric dipole. Thus, the potential of the dipole becomes

$$V = \frac{\mathbf{p}\cdot\hat{\mathbf{r}}}{4\pi\varepsilon_0 r^2}$$

In spherical coordinates,

$$\mathbf{E} = -\nabla V$$

$$= -\left(\hat{\mathbf{r}}\frac{\partial V}{\partial r} + \hat{\boldsymbol{\theta}}\frac{1}{r}\frac{\partial V}{\partial \theta} + \hat{\boldsymbol{\phi}}\frac{1}{r\sin\theta}\frac{\partial V}{\partial \phi}\right)$$

Substitution of V and its partial differentiation gives

$$\mathbf{E} = \frac{qd}{4\pi\varepsilon_0 r^3}(2\cos\theta\hat{\mathbf{r}} + \sin\theta\hat{\boldsymbol{\theta}})$$

The electric flux pattern of the electric dipole is shown in Figure 4.5.

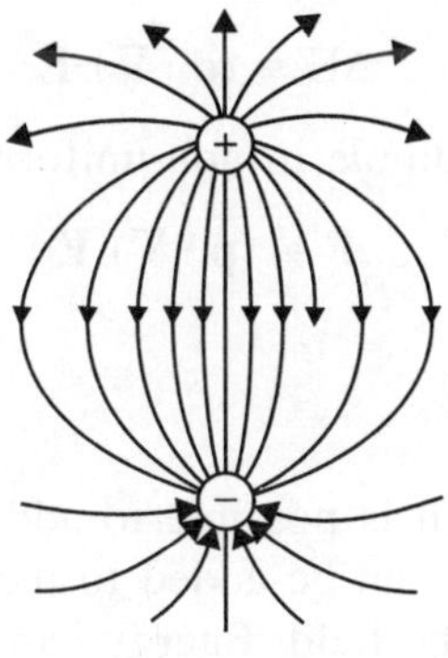

FIGURE 4.5 Electric flux pattern of electric dipole.

4.2.2 Torque and Force

Consider an electric dipole in a uniform electric field. The force on the positive end $\mathbf{F}_+ = q\mathbf{E}$ is exactly equal and opposite to the force on the negative end $\mathbf{F}_- = -q\mathbf{E}$, resulting in cancellation. But the points where the forces act are not the same and as a result there comes a torque into being (Figure 4.6) which is equal to

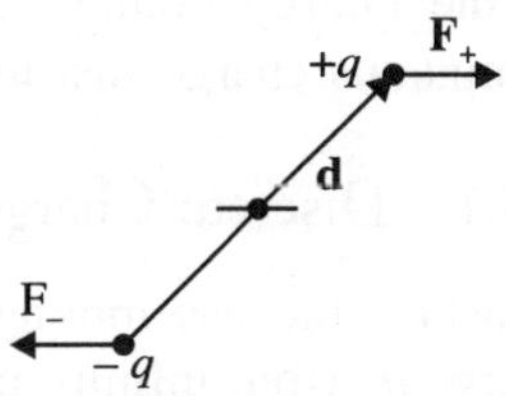

FIGURE 4.6 Torque on electric dipole placed in electric field.

$$\mathbf{T} = (\mathbf{r}_+ \times \mathbf{F}_+) + (\mathbf{r}_- \times \mathbf{F}_-)$$
$$= \left[\left(\frac{\mathbf{d}}{2}\right) \times (q\mathbf{E})\right] + \left[\left(\frac{-\mathbf{d}}{2}\right) \times (-q\mathbf{E})\right]$$
$$= q\mathbf{d} \times \mathbf{E} = \mathbf{p} \times \mathbf{E}$$

Thus, a dipole of moment $\mathbf{p} = q\mathbf{d}$ in a uniform field experiences a torque given by

$$\mathbf{T} = \mathbf{p} \times \mathbf{E} \tag{4.16}$$

When a dipole is placed in a non-uniform field, it experiences force as well as torque. The amount of torque can be computed using the above formula but it gives the torque about the centre of the dipole. If we need the torque about any other point, then the relation we have to use is

$$\mathbf{T} = (\mathbf{p} \times \mathbf{E}) + (\mathbf{r} \times \mathbf{F}) \tag{4.17}$$

Now we try to estimate the force on the dipole in a non-uniform field. It is the force on the positive end plus the force on the negative end and is given by

$$\mathbf{F} = \mathbf{F}_+ + \mathbf{F}_-$$
$$= q(\mathbf{E}_+ - \mathbf{E}_-) = q(\Delta\mathbf{E}) \tag{4.18}$$

where $\Delta\mathbf{E}$ represents the difference in the values of field intensity at the positive charge end and the negative charge end. But from the vector calculus, we can approximate a small change in E_x as

$$\Delta E_x \equiv (\nabla E_x) \cdot \mathbf{d} \tag{4.19}$$

Similar expressions can be written for other components of the field. More compactly, we can write

$$\Delta\mathbf{E} \equiv (\mathbf{d} \cdot \nabla)\,\mathbf{E} \tag{4.20}$$

and therefore the force on electric dipole in non-uniform electric field is given by

$$\mathbf{F} \equiv (\mathbf{p} \cdot \nabla)\,\mathbf{E} \tag{4.21}$$

4.3 ELECTRIC ENERGY

Fields are reservoirs of energy and it is possible to add energy to or extract energy from the field. One method by which energy can be added to the field is forcing a charged particle in opposition to the force exerted by the field. Energy can be taken out of the field by allowing a charged particle to get accelerated by the field. The energy of the fields is considered to be due to the conversion of work done into energy while assembling the source distribution. Now we try to estimate the energy in the electrostatic fields. Initially, we consider the finding of the energy content of discrete charge distributions and then we extend the result to continuous charge distributions.

4.3.1 Discrete Charge Distributions

Consider the assembling of n discrete charges in a charge-free region. To bring the first charge q_1 from infinity into this region does not require any work because the region initially is charge free and no force is applied against the movement of q_1. Next, consider the movement of the second charge q_2. To bring the second charge q_2 into the region, it is required that some work to be done to overcome the opposing force from the first charge which is equal to

$$W = \frac{1}{4\pi\varepsilon_0}\left(\frac{q_1 q_2}{r_{12}}\right) \tag{4.22}$$

Here r_{12} is the distance between two charges that exists in the region under consideration. Now consider the pulling of the third charge q_3 from infinity into the region which also requires some work to be done to counter the opposing force from the already existing charges, i.e. q_1 and q_2. This work is given by

$$W = \frac{1}{4\pi\varepsilon_0}\left[\left(\frac{q_1 q_3}{r_{13}}\right) + \left(\frac{q_2 q_3}{r_{23}}\right)\right] \tag{4.23}$$

Here r_{13} are r_{23} distances of q_3 from q_1 and q_2 respectively. Similarly, it is possible to find the work required to add a fourth charge q_4 to the charge constellation. It is given by

$$W = \frac{1}{4\pi\varepsilon_0}\left[\left(\frac{q_1 q_4}{r_{14}}\right) + \left(\frac{q_2 q_4}{r_{24}}\right) + \left(\frac{q_3 q_4}{r_{34}}\right)\right] \tag{4.24}$$

Now let us compute the total work done in assembling these four charges which is the sum of the works computed above. It is given by

$$W = \frac{1}{4\pi\varepsilon_0}\left[\left(\frac{q_1 q_2}{r_{12}}\right) + \left(\frac{q_1 q_3}{r_{13}}\right) + \left(\frac{q_1 q_4}{r_{14}}\right) + \left(\frac{q_2 q_3}{r_{23}}\right) + \left(\frac{q_2 q_4}{r_{24}}\right) + \left(\frac{q_3 q_4}{r_{34}}\right)\right] \tag{4.25}$$

The above expression for work can be put in a slightly different form useful in the present context.

$$W = \frac{1}{4\pi\varepsilon_0}\left[\left(\frac{q_1 q_2}{r_{21}}\right) + \left(\frac{q_1 q_3}{r_{31}}\right) + \left(\frac{q_1 q_4}{r_{41}}\right) + \left(\frac{q_2 q_3}{r_{32}}\right) + \left(\frac{q_2 q_4}{r_{42}}\right) + \left(\frac{q_3 q_4}{r_{43}}\right)\right] \tag{4.26}$$

Adding equations (4.25) and (4.26), we obtain

$$\begin{aligned}
2W = &\frac{1}{4\pi\varepsilon_0}\left[\left(\frac{q_1 q_2}{r_{12}}\right) + \left(\frac{q_1 q_3}{r_{13}}\right) + \left(\frac{q_1 q_4}{r_{14}}\right) + \left(\frac{q_2 q_3}{r_{23}}\right) + \left(\frac{q_2 q_4}{r_{24}}\right) + \left(\frac{q_3 q_4}{r_{34}}\right)\right. \\
&\left. + \left(\frac{q_1 q_2}{r_{21}}\right) + \left(\frac{q_1 q_3}{r_{31}}\right) + \left(\frac{q_1 q_4}{r_{41}}\right) + \left(\frac{q_2 q_3}{r_{32}}\right) + \left(\frac{q_2 q_4}{r_{42}}\right) + \left(\frac{q_3 q_4}{r_{43}}\right)\right] \\
= &\frac{1}{4\pi\varepsilon_0}\left[\left(\frac{q_1 q_2}{r_{12}}\right) + \left(\frac{q_1 q_3}{r_{13}}\right) + \left(\frac{q_1 q_4}{r_{14}}\right) + \left(\frac{q_1 q_2}{r_{21}}\right) + \left(\frac{q_2 q_3}{r_{23}}\right) + \left(\frac{q_2 q_4}{r_{24}}\right)\right. \\
&\left. + \left(\frac{q_1 q_3}{r_{31}}\right) + \left(\frac{q_2 q_3}{r_{32}}\right) + \left(\frac{q_3 q_4}{r_{34}}\right) + \left(\frac{q_1 q_4}{r_{41}}\right) + \left(\frac{q_2 q_4}{r_{42}}\right) + \left(\frac{q_3 q_4}{r_{43}}\right)\right] \\
= &\frac{1}{4\pi\varepsilon_0}\sum_{i=1}^{4}\sum_{\substack{j=1 \\ j\neq i}}^{4}\frac{q_i q_j}{r_{ij}}
\end{aligned}$$

or

$$W = \frac{1}{2}\frac{1}{4\pi\varepsilon_0}\sum_{i=1}^{4}\sum_{\substack{j=1 \\ j\neq i}}^{4}\frac{q_i q_j}{r_{ij}} \tag{4.27}$$

So this is the expression for the external work done to assemble the four number of discrete charges. Now we can extend this result to write down an expression for the required work to assemble n discrete charges as

$$W = \frac{1}{2}\frac{1}{4\pi\varepsilon_0}\sum_{i=1}^{n}\sum_{\substack{j=1 \\ j\neq i}}^{n}\frac{q_i q_j}{r_{ij}} = \frac{1}{2}\sum_{i=1}^{n} q_i \sum_{\substack{j=1 \\ j\neq i}}^{n}\left(\frac{1}{4\pi\varepsilon_0}\frac{q_j}{r_{ij}}\right) \tag{4.28}$$

We can observe that the term in the parentheses is an expression for the potential at the location of charge q_i due to all other charges which we can designate as V_i. In terms of this potential, the work done in assembling the charge distribution becomes

$$W = \frac{1}{2}\sum_{i=1}^{n} q_i V_i \rightarrow \text{Discrete charge distribution} \tag{4.29}$$

This expression gives us the work done on the charges to assemble them. As this work is considered to have gone into the field as energy, the expression can also be considered as the total energy stored in the field that surrounds the discrete charge distribution.

4.3.2 Continuous Charge Distributions

The above result can also be extended to the continuous charge distributions: line charge distributions, surface charge distributions and volume charge distributions. The line charge can be divided into several small length segments each of a differential length. The charge within each segment can be considered as a point charge and the above relation can be applied to the work done or energy of the field. As the charge distribution is continuous, the summation in the above expression becomes integration. The resultant expression for the energy of the line charge distribution then is

$$W = \frac{1}{2}\int \lambda V dl \rightarrow \text{Line charge distribution} \tag{4.30}$$

Similar expressions can be derived in case of other continuous charge distributions as follows:

$$W = \frac{1}{2}\int \sigma V da \rightarrow \text{Surface charge distribution} \tag{4.31}$$

$$W = \frac{1}{2}\int \rho V d\tau \rightarrow \text{Volume charge distribution} \tag{4.32}$$

In all the above expressions, the energy is given in terms of charge and potential functions. A much more general and useful relation for energy involves field quantities. It can be derived from the available relations using the Gauss theorem of electrostatics and the divergence theorem of vector calculus which connects surface and volume integrals. Applying the Gauss theorem $\nabla \cdot \mathbf{D} = \rho$ to the energy relation of energy of the volume charge distribution, we get the relation

$$W = \frac{1}{2}\int \rho V d\tau = \frac{1}{2}\int V\,(\nabla \cdot \mathbf{D})\, d\tau \tag{4.33}$$

Using the product rule

$$\nabla \cdot (f\mathbf{A}) = f(\nabla \cdot \mathbf{A}) + \mathbf{A} \cdot (\nabla f) \Rightarrow f(\nabla \cdot \mathbf{A}) = \nabla \cdot (f\mathbf{A}) - \mathbf{A} \cdot (\nabla f)$$

We get

$$W = \frac{1}{2}\left[\int (\nabla \cdot \mathbf{D}V) d\tau - \int \mathbf{D} \cdot \nabla V d\tau\right]$$

$$= \frac{1}{2}\left[\int (\nabla \cdot \mathbf{D}V) d\tau + \int \mathbf{D} \cdot \mathbf{E} d\tau\right] \tag{4.34}$$

In arriving at the above expression, we have used the relation $\mathbf{E} = -\nabla V$. Now we use the divergence theorem of vector calculus

$$\int (\nabla \cdot \mathbf{A}) d\tau = \oint \mathbf{A} \cdot d\mathbf{a}$$

In the first term of the right-hand side of Eq. (4.34) resulting in

$$W = \frac{1}{2}\left[\oint \mathbf{D}V \cdot d\mathbf{a} + \int \mathbf{D} \cdot \mathbf{E} d\tau\right] = \frac{1}{2}\left[\oint \mathbf{D}V \cdot d\mathbf{a} + \int \mathbf{D} \cdot \mathbf{E} d\tau\right] \tag{4.35}$$

The expression says that the energy has two components: one is given by a surface (4.35) integral and the other is a volume integral. Note that the integration surface of the first integral and the integration volume of the second integral are related: the volume of the second integral is one which is enclosed by the surface of the first integral. To evaluate the energy, we need to perform the integration over the entire region wherever the field exists. But the field due to any charge distribution almost extends up to infinity. Therefore, the region of integration becomes the entire space.

One mathematical technique used to evaluate the integral over the entire space is performing the integration over a spherical volume with the source distribution at its centre and then extending its radius to infinity. When the radius is extended to infinity, the surface integral tends to zero because the integrand, product of three factors, is inversely proportional to the radius. Therefore, the surface integral can be neglected and results in an expression for energy as

$$W = \frac{1}{2} \int_{\text{all space}} \mathbf{D} \cdot \mathbf{E} d\tau = \frac{\varepsilon}{2} \int_{\text{all space}} \mathbf{E} \cdot \mathbf{E} d\tau \tag{4.36}$$

In the case of free space $\varepsilon = \varepsilon_0$, the expression for energy of a charge distribution becomes

$$W = \frac{\varepsilon_0}{2} \int_{\text{all space}} E^2 d\tau \rightarrow \text{discrete and continuous distributions} \tag{4.37}$$

This is the expression for work which has been done to assemble the charge distribution and it is considered to have gone into the field and stored as energy. This energy is supposed to spread in the region of the entire field. The energy per unit volume then becomes $\varepsilon_0 E^2/2$ which is called *energy density* of the electrostatic fields.

Potential energies

Discrete charge distribution	$W = \frac{1}{2}\sum_{i=1}^{n} q_i V_i$
Surface charge distribution	$W = \frac{1}{2}\int \sigma V da$
Line charge distribution	$W = \frac{1}{2}\int \lambda V dl$

Volume charge distribution	$W = \frac{1}{2}\int \rho V d\tau$
General field expression	$W = \frac{\varepsilon_0}{2}\int_{\text{all space}} E^2 d\tau$
Dielectric systems	$W = \frac{1}{2}\int_{\text{all space}} \mathbf{D}\cdot\mathbf{E} d\tau$

One important aspect about electrostatic energy worth considering here is its quadratic relation with the field intensity. Hence, it does not obey the superposition principle, i.e. the energy of a compound field is not the sum of the energies of the individual fields.

4.4 POISSON'S LAW AND LAPLACE'S EQUATION

When a source charge distribution is given, it is possible to find the field intensity and the potential functions using

$$\mathbf{E}(\mathbf{r}) = \frac{1}{4\pi\varepsilon}\int \frac{\rho(\mathbf{r}')d\tau'}{R^2}\hat{\mathbf{R}} \tag{4.38}$$

and

$$V(\mathbf{r}) = \frac{1}{4\pi\varepsilon}\int \frac{\rho(\mathbf{r}')d\tau'}{R} \tag{4.39}$$

respectively.

However, in electrostatics, we come across a class of problems in which the charge distribution is not known and the thing that is available is potential over the boundary of the problem region. These types of problems are known as *boundary value problems* and as is evident, the above-mentioned equations or relations are not useful to determine **E** or V because the source charge distribution is not known. Instead, two equations known as Poisson's law and Laplace's equations are highly useful.

The boundary condition which specifies potential V at each and every point on the boundary of the problem region is called Dirichlet boundary condition. Sometimes the rate of change of potential V normal to boundary is specified. In that case, the boundary condition is called Neumann condition. The boundary condition which is a combination of the above two is called mixed boundary condition.

Considering divergence in both sides of $\mathbf{E} = -\nabla V$, we get

$$\nabla\cdot\mathbf{E} = -\nabla\cdot\nabla V = -\nabla^2 V \tag{4.40}$$

and using Gauss law $\nabla\cdot\mathbf{D} = \rho$, we have

$$\nabla^2 V = -\frac{\rho}{\varepsilon} \tag{4.41}$$

which is called Poisson's law.

In the case of situations where the field may be due to point charges and/or line charges and/or surface charges but with no volume charges, then the volume charge density $\rho = 0$ and the Poisson's equation becomes

$$\nabla^2 V = 0 \tag{4.42}$$

which is known as Laplace's equation.

4.4.1 Uniqueness Theorem

Uniqueness theorem has special importance in boundary value problems in assuring the uniqueness of their solutions. It states "within some close region, a solution to Laplace's equation to Poisson's equation for the potential function V is the only solution that satisfies the potential specified over the boundaries of the region". Therefore, it assures the uniqueness of the solution obtained by solving Laplace's equation or Poisson's equation for a given boundary condition.

We consider a few configurations of conductors to illustrate the utility of Laplace's equation in estimating the potential distribution and consequently field intensity. To begin with, consider two large parallel plates separated by a distance of d and placed parallel to the xy-plane. They have a potential difference applied across them and we are interested to find the potential distribution and field intensity in between the plates.

From the symmetry, we can observe that potential does not vary either with x or with y. The Laplace's equation therefore for this case is

$$\nabla^2 V = \frac{\partial^2 V}{\partial z^2} = 0 \tag{4.43}$$

Its solution is

$$V = k_1 z + k_2 \tag{4.44}$$

where k_1 and k_2 are arbitrary constants. From the available boundary conditions,

$$k_1 = \frac{V_1 - V_0}{d} \text{ and } k_2 = V_0$$

resulting in

$$V = \frac{V_1 - V_0}{d} z + V_0 \tag{4.45}$$

The field intensity can be computed as

$$\mathbf{E} = -\nabla V = -\frac{\partial V}{\partial z}\hat{\mathbf{z}} = -\frac{V_1 - V_0}{d}\,\hat{\mathbf{z}} \tag{4.46}$$

As a second example, we consider finding the potential distribution between concentric cylindrical conductors. As the configuration exhibits cylindrical symmetry, we need to use the Laplace's equation in the cylindrical coordinate system. Thus,

$$\nabla^2 V = \frac{1}{\rho}\frac{\partial}{\partial \rho}\left(\rho \frac{\partial V}{\partial \rho}\right) + \frac{1}{\rho^2}\left(\frac{\partial^2 V}{\partial \phi^2}\right) + \frac{\partial^2 V}{\partial z^2} \tag{4.47}$$

In the present case, the cylinders are quite long and the potential cannot have any variation either along ϕ or along z direction.

Therefore,

$$\nabla^2 V = \frac{1}{\rho}\frac{\partial}{\partial \rho}\left(\rho \frac{\partial V}{\partial \rho}\right) = 0 \tag{4.48}$$

The appropriate solution to this equation is

$$V = k_1 \ln \rho + k_2 \tag{4.49}$$

where k_1 and k_2 are arbitrary constants. The field intensity in the region between the cylinders can be computed as

$$\mathbf{E} = -\nabla V = -\frac{\partial V}{\partial \rho}\hat{\boldsymbol{\rho}} = -\frac{k_1}{\rho}\hat{\boldsymbol{\rho}} \tag{4.50}$$

As a third example, we consider two concentric spherical shells. As the configuration exhibits spherical symmetry, we need to use the Laplace's equation in spherical coordinate system. Thus,

$$\nabla^2 V = \frac{1}{r^2}\frac{\partial}{\partial r}\left(r^2 \frac{\partial V}{\partial r}\right) + \frac{1}{r^2 \sin\theta}\frac{\partial}{\partial \theta}\left(\sin\theta \frac{\partial V}{\partial \theta}\right) + \frac{1}{r^2 \sin^2\theta}\left(\frac{\partial^2 V}{\partial \phi^2}\right) \tag{4.51}$$

From the symmetry considerations, the potential cannot have any variation either along θ or along ϕ direction.

Therefore,

$$\nabla^2 V = \frac{1}{r^2}\frac{\partial}{\partial r}\left(r^2 \frac{\partial V}{\partial r}\right) = 0 \tag{4.52}$$

The appropriate solution to this equation is

$$V = \frac{k_1}{r} + k_2 \tag{4.53}$$

where k_1 and k_2 are arbitrary constants. The field intensity in the region between the spherical shells can be computed as

$$\mathbf{E} = -\nabla V = -\frac{\partial V}{\partial r}\hat{\mathbf{r}} = -\frac{k_1}{r^2}\hat{\mathbf{r}} \tag{4.54}$$

In all the above three examples, we can observe the utility and simplicity of the Laplace's equation in giving solutions to the boundary value problems. However, it should be kept in mind that the simplicity of the solution is possible only when the configuration exhibits high level of symmetry.

SOLVED PROBLEMS

1. Two point charges each of charge equal to q C is separated by a distance d m. Find the potential at a point which is at a distance of z m from the centre of the line joining the point charges.

Solution: The potential obeys the superposition principle. The total potential at an arbitrary point due to two point charges is equal to the sum of the potentials at the arbitrary point due to individual point charges.

Charges $-q$ and q are separated by a distance of d and the observation point P is at a distance of z above the charge distribution.

Potential due to $-q$ is given by

$$V_{-}(P) = \frac{-q}{4\pi\varepsilon_0\sqrt{z^2 + \left(\frac{d}{2}\right)^2}}$$

and that due to q is given by

$$V_{+}(P) = \frac{q}{4\pi\varepsilon_0\sqrt{z^2 + \left(\frac{d}{2}\right)^2}}$$

Then the potential due to both the charges will be

$$V(P) = V_{-}(P) + V_{+}(P) = \frac{-q}{4\pi\varepsilon_0\sqrt{z^2 + \left(\frac{d}{2}\right)^2}} + \frac{q}{4\pi\varepsilon_0\sqrt{z^2 + \left(\frac{d}{2}\right)^2}} = 0$$

2. A straight line charge of length $2L$ is carrying a uniform charge density of λ C/m^2. Find the potential at a point which is at a distance of z m from the centre of the line charge.

Solution: The observation point is at a distance of z above the uniform line charge of length $2L$ and charge density λ.

$$V(P) = V(P) = \frac{1}{4\pi\varepsilon_0}\int_{-L}^{L}\frac{\lambda dx}{\sqrt{z^2 + x^2}} = \frac{\lambda}{4\pi\varepsilon_0}\ln\left(x + \sqrt{z^2 + x^2}\right)\Bigg|_{-L}^{L}$$

$$= \frac{\lambda}{4\pi\varepsilon_0}\ln\left(\frac{L + \sqrt{z^2 + L^2}}{-L + \sqrt{z^2 + L^2}}\right) = \frac{\lambda}{2\pi\varepsilon_0}\ln\left(\frac{L + \sqrt{z^2 + L^2}}{z}\right)$$

3. A circular line charge of radius a is carrying a uniform charge density of λ C/m. Find the potential at a point which is at a distance of z m from the centre of the circular charge.

Solution: The observation point P is at a distance of z above the circular charge ring with radius R and line charge density λ C/m. Let it be over the xy-plane with its centre over the origin.

$$dQ = \lambda a d\phi$$

$$V(P) = \frac{1}{4\pi\varepsilon_0}\int_0^{2\pi}\frac{\lambda a d\phi}{\sqrt{a^2+z^2}}$$

$$= \frac{1}{4\pi\varepsilon_0}\int_0^{2\pi}\frac{\lambda a d\phi}{\sqrt{a^2+z^2}} = \frac{\lambda a}{2\varepsilon_0\sqrt{a^2+z^2}}$$

4. A circular charged disc of radius a is carrying a uniform surface charge density of σ C/m^2. Find the potential at a point which is at a distance of z m from the centre of the circular charge disc.

Solution: The observation point P is at a distance of z above the circular charge disc with radius a and charge density σ. Thus,

$$dQ = \sigma\rho d\rho d\phi$$

$$V(P) = \frac{1}{4\pi\varepsilon_0}\int_0^a\int_0^{2\pi}\frac{\sigma\rho d\rho d\phi}{\sqrt{\rho^2+z^2}}$$

$$= \frac{1}{2\varepsilon_0}\int_0^a\frac{\sigma\rho d\rho}{\sqrt{\rho^2+z^2}}$$

$$= \frac{\sigma}{2\varepsilon_0}\sqrt{\rho^2+z^2}\Big|_0^a = \frac{\sigma}{2\varepsilon_0}\left(\sqrt{a^2+z^2}-z\right)$$

5. A straight line charge of infinite length is carrying a uniform charge density of λ C/m. Find the potential at a point which is at a distance of ρ m from the line charge.

Solution: First let us try to find the absolute potential, i.e. with respect to zero potential point at infinity. The field intensity at a distance of ρ from the line charge is

$$\mathbf{E} = \frac{2\lambda}{4\pi\varepsilon_0\rho}\hat{\boldsymbol{\rho}}$$

Using $V = -\int_\infty^P \mathbf{E}\cdot d\mathbf{l}$, we have

$$V(\rho) = -\frac{1}{4\pi\varepsilon_0}\int_\infty^\rho\frac{2\lambda}{\rho}\hat{\boldsymbol{\rho}}\cdot(d\rho\hat{\boldsymbol{\rho}}+\rho d\phi\hat{\boldsymbol{\phi}}+dz\hat{\mathbf{z}}) = -\frac{1}{4\pi\varepsilon_0}\int_a^\rho\frac{2\lambda}{\rho}d\rho$$

$$= -\frac{1}{4\pi\varepsilon_0}2\lambda\ln\left(\frac{\rho}{a}\right)$$

Now we can observe, if $a \to \infty$, the expression for the potential is indeterminate. Therefore, absolute potential does not exist for infinite line charges. The expression can give a finite determinate value only when is non-zero finite. Hence, we can have only relative potentials with reference to points set at a finite distance from the charge distribution for infinite line charge distributions.

Now we consider a numerical example. A line charge $\lambda = 400$ pC/m lies along the x-axis. The reference point is (0, 5, 12) m. We need to find the potential at the point (2, 3, −4) m. From the given values, we have

$$a = \sqrt{5^2 + 12^2} = 13$$

$$\rho = \sqrt{3^2 + 4^2} = 5$$

$$V = \frac{1}{4\pi\varepsilon_0} 2\lambda \ln\left(\frac{\rho}{a}\right)$$

$$= -\frac{36\pi}{4\pi \times 10^{-9}} \times 2 \times 40 \times 10^{-12} \ln\left(\frac{5}{13}\right)$$

$$= -0.32 \ln\left(\frac{5}{13}\right) = 0.306$$

6. Determine the potential distribution about a long parallel pair of wires when the wires have equal and opposite charges distributed along their length.

Solution: Potential function obeys the superposition principle. The relative potential due to a long conductor carrying positive line charge is

$$V(\rho) = -\frac{1}{4\pi\varepsilon_0} 2\lambda \ln\frac{\rho}{a}$$

And the potential due to a long conductor carrying negative line charge becomes

$$V(\rho) = \frac{1}{4\pi\varepsilon_0} 2\lambda \ln\frac{\rho}{a}$$

Let us suppose the arbitrary point P (Figure 4.7), where potential is required, is at a normal distance of ρ_+ from the positive conductor and ρ_- from the negative conductor, a_+ and a_- are the normal distances of the reference point from the positive and negative conductors. Then the potential at the point P due to the conductors becomes the sum due to individual conductors.

$$V(P) = \frac{1}{4\pi\varepsilon_0} 2\lambda \ln\frac{\rho_-}{a_-} - \frac{1}{4\pi\varepsilon_0} 2\lambda \ln\frac{\rho_+}{a_-} = \frac{1}{4\pi\varepsilon_0} 2\lambda \ln\frac{\rho_-}{\rho_+} + \frac{1}{4\pi\varepsilon_0} 2\lambda \ln\frac{a_+}{a_-}$$

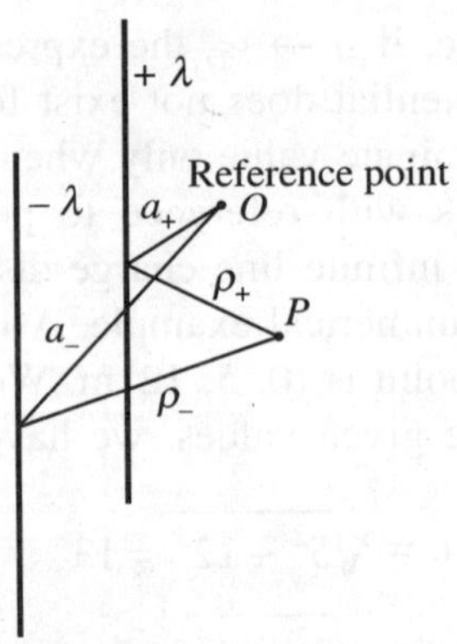

FIGURE 4.7 Electric potential function of parallel long wire system.

As the reference point moves towards infinity, the ratio a_+/a_- tends to one, resulting in a value of zero to the second term of the right-hand side of the above equation. Therefore, the potential of the given parallel wire conductor system, with reference point at infinity, is

$$V(P) = \frac{1}{4\pi\varepsilon_0} 2\lambda \ln \frac{\rho_-}{\rho_+}$$

7. A plane sheet charge of infinite extent is carrying a uniform surface charge density of σ C/m². Find the potential at a point which is at a distance of z m from the plane sheet charge.

Solution: In this case also, first let us try to find the absolute potential, i.e. with respect to zero potential point at infinity. The field intensity due to the infinite sheet charge is

$$\mathbf{E} = \frac{\sigma}{2\varepsilon_0} \hat{\mathbf{z}}$$

Using

$$V = -\int_{\infty}^{P} \mathbf{E} \cdot d\mathbf{l}$$

we have

$$V(\rho) = -\frac{\sigma}{2\varepsilon_0} \int_a^z \hat{\mathbf{z}} \cdot (d\rho\hat{\boldsymbol{\rho}} + \rho d\phi\hat{\boldsymbol{\phi}} + dz\hat{\mathbf{z}}) = -\frac{\sigma}{2\varepsilon_0} \int_a^z dz = -\frac{\sigma}{2\varepsilon_0}(z - a) = -\frac{\sigma z}{2\varepsilon_0}$$

Now it is clear why reference at infinity is unacceptable as we can observe, if $a \to \infty$, the expression for the potential is indeterminate. Therefore, absolute potential cannot be calculated with zero potential point at infinity for infinite sheet charges. In fact, we can observe that the zero potential point for this distribution is at the origin, i.e. a is zero. Hence, for this charge distribution, we can have absolute potential with respect to origin or relative potential with reference point set at a finite distance from the charge distribution.

8. A spherically charged shell of radius R is carrying a uniform surface charge density of σ C/m^2. Find the potential at a point which is both inside and outside the shell.

Solution: To find the required quantities, first we need to obtain the expressions for field intensities at points both inside and outside. With the help of Gauss law, for points outside the shell, the field intensity can be found as

$$\mathbf{E} = \frac{1}{4\pi\varepsilon}\frac{Q}{r^2}\hat{\mathbf{r}}$$

and inside the shell it is zero. Here $\hat{\mathbf{r}}$ is the vector joining the shell centre with the observation point P and Q is the total charge over the shell.
The potential at point P now is calculated using the defining relation

$$V = \int_{\infty}^{P} \mathbf{E}\cdot d\mathbf{l}$$

For the spherical shell and when the observation point is outside, the field intensity $\mathbf{E}$ is given by

$$\mathbf{E} = \frac{1}{4\pi\varepsilon}\frac{Q}{r^2}\hat{\mathbf{r}}$$

Here $\hat{\mathbf{r}}$ is the position vector of the observation point P. And the differential length in the spherical coordinate system is

$$d\mathbf{l} = dr\hat{\mathbf{r}} + rd\theta\hat{\boldsymbol{\theta}} + r\sin\theta\, d\phi\hat{\boldsymbol{\phi}}$$

Substituting the above two relations in the defining relation of potential, we get

$$V = -\frac{Q}{4\pi\varepsilon}\int_{\infty}^{P}\frac{1}{r^2}\hat{\mathbf{r}}\cdot(dr\hat{\mathbf{r}} + rd\theta\hat{\boldsymbol{\theta}} + r\sin\theta\, d\phi\hat{\boldsymbol{\phi}})$$

$$= -\frac{Q}{4\pi\varepsilon}\int_{\infty}^{P}\frac{dr}{r^2} = -\frac{Q}{4\pi\varepsilon}\left.\frac{-1}{r}\right|_{\infty}^{P} = \frac{Q}{4\pi\varepsilon r}$$

When it comes to inside points, i.e. if the point is inside the shell, then the defining integral can be split as the sum of two integrals.

$$V = \int_{\infty}^{P}\mathbf{E}\cdot d\mathbf{l} = \int_{\infty}^{R}\mathbf{E}\cdot d\mathbf{l} + \int_{R}^{P}\mathbf{E}\cdot d\mathbf{l}$$

$$= \int_{\infty}^{R}\mathbf{E}\cdot d\mathbf{l} = \frac{1}{4\pi\varepsilon}\frac{Q}{R}$$

Therefore, for a spherical shell with radius R, carrying uniform surface charge density σ with respect to reference point at infinity, the potential functions are given by

(i) For points outside the spherical shell,

$$V(r) = \frac{1}{4\pi\varepsilon}\frac{Q}{r}$$

(ii) For points inside the spherical shell,

$$V(r) = \frac{1}{4\pi\varepsilon}\frac{Q}{R}$$

Note that the potential inside the shell is not zero but it is a constant.

9. A solid spherical charge of radius R is carrying a uniform volume charge density of ρ C/m^3. Find the potential at a point which is both outside and inside the sphere.

Solution: The solid sphere is with radius R and is carrying uniform volume charge density ρ. With reference point at infinity, we know

$$V = -\int_{\infty}^{P} \mathbf{E}\cdot d\mathbf{l}$$

(i) For points outside the solid sphere ($r > R$),

$$\mathbf{E}(r) = \frac{1}{4\pi\varepsilon_0}\frac{q}{r^2}\hat{\mathbf{r}}$$

$$V(r) = -\int_{\infty}^{r} \frac{1}{4\pi\varepsilon_0}\frac{q}{r^2}dr = \frac{1}{4\pi\varepsilon_0}\frac{q}{r}$$

(ii) For points inside the solid sphere ($r < R$),

$$\mathbf{E} = \frac{q}{4\pi\varepsilon_0}\frac{r}{R^3}\hat{\mathbf{r}}$$

$$V(r) = -\int_{\infty}^{R} \frac{1}{4\pi\varepsilon_0}\frac{q}{r^2}dr - \int_{R}^{r} \frac{1}{4\pi\varepsilon_0}\frac{qr}{R^3}dr = \frac{q}{4\pi\varepsilon_0}\left[\frac{1}{R} - \frac{(r^2 - R^2)}{2R^3}\right]$$

$$= \frac{q}{4\pi\varepsilon_0}\frac{1}{2R}\left[3 - \frac{r^2}{R^2}\right]$$

10. Find the energy of a discrete charge distribution consisting of four discrete charges, each carrying a charge of q C at the corners of a square of a side of a m (see Figure 4.8).

Solution: The work done in assembling or the energy of a discrete charge distribution is

$$W = \frac{1}{2}\sum_{i=1}^{n} q_i V_i$$

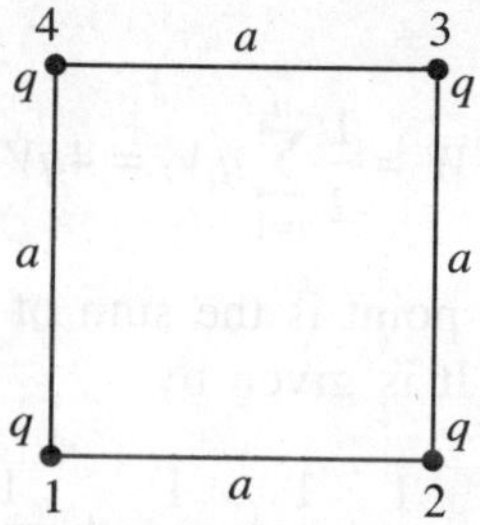

FIGURE 4.8 Energy of discrete charge distribution—charges at the corners of a square.

In the present case, $n = 4$, $q_1 = q_2 = q_3 = q_4 = q$ and $V_1 = V_2 = V_3 = V_4$, let us say V. Therefore,

$$W = \frac{1}{2}\sum_{i=1}^{n} q_i V_i = 2qV$$

The potential at any corner point is the sum of the potentials due to charges located at the other three corners. It is given by

$$V = \frac{1}{4\pi\varepsilon_0}\left(\frac{1}{a} + \frac{1}{a} + \frac{1}{\sqrt{2}a}\right) = \frac{(2\sqrt{2}+1)}{4\pi\varepsilon_0 a\sqrt{2}}$$

$$W = 2qV = \frac{\sqrt{2}(2\sqrt{2}+1)q}{4\pi\varepsilon_0 a}$$

11. Find the energy of a discrete charge distribution consisting of eight discrete charges, each carrying a charge of q C at the corners of a cube of a side of a m (see Figure 4.9).

Solution: Energy of a discrete charge distribution is given by
In the present case, $n = 8$, $q_1 = q_2 = \cdots = q_8 = q$ and $V_1 = V_2 = ... = V_8$, let us say V.

$$W = \frac{1}{2}\sum_{i=1}^{n} q_i V_i$$

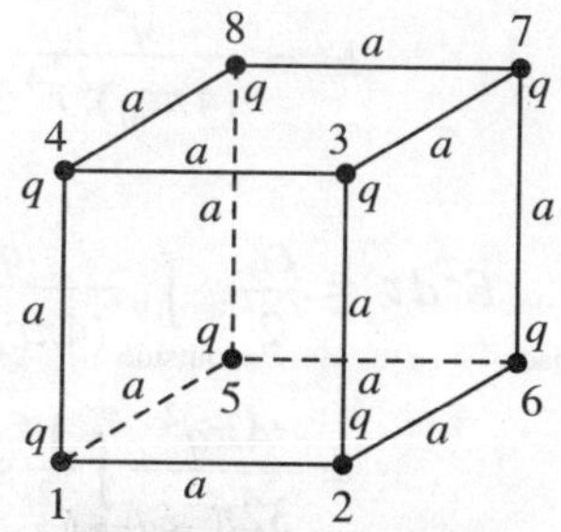

FIGURE 4.9 Energy of discrete charge distribution—charges at the corners of a cube.

Therefore,

$$W = \frac{1}{2}\sum_{i=1}^{n} q_i V_i = 4qV$$

The potential at any corner point is the sum of the potentials due to charges located at the other seven corners. It is given by

$$V = \frac{1}{4\pi\varepsilon_0}\left(\frac{1}{a} + \frac{1}{a} + \frac{1}{a} + \frac{1}{\sqrt{2}a} + \frac{1}{\sqrt{2}a} + \frac{1}{\sqrt{2}a} + \frac{1}{\sqrt{3}a}\right)$$

$$= \frac{1}{4\pi\varepsilon_0 a}\left(\frac{3\sqrt{6} + 3\sqrt{3} + \sqrt{2}}{\sqrt{6}}\right)$$

$$W = 4qV = \frac{q}{\pi\varepsilon_0 a}\left(\frac{3\sqrt{6} + 3\sqrt{3} + \sqrt{2}}{\sqrt{6}}\right)$$

12. Find the energy of a uniformly charged spherical shell of total charge q and radius R.

Solution: We can arrive at the result using either of the two expressions for the electric energy: the first one gives the expression for energy in terms of charge distribution and the second one gives energy in terms of the field intensity.

$$W = \frac{1}{2}\int \sigma V da \text{ or } W = \frac{\varepsilon_0}{2}\int_{\text{all space}} E^2 d\tau$$

We know that the potential at the surface of this sphere is $V = q/4\pi\varepsilon_0 R$, a constant, so

$$W = \frac{1}{2}\int \sigma V da = \frac{q}{8\pi\varepsilon_0 R}\int \sigma da = \frac{q^2}{8\pi\varepsilon_0 R}$$

The second method requires the expression for field and we also know that the field inside the sphere is zero and outside the sphere, it is

$$\mathbf{E} = \frac{q}{4\pi\varepsilon_0 r^2}\hat{\mathbf{r}}$$

$$E^2 = \frac{q^2}{(4\pi\varepsilon_0)^2 r^4}$$

Therefore,

$$W = \frac{\varepsilon_0}{2}\int_{\text{all space}} E^2 d\tau = \frac{\varepsilon_0}{2}\int_{\text{outside}} \frac{q^2}{(4\pi\varepsilon_0)^2 r^4} r^2 \sin\theta dr d\theta d\phi$$

$$= \frac{4\pi q^2}{32\pi^2\varepsilon_0}\int_R^{\infty}\frac{1}{r^2}\sin\theta dr = \frac{q^2}{8\pi\varepsilon_0 R}$$

Observe that the results obtained by both the methods agree.

13. Fing the energy stored in a uniformly charged solid sphere of total charge q and radius R.

Solution: We can have the solution to the problem and arrive at the result using either of the two expressions for the electric energy: the first one gives the expression for energy in terms of charge distribution and the second one gives energy in terms of the field intensity.

$$W=\frac{1}{2}\int \rho V d\tau \text{ and } W=\frac{\varepsilon_0}{2}\int_{\text{all space}} E^2 d\tau$$

We know that the potential at a point located at a distance of $r < R$ from the centre of the sphere is

$$V(r)=\frac{q}{4\pi\varepsilon_0}\frac{1}{2R}\left[3-\frac{r^2}{R^2}\right]$$

Substituting this relation in the equation for energy, we get

$$W=\frac{1}{2}\int \rho V d\tau=\frac{1}{2}\frac{q}{4\pi\varepsilon_0}\frac{\rho}{2R}\int_0^R\int_0^{2\pi}\int_0^{\pi}\left[3-\frac{r^2}{R^2}\right]r^2 \sin\theta d\theta d\phi dr$$

$$=\frac{q\rho}{4\varepsilon_0 R}\left[3\frac{r^3}{3}-\frac{r^5}{5R^2}\right]_0^R=\frac{q\rho R^2}{5\varepsilon_0}=\frac{qR^2}{5\varepsilon_0}\frac{3q}{4\pi R^3}=\frac{q^2}{4\pi\varepsilon_0}\frac{3}{5R}$$

We also calculate the energy using the second relation which is in terms of field. We know that the field of the sphere is

$$\mathbf{E}=\begin{cases}\dfrac{q}{4\pi\varepsilon_0}\dfrac{r}{R^3}\,\hat{\mathbf{r}} & \text{Inside}\\[2ex] \dfrac{q}{4\pi\varepsilon_0 r^2}\,\hat{\mathbf{r}} & \text{Outside}\end{cases}$$

Therefore, the energy stored in the system becomes

$$W=\frac{\varepsilon_0}{2}\int_{\text{all space}} E^2 d\tau=\frac{\varepsilon_0}{2}\frac{q^2}{(4\pi\varepsilon_0)^2}\left[\int_0^R\left(\frac{r}{R^3}\right)^2 4\pi r^2 dr+\int_R^{\infty}\frac{1}{r^4}4\pi r^2 dr\right]$$

$$=\frac{q^2}{4\pi\varepsilon_0}\frac{3}{5R}$$

14. Find the energy stored in a uniformly charged spherical shell of total charge q and radius a and surrounded by a linear dielectric material of susceptibility χ_e out to radius b.

Solution: From the given information, the field intensity can be expressed as

$$\mathbf{E} = \begin{cases} 0 & \text{for } r < a \\ \dfrac{q}{4\pi\varepsilon r^2}\hat{\mathbf{r}} & \text{for } a < r < b \\ \dfrac{q}{4\pi\varepsilon_0 r^2}\hat{\mathbf{r}} & \text{for } r > b \end{cases}$$

And the displacement density **D** becomes

$$\mathbf{D} = \begin{cases} 0 & \text{for } r < a \\ \varepsilon\mathbf{E} & \text{for } a < r < b \\ \varepsilon_0\mathbf{E} & \text{for } r > b \end{cases}$$

The energy then is

$$W = \frac{1}{2}\int_{\text{all space}} \mathbf{D}\cdot\mathbf{E}\,d\tau$$

$$d\tau = r^2 \sin\theta dr\, d\theta\, d\phi$$

Integration with respect to ϕ results in 2π and with respect θ results in 2.

$$W = \frac{1}{2}\left(\frac{q}{4\pi}\right)^2 4\pi\left(\frac{1}{\varepsilon}\int_a^b \frac{1}{r^2}\frac{1}{r^2}r^2 dr + \frac{1}{\varepsilon_0}\int_b^\infty \frac{1}{r^2}\frac{1}{r^2}r^2 dr\right)$$

$$= \frac{q^2}{8\pi}\left[\frac{1}{\varepsilon}\left(-\frac{1}{r}\right)\bigg|_a^b + \frac{1}{\varepsilon_0}\left(-\frac{1}{r}\right)\bigg|_b^\infty\right]$$

$$= \frac{q^2}{8\pi\varepsilon_0}\left[\frac{1}{1+\chi_e}\left(\frac{1}{a}-\frac{1}{b}\right)+\frac{1}{b}\right]$$

$$= \frac{q^2}{8\pi\varepsilon_0\,(1+\chi_e)}\left[\left(\frac{1}{a}+\frac{\chi_e}{b}\right)\right]$$

EXERCISE SECTION

Essay Type Questions

1. Define the conservative property and show that electrostatic field is conservative.
2. Derive the integral expression for the electric scalar potential in terms of the field intensity. Also derive differential expression of field intensity in terms of the scalar potential.
3. Differentiate the relative potential from the absolute potential. Explain why absolute potentials do not exist for infinite straight line charge and infinite surface charge.

4. Find an expression for the potential of a point charge and extend it with the aid of the superposition principle to the other continuous charge distributions.
5. Sketch an electric dipole and define its dipole moment. Show that the potential of the electric dipole is inversely proportional to the square of the distance to the observation point.
6. Consider an electric field associated with a volume charge distribution and derive a relation for the divergence and curl of the field.
7. Consider a set of discrete charges at infinity and derive an expression for the energy required to assemble these charges.
8. Derive an expression for the energy stored in a line charge distribution and extend it to the surface and volume charge distributions.
9. Show that the energy density in the electric field is proportional to the square of the field intensity.

Numerical Problems

1. Find the work done in moving a point charge 5 mC from the point (2, 1, 3) to (4, – 2, 5) in the electric field $\mathbf{E} = 4\hat{\mathbf{x}} + 3x^2\hat{\mathbf{y}} + 2yz^2\hat{\mathbf{z}}$ V/m .
2. Two point charges 5 mC and –10 mC are located at (1, 1, 1) and (0, –1, 1) respectively. Assuming zero potential point at infinity, find the potentials at the points (a) (1, 1, 0), (b) (0, 0, –1) and (c) (0, 1, 0).
3. A uniform line charge of length 2 m with a total charge of 3 nC is situated coincident to the z-axis with its centre point 2 m from the origin. At a point on the x-axis, 2 m from the origin, find V and $\mathbf{E}$.
4. Find the potential corresponding to the following electric field intensities:

   (a) $\mathbf{E} = 2yz\hat{\mathbf{x}} + z^2\hat{\mathbf{y}} - y\hat{\mathbf{z}}$

   (b) $\mathbf{E} = e^{-z}\,(\sin\phi\hat{\boldsymbol{\rho}} + \cos\phi\hat{\boldsymbol{\phi}} - \rho\sin\phi\hat{\mathbf{z}})$
5. A line charge is $2a$ m long and has a uniform charge λ C/m. Find the potential at a point r m from the line and located on the plane which bisects the line.

Fill in the Blanks

1. Electric potential is work done by __________ agency.
2. To compute the absolute potential, the reference is set at __________.
3. To compute the relative potential, the reference is at a __________ distance.
4. Potential due to point charge is inversely proportional to __________.
5. The potential inside a charged spherical shell is __________.
6. Equipotential surfaces of a point charge are __________.
7. Equipotential surfaces of an infinite straight line charge are __________.
8. Energy stored in an assembly of discrete charged particles is __________.

9. Energy density in electrostatic fields is given by ________.
10. Equipotential lines are always ________ to flux lines.
11. The potential due to electric dipole is inversely proportional to the ________ of distance.
12. The field intensity due to electric dipole is inversely proportional to the ________ of distance.

Answers

1. external	2. infinity
3. finite	4. distance
5. constant	6. concentric spherical shells
7. coaxial cylindrical surfaces	8. $\sum q_i V_i$
9. $\dfrac{\varepsilon_0 E^2}{2}$	10. perpendicular
11. square	12. cube

Match the Following

Set I

1. Absolute potential	(a) scalar potential
2. Relative potential	(b) inverse proportionality with distance
3. Equipotential contours	(c) normal to flux lines
4. Potential of point charge	(d) finite potential reference
5. Electric potential	(e) zero potential reference

Set II

1. Point charge	(a) $V = \dfrac{1}{4\pi\varepsilon} \sum_{k=1}^{n} \dfrac{Q_k}{R_k}$
2. Discrete charge distribution	(b) $V = \dfrac{1}{4\pi\varepsilon} \dfrac{Q}{R}$
3. Line charge	(c) $V(\mathbf{r}) = \dfrac{1}{4\pi\varepsilon} \int \dfrac{\sigma(\mathbf{r}')da'}{R}$
4. Surface charge	(d) $V = \dfrac{1}{4\pi\varepsilon} \int \dfrac{\lambda(\mathbf{r}')dl'}{R}$
5. Volume charge	(e) $V(\mathbf{r}) = \dfrac{1}{4\pi\varepsilon} \int \dfrac{\rho(\mathbf{r}')d\tau'}{R}$

Set III

1. Absolute potential	(a) $V_{ab} = -\int_b^a \mathbf{E}\cdot d\mathbf{l}$
2. Relative potential	(b) $V = -\int_\infty^P \mathbf{E}\cdot d\mathbf{l}$
3. Field intensity	(c) $\nabla^2 V = -\dfrac{\rho_v}{\varepsilon}$
4. Poisson's equation	(d) $-\nabla V$
5. Laplace's equation	(e) $\nabla^2 V = 0$

Set IV

1. Potential of point charge Q	(a) $\dfrac{1}{2}\int \rho V d\tau$
2. Discrete charge distribution	(b) $\dfrac{1}{2}\int \lambda V dl$
3. Surface charge	(c) $\dfrac{1}{2}\int \sigma V da$
4. Line charge	(d) $\dfrac{1}{2}\sum_{i=1}^{n} q_i V_i$
5. Volume charge	(e) QV

Set V

1. Conservative field	(a) $\dfrac{\varepsilon E^2}{2}$
2. Energy density	(b) $\nabla \times \mathbf{E} = 0$
3. Potential	(c) joules
4. Potential energy	(d) volts
5. Divergence	(e) volume charge density

Answers

Set I:	1. (e)	2. (d)	3. (c)	4. (b)	5. (a)
Set II:	1. (b)	2. (a)	3. (d)	4. (c)	5. (e)
Set III:	1. (b)	2. (a)	3. (d)	4. (c)	5. (e)
Set IV:	1. (e)	2. (d)	3. (c)	4. (b)	5. (a)
Set V:	1. (b)	2. (a)	3. (d)	4. (c)	5. (e)

Multiple Choice Questions

1. The potential is the work done on the unit positive charge by
   (a) external agency (b) field
   (c) both (a) and (b) (d) none of these
2. The relation between potential V and field intensity $\mathbf{E}$ is
   (a) $\mathbf{E} = -\nabla V$ (b) $V = -\int \mathbf{E} \cdot d\mathbf{l}$
   (c) both (a) and (b) (d) none of these
3. The relation $\mathbf{E} = -\nabla V$ is independent of
   (a) charge distribution (b) reference point of potential
   (c) both (a) and (b) (d) none of these
4. The electric scalar potential V due to point charges is
   (a) inversely proportional to distance (b) directly proportional to distance square
   (c) independent of the distance (d) none of these
5. The electric scalar potential V inside the spherical charge shell is
   (a) non-zero (b) constant
   (c) both (a) and (b) (d) none of these
6. The electric scalar potential V outside the spherical charge shell is
   (a) non-zero (b) a distance-dependent variable
   (c) both (a) and (b) (d) none of these
7. The reference point of scalar potential V due to charge distributions extending to infinity cannot be
   (a) at infinity (b) at a point away from the charge
   (c) at the centre (d) none of these
8. The reference point of scalar potential V due to a point charge can be
   (a) at infinity (b) at a point away from the charge
   (c) both (a) and (b) (d) none of these
9. The reference point of scalar potential V due to infinite line charges can be
   (a) at infinity
   (b) at the centre of line charge
   (c) at a point away from the line charge
   (d) none of these
10. The reference point of scalar potential V due to infinite sheet charges can be
    (a) at a point away from the sheet charge
    (b) at the centre of sheet charge
    (c) both (a) and (b)
    (d) at infinity

11. The electric scalar potential V outside the spherical charge shell is
    (a) non-zero
    (b) a distance-dependent variable
    (c) both (a) and (b)
    (d) none of these
12. The direction of the electric dipole is from
    (a) positive to negative charge
    (b) negative to positive charge
    (c) both (a) and (b)
    (d) none of these
13. The field intensity **E** due to dipole is
    (a) proportional to dipole moment qd
    (b) inversely proportional to the cube of distance
    (c) both (a) and (b)
    (d) none of these
14. The scalar potential V due to dipole is
    (a) proportional to dipole moment qd
    (b) inversely proportional to the square of distance
    (c) both (a) and (b)
    (d) none of these
15. The torque on dipole due to electric field is
    (a) $\mathbf{T} = \mathbf{p} \times \mathbf{E}$
    (b) $\mathbf{T} = (\mathbf{p} \cdot \nabla)\, \mathbf{E}$
    (c) both (a) and (b)
    (d) none of these
16. The force on electric dipole due to non-uniform electric field is
    (a) $\mathbf{F} = \mathbf{p} \times \mathbf{E}$
    (b) $\mathbf{F} = (\mathbf{p} \cdot \nabla)\, \mathbf{E}$
    (c) both (a) and (b)
    (d) none of these
17. Identify the wrong combination
    (a) $\nabla^2 V = -\dfrac{\rho_v}{\varepsilon}$—Poisson's law
    (b) $\nabla^2 \cdot V = 0$—Laplace's equation
    (c) $\nabla \cdot \mathbf{D} = \rho$—Gauss law
    (d) none of these
18. The electrostatic field **E** is
    (a) conservative
    (b) non-conservative
    (c) both (a) and (b)
    (d) none of these
19. For electrostatic field **E** obeys
    (a) $\oint \mathbf{E} \cdot d\mathbf{l} = 0$
    (b) $\nabla \times \mathbf{E} = 0$
    (c) both (a) and (b)
    (d) none of these
20. The energy density in electrostatic field **E** is
    (a) $\dfrac{\varepsilon_0 E^2}{2}$
    (b) $\mathbf{D} \cdot \dfrac{\mathbf{E}}{2}$
    (c) both (a) and (b)
    (d) none of these

21. The electrical energy stored in various charge distributions is

(a) $\frac{1}{2}\sum_{i=1}^{n} q_i V_i$ (b) $\frac{1}{2}\int \sigma V da$

(c) $\frac{1}{2}\int \rho V d\tau$ (d) all of these

Answers

1. (a)	2. (c)	3. (c)	4. (a)	5. (c)
6. (c)	7. (a)	8. (c)	9. (c)	10. (c)
11. (c)	12. (a)	13. (c)	14. (c)	15. (a)
16. (b)	17. (d)	18. (a)	19. (c)	20. (c)
21. (d)				

5

Conductors, Dielectrics and Capacitors

Success is to be measured not so much by the position that one has reached in life as by the obstacles which he has overcome.

—Booker T. Washington

The materials and media that we come across while discussing the various applications of fields can be divided into two basic types: conductors and dielectrics. Conductors allow free flow of electricity, whereas dielectrics do not allow the electric current to pass through them.

In this chapter, conductors are defined. Their properties and the reasons behind their behaviour are explained. Then the definition and properties of dielectrics are given. The polarization which is the phenomenon that occurs when a dielectric is placed in field is analysed thoroughly.

Capacitors are energy storage electrical elements. Usually, their structure consists of two conductors separated by a dielectric. They are highly useful for a variety of purposes in electrical circuits. They are defined and the capacitance of a variety of capacitors is calculated.

5.1 CONDUCTORS

Conductors are a class of materials characterized by the existence of plenty of free electrons and consequently good conductivity. Silver, copper, aluminium, etc. are examples of such materials. They exhibit certain interesting and useful properties when placed in the electrostatic fields which are listed below.

The electric field is always zero inside a conductor, i.e. $\mathbf{E} = 0$ inside a conductor. If it is non-zero initially, let us say due to some external applied field, then the free charges inside experience a force due to this non-zero field and move over to the surface resulting in their redistribution. The redistribution is such that the field causing the redistribution gets cancelled by the field which is the result of the redistribution of free charge resulting in nil field inside. Therefore, the electric field in the interior of the conductor is always zero regardless of the externally applied field or shape of the conductor.

The net charge density in the interior of a conductor is zero, i.e. $\rho = 0$ inside a conductor. This fact can be understood easily by applying the Gauss law $\nabla \cdot \mathbf{E} = (\rho/\varepsilon)$ to the interior of the conductor. As the field intensity is zero inside the conductor, i.e. $\mathbf{E} = 0$, the Gauss law says that the charge density must be zero. If we throw some charge onto the conductor, then, due to mutual attraction or repulsion, the added charge immediately moves over to the surface resulting in nil net charge in the interior of the conductor.

The net charge density inside a conductor is zero but there exists a surface charge over its surface. This surface charge is the result of mutual repulsion that exists between the free charges within the conductor. The surface charge results in field intensity outside the conductor which can be found to be as $\mathbf{E} = (\sigma/\varepsilon_0)\,\hat{\mathbf{n}}$ by applying the Gauss law.

The surface charge can be estimated in terms of the gradient of the potential. We know that

$$\mathbf{E} = -\frac{\partial V}{\partial n}\hat{\mathbf{n}} \tag{5.1}$$

But it is found that at the surface of the conductor,

$$\mathbf{E} = \left(\frac{\sigma}{\varepsilon_0}\right)\hat{\mathbf{n}} \tag{5.2}$$

Comparing both the equations, we can write the relation for surface density as

$$\sigma = -\,\varepsilon_0\frac{\partial V}{\partial n} \tag{5.3}$$

At various points outside as well as inside a conductor, the potential always remains the same, i.e. V is constant throughout the conductor. This aspect can be understood easily from the fact that the electric field on the surface of the conductor (external to the surface) is perpendicular to the surface and due to this, the potential on the surface of any conductor is constant.

The electric field **E** that exists outside the conductor can only be perpendicular to the surface. If it has a tangential component to the surface, the resultant force on the surface charge leads to its redistribution in such a way that, ultimately, there is no tangential component.

The surface of the conductor experiences certain amount of force due to the fields set up by the surface charge that exists over other parts of the surface of the conductor and it can be shown, per unit area, that it is equal to

$$\mathbf{f} = \frac{\sigma^2}{2\varepsilon_0}\hat{\mathbf{n}} \tag{5.4}$$

- Electric field is zero inside the conductor.
- Charge density is zero inside the conductor.
- Any net charge resides on the surface.
- The conductor is always equipotential.
- Electric field is always normal to the surface of the conductor.
- Conductors experience an outward electrostatic pressure equal to $\varepsilon_0 E^2/2$.

5.1.1 Types of Currents

In electromagnetic field theory, we come across three types of currents: conduction current, convection current and displacement current.

Conduction current results when the atoms or matters of the conducting medium do not move. In metal wire, for example, all the positive charge and most of the negative charge cannot move; only the electrons in the outermost shells move resulting in conduction current. The conduction current obeys the Ohm's law.

Convection current is generated by the actual movement of electrically charged matter. A wind-driven charged cloud, electron beam of cathode ray tube, etc. are examples of convection current. The convection current does not obey the Ohm's law.

The concept of *displacement current* was first introduced by James Clerk Maxwell in 1873 in his successful attempt to establish a unified connection between electric and magnetic fields under time-varying conditions. It is related to electric field through

$$\mathbf{J}_D = \frac{\partial \mathbf{D}}{\partial t} \tag{5.5}$$

Here, we confine our discussion to conduction currents only.

5.1.2 Ohm's Law in Point Form

When an external field is applied, there results a conduction current in conductors and semi-conductors. This current is due to the flow of electrons in the case of conductors and due to electrons as well as holes in the case of semi-conductors. In either case, the resultant current is proportional to the applied external field.

The charge carriers of the medium drift with certain velocity v_d which depends upon the externally applied field E.

$$v_d = \pm\mu E \tag{5.6}$$

Here μ is the proportionality constant known as the mobility of the carriers, i.e. electrons or holes. The '+' sign is for holes and the '–' sign is for electrons. The electrons and holes have different mobilities and in general it is the function of effective mass of the carrier as well as the mean distance the carrier moves in between collisions.

Let us consider the case of a conductor in which the current mechanism is entirely due to electrons. By definition, the current density J is the charge crossing the unit cross-sectional area in unit time and it is equal to the product of volume charge density ρ and the drift velocity v_d. Thus,

$$J = \rho v_d \tag{5.7}$$

But the volume charge density is the product of the number of electrons n per unit volume and the charge over a single electron $-e$. Therefore,

$$J = -nev_d \tag{5.8}$$

But we know that $v_d = -\mu E$. So

$$J = ne\mu E \tag{5.9}$$

Hence, the expression for current in conductors becomes

$$J = ne\mu E = \sigma E \tag{5.10}$$

The constant $\sigma = ne\mu$ is called the conductivity of the conductor medium. The conductivities of different conductors and insulators are given in Table 5.1.

Table 5.1 Conductivity σ in mhos per metre

Conductor	σ	Insulator	σ
Silver	6.17×10^7	Quartz	10^{-17}
Copper	5.80×10^3	Polystyrene	10^{-16}
Gold	4.10×10^7	Rubber (hard)	10^{-15}
Aluminium	3.82×10^7	Mica	10^{-14}
Tungsten	1.82×10^7	Porcelain	10^{-13}
Zinc	1.67×10^7	Diamond	10^{-13}
Brass	1.50×10^7	Glass	10^{-12}
Nickel	1.45×10^7	Bakelite	10^{-9}
Iron	1.03×10^7	Marble	10^{-8}
Bronze	1.00×10^7	Soil (sandy)	10^{-5}
Solder	0.70×10^7	Sands (dry)	2×10^{-4}
Steel (stainless)	0.11×10^7	Clay	10^{-4}
Nichrome	0.10×10^7	Ground (dry)	$10^{-4} - 10^{-5}$
Graphite	7.00×10^4	Ground (wet)	$10^{-2} - 10^{-3}$
Silicon	1.20×10^3	Water (distilled)	2×10^{-4}
Water (sea)	3 – 5	Water (fresh)	10^{-3}
		Femite (typical)	10^{-2}

In the case of semi-conductors, the current is partly due to electrons and partly due to holes, so the expression for current becomes

$$J = ne\mu_e E + pe\mu_p E = (ne\mu_e + pe\mu_p) E = \sigma E \tag{5.11}$$

The constant $\sigma = e(n\mu_e + p\mu_p)$ is the conductivity of the semi-conductor medium. Here n and p are electron and hole densities and μ_c and μ_p are the electron and hole mobilities respectively. In either case, the current density $J = \sigma E$. This relation is known as the Ohm's law in pointform. It says that the conduction current density at an arbitrary point in a medium is directly proportional to the externally applied field intensity at that point.

An important limitation of the law is that it is applicable only to the conduction currents.

5.1.3 Joule's Law

This law gives an expression for the power dissipated in a conducting medium in the presence of an electric field **E**. Under the influence of the applied field, the free electrons acquire drift velocities increasing their collision rate. As a result of all the collisions, the work done by the electrical force is converted into heat and the Joule's law is an expression for the dissipated power.

Consider a conductor with free electron volume charge density of ρ. Then the charge in a differential volume $d\tau$ becomes $\rho d\tau$. Under the influence of the externally applied electric field **E**, this charge experiences a force

$$d\mathbf{F} = \mathbf{E}\rho d\tau \tag{5.12}$$

Let us suppose that because of the force, the charge moved a distance dl in a duration of dt. Then the work done dW on the charge is

$$\begin{aligned} dW &= d\mathbf{F} \cdot d\mathbf{l} = dFdl \\ &= E\rho d\tau dl \end{aligned} \tag{5.13}$$

This is the work done or energy spent by the external field in moving the charge. The power dP expended by the field then is

$$dP = \frac{dW}{dt} = E\rho d\tau \frac{dl}{dt} = E\rho d\tau v_d \tag{5.14}$$

In the above expression, v_d stands for drift velocity of the electrons. The total power P being spent moving the electrons becomes

$$\begin{aligned} P &= \int_v dP = \int_v E\rho v_d d\tau \\ &= \int_v \mathbf{E} \cdot \mathbf{J} d\tau \end{aligned} \tag{5.15}$$

This equation is a mathematical statement of the Joule's law.

If we consider the Ohm's law in point form $\mathbf{J} = \sigma\mathbf{E}$, then

$$P = \int_v \mathbf{E} \cdot \mathbf{J} d\tau = \int_v \sigma E^2 \, d\tau \tag{5.16}$$

5.1.4 Relaxation Time Constant

The conductors are electrically neutral because the charge on the electrons is balanced by the positive charge in the atoms' nuclei. When an excess free charge q is introduced at some interior point in the conductor, it gives rise to an electric field which forces the charges of the host material nearest to the excess charge to rearrange their locations, which in turn cause other charges to move and the process continues until neutrality is reestablished in the conductor material and a charge equal to q is supplied to the conductor's surface.

To find the rate at which the excess charge dissipates, let us introduce a volume charge density ρ_{vol} at the interior of a conductor and consider the equation of continuity

$$\nabla \cdot \mathbf{J} = -\frac{\partial \rho_v}{dt} \tag{5.17}$$

But according to the Ohm's law,

$$\mathbf{J} = \sigma \mathbf{E} \tag{5.18}$$

Hence,

$$\sigma \nabla \cdot \mathbf{E} = -\frac{\partial \rho_v}{dt} \tag{5.19}$$

Now use the Gauss law $\nabla \cdot \mathbf{E} = \rho_v/\varepsilon$ to obtain the partial differential equation

$$\frac{\partial \rho_v}{dt} + \frac{\sigma}{\varepsilon} \rho_v = 0 \tag{5.20}$$

Given that $\rho_v = \rho_{\text{vol}}$ at $t = 0$, the solution of the equation is

$$\rho_v(t) = \rho_{\text{vol}}\, e^{-(\sigma/\varepsilon)t} = \rho_{\text{vol}}\, e^{-t/\tau_r} \tag{5.21}$$

where $\tau_r = \varepsilon/\sigma$ is called the relaxation time constant.

We see from the equation that the initial excess charge ρ_{vol} decays exponentially at a rate τ_r. The relaxation time constant is very small for conductors and it is only $\tau_r = 1.53 \times 10^{-19}$ s for copper. Thus, the charge dissipation process in a conductor is extremely fast. In contrast, the decay process is very slow in a good insulator. For mica, the relaxation time constant $\tau_r = 5.31 \times 10^4$ s. So the charge dissipation process is very slow in insulators. For certain types of semi-conductors like GaAs, the relaxation time constant is negative resulting in dynamic negative resistance over certain portion of its *E-J* characteristic.

5.2 DIELECTRICS

Dielectrics are materials characterized by very poor conductivity. They can be of any state—solid, liquid or gaseous and of any form—crystalline or amorphous. Their atomic structure is characterized by tightly bound valence electrons, resulting in the absence of free or mobile electrons. In the case of perfect dielectrics, which do not exist in the real world, free electron content is nil resulting in zero conductivity. In the dielectrics that are available in the real world, free electron concentration is not zero and it results in small amount of non-zero conductivity.

The dielectric nature of the materials is characterized by one parameter known as *permittivity* ε. In the case of ordinary materials, the ratio of conductivity to permittivity is an indication of its conducting or dielectric nature. In fact, the ratio $\sigma/\omega\varepsilon$, called the dissipation factor of the dielectric, equal to one is considered as the dividing line between the conductors and the dielectrics.

The dielectrics can be divided into linear and non-linear. Within the material, if the ratio of displacement density to field intensity is constant irrespective of the field intensity, then that material is considered as linear dielectric. otherwise non-linear dielectric. They can also be classified as homogeneous and non-homogeneous. Homogeneous dielectrics are characterized by uniform nature of properties throughout the material. In the case of non-homogeneous dielectrics, the properties vary from point to point.

When a dielectric is placed in an external field, two things usually happen. The material gets polarized resulting in the induction of some bound charge which in turn gives rise to its own field. It can be shown that when a homogeneous linear dielectric is placed in a field, the induced bound volume charge density ρ_b is proportional to the free charge density ρ_f.

The second change is that due to the field of the induced bound charge, the existing original field gets modified. Now the net field is the vector sum of the external field and the field due to the polarization. If a homogeneous linear dielectric is placed in the field, the field intensity in that region gets reduced by a factor of one over the dielectric constant, i.e $1/\varepsilon_r$.

5.2.1 Polarization of Dielectrics

Dielectric materials are electrically neutral. But when an external field is applied, the positive and negative charges within shift from their mean position thereby creating dipoles. This phenomenon in which the creation of electric dipoles takes place is called *polarization*.

The polarization that takes place upon the application of external field may be ascribed to one or more of the three factors: electron polarization, ionic polarization and orientational polarization. The electron polarization occurs due to the shift of the neutral atom's positive and negative charges, whereas the ionic polarization is due to the shift of positive and negative ions from their mean positions in a molecule upon the application of the field. The last one orientational polarization is due to the tendency of randomly oriented existing dipoles to align with the externally applied field.

The dielectrics can also be classified into polar and non-polar. In polar dielectrics, the positive and negative charges in the molecules remain shifted permanently from their mean positions resulting in several pairs acting as dipoles. In general, these dipoles are oriented randomly with the body of the dielectric and the application of the electric field tries to align them in one direction. In non-polar materials, the dipoles do not exist prior to the application of the electric field but they come into being due to the application of the field.

When a piece of dielectric material is placed in an electric field:

- The field will induce in each atom a tiny dipole moment pointing in the same direction as the field, if the material is made up of non-polar molecules.
- Each permanent dipole will experience a torque, tending to line it up along the field direction, if the material is made up of polar molecules.

In either case, the result is a lot of little dipoles pointing along the direction of the field and the dielectric is said to be *polarized*. A convenient measure of this effect is polarization **P**, the dipole moment per unit volume.

5.2.2 Field of a Polarized Object

Here we try to estimate the field due to a polarized object. A polarized object consists of, at microscopic level, several electric dipoles and each dipole gives rise to its own field. The total field due to the object is then the vector sum of the fields due to the individual dipoles. We have already seen that a single dipole with a dipole moment **p** gives rise to a potential given by

$$V = \frac{1}{4\pi\varepsilon_0}\frac{\hat{\mathbf{r}}\cdot\mathbf{p}}{r^2} \tag{5.22}$$

Now we are going to extend the result to find the potential due to a polarized object. Let us suppose that the polarization of the object, i.e. the dipole moment per unit volume, be **P**. The dipole moment due to a differential volume $d\tau'$ becomes equal to $\mathbf{P}\,d\tau'$. Then the potential due to this dipole moment is

$$dV = \frac{1}{4\pi\varepsilon_0}\left[\frac{\hat{\mathbf{R}}\cdot\mathbf{P}}{R^2}\,d\tau'\right] \tag{5.23}$$

In the above expression, $\hat{\mathbf{R}}$ is the vector joining the source point $\mathbf{r}'$ within $d\tau'$ to the field point **r**, i.e. $\mathbf{R} = \mathbf{r} - \mathbf{r}'$. The electric potential due to polarized object in shown in Figure 5.1.

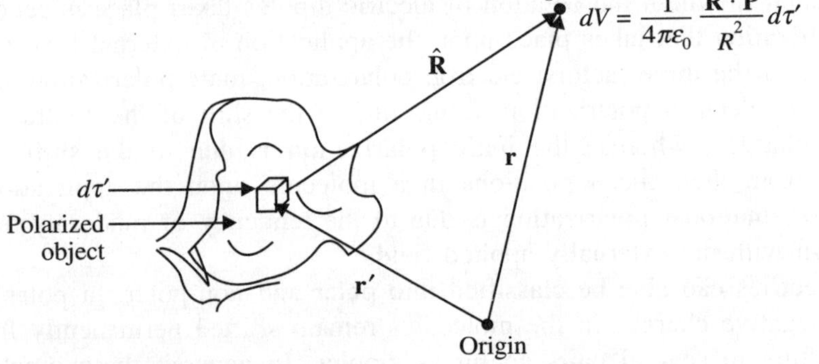

FIGURE 5.1 Electric potential due to polarized object.

The total potential due to the entire object is

$$V = \frac{1}{4\pi\varepsilon_0}\int_{\text{vol}}\frac{\hat{\mathbf{R}}\cdot\mathbf{P}}{R^2}\,d\tau' = \frac{1}{4\pi\varepsilon_0}\int_{\text{vol}}\mathbf{P}\cdot\nabla'\left(\frac{1}{R}\right)d\tau' \tag{5.24}$$

Using the vector identity

$$\nabla\cdot(f\mathbf{A}) = f(\nabla\cdot\mathbf{A}) + \mathbf{A}\cdot(\nabla f) \Rightarrow \mathbf{A}\cdot(\nabla f) = \nabla\cdot(f\mathbf{A}) - f(\nabla\cdot\mathbf{A})$$

The potential can be expressed as

$$V = \frac{1}{4\pi\varepsilon_0}\int_{\text{vol}}\left[\nabla'\cdot\left(\frac{\mathbf{P}}{R}\right) - \frac{1}{R}\nabla'\cdot\mathbf{P}\right]d\tau' \tag{5.25}$$

With the help of the divergence theorem, the first term under the integral may be converted into a surface integral. Then the expression for potential becomes

$$V = \frac{1}{4\pi\varepsilon_0}\oint_{\text{sur}}\frac{\mathbf{P}\cdot d\mathbf{a}'}{R} - \frac{1}{4\pi\varepsilon_0}\int_{\text{vol}}\frac{1}{R}\nabla'\cdot\mathbf{P}\,d\tau'$$

or

$$V = \frac{1}{4\pi\varepsilon_0}\oint_{\text{sur}}\frac{\sigma_b da'}{R} - \frac{1}{4\pi\varepsilon_0}\int_{\text{vol}}\frac{\rho_b}{R}d\tau' \tag{5.26}$$

where

$$\sigma_b = \mathbf{P}\cdot\hat{\mathbf{n}}$$

and

$$\rho_b = -\nabla\cdot\mathbf{P}$$

It means the potential and hence also the field of a polarized object is the same as that produced by a volume charge density $\rho_b = -\nabla\cdot\mathbf{P}$ plus a surface charge density $\sigma_b = \mathbf{P}\cdot\hat{\mathbf{n}}$.

Also note that the net charge of the object is zero. It leads to the fact that the total. surface charge must be equal and opposite in polarity to the volume charge. Thus, if

$$Q_\rho = \int_{\text{vol}}(-\nabla\cdot\mathbf{P})\,d\tau \tag{5.27}$$

$$Q_\sigma = \int_{\text{sur}}\mathbf{P}\cdot d\mathbf{a} \tag{5.28}$$

then

$$Q_\rho + Q_\sigma = 0 \tag{5.29}$$

5.2.3 Gauss Law

Within the dielectric, the total charge density consists of two parts: one is the bound charge which is a result of polarization and the other is the charge which may be placed over it by an external agency called free charge. The total charge then can be expressed as

$$\rho = \rho_b + \rho_f \tag{5.30}$$

where ρ_b is a result of polarization, bound charge density and ρ_f is not a result of polarization, free charge.

Now the Gauss law reads

$$\varepsilon_0\nabla\cdot\mathbf{E} = \rho = \rho_b + \rho_f = -\nabla\cdot\mathbf{P} + \rho_f \tag{5.31}$$

where **E** is the total field

$$\Rightarrow\quad \nabla\cdot(\varepsilon_0\mathbf{E} + \mathbf{P}) = \rho_f \quad\Rightarrow\quad \mathbf{D} \equiv \varepsilon_0\mathbf{E} + \mathbf{P} \tag{5.32}$$

In terms of **D**, the electric displacement, the Gauss law reads

$$\nabla\cdot\mathbf{D} = \rho f \rightarrow \oint\mathbf{D}\cdot d\mathbf{a} = Q_{\text{enc}} \tag{5.33}$$

It is worth mentioning that 'the source of **D** can be considered as free charge and the source of **P** is the bound charge and the source of **E** can be the charges of either type'.

Electric dipole: $V = \dfrac{\mathbf{p}\cdot\hat{\mathbf{r}}}{4\pi\varepsilon_0 r^2}$

$$\mathbf{E} = \frac{qd}{4\pi\varepsilon_0 r^3}(2\cos\theta\hat{\mathbf{r}} + \sin\theta\hat{\boldsymbol{\theta}})$$

Charge over a polarized object:

$$\sigma_b = \mathbf{P}\cdot\hat{\mathbf{n}}$$

$$\rho_b = -\nabla\cdot\mathbf{P}$$

Linear dielectrics: $\mathbf{P} = \varepsilon_0\chi_e\mathbf{E}$

$$\rho_b = \frac{-\chi_e}{1+\chi_e}\rho_f$$

$$\varepsilon_r = 1 + \chi_e$$

5.2.4 Linear Dielectrics

The relationship between the externally applied electric field **E** and the resultant polarization **P** depends upon the nature of the dielectric material. In isotropic materials, **E** and **P** are always parallel, irrespective of the direction of the field. Most of the widely used dielectrics in engineering are isotropic. Not only that, they are also linear, i.e. the relationship between the applied electric field **E** and the resultant polarization **P** is linear as long as the applied field is neither too low nor too high. Some of the dielectrics do exhibit non-linear behaviour and one class of dielectric materials, known as ferroelectrics, belongs to this category. Barium titanate and Rochelle salt are two prominent examples of non-linear dielectric materials.

In linear dielectrics, the polarization is proportional to the field, i.e.

$$\mathbf{P} = \varepsilon_0\chi_e\mathbf{E} \tag{5.34}$$

The proportionality constant χ_e is called *electric susceptibility*. So in linear media, we have

$$\mathbf{D} = \varepsilon_0\mathbf{E} + \mathbf{P} = \varepsilon_0\mathbf{E} + \varepsilon_0\chi_e\mathbf{E} = \varepsilon_0(1+\chi_e)\mathbf{E} \tag{5.35}$$

Therefore,

$$\mathbf{D} = \varepsilon\mathbf{E} \tag{5.36}$$

where $\varepsilon = \varepsilon_0(1 + \chi_e)$ is called the permittivity of the material. And $\varepsilon_r = (1 + \chi_e)$ is called the dielectric constant or relative permittivity of the dielectric material.

Most of the real-world dielectrics have certain amount of non-zero conductivity making them lossy ones and in that case the relative permittivity becomes complex. For lossy dielectrics,

$$\dot{\varepsilon}_r = \varepsilon_r - j\frac{\sigma}{\omega\varepsilon_0} = \varepsilon' - j\varepsilon'' = \varepsilon'(1 - j\tan\theta) \tag{5.37}$$

where

$$\tan\delta = \text{loss tangent} = \frac{\varepsilon''}{\varepsilon'} = \frac{\sigma}{\omega\varepsilon_0\varepsilon_r}$$

In the case of isotropic materials, **D** and **E** are not in the same direction, and in these materials, a component of **D** may depend upon one or more components of **E**. The relation $\mathbf{D} = \varepsilon\mathbf{E}$ for these materials becomes a matrix equation in which ε is a 3×3 square matrix.

An interesting property exhibited by linear dielectrics is that bound charge is proportional to free charge, i.e.

$$\rho_b \propto \rho_f \tag{5.38}$$

It can be proved as follows:

$$\rho_b = -\nabla \cdot \mathbf{P} = -\nabla \cdot \varepsilon_0 \chi_e \mathbf{E}$$

$$= -\varepsilon_0 \chi_e \nabla \cdot \mathbf{E} = -\varepsilon_0 \chi_e \nabla \cdot \frac{\mathbf{D}}{\varepsilon}$$

$$= -\chi_e \nabla \cdot \frac{\mathbf{D}}{\varepsilon_r} = -\frac{\chi_e}{1+\chi_e}\rho_f \tag{5.39}$$

As χ_e is constant, ρ_b is proportional to ρ_f. Hence proved.

5.2.5 Anisotropic Dielectrics

In the case of isotropic dielectrics, the direction of **D** is the same as the direction of the applied electric field **E**. Against this class, there exists another type of dielectrics known as *anisotropic dielectrics*. In this type of dielectrics, the induced polarization vector **P** is not in the direction of **E**. The relationship between **D** and **E** in such type of dielectrics can be expressed as

$$\begin{bmatrix} D_x \\ D_y \\ D_z \end{bmatrix} = \begin{bmatrix} \varepsilon_{xx} & \varepsilon_{xy} & \varepsilon_{xz} \\ \varepsilon_{yx} & \varepsilon_{yy} & \varepsilon_{yz} \\ \varepsilon_{zx} & \varepsilon_{zy} & \varepsilon_{zz} \end{bmatrix} \begin{bmatrix} E_x \\ E_y \\ E_z \end{bmatrix}$$

Thus, the permittivity of an anisotropic dielectric is not a single constant, instead it is a tensor or a 3×3 matrix. It indicates that how the three components of **D** depend upon the various components of **E**. Even though in anisotropic dielectrics, in general, **D** is not parallel to **E**, there exist certain crystals in which they are parallel to each other in certain directions. These directions are called characteristic polarizations of the crystal. For certain crystals, their electric behaviour is such that their permittivity tensor is simple shown as follows:

$$\begin{bmatrix} D_x \\ D_y \\ D_z \end{bmatrix} = \begin{bmatrix} \varepsilon_x & 0 & 0 \\ 0 & \varepsilon_y & 0 \\ 0 & 0 & \varepsilon_z \end{bmatrix} \begin{bmatrix} E_x \\ E_y \\ E_z \end{bmatrix}$$

If the three non-zero elements of the tensor are all different, then the crystal is called *biaxial* and if two of the three are equal, then it is called *uniaxial*.

5.2.6 Energy and Force

Dielectric materials are characterized by one important property: capability of storing the electric energy. They are able to store the energy due to the shift of the positive and negative internal bound charges from their mean positions when an external field is applied. Therefore, the energy storage is in the shift of bound charges and the source of energy is applied field. The energy in the dielectric systems is given by

$$W = \frac{1}{2}\int \mathbf{D}\cdot\mathbf{E}\,d\tau \tag{5.40}$$

Just like conductors, the dielectrics also get attracted into the electric fields. The cause of this drag is the accumulation of bound charge near the free charge of the opposite sign. It is the fringing field that actually causes the pull. The force on the dielectric can be given in an empirical form as

$$F = \frac{1}{2}V^2\frac{dC}{dx} \tag{5.41}$$

Derivation

Consider a parallel plate capacitor which has a dielectric inserted partly. Let us suppose that the insertion is along the x-direction and the insertion is up to a distance of x. Now the dielectric is pulled out a distance of dx by applying a force F_{ext}. The work done by the external agency or the energy change in the capacitor system in pulling the dielectric is

$$dW = F_{ext}\,dx$$

$$F_{ext} = \frac{dW}{dx}$$

The electric force F on the dielectric must be equal to the external force with the sign reversed. Thus,

$$F_{ext} = -F$$

$$F = -\frac{dW}{dx}$$

We know that the energy stored in the capacitor is

$$W = \frac{1}{2}CV^2 = \frac{1}{2}\frac{Q^2}{C}$$

Therefore,

$$F = -\frac{dW}{dx} = \frac{1}{2}\frac{Q^2}{C^2}\frac{dC}{dx} = \frac{1}{2}V^2\frac{dC}{dx}$$

One important application of dielectrics in electrostatics is that they are used as insulating materials between the plates of a capacitor. When a parallel plate capacitor is filled with a material of dielectric constant ε_r, then the field gets reduced to E/ε_r, the potential gets reduced to V/ε_r and the capacitance increases to $\varepsilon_r C$.

5.3 ELECTRIC BOUNDARY CONDITIONS

It is possible to derive a set of relations connecting field quantities on either side of the dielectric interface or boundary. These are usually called boundary conditions and are useful to find quickly and easily the field on one side of the interface if the field on the other side is given.

Tangential components: Consider an interface of two dielectrics, one above the other, having permittivities ε_1 and ε_2 occupying regions 1 and 2 respectively as shown in Figure 5.2(a). Now apply the equation

$$\oint \mathbf{E} \cdot d\mathbf{l} = 0 \tag{5.42}$$

around a closed rectangular path encompassing a part of the interface as shown in Figure 5.2(a). Now,

$$\oint \mathbf{E} \cdot d\mathbf{l} = \int_{\text{top side}} \mathbf{E} \cdot d\mathbf{l} + \int_{\text{right side}} \mathbf{E} \cdot d\mathbf{l} + \int_{\text{bottom side}} \mathbf{E} \cdot d\mathbf{l} + \int_{\text{left side}} \mathbf{E} \cdot d\mathbf{l} \tag{5.43}$$

As the width of the loop tends to zero, the integrals over the left and right sides become zeros resulting in

$$E_{y1}\Delta z - E_{y2}\Delta z = 0$$

or

$$E_{y1} = E_{y2} \tag{5.44}$$

As the y-components are parallel to the interface, we can say that the parallel electric field is continuous across the boundary, i.e.

$$E_{t1} = E_{t2} \tag{5.45}$$

Another aspect of the tangential components is that as the intensities are continuous, the displacements **D** must be discontinuous.

$$E_{t1} = E_{t2} \Rightarrow \frac{D_{t1}}{\varepsilon_1} = \frac{D_{t2}}{\varepsilon_2} \quad \text{or} \quad \frac{D_{t1}}{D_{t2}} = \frac{\varepsilon_1}{\varepsilon_2} \tag{5.46}$$

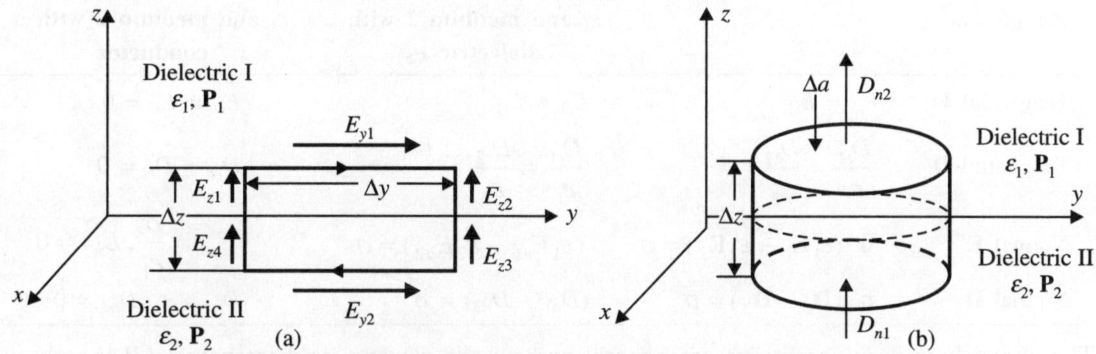

FIGURE 5.2 Boundary conditions—interface of dielectric media.

Normal components: To explore the relation that exists between the normal field components, we now consider a small Gaussian pillbox enclosing the interface as shown in Figure 5.2(b) and then apply the Gauss law

$$\oint \mathbf{D} \cdot d\mathbf{a} = Q_{\text{enc}} = \int \rho d\tau$$

$$\oint \mathbf{D} \cdot d\mathbf{a} = \int_{\text{top face}} \mathbf{D} \cdot d\mathbf{a} + \int_{\text{bottom face}} \mathbf{D} \cdot d\mathbf{a} + \int_{\text{side face}} \mathbf{D} \cdot d\mathbf{a}$$

Therefore,

$$Q_{enc} = \int \rho d\tau = \oint \mathbf{D} \cdot d\mathbf{a} = \int_{\text{top face}} \mathbf{D} \cdot d\mathbf{a} + \int_{\text{bottom face}} \mathbf{D} \cdot d\mathbf{a} + \int_{\text{side face}} \mathbf{D} \cdot d\mathbf{a} \tag{5.47}$$

If ρ is the volume density of the charge that exists over the interface, the net charge enclosed becomes $Q_{enc} = \rho\Delta z\Delta a$. As the thickness of the box tends to zero, the side wall area and consequently, the flux through the side wall become zero and hence the third term on the right hand side becomes zero. The above relation then becomes

$$D_{n1}\Delta a - D_{n2}\Delta a = \rho\Delta z\Delta a = \sigma\Delta a \tag{5.48}$$

It results in

$$D_{n1} - D_{n2} = \sigma \tag{5.49}$$

Note that σ (= $\rho\Delta z$) is the surface charge density and it represents the charge that has been placed deliberately over the interface and not the bound charge. In general, $\sigma = 0$ resulting in

$$D_{n1} - D_{n2} = 0 \Rightarrow D_{n1} = D_{n2} \tag{5.50}$$

which indicates the continuity of the normal components of the displacement density **D**. Note that the continuity of the normal components of **D** implies the discontinuity in the normal components of **E**.

$$D_{n1} = D_{n2} \Rightarrow \varepsilon_1 E_{n1} = \varepsilon_2 E_{n2} \Rightarrow \frac{E_{n1}}{E_{n2}} = \frac{\varepsilon_2}{\varepsilon_1} \tag{5.51}$$

Table 5.2 Electric boundary conditions

Field component	Any two media	Medium 1 with dielectric ε_1 and medium 2 with dielectric ε_2	Medium 1 with dielectric ε_1 and medium 2 with a conductor
Tangential **E**	$E_{t1} = E_{t1}$	$E_{t1} = E_{t1}$	$E_{t1} = E_{t2} = 0$
Tangential **D**	$\frac{D_{t1}}{\varepsilon_1} = \frac{D_{t2}}{\varepsilon_2}$	$\frac{D_{t1}}{\varepsilon_1} = \frac{D_{t2}}{\varepsilon_2}$	$D_{t1} = D_{t2} = 0$
Normal **E**	$\hat{\mathbf{n}} \cdot (\varepsilon_1\mathbf{E}_1 - \varepsilon_2\mathbf{E}_2) = \sigma$	$(\varepsilon_1 E_{n1} - \varepsilon_2 E_{n2}) = \sigma$	$E_{n1} = \frac{\sigma}{\varepsilon_1}, E_{n2} = 0$
Normal **D**	$\hat{\mathbf{n}} \cdot (\mathbf{D}_1 - \mathbf{D}_2) = \sigma$	$(D_{n1} - D_{n2}) = \sigma$	$D_{n1} = \sigma, D_{n2} = 0$

The above boundary conditions are general and on the electric field intensity. All the above derived information is provided in Table 5.2.

Direction and magnitude of field in medium 2: Sometimes it is required to find the field intensity in medium 2 from the information of the field in medium 1. The necessary relations are derived hereunder.

Let $\mathbf{D}_1$ and $\mathbf{E}_1$ make an angle of θ_1 with the normal to the interface and $\mathbf{D}_2$ and $\mathbf{E}_2$ make an angle of θ_2. Then

$$\frac{D_{t1}}{D_{t2}} = \frac{\varepsilon_1}{\varepsilon_2} = \frac{D_1 \sin\theta_1}{D_2 \sin\theta_2} \tag{5.52}$$

$$D_{n1} = D_{n2} \Rightarrow D_1 \cos \theta_1 = D_2 \cos \theta_2 \tag{5.53}$$

Combining Eqs. (5.52) and (5.53), we get

$$\frac{\tan \theta_1}{\tan \theta_2} = \frac{\varepsilon_1}{\varepsilon_2} \tag{5.54}$$

It can also be found that

$$D_2 = D_1 \sqrt{\cos^2 \theta_1 + \left(\frac{\varepsilon_2}{\varepsilon_1}\right)^2 \sin^2 \theta_1} \tag{5.55}$$

and

$$E_2 = E_1 \sqrt{\sin^2 \theta_1 + \left(\frac{\varepsilon_1}{\varepsilon_2}\right)^2 \cos^2 \theta_1} \tag{5.56}$$

It can be observed that D is larger in the medium of larger permittivity except when $\theta_1 = \theta_2 = 0$ and E is larger in the medium of smaller permittivity except when $\theta_1 = \theta_2 = 90°$.

5.4 CAPACITORS

By definition, capacitor is an arrangement of two conductors, of any shape and size, called plates and separated by a distance, even infinity. They are widely used in various electrical and electric circuits. They are one of basic electrical elements and capable of storing energy in electrical form.

Let us suppose, the two conductors or plates are connected with a battery of voltage V across them. If a charge Q gets deposited on the plate connected to the positive terminal of the battery, then an equal amount of negative charge, i.e. $-Q$, gets deposited on the plate connected to the negative terminal of the battery.

The ratio of charge to voltage Q/V is a constant depending only upon the geometrical arrangement of the plates and is called *capacitance*. Its value is determined by the sizes, shapes and separation of the two conductors.

5.4.1 Parallel Plate Capacitor

The parallel plate capacitor is used widely in various electrical circuits. Structurally, it consists of two parallel conducting plates with a dielectric slab in between. In the absence of any physical dielectric, the capacitor is called air-dielectric capacitor. Usually the spacing of the plates is very small when compared to any other dimension in the structure.

When a voltage source is connected across the capacitor, the positive charge gets deposited on the plate connected to the positive terminal of the source and an equal amount of negative charge gets deposited on the other plate. As a result of the existence of charge on the plates, there comes into being a field which is uniform in between the plates whose direction is from the positive plate to the negative plate. The capacitance of the structure can be evaluated as

$$C = \frac{\varepsilon A}{d} \tag{5.57}$$

Proof: Let us suppose that the area of the plates is A and the plates are separated by a small distance d (Figure 5.3). Also suppose that a source of voltage V is connected across the plates resulting in the deposition of charge Q on the bottom plate and $-Q$ on the top plate. With the help of the Gauss law, it can be found that the field in between the shells is

$$\mathbf{E} = \frac{Q}{A\varepsilon}\hat{\mathbf{n}}$$

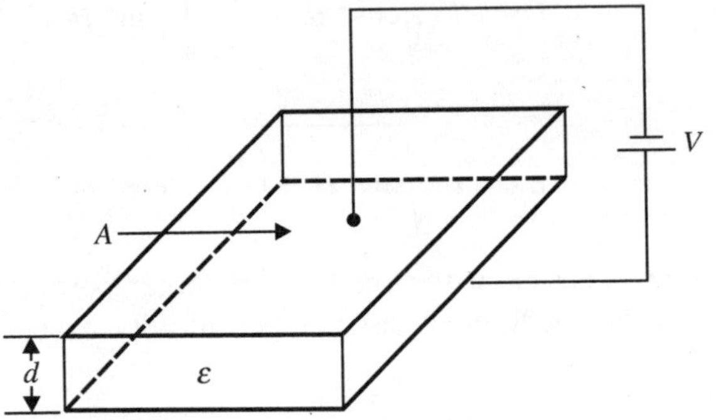

FIGURE 5.3 Parallel plate capacitor.

Now using the basic relation connecting the potential with the field intensity

$$V = -\frac{1}{4\pi\varepsilon}\int E \cdot d\mathbf{l}$$

we calculate the voltage across the plates V_{bt} and it must be equal to the applied voltage V. To move further, we need to select a coordinate system, and go for the Cartesian system and place the capacitor over the xy-plane with the bottom plate at $z = 0$ and the top plate at $z = d$. With reference to this system, the field intensity becomes

$$\mathbf{E} = \frac{Q}{A\varepsilon}\hat{\mathbf{z}}$$

And the potential of the bottom plate with respect to the top plate is

$$V_{bt} = -\frac{Q}{A\varepsilon}\int_t^b \hat{\mathbf{z}} \cdot d\mathbf{l} = -\frac{Q}{A\varepsilon}\int_d^0 dz = \frac{Qd}{A\varepsilon} = V$$

By definition, capacitance $C \triangleq Q/V$. Therefore,

$$C = \frac{Q}{V} = \frac{\varepsilon A}{d}$$

If the plates are separated by free space, then we need to replace ε with ε_0.

5.4.2 Spherical Shell Capacitor

We have applications in circuit theory as well as in field theory where a spherical conductor surface encloses another similar surface. When they are charged, by definition, there exists certain amount of capacitance between them. In the case of two spherical concentric shells, one enclosing the other, with a dielectric in between, the capacitance can be found to be

$$C = 4\pi\varepsilon\frac{ab}{(b-a)} \tag{5.58}$$

Note in the present case that when a source is connected across the shells, the net charge deposited over the shells remains the same but the charge densities over the shells are not the same.

Proof: Let us suppose that the radius of the inner spherical shell is a and that of the outer shell is b (Figure 5.4) and a source of voltage V is connected across the shells resulting in the deposition of charge Q on the inner shell and $-Q$ on the outer shell. With the help of the Gauss law, it can be found that the field in between the shells is

$$\mathbf{E} = \frac{1}{4\pi\varepsilon}\frac{Q}{r^2}\hat{\mathbf{r}}$$

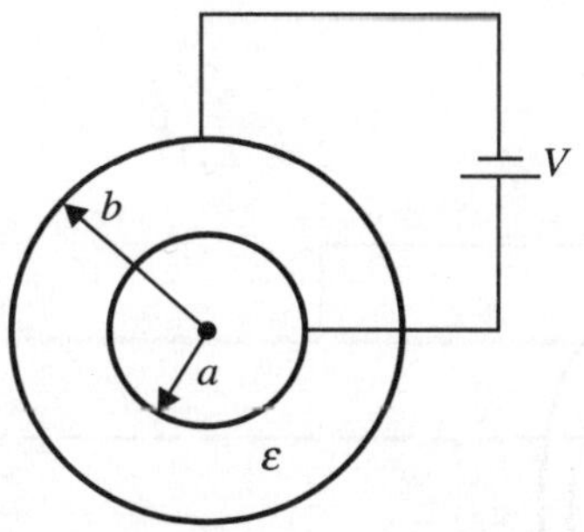

FIGURE 5.4 Spherical shell capacitor.

Now using the basic relation connecting the potential with the field intensity

$$V = -\frac{1}{4\pi\varepsilon}\int \mathbf{E}\cdot d\mathbf{l}$$

we calculate the voltage across the shells V_{ba} and it must be equal to the applied voltage V.

$$V = V_{ba} = -\frac{1}{4\pi\varepsilon}\int_b^a \frac{Q}{r^2}\hat{\mathbf{r}}\cdot d\mathbf{l} = \frac{-1}{4\pi\varepsilon}\int_b^a \frac{Q}{r^2}\,dr$$

$$= \frac{Q}{4\pi\varepsilon}\frac{1}{r}\bigg|_b^a = \frac{Q}{4\pi\varepsilon}\left(\frac{1}{a}-\frac{1}{b}\right) = \frac{Q}{4\pi\varepsilon}\left(\frac{b-a}{ab}\right)$$

By definition, capacitance $C \triangleq Q/V$. Therefore,

$$C = \frac{Q}{V} = 4\pi\varepsilon\frac{ab}{(b-a)}$$

If the shells are separated by free space, then replace ε with ε_0.

5.4.3 Cylindrical Capacitor

Coaxial cables are an important type of transmission lines widely used to transmit high frequency telecom signals. They are also used as power cables to transfer electrical energy. Structurally they consist of an inner conductor rod of cylindrical cross section and an outer conductor of cylindrical cross section enclosing the inner conductor. The conductors are separated by dielectric material. There exists capacitance between the inner and outer conductors and it is given by

$$C = 2\pi\varepsilon \frac{L}{\ln\left(\frac{b}{a}\right)} \tag{5.59}$$

Proof: Let us suppose that the radius of the inner cylindrical shell is a and that of the outer one is b (Figure 5.5) and the length of the capacitor is L. Also suppose that a source of voltage V is connected across the shells resulting in the deposition of charge Q on the inner shell and $-Q$ on the outer shell. With the help of the Gauss law, it can be found that the field in between the shells is

$$\mathbf{E} = \frac{1}{2\pi\varepsilon}\frac{Q}{L\rho}\hat{\rho}$$

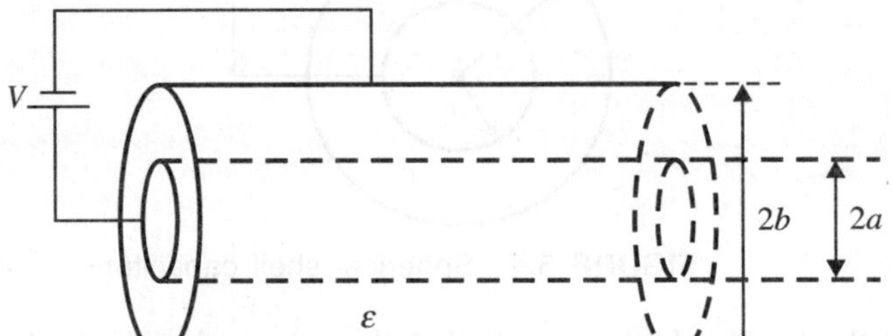

FIGURE 5.5 Cylindrical shell capacitor.

Now we calculate the voltage across the shells V_{ba} and it must be equal to the applied voltage V.

$$V_{ab} = \frac{-Q}{2\pi\varepsilon L}\int_b^a \frac{1}{\rho}\hat{\rho}\cdot d\mathbf{l} = \frac{-Q}{2\pi\varepsilon L}\int_b^a \frac{1}{\rho}\,d\rho$$

$$= \frac{Q}{2\pi\varepsilon L}\ln\left(\frac{b}{a}\right) = V$$

By definition, capacitance $C \triangleq Q/V$. Therefore,

$$C = \frac{Q}{V} = \frac{2\pi\varepsilon L}{\ln\left(\frac{b}{a}\right)}$$

This is the capacitance within a length L. If it is required per unit length, then we have to divide it by L. In the derivation, it is assumed that the space between the shells is filled with a medium of permittivity ε. If it is free space, then replace ε with ε_0.

5.4.4 Parallel Cylindrical Conductors

Here we try to find out the capacitance that exists across the conductors of parallel wire transmission lines. Let us suppose, the spacing is b and the radius of the conductors is a (Figure 5.6). We assume that the conductors are long, both carrying equal charge density of λ C/m but one with positive polarity and the other with negative polarity. We also assume that the charge which is actually spread over the cross section of the conductor is concentrated over a thin line. It results in a configuration of two long parallel straight line charges separated by a distance different from that of the actual conductor spacing. The potential distribution for this dual conductor system is

$$V(P) = \frac{\lambda}{2\pi\varepsilon_0} \ln \frac{\rho_-}{\rho_+} \tag{5.60}$$

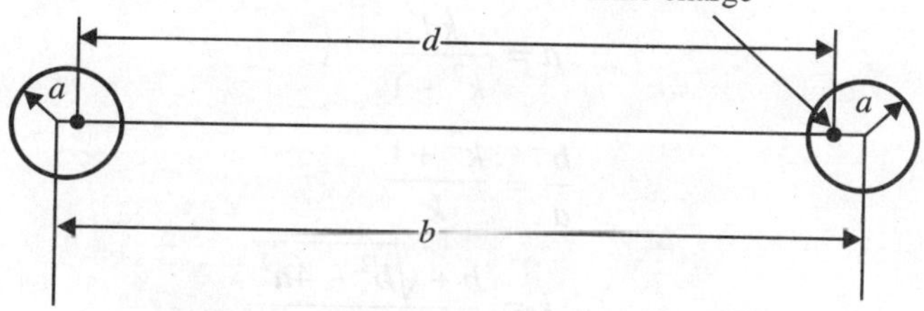

FIGURE 5.6 Parallel cylindrical conductors.

The surfaces of the conductors are equipotential surfaces. Therefore, for points over the surfaces of the conductors,

$$V(P) = \frac{\lambda}{2\pi\varepsilon_0} \ln \frac{\rho_-}{\rho_+} = k_1 \tag{5.61}$$

or

$$\frac{\rho_-}{\rho_+} = k \tag{5.62}$$

where k and k_1 are constants. From the geometry of the configuration, with d as the separation distance between the line charges,

$$\rho_+^2 = \left(x + \frac{d}{2}\right)^2 + y^2 \tag{5.63}$$

$$\rho_-^2 = \left(x - \frac{d}{2}\right)^2 + y^2 \tag{5.64}$$

Therefore,

$$\frac{\rho_-^2}{\rho_+^2} = \frac{\left(x - \frac{d}{2}\right)^2 + y^2}{\left(x + \frac{d}{2}\right)^2 + y^2} = k^2 \tag{5.65}$$

Manipulation of the above equation results in

$$\left[x + \frac{d}{2}\left(\frac{k^2+1}{k^2-1}\right)\right]^2 + y^2 = \frac{k^2 d^2}{(k^2-1)^2} \tag{5.66}$$

which is the equation of the family of circles with radius $kd/(k^2 - 1)$ and centre at $(d/2)\,[(k^2 + 1)/(k^2 - 1)]$, 0. Because of invariance in the z-direction, the equipotential surfaces are cylinders.

If the radius of the conductors is a with a separation of b between their axes, then

$$b = d\,\frac{k^2+1}{k^2-1} \tag{5.67}$$

$$a = \frac{kd}{k^2-1} \tag{5.68}$$

$$\frac{b}{a} = \frac{k^2+1}{k} \tag{5.69}$$

$$k = \frac{b + \sqrt{b^2 - 4a^2}}{2a} \tag{5.70}$$

But $V(P) = \dfrac{1}{4\pi\varepsilon_0}\, 2\lambda \ln \dfrac{\rho_-}{\rho_+}$ with $k = \dfrac{\rho_-}{\rho_+}$

Therefore, the potential at the surface of any one conductor is

$$V_1 = \frac{\lambda}{2\pi\varepsilon_0} \ln \frac{b + \sqrt{b^2 - 4a^2}}{2a} \tag{5.71}$$

When the separation is large compared with the radius $b \gg a$, it becomes

$$V_1 = \frac{\lambda}{2\pi\varepsilon_0} \ln \frac{b + \sqrt{b^2 - 4a^2}}{2a} = \frac{\lambda}{2\pi\varepsilon_0} \ln \frac{b}{a} \tag{5.72}$$

The potential at the surface of other conductor must be equal but opposite. Hence,

$$V = V_1 - V_2 = \frac{\lambda}{\pi\varepsilon_0} \ln \frac{b + \sqrt{b^2 - 4a^2}}{2a} \tag{5.73}$$

The capacitance per unit length is

$$C = \frac{\lambda}{V} = \frac{\pi\varepsilon_0}{\ln\left[\dfrac{\left(b + \sqrt{b^2 - 4a^2}\right)}{2a}\right]} \tag{5.74}$$

When the separation is large compared with the radius $b \gg a$, it becomes

$$C = \frac{\lambda}{V} = \frac{\pi\varepsilon_0}{\ln\dfrac{b}{a}} \tag{5.75}$$

5.4.5 Multiple Dielectric Capacitors

Consider a parallel plate capacitor consisting of two different dielectrics with the interface parallel to the electric the field. In this case, the equivalent capacitance can be obtained by treating the configuration as two capacitors in parallel [Figure 5.7(a)]: one with a capacitance of $C_1 = \varepsilon_1 A_1/d$ and the other with a capacitance of $C_2 = \varepsilon_2 A_2/d$

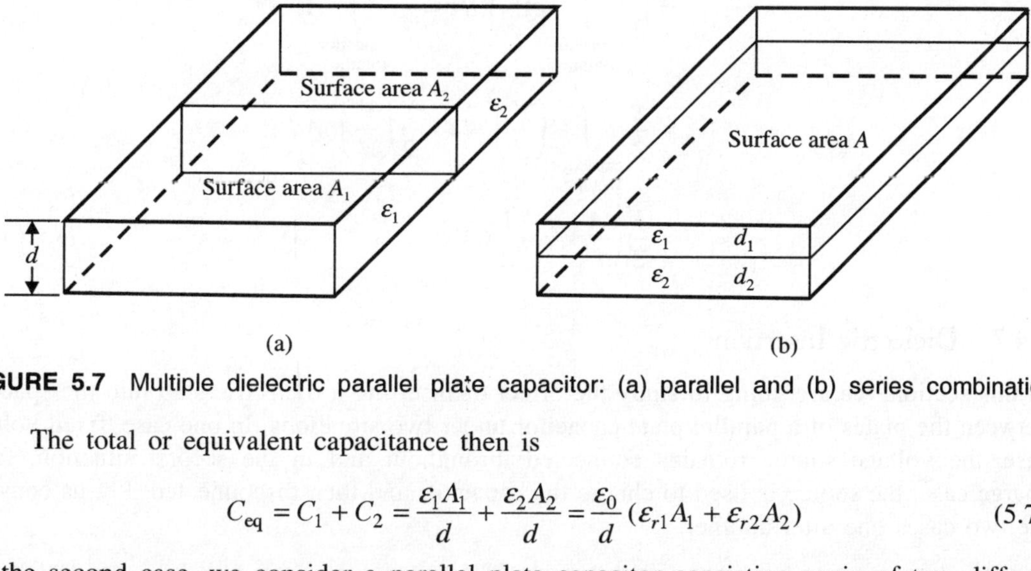

FIGURE 5.7 Multiple dielectric parallel plate capacitor: (a) parallel and (b) series combination.

The total or equivalent capacitance then is

$$C_{eq} = C_1 + C_2 = \frac{\varepsilon_1 A_1}{d} + \frac{\varepsilon_2 A_2}{d} = \frac{\varepsilon_0}{d}(\varepsilon_{r1}A_1 + \varepsilon_{r2}A_2) \tag{5.76}$$

In the second case, we consider a parallel plate capacitor consisting again of two different dielectrics but with the interface normal to the electric field. In this situation, the equivalent capacitance is that of the two capacitors in series [Figure 5.7(b)]: one with a capacitance of $C_1 = \varepsilon_1 A/d_1$ and the other with a capacitance of $C_2 = \varepsilon_2 A/d_2$.

The total or equivalent capacitance then is

$$\frac{1}{C_{eq}} = \frac{1}{C_1} + \frac{1}{C_2} = \frac{1}{\varepsilon_1 \dfrac{A}{d_1}} + \frac{1}{\varepsilon_2 \dfrac{A}{d_2}} \tag{5.77}$$

Therefore, we can say that if the field is parallel to the interface, the component capacitors are to be considered as parallel and if the field is normal, the component capacitors are to be considered as series in computing the equivalent capacitance of a capacitor with multiple dielectrics.

5.4.6 Energy Stored

The energy stored in a dielectric system is given by

$$W = \frac{1}{2} \int_{\text{all space}} \mathbf{D} \cdot \mathbf{E} d\tau \tag{5.78}$$

We need to perform the integration wherever the field exists. In the case of a parallel plate capacitor, the field exists only in between the plates. And the displacement density there is

$$\mathbf{D} = \varepsilon \mathbf{E} \tag{5.79}$$

Substituting Eq. (5.79) in Eq. (5.78), we get

$$= \frac{1}{2} \int_{\substack{\text{capacitor}\\\text{volume}}} \varepsilon \mathbf{E} \cdot \mathbf{E} d\tau = \frac{\varepsilon}{2} \int_{\substack{\text{capacitor}\\\text{volume}}} \mathbf{E}^2 d\tau$$

$$= \frac{\varepsilon}{2} \int_{\substack{\text{capacitor}\\\text{volume}}} \left(\frac{V}{d}\right)^2 d\tau = \frac{\varepsilon}{2}\left(\frac{V}{d}\right)^2 Ad$$

$$= \frac{1}{2}\frac{\varepsilon A}{d} V^2 = \frac{1}{2} C V^2 \tag{5.80}$$

5.4.7 Dielectric Insertion

In this section, we are going to study the effect of inserting a dielectric slab into the space in between the plates of a parallel plate capacitor under two situations. In one case, fixed voltage case, the voltage source remains connected throughout and in the second situation, fixed charge case, the source is used to charge the capacitor and then disconnected. Let us consider the two cases one after another.

Fixed voltage case: Let us suppose that a parallel plate capacitor with free space dielectric remains permanently connected to a source of constant voltage V. If a dielectric with a dielectric constant of ε_r is inserted into the entire space in between the plates, then we need to know how the field intensity and displacement density change with insertion neglecting the fringing fields.

With free space as dielectric, the field intensity and the displacement, respectively, are

$$\mathbf{E}_0 = \frac{V}{d}\hat{\mathbf{n}} \tag{5.81}$$

and

$$\mathbf{D}_0 = \varepsilon_0 \mathbf{E}_0 = \frac{\varepsilon_0 V}{d}\,\hat{\mathbf{n}} \tag{5.82}$$

With the dielectric filling the space in between the plates, the field intensity and the displacement, respectively, become

$$\mathbf{E} = \frac{V}{d}\,\hat{\mathbf{n}} = \mathbf{E}_0 \tag{5.83}$$

and

$$\mathbf{D} = \varepsilon_0 \varepsilon_r \mathbf{E}_0 = \frac{\varepsilon_0 \varepsilon_r V}{d}\,\hat{\mathbf{n}} = \varepsilon_r \mathbf{D}_0 \tag{5.84}$$

In this case, which is called 'fixed voltage case', therefore, insertion of dielectric does not change the field intensity anyway, but the displacement density increases.

It can also be observed that the energy, capacitance, surface charge density and also the total charge increase to ε_r times their respective values before insertion.

Fixed charge case: In this case, the parallel plate capacitor with free space dielectric is charged until it holds Q on the positive plate and $-Q$ on the negative plate and the source is then disconnected. Now if a dielectric with a dielectric constant of ε_r is inserted into the entire space in between the plates, we need to know the changes in the field intensity and displacement density after the dielectric insertion. Here also we neglect the fringing fields.

With free space as dielectric, the displacement and the field intensity, respectively, are

$$\mathbf{D}_0 = \frac{Q}{A}\,\hat{\mathbf{n}} \tag{5.85}$$

and

$$\mathbf{E}_0 = \frac{\mathbf{D}_0}{\varepsilon_0} = \frac{Q}{\varepsilon_0 A}\,\hat{\mathbf{n}} \tag{5.86}$$

With the dielectric filling the space in between the plates, the displacement and the field intensity, respectively, become

$$\mathbf{D} = \frac{Q}{A}\,\hat{\mathbf{n}} = \mathbf{D}_0 \tag{5.87}$$

and

$$\mathbf{E} = \frac{\mathbf{D}_0}{\varepsilon_0 \varepsilon_r} = \frac{Q}{\varepsilon_0 \varepsilon_r A}\,\hat{\mathbf{n}} = \frac{\mathbf{E}_0}{\varepsilon_r} \tag{5.88}$$

In the present case called 'fixed charge case', therefore, the insertion of dielectric does not change the displacement density anyway, but the field intensity decreases. It can be noted that, in this case, the energy and voltage decrease to $1/\varepsilon_r$ times their previous values but the capacitance increases to ε_r times its value before the insertion.

SOLVED PROBLEMS

1. Two large metal plates, each of area A, are held at a distance d apart. Suppose we put a charge Q on each plate. What is the electrostatic pressure on the plates?

Solution: Force on the inner surface of each conductor per unit area is

$$\mathbf{f} = \frac{\sigma^2}{2\varepsilon_0}\hat{\mathbf{n}}$$

As the charge on each plate is Q, the surface charge density is

$$\sigma = \frac{Q}{A}$$

and the field intensity is

$$E = \frac{\sigma}{\varepsilon_0} = \frac{Q}{A\varepsilon_0} \quad \Rightarrow \quad Q = EA\varepsilon_0$$

The electrostatic pressure on each of the plates is

$$f = \frac{(E\varepsilon_0)^2}{2\varepsilon_0} = \frac{1}{2}\varepsilon_0 E^2$$

So, the total force on each of the plates is

$$F = fA = \frac{(E\varepsilon_0)^2}{2\varepsilon_0}A = \frac{1}{2}\varepsilon_0 E^2 A$$

2. A metal sphere of radius R carries a total charge Q. What is the force of repulsion between the northern hemisphere and the southern hemisphere?

Solution: The field inside a uniformly charged sphere is

$$\mathbf{E} = \frac{1}{4\pi\varepsilon_0}\frac{Q}{R^3}\hat{\mathbf{r}}$$

So the force per unit volume

$$\mathbf{f}_r = \rho\mathbf{E} = \frac{3Q}{4\pi R^3}\frac{1}{4\pi\varepsilon_0}\frac{Q}{R^3}\hat{\mathbf{r}} = \frac{3}{\varepsilon_0}\frac{Q^2}{(4\pi R^3)^2}\hat{\mathbf{r}}$$

So the force in the z-direction on $d\tau$ is

$$d\mathbf{F}_z = \mathbf{f}_z d\tau = \mathbf{f}_r \cos\theta d\tau = \frac{3}{\varepsilon_0}\frac{Q^2}{(4\pi R^3)^2} r\cos\theta(r^2 \sin\theta dr d\theta d\phi)$$

So the total force on the northern hemisphere is

$$\mathbf{F}_z = \int d\mathbf{F}_z = \frac{3}{\varepsilon_0}\frac{Q^2}{(4\pi R^3)^2}\int_0^{2\pi}\int_0^{\pi/2}\int_0^{R} r\cos\theta(r^2\sin\theta dr d\theta d\phi) = \frac{3}{64\pi\varepsilon_0}\frac{Q^2}{R^2}$$

3. A metal sphere of radius R carrying charge q is surrounded by a thick concentric metal shell (inner radius a and outer radius b). The shell carries no net charge. Find

(a) surface charge density σ at R, at a and at b and

(b) the potential at the centre

(c) if the outer surface is grounded, what happens to (a) and (b) if $\sigma_b = 0$ (the charge drains off)?

Solution:

(a) The surface charge density σ at R, at a and at b:

$$\sigma_R = \frac{q}{4\pi R^2};\ \sigma_a = \frac{-q}{4\pi a^2};\ \sigma_b = \frac{q}{4\pi b^2}$$

(b) Potential at the centre, using infinity as the reference point:

$$V(0) = -\int_{\infty}^{0}\mathbf{E}\cdot d\mathbf{l} = -\int_{\infty}^{b}\frac{1}{4\pi\varepsilon_0}\frac{q}{r^2}dr - \int_{b}^{a}(0)\,dr - \int_{a}^{R}\frac{1}{4\pi\varepsilon_0}\frac{q}{r^2}dr - \int_{R}^{\infty}(0)\,dr$$

$$= \frac{1}{4\pi\varepsilon_0}\left[\frac{q}{b}+\frac{q}{R}-\frac{q}{a}\right]$$

(c) If the outer surface is grounded, let us see what happens to (a) and (b) if $\sigma_b = 0$ (the charge drains off):

$$V(0) = -\int_{\infty}^{0}\mathbf{E}\cdot d\mathbf{l} = -\int_{\infty}^{b}(0)\,dr - \int_{b}^{a}(0)\,dr - \int_{a}^{R}\frac{1}{4\pi\varepsilon_0}\frac{q}{r^2}dr - \int_{R}^{\infty}(0)\,dr$$

$$= \frac{1}{4\pi\varepsilon_0}\left[\frac{q}{R}-\frac{q}{a}\right]$$

4. Two spherical cavities, of radii a and b, are hallowed out from the interior of a neutral conducting sphere of radius R. At the centre of each cavity, a point charge is placed, call these charges q_a and q_b. Find

(a) the surface charge densities σ_a, σ_b, and σ_R:

(b) field outside the conductor,

(c) field within each cavity,

(d) force on q_a and q_b,

(e) which of these answers would change if a third charge q_c were brought near the conductor?

Solution:

(a) The surface charge densities, σ_a, σ_b and σ_R :

$$\sigma_a = \frac{-q_a}{4\pi a^2}; \quad \sigma_b = \frac{-q_b}{4\pi b^2}; \quad \sigma_R = \frac{q_a + q_b}{4\pi R^2}$$

(b) The field outside the conductor :

$$\mathbf{E}_{\text{ot}} = \frac{1}{4\pi\varepsilon_0}\frac{q_a + q_b}{r^2}\hat{\mathbf{r}}$$

where $\hat{\mathbf{r}}$ is the vector from the centre of a large sphere.

(c) The field within each cavity:

$$\mathbf{E}_a = \frac{1}{4\pi\varepsilon_0}\frac{q_a}{r_a^2}\hat{\mathbf{r}}_a \text{, where } \hat{\mathbf{r}}_a \text{ is the vector from the centre of cavity } a.$$

$$\mathbf{E}_b = \frac{1}{4\pi\varepsilon_0}\frac{q_b}{r_a^2}\hat{\mathbf{r}}_b \text{, where } \hat{\mathbf{r}}_b \text{ is the vector from the centre of cavity } b.$$

(d) The force on q_a and q_b:
Zero

(e) Let us see which of these answers would change if a third charge q_c were brought near the conductor:
The quantity σ_R would change (not σ_a or σ_b).
The quantity $\mathbf{E}_{\text{ot}}$ would change (not $\mathbf{E}_a$ or $\mathbf{E}_b$).
The force on q_a and q_b is still zero.

5. Two coaxial long cylinders (radii a and b) are separated by the material of conductivity σ. If they are maintained at potential difference V, find the current that flows from one to the other in a length L?

Solution: The field between the cylinders, using the Gauss law, is

$$\mathbf{E} = \frac{1}{4\pi\varepsilon_0}\frac{\lambda}{\rho}\hat{\boldsymbol{\rho}}$$

where λ is the charge per unit length on the inner cylinder. The current is therefore

$$I = \int \mathbf{J} \cdot d\mathbf{a} = \sigma \int \mathbf{E} \cdot d\mathbf{a} = \sigma \frac{\lambda L}{\varepsilon_0}$$

(The integral is over any surface enclosing the inner cylinder). Meanwhile the potential difference between the cylinders is

$$V_{ab} = -\int_b^a \mathbf{E} \cdot d\mathbf{l} = \frac{\lambda}{2\pi\varepsilon_0} \ln\left(\frac{b}{a}\right)$$

So,

$$I = \frac{2\pi\sigma L}{\ln\left(\frac{b}{a}\right)} V$$

6. Two concentric metal spherical shells, of radii a and b, respectively, are separated by a weakly conducting material of conductivity σ. If the potential difference is maintained at V, which current flows from one to the other and what is the resistance between the shells?

Solution: Let Q be the charge on the inner shell. Then in the space between them, the electric field is

$$\mathbf{E} = \frac{1}{4\pi\varepsilon_0}\frac{Q}{r^2}\hat{\mathbf{r}}$$

and the potential is

$$V_{ab} = -\int_b^a \mathbf{E}\cdot d\mathbf{l} = -\frac{Q}{4\pi\varepsilon_0}\int_b^a \frac{1}{r^2}\,dr = \frac{Q}{4\pi\varepsilon_0}\left(\frac{1}{a}-\frac{1}{b}\right)$$

$$I = \int \mathbf{J}\cdot d\mathbf{a} = \sigma\int \mathbf{E}\cdot d\mathbf{a} = \sigma\frac{Q}{\varepsilon_0} = \frac{\sigma}{\varepsilon_0}\frac{4\pi\varepsilon_0\,(V_a - V_b)}{\left(\frac{1}{a}-\frac{1}{b}\right)} = \frac{4\pi\sigma(V_a - V_b)}{\left(\frac{1}{a}-\frac{1}{b}\right)}$$

$$R = \frac{V_a - V_b}{I} = \frac{1}{4\pi\sigma}\left(\frac{1}{a}-\frac{1}{b}\right)$$

For large b ($b \gg a$), the second term is negligible, and

$$R = \frac{1}{4\pi\sigma a}$$

7. Two metal objects are embedded in a weakly conducting material of conductivity σ. Show that the resistance between them is related to the capacitance of the arrangement by $R = \varepsilon_0/\sigma C$.

Solution: The conduction current is related to its density and field intensity through

$$I = \int \mathbf{J}\cdot d\mathbf{a} = \sigma\int \mathbf{E}\cdot d\mathbf{a}$$

By the application of the Gauss law,

$$\sigma\int \mathbf{E}\cdot d\mathbf{a} = \sigma\frac{Q}{\varepsilon_0}$$

Therefore,

$$I = \int \mathbf{J}\cdot d\mathbf{a} = \sigma\int \mathbf{E}\cdot d\mathbf{a} = \sigma\frac{Q}{\varepsilon_0}$$

where the integral is taken over a surface enclosing the positively charged conductor. But

$$Q = CV$$

and

$$V = IR$$

so

$$I = \frac{\sigma}{\varepsilon_0} CIR$$

or

$$R = \frac{\varepsilon_0}{\sigma C}$$

8. Show that the electric field of a dipole in coordinate-free form is

$$\mathbf{E}_{\text{dip}}(\mathbf{r}) = \frac{1}{4\pi\varepsilon_0}\frac{1}{r^3}[3(\mathbf{p}\cdot\hat{\mathbf{r}})\,\hat{\mathbf{r}} - \mathbf{p}]$$

Solution: The field due to dipole is

$$\mathbf{E}_{\text{dip}}(\mathbf{r}) = \frac{p}{4\pi\varepsilon_0}\frac{1}{r^3}(2\cos\theta\hat{\mathbf{r}} + \sin\theta\hat{\boldsymbol{\theta}})$$

A vector can be resolved into two normal components. So

$\mathbf{p} = (\mathbf{p}\cdot\hat{\mathbf{r}})\,\hat{\mathbf{r}} + (\mathbf{p}\cdot\hat{\boldsymbol{\theta}})\,\hat{\boldsymbol{\theta}} = p\cos\theta\hat{\mathbf{r}} - p\sin\theta\hat{\boldsymbol{\theta}}$. Then

$3(\mathbf{p}\cdot\hat{\mathbf{r}})\,\hat{\mathbf{r}} - \mathbf{p} = 3p\cos\theta\hat{\mathbf{r}} - p\cos\theta\hat{\mathbf{r}} + p\sin\theta\hat{\boldsymbol{\theta}} = 2p\cos\theta\hat{\mathbf{r}} + p\sin\theta\hat{\boldsymbol{\theta}}$

Hence,

$$\mathbf{E}_{\text{dip}}(\mathbf{r}) = \frac{p}{4\pi\varepsilon_0}\frac{1}{r^3}(2\cos\theta\hat{\mathbf{r}} + \sin\theta\hat{\boldsymbol{\theta}}) = \frac{1}{4\pi\varepsilon_0}\frac{1}{r^3}[3(\mathbf{p}\cdot\hat{\mathbf{r}})\hat{\mathbf{r}} - \mathbf{p}]$$

9. An infinite length line charge of uniform charge density λ C/m lying along the z-axis is surrounded by a shell of dielectric material with permittivity $4\varepsilon_0$ in the region $a < r < b$. Find

(a) **D** and **E** in the regions $0 < r < a$, $a < r < b$, $r > b$ and
(b) the polarization vector inside the cylindrical shell.

Solution:

(a) With the help of the Gauss law, it can be found that $\mathbf{D} = \frac{\lambda}{2\pi\rho}\hat{\boldsymbol{\rho}}$ everywhere.

Now, after substituting the concerned values of permittivity

$$\mathbf{E} = \frac{\mathbf{D}}{\varepsilon} = \begin{cases} \dfrac{\lambda}{2\pi\varepsilon_0\rho}\hat{\boldsymbol{\rho}} & \text{for } 0 < \rho < a \\ \dfrac{\lambda}{8\pi\varepsilon_0\rho}\hat{\boldsymbol{\rho}} & \text{for } a < \rho < b \\ \dfrac{\lambda}{2\pi\varepsilon_0\rho}\hat{\boldsymbol{\rho}} & \text{for } \rho > b \end{cases}$$

(b) The polarization vector inside the cylindrical shell. From the relation $\mathbf{D} = \varepsilon_0\mathbf{E} + \mathbf{P}$,

$$\mathbf{P} = \mathbf{D} - \varepsilon_0\mathbf{E} = \frac{\lambda}{2\pi\rho}\hat{\rho} - \frac{\lambda}{8\pi\rho}\hat{\rho} = \frac{3\lambda}{8\pi\rho}\hat{\rho}$$

10. Let us suppose that the electric field intensity in medium '1', free space, is given by $E_1 = 3\hat{\mathbf{x}} + 2\hat{\mathbf{y}} + 4\hat{\mathbf{z}}$ V/m. Find the displacement density $\mathbf{D}_2$ in medium '2' whose dielectric constant is $\varepsilon_r = 3$, at the charge-free interface

(a) x = constant,

(b) y = constant and the angles made by $\mathbf{E}_1$ and $\mathbf{E}_2$ with the interfaces.

Solution:

(a) x = constant

We can estimate the y- and z-components of $\mathbf{E}_2$ by using the fact that the tangential components of the electric field must be the same across the interface.

$$\mathbf{E}_1 = 3\hat{\mathbf{x}} + 2\hat{\mathbf{y}} + 4\hat{\mathbf{z}}$$
$$\mathbf{E}_2 = E_{x2}\hat{\mathbf{x}} + 2\hat{\mathbf{y}} + 4\hat{\mathbf{z}}$$

The displacement densities in the media '1' and '2' can be found using $\mathbf{D} = \varepsilon\mathbf{E}$.

$$\mathbf{D}_1 = \varepsilon_1\mathbf{E}_1 = 3\varepsilon_1\hat{\mathbf{x}} + 2\varepsilon_1\hat{\mathbf{y}} + 4\varepsilon_1\hat{\mathbf{z}}$$
$$\mathbf{D}_2 = \varepsilon_2\mathbf{E}_2 = E_{x2}\varepsilon_2\hat{\mathbf{x}} + 2\varepsilon_2\hat{\mathbf{y}} + 4\varepsilon_2\hat{\mathbf{z}}$$

Now we can estimate the x-component of $\mathbf{D}_2$ by using the fact that the normal components of the electric displacement must be the same across the interface in the absence of the surface charge.

By comparing the x-components of displacements, we get

$3\varepsilon_1 = E_{x2}\varepsilon_2$ or $E_{x2} = 3\varepsilon_1/\varepsilon_2$. It is given that $\varepsilon_1/\varepsilon_2 = 1/3 \Rightarrow E_{x2} = 1$

Therefore,

$$\mathbf{D}_2 = \varepsilon_2\mathbf{E}_2 = 3\varepsilon_0\hat{\mathbf{x}} + 6\varepsilon_0\hat{\mathbf{y}} + 12\varepsilon_0\hat{\mathbf{z}}$$

Angles:

$$\mathbf{E}_1 \cdot \hat{\mathbf{x}} = |E_1| \cos(90 - \theta_1)$$
$$3 = \sqrt{29}\sin\theta_1 \Rightarrow \theta_1 = \sin^{-1}\left(\frac{3}{\sqrt{29}}\right) = 33.85°$$
$$\mathbf{E}_2 \cdot \hat{\mathbf{x}} = |E_2| \cos(90 - \theta_2)$$
$$1 = \sqrt{21}\sin\theta_2 \Rightarrow \theta_2 = \sin^{-1}\left(\frac{1}{\sqrt{21}}\right) = 12.6°$$

(b) y = constant

To begin with, we estimate the x- and z-components of $\mathbf{E}_2$ which are tangential components of the electric field.

$$\mathbf{E}_1 = 3\hat{\mathbf{x}} + 2\hat{\mathbf{y}} + 4\hat{\mathbf{z}}$$
$$\mathbf{E}_2 = 3\hat{\mathbf{x}} + E_{y2}\hat{\mathbf{y}} + 4\hat{\mathbf{z}}$$

The displacement densities in the media '1' and '2' can be found using $\mathbf{D} = \varepsilon\mathbf{E}$.

$$\mathbf{D}_1 = \varepsilon_1\mathbf{E}_1 = 3\varepsilon_1\hat{\mathbf{x}} + 2\varepsilon_1\hat{\mathbf{y}} + 4\varepsilon_1\hat{\mathbf{z}}$$
$$\mathbf{D}_2 = \varepsilon_2\mathbf{E}_2 = 3\varepsilon_2\hat{\mathbf{x}} + E_{y2}\varepsilon_2\hat{\mathbf{y}} + 4\varepsilon_2\hat{\mathbf{z}}$$

Now we can estimate the y-component of $\mathbf{D}_2$ which is normal to the interface. By comparing the y-components of displacement densities, we get $3\varepsilon_1 = E_{y2}\varepsilon_2$ or $E_{y2} = 2\varepsilon_1/\varepsilon_2$. It is given that $\varepsilon_1/\varepsilon_2 = 1/3 \Rightarrow E_{y2} = 2/3$ Therefore,

$$\mathbf{D}_2 = \varepsilon_2\mathbf{E}_2 = 9\varepsilon_0\hat{\mathbf{x}} + 2\varepsilon_0\hat{\mathbf{y}} + 12\varepsilon_0\hat{\mathbf{z}}$$

Angles:

$$\mathbf{E}_1 \cdot \hat{\mathbf{y}} = |E_1| \cos(90 - \theta_1)$$
$$2 = \sqrt{29}\sin\theta_1 \quad \Rightarrow \quad \theta_1 = \sin^{-1}\left(\frac{2}{\sqrt{29}}\right) = 21.8°$$
$$\mathbf{E}_2 \cdot \hat{\mathbf{y}} = |E_2| \cos(90 - \theta_2)$$
$$\frac{2}{3} = \sqrt{21}\sin\theta_2 \Rightarrow \theta_2 = \sin^{-1}\left(\frac{2}{3\sqrt{21}}\right) = 8.36°$$

11. Show that at the interface of two dielectrics, the tangential components of displacement density and polarization are related through

$$D^{||}_{\text{above}} - D^{||}_{\text{below}} = P^{||}_{\text{above}} - P^{||}_{\text{below}}$$

Solution: Consider the displacement density in a dielectric and its curl.

$$\nabla \times \mathbf{D} = \varepsilon_0 (\nabla \times \mathbf{E}) + \nabla \times \mathbf{P}$$

As

$$\nabla \times \mathbf{E} = 0$$
$$\nabla \times \mathbf{D} = \nabla \times \mathbf{P}$$

Considering surface integral both sides results in

$$\int (\nabla \times \mathbf{D}) \cdot d\mathbf{a} = \int (\nabla \times \mathbf{P}) \cdot d\mathbf{a}$$

Applying the curl theorem results in

$$\oint \mathbf{D} \cdot d\mathbf{l} = \oint \mathbf{P} \cdot d\mathbf{l}$$

Applying this relation to the interface under consideration, we get

$$D^{||}_{\text{above}} - D^{||}_{\text{below}} = P^{||}_{\text{above}} - P^{||}_{\text{below}}$$

12. If θ_1 and θ_2 are angles made by $\mathbf{E}_1$ and $\mathbf{E}_2$ with the interface defined by z = constant, then show that

$$\tan\theta_1 = \frac{E_{z1}}{\sqrt{E_{x1}^2 + E_{y1}^2}} = \frac{\dfrac{D_{z1}}{\varepsilon_1}}{\sqrt{E_{x1}^2 + E_{y1}^2}}$$

$$\tan\theta_2 = \frac{E_{z2}}{\sqrt{E_{x2}^2 + E_{y2}^2}} = \frac{\dfrac{D_{z2}}{\varepsilon_1}}{\sqrt{E_{x2}^2 + E_{y2}^2}}$$

Solution: Given that the electric vector $\mathbf{E}_1$ makes and angle of θ_1 with z = constant surface. The vector now can be resolved into two components, one along the z-direction and the other along normal to the z-directions, respectively, then the component parallel and normal to the z-direction becomes

E_{z1} and $\sqrt{E_{x1}^2 + E_{y1}^2}$, respectively. Therefore,

$$\tan\theta_1 = \frac{E_{z1}}{\sqrt{E_{x1}^2 + E_{y1}^2}}$$

In terms of the displacement, it becomes

$$\frac{E_{z1}}{\sqrt{E_{x1}^2 + E_{y1}^2}} = \frac{\dfrac{D_{z1}}{\varepsilon_1}}{\sqrt{E_{x1}^2 + E_{y1}^2}}$$

Hence,

$$\tan\theta_1 = \frac{\dfrac{D_{z1}}{\varepsilon_1}}{\sqrt{E_{x1}^2 + E_{y1}^2}}$$

Similarly it can be shown very easily that

$$\tan\theta_2 = \frac{E_{z2}}{\sqrt{E_{x2}^2 + E_{y2}^2}} = \frac{\dfrac{D_{z2}}{\varepsilon_2}}{\sqrt{E_{x2}^2 + E_{y2}^2}}$$

13. Find the capacitance of an isolated sphere of radius a and charge Q.

Solution: By definition

$$C = \frac{Q}{V}$$

We know that the potential of an isolated sphere is

$$V = \frac{1}{4\pi\varepsilon_0}\frac{Q}{a}$$

The ratio of Q to V, which is capacitance, then becomes

$$C = \frac{Q}{V} = 4\pi\varepsilon_0 a$$

14. Find the capacitance of two spherical shells of radius a separated by a distance d.

Solution: The capacitances due to individual shells are in series and the capacitance of each one is

$$C = 4\pi\varepsilon_0 a$$

Therefore, the capacitance of the configuration becomes

$$C_{eq} = \frac{C_1 C_2}{C_1 + C_2} = \frac{CC}{C + C} = \frac{C}{2} = 2\pi\varepsilon_0 a$$

15. The space between the plates of a parallel plate capacitor is filled with two slabs of linear dielectric material. Each slab has thickness a, so the total distance between the plates is $2a$. Slab 1 has a dielectric constant of 2, and slab 2 has a dielectric constant of 1.5. The free charge density on the top plate is σ and on the bottom plate it is $-\sigma$. Find

(a) **D** in each slab,
(b) **E** in each slab,
(c) **P** in each slab,
(d) Potential difference between the plates,
(e) Location and amount of all bound charge and
(f) Recalculate the field in each slab by using the known free and bound charge.

Solution:

(a) The electric displacement **D** in each slab:
It can be found by applying the Gauss law to a thin pillbox enclosing one of the plates of the capacitor. Thus,

$$D = \sigma$$

Its direction is from the top plate to the bottom plate.

(b) Electric field **E** in each slab:
It can be found by using $E = D/\varepsilon$

$$E_1 = \frac{\sigma}{2\varepsilon_0};\ E_2 = \frac{2\sigma}{3\varepsilon_0}$$

(c) Polarization **P** in each slab:
It can be found by using

$$D = \varepsilon_0 E + P \Rightarrow P = D - \varepsilon_0 E$$

It results in

$$P_1 = \frac{\sigma}{2};\ P_2 = \frac{\sigma}{3}$$

(d) Potential difference between the plates (see Figure 5.8):
It can be found by using

$$V = -\int \mathbf{E} \cdot d\mathbf{l}$$

$$= \int Edl = E_1 a + E_2 a + = \frac{7}{6}\frac{\sigma a}{\varepsilon_0}$$

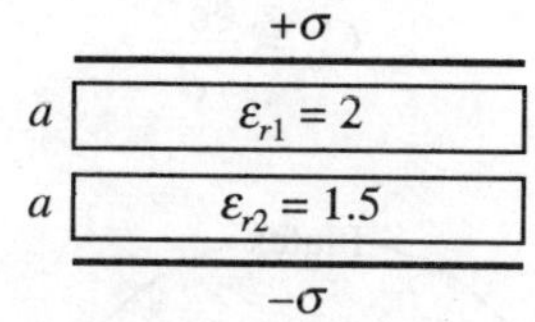

FIGURE 5.8 Parallel plate capacitor with multiple dielectrics.

(e) Location and amount of all bound charge:

$$\sigma_b = P_1 \text{ at bottom of slab 1} = \frac{\sigma}{2}$$

$$\sigma_b = -P_1 \text{ at top of slab 1} = -\frac{\sigma}{2}$$

$$\sigma_b = P_2 \text{ at bottom of slab 2} = \frac{\sigma}{3}$$

$$\sigma_b = -P_2 \text{ at top of slab 2} = -\frac{\sigma}{3}$$

(f) Recalculate the field in each slab by using the known free and bound charge:
In slab 1:

Total surface charge above: $\sigma - \frac{\sigma}{2} = \frac{\sigma}{2}$

Total surface charge below: $\frac{\sigma}{2} - \frac{\sigma}{3} + \frac{\sigma}{3} - \sigma = -\frac{\sigma}{2}$

Therefore, $E_1 = \dfrac{\sigma}{2\varepsilon_0}$

In slab 2:

Total surface charge above: $\sigma - \frac{\sigma}{2} + \frac{\sigma}{2} - \frac{\sigma}{3} = \frac{2\sigma}{3}$

Total surface charge below: $\frac{\sigma}{3} - \sigma = -\frac{2\sigma}{3}$

Therefore, $E_2 = \dfrac{2\sigma}{3\varepsilon_0}$

16. Suppose there is enough linear dielectric material, of dielectric constant ε_r, to half fill a parallel plate capacitor. By what fraction, is the capacitance increased when you distribute the material throughout and to half of the capacitor as shown in Figure 5.9? For a given potential difference V between the plates, find **E**, **D** and **P** in each region and the free and bound charge on all surfaces, for both the cases.

Solution: With no dielectric, the capacitance is

$$C_0 = \frac{\varepsilon_0 A}{d}$$

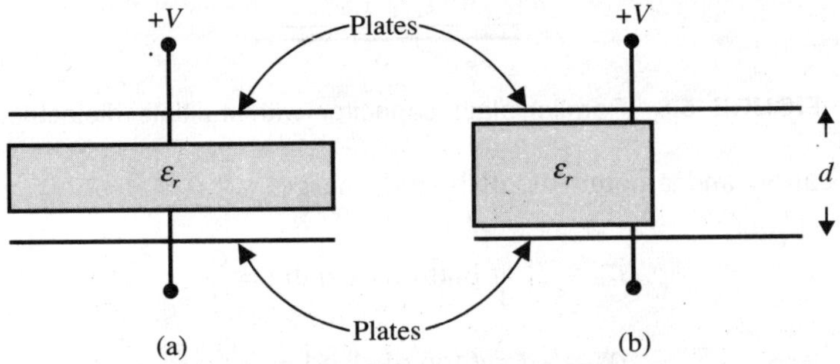

FIGURE 5.9 Half filled parallel plate capacitor.

In configuration (a) with $+\sigma$ on the upper plate, $-\sigma$ on lower, $D = \sigma$ between the plates. It can be found by applying the Gauss law to a thin pillbox enclosing one of the plates of the capacitor. Therefore, by the application of $E = D/\varepsilon$, it can be found that

$$E_1 = \frac{\sigma}{\varepsilon_0} \text{ (in air) and } E_1 = \frac{\sigma}{\varepsilon} \text{ (in dielectric)}$$

And the voltage can be computed using

$$\begin{aligned} V &= -\int \mathbf{E} \cdot d\mathbf{l} \\ &= \int Edl = E_1\left(\frac{d}{4}\right) + E_2\left(\frac{d}{2}\right) + E_1\left(\frac{d}{4}\right) \\ &= \frac{\sigma}{\varepsilon_0}\frac{d}{2} + \frac{\sigma}{\varepsilon}\frac{d}{2} = \frac{Qd}{2\varepsilon_0 A}\left(1 + \frac{\varepsilon_0}{\varepsilon}\right) \end{aligned}$$

$$C_a = \frac{Q}{V} = \frac{\varepsilon_0 A}{d}\left(\frac{2}{1 + \dfrac{1}{\varepsilon_r}}\right)$$

$$\frac{C_a}{C_0} = \left(\frac{2\varepsilon_r}{\varepsilon_r + 1}\right)$$

In configuration (b), the field intensity with potential difference V is $E = V/d$. So $\sigma = \varepsilon_0 E = \varepsilon_0 V/d$ (in air). Therefore,

$$P = \varepsilon_0 \chi_e E = \varepsilon_0 \chi_e \frac{V}{d} \text{ (in dielectric), so } \sigma_b = -\varepsilon_0 \chi_e \frac{V}{d} \text{ (on top of the dielectric)}$$

$$\sigma_{\text{tot}} = \varepsilon_0 \frac{V}{d} = \sigma_f - \sigma_b = \sigma_f - \varepsilon_0 \chi_e \frac{V}{d}$$

Thus,

$$\sigma_f = \varepsilon_0 \frac{V}{d}(1+\chi_e) = \varepsilon_0 \frac{V}{d} \varepsilon_r$$

$$C_b = \frac{Q}{V} = \frac{1}{V}\left(\sigma \frac{A}{2} + \sigma_f \frac{A}{2}\right)$$

$$= \frac{A}{2V}\left(\varepsilon_0 \frac{V}{d} + \varepsilon_0 \varepsilon_r \frac{V}{d}\right) = \frac{\varepsilon_0 A}{d}\left(\frac{\varepsilon_r + 1}{2}\right)$$

$$\frac{C_b}{C_0} = \left(\frac{\varepsilon_r + 1}{2}\right)$$

It can be observed that $C_b > C_a$.

17. A parallel plate capacitor with free space dielectric is connected to a source of constant voltage V. If a dielectric with a dielectric constant of $\varepsilon_r = 3$ is inserted into the entire space in between the plates, then estimate the field intensity and displacement density. Neglect the fringing fields.

Solution: If $\mathbf{E}_0$ and $\mathbf{D}_0$ are the field intensity and the displacement, respectively, with free space as dielectric, then with dielectric filling the space in between the plates, the field intensity and the displacement, respectively, become

$$\mathbf{E} = \frac{V}{d}\hat{\mathbf{n}} = \mathbf{E}_0 \text{ and } \mathbf{D} = \varepsilon_0 \times 3 \times \mathbf{E}_0 = \frac{\varepsilon_0 \times 3 \times V}{d}\hat{\mathbf{n}} = 3\mathbf{D}_0$$

It can also be observed that the energy, capacitance, surface charge density and also the total charge increase to 3 times their respective values before insertion.

18. A parallel plate capacitor with free space dielectric is charged until it holds Q on the positive plate and $-Q$ on the negative plate and the source is then disconnected. If a dielectric with a dielectric constant of 4 is inserted into the entire space in between the plates, then estimate the field intensity and displacement density before and after the dielectric insertion. Neglect the fringing fields.

Solution: If $\mathbf{E}_0$ and $\mathbf{D}_0$ are the field intensity and the displacement, respectively, with free space as dielectric, then with dielectric filling the space in between the plates, the field intensity and the displacement, respectively, become

$$\mathbf{D} = \frac{Q}{A}\hat{\mathbf{n}} = \mathbf{D}_0 \quad \text{and} \quad \mathbf{E} = \frac{\mathbf{D}_0}{4\varepsilon_0} = \frac{Q}{4\varepsilon_0 A}\hat{\mathbf{n}} = \frac{\mathbf{E}_0}{4}$$

In the present case called 'fixed charge case', therefore, the insertion of dielectric does not change the displacement density anyway but the field intensity decreases.

It can be noted that, in this case, the energy and voltage decrease to 0.25 times their previous values but the capacitance increases to 4 times its value before the insertion.

19. A sphere of linear dielectric material has embedded in it a uniform free charge density ρ. Find the potential at the centre of the sphere if its radius is R and dielectric constant is ε_r.

Solution: Using the Gauss law $\oint \mathbf{D} \cdot d\mathbf{a} = Q_{\text{fenc}}$ find the electric field intensity as follows:

$$\mathbf{E} = \frac{\rho r}{3\varepsilon}\hat{\mathbf{r}} \quad \text{for } r \le R$$

and

$$\mathbf{E} = \frac{\rho R^3}{3\varepsilon_0 r^2}\hat{\mathbf{r}} \quad \text{for } r > R$$

And then find out the potential.

$$V = -\int_{\infty}^{a} \mathbf{E} \cdot d\mathbf{l} = \frac{\rho R^3}{3\varepsilon_0} \frac{1}{r}\bigg|_{\infty}^{R} - \frac{\rho}{3\varepsilon}\int_{R}^{0} r dr$$

$$= \frac{\rho R^2}{3\varepsilon_0} + \frac{\rho}{3\varepsilon}\frac{R^2}{2} = \frac{\rho R^2}{3\varepsilon_0}\left(1 + \frac{1}{2\varepsilon_r}\right)$$

20. Find the capacitance of a cylindrical capacitor with two different dielectrics occupying equal volumes with interface parallel to the field.

Solution: As the interface is parallel to the field, the arrangement can be considered as two capacitors in parallel.

$$C_{\text{eq}} = C_1 + C_2 = \pi\varepsilon_1 \frac{L}{\ln\left(\frac{b}{a}\right)} + \pi\varepsilon_2 \frac{L}{\ln\left(\frac{b}{a}\right)} = \frac{\pi L}{\ln\left(\frac{b}{a}\right)}(\varepsilon_1 + \varepsilon_2)$$

Note that in the expression for the capacitance, factor '2' is missing. It is because each dielectric is filling only half of the volume of the capacitor. So the capacitance becomes half of that due to the complete volume.

21. A coaxial cable consists of a copper wire, radius a surrounded by a concentric copper tube of inner radius c. The space between then is partially filled from b to c with a material of dielectric constant ε_r. Find the capacitance per unit length of the cable.

Solution: Let Q be the charge of length l of the inner conductor.

Using the Gauss law $\oint \mathbf{D} \cdot d\mathbf{a} = Q_{\text{fenc}}$ find the electric field intensity as follows:

$$D2\pi rl = Q$$

$$E = \frac{Q}{2\pi\varepsilon_0 \rho l} \qquad (a < \rho < b)$$

$$E = \frac{Q}{2\pi\varepsilon\rho l} \qquad (b < \rho < c)$$

Hence

$$V_{ac} = -\int_c^a \mathbf{E}\cdot d\mathbf{l} = \int_a^b \frac{Q}{2\pi\varepsilon_0 l}\frac{d\rho}{\rho} + \int_b^c \frac{Q}{2\pi\varepsilon l}\frac{d\rho}{\rho} = \frac{Q}{2\pi\varepsilon_0 l}\left[\ln\left(\frac{b}{a}\right) + \frac{\varepsilon_0}{\varepsilon}\ln\left(\frac{c}{b}\right)\right]$$

Therefore

$$\frac{C}{l} = \frac{Q}{Vl} = \frac{2\pi\varepsilon_0}{\left[\ln\left(\frac{b}{a}\right) + \left(\frac{l}{k}\right)\ln\left(\frac{c}{b}\right)\right]}$$

22. A parallel plate capacitor of length l with area a is inserted with a dielectric up to a distance of $l - x$. Find the electric force on the dielectric slab.

Solution: The force on the slab is given by

$$F = -\frac{dW}{dx} = \frac{1}{2}\frac{Q^2}{C^2}\frac{dC}{dx} = \frac{1}{2}V^2\frac{dC}{dx}$$

And the capacitance of the partially dielectric filled capacitor is

$$C = \frac{\varepsilon_0 a}{d}(\varepsilon_r l - \chi_e x)$$

and

$$\frac{dC}{dx} = -\frac{\varepsilon_0 \chi_e a}{d}$$

Therefore,

$$F = -\frac{dW}{dx} = \frac{1}{2}\frac{Q^2}{C^2}\frac{dC}{dx}$$

$$= -\frac{\varepsilon_0 \chi_e a}{2d}V^2$$

This gives an expression for the electric force on the dielectric. The negative sign indicates that the force is a pull into the capacitor.

23. Suppose the plates of a parallel plate capacitor move closer together by an infinitesimal distance ε, as a result of mutual attraction. Find the amount of work done by the electrostatic forces and the energy lost by the field in this process.

Solution: The amount of work done by the electrostatic forces is

$$W = (\text{force}) \times (\text{distance}) = (\text{pressure}) \times (\text{area}) \times (\text{distance}) = \frac{\varepsilon_0}{2} E^2 A\varepsilon$$

The energy lost in this process is

$$W = (\text{energy per unit volume}) \times (\text{decrease in volume}) = \frac{\varepsilon_0}{2} E^2 A\varepsilon$$

EXERCISE SECTION

Essay Type Questions

1. Define conductor and list out all its important features.
2. What are dielectrics? Explain qualitatively the reason for their conductivity being zero.
3. Define the terms 'linear', 'homogeneous' and 'isotropic' with respect to dielectrics.
4. Derive boundary conditions in the presence of dielectrics.
5. Show that the field of a polarized object is the same as the field due to a surface charge plus due to a volume charge.
6. Explain qualitatively why a dielectric gets attracted into an electric field. Derive an expression for the force on the dielectric exerted by the field.
7. Define capacitance mentioning its nature and units. Explain the standard procedures followed in computing the capacitance.
8. Derive expressions for the capacitance of
   (a) parallel plate capacitor,
   (b) concentric spherical shell capacitor and
   (c) concentric cylindrical shell capacitor.

Numerical Problems

1. A sphere of radius a has the charge distribution $\rho(r)$ C/m^3 which produces an electric field intensity given by,
$$E(r) = Ar^4 \quad \text{for } r \le a$$
$$= Ar^{-2} \quad \text{for } r > a$$
where A is a constant. Find the corresponding charge distribution $\rho(r)$ C/m^3.
2. In a medium of $\varepsilon = 5\varepsilon_0$, $\mu = 2.5\mu_0$, $\sigma = 0.2$ m℧ and $E = 20$ μV/m. Find the conduction current density. If this current density exists in a cylindrical rod of 2 cm diameter, evaluate the current that can flow through the rod.
3. A parallel plate capacitor with air as dielectric has a plate area of 36π cm^2 and a separation between the plates of 1 mm. It is charged to a potential difference of 100 V by connecting it across a battery. If the battery is disconnected and the plate separation is increased to 2 mm, calculate the charge in

(a) potential difference across the plates and

(b) energy stored. How do you account for the change in the energy stored?

4. A parallel plate capacitor consists of two square metal plates with 50 mm side separated by 10 mm. A slab of sulphur ($\varepsilon_r = 4$) 6 mm thick is placed on the lower plate and the air gap is of 4 mm. Find the capacitance of the capacitor.
5. (a) Determine the capacitance of a capacitor consisting of two parallel metal plates 30 cm by 30 cm surface separated by 5 mm in air with $\varepsilon_0 = 8.854 \times 10^{-12}$.

   (b) What is the energy stored by the capacitor if it is charged to a potential difference of 500 V?

   (c) What is energy density?

Fill in the Blanks

1. Perfect conductor is one for which the conductivity is __________.
2. The field is always __________ to the surface of the perfect conductor.
3. Perfect dielectric is one for which the conductivity is __________.
4. The separation line from the conductor to the dielectric is __________.
5. Charge density and field intensity inside a conductor are __________.
6. The potential over a conductor are __________.
7. The pressure over the surface of the conductor is __________.
8. The relaxation constant is given by __________.
9. Linear dielectric is one for which __________.
10. The conductivity in the case of semiconductor is __________.
11. Joules law is given by __________.
12. In the case of dielectrics, the Gauss law is __________.
13. The field due to dielectric is the sum of fields due to ______ charge and ______ charge.
14. In the case of linear dielectric, the relation between bound charge and free charge is __________.
15. The presence of dielectric reduces the field intensity by a factor equal to ________.
16. The defining relation of capacitance is __________ and its unit is __________.
17. The capacitance of the parallel plate capacitor is __________.
18. The capacitance of the spherical shell capacitor is __________.
19. The capacitance of the cylindrical capacitor is __________.
20. Dielectric insertion __________ the size of the capacitor for a given capacitance.

Answers

1. infinity
2. normal
3. zero
4. $\sigma = \omega\varepsilon$

5. zero

6. constant

7. $\mathbf{f} = \frac{\sigma^2}{2\varepsilon_0}\hat{\mathbf{n}}$

8. $\tau_r = \frac{\varepsilon}{\sigma}$

9. $\mathbf{P} = \varepsilon_0 \chi \mathbf{E}$

10. $\sigma = e(n\mu_e + p\mu_p)$

11. $P = \int_v \mathbf{E} \cdot \mathbf{J}\, d\tau$

12. $\oint \mathbf{D} \cdot d\mathbf{a} = Q_f$

13. surface, volume

14. $\rho_b = \frac{-\chi_e}{1+\chi_e}\rho_f$

15. $\frac{1}{\varepsilon_r}$

16. $C = \frac{Q}{V}$, farad

17. $\frac{\varepsilon A}{d}$

18. $C = 4\pi\varepsilon \frac{ab}{(b-a)}$

19. $C = 2\pi\varepsilon \frac{L}{\ln\left(\frac{b}{a}\right)}$

20. reduces

Match the Following

Set I

1. E and ρ inside conductor	(a) conductor
2. Equipotential	(b) zero
3. Outward pressure	(c) $\varepsilon_0 \frac{E^2}{2}$
4. Perfect conductor	(d) normal to surface
5. Direction of field	(e) infinite conductivity

Set II

1. Relaxation time	(a) $P = \int_v \mathbf{E} \cdot \mathbf{J}\, d\tau$
2. Joules law	(b) $\mathbf{P} = \varepsilon_0 \chi \mathbf{E}$
3. Linear dielectric	(c) $\frac{\varepsilon}{\sigma}$
4. Gauss law for dielectrics	(d) $\frac{-\chi_e}{1+\chi_e}\rho_f$
5. Bound charge in linear dielectrics	(e) $\oint \mathbf{D} \cdot d\mathbf{a} = Q_f$

Set III

1. Net bound charge	(a) $\sigma_b = \mathbf{P} \cdot \hat{\mathbf{n}}$
2. Surface bound charge	(b) zero
3. Volume bound charge	(c) $\rho_b = -\nabla \cdot \mathbf{P}$
4. Relative permittivity ε_r	(d) $\sigma = 0$
5. Perfect dielectric	(e) $\varepsilon_0(1 + \chi_e)$

Set IV

1. Capacitance	(a) $\frac{CV^2}{2} \frac{Q}{V}$
2. Energy	(b) $\frac{\varepsilon A}{d}$
3. Parallel plate capacitor	(c) $\frac{Q}{V}$
4. Cylindrical shell capacitor	(d) $4\pi\varepsilon \frac{ab}{(b-a)}$
5. Spherical shell capacitor	(e) $2\pi\varepsilon \frac{L}{\ln\left(\frac{b}{a}\right)}$

Answers

Set I:	1. (b)	2. (a)	3. (c)	4. (e)	5. (d)
Set II:	1. (c)	2. (a)	3. (b)	4. (e)	5. (d)
Set III:	1. (b)	2. (a)	3. (c)	4. (e)	5. (d)
Set IV:	1. (c)	2. (a)	3. (b)	4. (e)	5. (d)

Multiple Choice Questions

1. Inside the conductor,
   (a) electric field is zero (b) net charge density is zero
   (c) both (a) and (b) (d) none of these
2. Outside the conductor, tangential and normal electric fields are
   (a) zero, non-zero respectively
   (b) non-zero, zero respectively
   (c) zero, zero respectively
   (d) none of these

3. The average force per unit area on the surface of conductor is
   (a) $\frac{\varepsilon_0}{2}E^2$ (b) $\frac{1}{2\varepsilon_0}\sigma$
   (c) both (a) and (b) (d) none of these
4. The conductor is
   (a) equipotential
   (b) one where net charge can be non-zero on the surface
   (c) the surface charge can be $\sigma = -\varepsilon_0 \frac{\partial V}{\partial n}$
   (d) all of these
5. The capacitance C is
   (a) always positive (b) a geometrical quantity
   (c) equal to charge to voltage ratio (d) all of these
6. The energy stored in the capacitor is
   (a) $\frac{CV^2}{2}$ (b) $\frac{Q^2}{2C}$
   (c) both (a) and (b) (d) none of these
7. The capacitance of a spherical conductor is
   (a) $\frac{\varepsilon A}{d}$ (b) $\frac{Q^2}{2C}$
   (c) both (a) and (b) (d) none of these
8. The capacitance of a parallel plate capacitor is
   (a) $\frac{\varepsilon A}{d}$ (b) $4\pi\varepsilon r$
   (c) $4\pi\varepsilon\frac{ab}{b-a}$ (d) none of these
9. The capacitance between two concentric spherical shells is
   (a) $\frac{\varepsilon A}{d}$ (b) $4\pi\varepsilon r$
   (c) $4\pi\varepsilon\frac{ab}{b-a}$ (d) none of these
10. The field of these polarized object is equal to
    (a) due to volume charge density $\rho_b = -\nabla \cdot \mathbf{P}$
    (b) due to surface charge density $\sigma_b = \mathbf{P} \cdot \hat{\mathbf{n}}$
    (c) the sum of (a) and (b)
    (d) none of these

11. In the presence of dielectrics,
    (a) $\mathbf{D} = \varepsilon_0\mathbf{E} + \mathbf{P}$ (b) $\rho_f = \nabla \cdot \mathbf{D}$
    (c) both (a) and (b) (d) none of these
12. In the case of linear dielectrics, the permittivity is
    (a) $\varepsilon = \varepsilon_0\varepsilon_r$ (b) $\varepsilon = \varepsilon_0(1 + \chi_e)$
    (c) both (a) and (b) (d) none of these
13. In the case of linear dielectrics, the polarization **P** is
    (a) proportional to the field intensity (b) $\mathbf{P} = \varepsilon_0\chi_e\mathbf{E}$
    (c) both (a) and (b) (d) none of these
14. In the case of linear dielectrics,
    (a) loss tangent $= \dfrac{\sigma}{\omega\varepsilon}$ (b) dielectric constant $\varepsilon_r = (1 + \chi_e)$
    (c) both (a) and (b) (d) none of these

Answers

1. (c)	2. (a)	3. (c)	4. (d)	5. (d)
6. (c)	7. (b)	8. (a)	9. (c)	10. (c)
11. (c)	12. (c)	13. (c)	14. (c)	

UNIT III

STEADY MAGNETIC FIELDS

6

Magnetostatics

Success isn't a result of spontaneous combustion. You must set yourself on fire.
—Arnold H. Glasow

The regions that surround the conductors carrying steady currents (direct currents) can exert force on current-carrying conductors and moving charges. As these regions, called steady magnetic fields, are of special and useful properties, we study about them in the coming sections under the title magnetostatics.

There exists a connection between electricity and magnetism which was discovered by the Danish scientist Hans Oersted in 1819. He found that an electric current in a wire caused a compass needle placed in its vicinity to deflect and the deflection was in such a direction that the needle's direction was always perpendicular to the wire and the radial line connecting the wire to the needle.

From these experiments, it was found that the current-carrying wire induced a magnetic field that formed closed circular loops around the wire. Shortly after Oersted's discovery, two French scientists, Jean-Baptiste Biot and Félix Savart, developed a relation for magnetic field intensity at a point in space due to current in a conductor.

6.1 Current Distributions

Currents are sources of magnetic fields and, usually, they exist in one of the three configurations: line or filamentary currents, sheet currents and volume currents. Sometimes the current exists in the form of a thin line like the current in a single strand of copper wire and in such a case the current is called *line current*. Note that line currents are characterized by negligible cross-sectional area normal to the current direction. The current also exists in the form of a thin sheet, which then is called *surface current* or *sheet current*. The sheet current normally exists in a finite width conductor of negligible thickness. The current in thin copper plate is an example of this type of current. The third configuration of the current is called *volume current* and it exists in conductors which are thick having a finite cross-sectional area normal to the direction of the current.

In electromagnetics, we come across three quantities useful to describe different types of current distributions. They are: current **I** measured in amperes, surface current density **K** with units of amperes per metre and volume current density **J** in amperes per square metre. All these three quantities are vectors and are also interrelated. And now we try to explore the relation among them.

By definition, the current **I** through a conductor is the charge crossing a cross section of the conductor in unit time. If the charge flow is in the *x*-direction and the charge crossing is *dQ* through a cross section during time *dt*, then the current **I** is given by

$$\mathbf{I} = \frac{dQ}{dt}\,\hat{\mathbf{x}} \tag{6.1}$$

This quantity, sometimes called line current, exists or can be defined for all the three types of current distributions: line, surface as well as volume currents.

The second current describing quantity **K** can be defined only for the surface or volume current distribution. It is the current **I** per unit normal width of the conductor. Thus,

$$\mathbf{K} = \frac{d\mathbf{I}}{dl_\perp} \tag{6.2}$$

In integral form, this relation is

$$\mathbf{I} = \int_L \mathbf{K}\,dl_\perp \tag{6.3}$$

The third quantity is volume current density **J**. It can be defined only for the volume current distribution. It is actually the current **I** per unit normal cross-sectional area of the current-carrying conductor. Thus,

$$\mathbf{J} = \frac{d\mathbf{I}}{da_\perp} \tag{6.4}$$

The relation can also be expressed in integral form as follows:

$$\mathbf{I} = \int_A \mathbf{J}\,da_\perp \tag{6.5}$$

Various types of current distributions are shown in Figure 6.1.

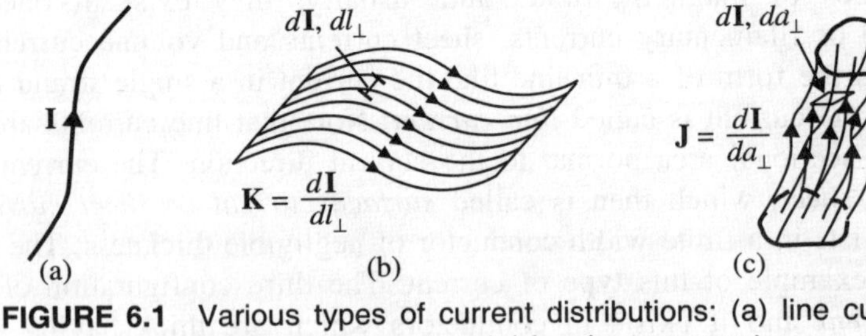

FIGURE 6.1 Various types of current distributions: (a) line current, (b) sheet current and (c) volume current.

Many of the relations in magnetostatics are given in terms of differential current elements. These are three in number and they are given by **I***dl*, **K***da* and **J***dτ*. The first one is the product of line current with differential length, **I***dl*, called *line current element*, the second one is the product of surface current density with differential area, **K***da*, called surface current element and finally the product of volume current density with differential volume, **J***dτ*, is called *volume current element.*

All the three current elements are related and the relation can be expressed as

$$\sum_{i=1}^{n} ()\, q_i \mathbf{v}_i \sim \int_{\text{line}} ()\, \mathbf{I}dl \sim \int_{\text{sur}} ()\, \mathbf{K}da \sim \int_{\text{vol}} ()\, \mathbf{J}d\tau \tag{6.6}$$

We are going to use the above equivalence relation frequently, hence it requires our attention a little more than that is given. We do the mathematical analysis to prove the equivalence.

Consider, to start with, a filamentary line current. The differential charge dQ within the differential length dl can be considered a point charge. If the charge is travelling with a velocity v and the current is I, then the charge crossing the cross section in unit time I coulombs must be within a length of v metres. Therefore, the charge within differential length must be

$$dQ = \frac{I}{v}\, dl \tag{6.7}$$

Now, consider the sheet current with current density K. The sheet can be divided into strips of width dw normal to current flow and the differential charge dQ in differential area $dldw$ can be considered a point charge. Thus,

$$dQ = \frac{K}{v}\, dldw = \frac{K}{v}\, da \tag{6.8}$$

Lastly, consider the volume current with current density J. The current can be divided into pipes of cross-sectional area da normal to current flow and the differential charge dQ in differential volume $dlda$ can be considered a point charge. Thus,

$$dQ = \frac{J}{v}\, dlda = \frac{J}{v}\, d\tau \tag{6.9}$$

By comparing the differential charges of various current distributions, we can arrive at the following relation:

$$\sum_{i=1}^{n} ()\, q_i \mathbf{v}_i \sim \int_{\text{line}} ()\, \mathbf{I}dl \sim \int_{\text{sur}} ()\, \mathbf{K}da \sim \int_{\text{vol}} ()\, \mathbf{J}d\tau \qquad \text{[as given by Eq. (6.6)]}$$

6.2 BIOT–SAVART LAW

The current-carrying conductors are surrounded by magnetic fields. The intensity of the magnetic field is specified by the Biot–Savart law. It gives the dependence of the magnetic field intensity on the current source distribution.

The magnetic field intensity $d\mathbf{H}$ at an arbitrary field point P due to a differential current element $Id\mathbf{l}$ is

1. directly proportional to the product of the current, magnitude of differential length and the sine of the angle between the filament and a line connecting the differential element to the field point P,
2. inversely proportional to R^2, R being the distance between the current element and the field point P and
3. perpendicular to both the differential current element and the distance vector joining the differential element with the field point P.

The proportionality constant in rationalized MKS units is $1/4\pi$. Mathematically, it can be put for line currents as

$$d\mathbf{H}=\frac{1}{4\pi}\frac{Id\mathbf{l}\times\hat{\mathbf{R}}}{R^2}$$

or

$$\mathbf{H}=\int d\mathbf{H}=\frac{I}{4\pi}\int\frac{d\mathbf{l}\times\hat{\mathbf{R}}}{R^2} \tag{6.10}$$

Observe the similarity of this relation with the Coulomb's law for electric field intensity. The role of point charge here is taken over by the current element $Id\mathbf{l}$. However, note that the magnetic field intensity is independent of the medium characteristics, whereas the electric field intensity depends upon the medium's permittivity. The Biot–Savart law for various current distributions is shown in Figure 6.2.

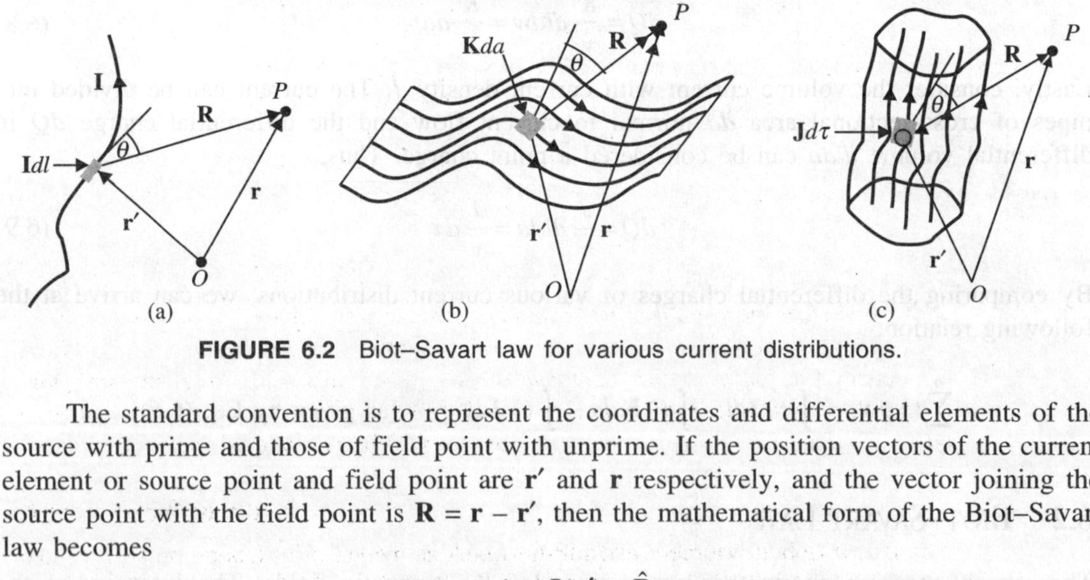

FIGURE 6.2 Biot–Savart law for various current distributions.

The standard convention is to represent the coordinates and differential elements of the source with prime and those of field point with unprime. If the position vectors of the current element or source point and field point are $\mathbf{r}'$ and $\mathbf{r}$ respectively, and the vector joining the source point with the field point is $\mathbf{R} = \mathbf{r} - \mathbf{r}'$, then the mathematical form of the Biot–Savart law becomes

$$\mathbf{H}(\mathbf{r})=\frac{1}{4\pi}\int\frac{\mathbf{I}(\mathbf{r}')\times\hat{\mathbf{R}}}{R^2}\,dl' \tag{6.11}$$

Even though it is given for filamentary line currents, it can be adapted for surface currents as well as volume currents. We have seen that the current elements are related through $\mathbf{I}dl = \mathbf{K}da = \mathbf{J}d\tau$.

With the help of this equivalence, we can write down expressions for the field intensity due to other current distributions as follows:

$$\mathbf{H}(\mathbf{r}) = \frac{1}{4\pi}\int \frac{\mathbf{K}(\mathbf{r}') \times \hat{\mathbf{R}}}{R^2}\, da' \tag{6.12}$$

for surface or sheet current distribution, and

$$\mathbf{H}(\mathbf{r}) = \frac{1}{4\pi}\int \frac{\mathbf{J}(\mathbf{r}') \times \hat{\mathbf{R}}}{R^2}\, d\tau' \tag{6.13}$$

for the volume current distribution.

> Biot–Savart Law:
>
> $$\mathbf{H}(\mathbf{r}) = \frac{1}{4\pi}\int \frac{\mathbf{I}(\mathbf{r}') \times \hat{\mathbf{R}}}{R^2}\, dl'$$
>
> $$\mathbf{H}(\mathbf{r}) = \frac{1}{4\pi}\int \frac{\mathbf{K}(\mathbf{r}') \times \hat{\mathbf{R}}}{R^2}\, da'$$
>
> $$\mathbf{H}(\mathbf{r}) = \frac{1}{4\pi}\int \frac{\mathbf{J}(\mathbf{r}') \times \hat{\mathbf{R}}}{R^2}\, d\tau'$$

6.2.1 Finite Straight Line Current

Here we try to find out the field intensity due to a finite length current-carrying conductor by using the Biot–Savart law which is a canonical problem. The solution of this problem can be applied to find the field intensities of various other different configurations. Therefore, it is worth considering the various aspects of the solution seriously.

Arrange the finite length along the z-axis as shown in Figure 6.3 such that the observation falls over the xy-plane. Locate θ and α as shown and let θ_1, α_1 correspond to the lower end and θ_2, α_2 correspond to the upper end. Consider a differential length over the conductor at a height of z with angles θ and α.

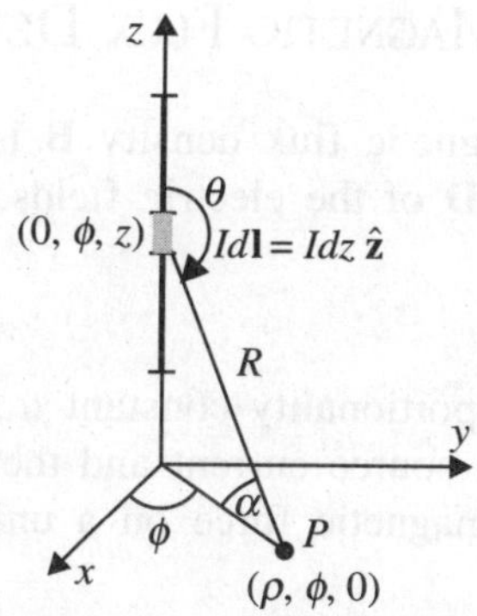

FIGURE 6.3 Magnetic field intensity due to finite length conductor.

From the figure,

$$z = \rho \tan \alpha \Rightarrow dz = \rho \sec^2 \alpha d\alpha$$

The distance between the differential length and the observation point is $r = \rho \sec \alpha$.
And $d\mathbf{l} \times \hat{\mathbf{r}} = dz \sin (90 - \alpha)\, \hat{\boldsymbol{\phi}} = dz \cos \alpha\, \hat{\boldsymbol{\phi}}$

Substituting all these values in the following expression

$$\mathbf{H} = \frac{I}{4\pi} \int \frac{d\mathbf{l} \times \hat{\mathbf{r}}}{r^2}$$

we obain

$$\mathbf{H} = \frac{I}{4\pi} \int_{\theta_1}^{\theta_2} \frac{dz \cos \alpha\, \hat{\boldsymbol{\phi}}}{(\rho \sec \alpha)^2} = \frac{I}{4\pi} \int_{\theta_1}^{\theta_2} \frac{\rho \sec^2 \alpha d\alpha \cos \alpha\, \hat{\boldsymbol{\phi}}}{(\rho \sec \alpha)^2}$$

$$= \frac{I}{4\pi} \int_{\alpha_1}^{\alpha_2} \frac{\rho \sec^2 \alpha d\alpha \cos \alpha\, \hat{\boldsymbol{\phi}}}{(\rho \sec \alpha)^2} = \frac{I}{4\pi} \int_{\alpha_1}^{\alpha_2} \frac{\cos \alpha d\alpha\, \hat{\boldsymbol{\phi}}}{\rho}$$

$$= \frac{I}{4\pi\rho} (\sin \alpha_2 - \sin \alpha_1)\, \hat{\boldsymbol{\phi}} \tag{6.14}$$

From the geometry shown above, we can observe that $\theta = 90° + \alpha$ and if θ_1, α_1 corresponds to the lower end and θ_2, α_2 corresponds to the upper end, then $\alpha_1 = \theta_1 - 90°$ and $\alpha_2 = \theta_2 - 90°$. With these relations, it is possible to express the field intensity in terms of θ_1 and θ_2 as

$$\mathbf{H} = \frac{I}{4\pi\rho} (\cos \theta_1 - \cos \theta_2)\, \hat{\boldsymbol{\phi}} \tag{6.15}$$

6.3 MAGNETIC FLUX DENSITY

The magnetic flux density **B** is a quantity which is analogous to the electric displacement density **D** of the electric fields. By definition, it is related to the magnetic field intensity **H** through

$$\mathbf{B} = \mu \mathbf{H} \tag{6.16}$$

The proportionality constant μ is called permeability of the medium in which the conductor carrying source current and the field point are located. The flux density can also be defined as the magnetic force on a unit positive charge moving with unit velocity normally to the field.

Permeability of the medium: It is an important characteristic of the medium describing the nature of its magnetic properties. It is, in general, the product of two factors, $\mu = \mu_0 \mu_r$. The first factor is known as the permeability of free space and its value is $\mu_0 = 4\pi \times 10^{-7}$ H/m. The second one is called the relative permeability of the medium. And for free space, its value is one. The relative permeability of different materials is shown in Table 6.1.

Table 6.1 Relative permeability μ_r

Types of material	μ_r
Diamagnetic material	
Bismuth	0.99999860
Paraffin	0.99999942
Wood	0.99999950
Silver	0.99999981
Paramagnetic material	
Aluminium	1.00000065
Beryllium	1.00000079
Nickel chloride	1.00004
Manganese sulphate	1.0001
Ferromagnetic material	
Nickel	50
Cast iron	60
Cobalt	60
Mahine steel	300
Ferrite typical	1000
Transformer iron	3000
Silicon iron	4000
Iron (pure)	4000
Mumetal	20,000
Supermalloy	100,000

6.4 MAGNETIC FORCE

It has been observed that the magnetic fields exert force on moving charges and current-carrying conductors. The regions that surround the current-carrying conductors possess magnetic fields and if a moving charge or current-carrying conductor is introduced into these regions, they are exerted a force called *magnetic force*. The amount and direction of this force are described by two laws: one is the fundamental Lorentz force law and the second one is the Ampere's force law which describes the force between the current-carrying circuits.

6.4.1 Lorentz Force Law

When a charged particle is in a region where there exist fields, it experiences a force depending upon the state of the particle and type of the fields. The Lorentz force law is a fundamental law that gives a relation between the forces on the particle due to the combined impact of electric and magnetic fields.

According to this law, the force on a moving point charge Q with velocity $\mathbf{v}$ in a region having both electric field $\mathbf{E}$ and magnetic field $\mathbf{B}$ is

$$\begin{aligned}\mathbf{F} &= Q\mathbf{E} + Q(\mathbf{v} \times \mathbf{B}) \\ &= \mathbf{F}_{ele} + \mathbf{F}_{mag}\end{aligned} \tag{6.17}$$

The first term of the equation represents the electric force on the moving charge and the second term represents the magnetic force. Note the difference between the Lorentz electric force and the Coulomb's electric force: the former is on the moving charges, whereas the latter is on the stationary charges. As it is, this law can give the force on point charges only, but it can also be extended to various different current distributions using the superposition principle.

Let us consider the filamentary line current: the differential charge dQ within a differential length dl of the conductor can be considered a point charge and it is $dQ = (I/v)\, dl$. The differential force on the differential charge, according to the Lorentz's law, now is

$$d\mathbf{F}_{mag} = (\mathbf{v} \times \mathbf{B})\, dQ = (\mathbf{v} \times \mathbf{B}) \frac{I}{v}\, dl = (\mathbf{I} \times \mathbf{B})\, dl \tag{6.18}$$

The total force then becomes

$$\mathbf{F}_{mag} = \int d\mathbf{F}_{mag} = \int (\mathbf{I} \times \mathbf{B})\, dl = I \int d\mathbf{l} \times \mathbf{B} \tag{6.19}$$

Next, we consider the sheet current to extend the Lorentz's force law. The differential charge dQ within a differential area da can be considered a point charge and it is $dQ = (K/v)\, da$. It leads to

$$\mathbf{F}_{mag} = \int d\mathbf{F}_{mag} = K \int d\mathbf{a} \times \mathbf{B} = \int \mathbf{K} \times \mathbf{B}\, da \tag{6.20}$$

In a similar manner, the force law can also be extended to volume currents. The differential charge dQ in a differential volume $d\tau$ can be considered a point charge and it is $dQ = (J/v)\, d\tau$. It leads to

$$\mathbf{F}_{mag} = \int d\mathbf{F}_{mag} = \int \mathbf{J} \times \mathbf{B}\, d\tau \tag{6.21}$$

Therefore,

$$\begin{aligned}\mathbf{F}_{mag} &= I \int d\mathbf{l} \times \mathbf{B} \quad \text{for line currents} \\ &= K \int d\mathbf{a} \times \mathbf{B} \quad \text{for surface currents} \\ &= \int \mathbf{J} \times \mathbf{B}\, d\tau \quad \text{for volume currents}\end{aligned}$$

6.4.2 Cycloid

We consider the analysis of motion of a charged particle under the influence of both electric and magnetic fields when they are perpendicular to each other. It is not only an example of the application of the Lorentz force law, but also useful in the analysis of microwave tubes and particle accelerators. The salient features of a particle motion under the influence of normal fields are given as follows:

- Cycloid is the path traced by a spot on the rim of a rolling wheel. If the spot is on the spoke of the wheel, then the path is called *trochoid.*
- Under the influence of the normal fields, the initially stationary charged particle traces a cycloid path. Its equation is

$$(y - R\omega t)^2 + (z - R)^2 = R^2 \tag{6.22}$$

where the field is in the + x-direction, the field E is in the $+z$-direction, $\omega = (Q/m)B$ and $R = (E/\omega)B$. Here ω represents the angular velocity of the particle and ωt represents the velocity of the translation of the particle.
- The maximum displacement of the particle along the direction of the E field is $2R$ and the distance between the cusps is $2\pi R$. The velocity of the particle at the cusp is zero.
- If the particle has an initial velocity parallel to the magnetic field, then the motion of the particle is in the cycloid helical path. If the initial velocity is normal to the magnetic field, then the path is trochoid.

Derivation

Initially, the particle is at rest, so the magnetic force is zero, and the electric field accelerates the charge in the z-direction. As it picks up speed, a magnetic force develops which pulls the charge around to the right. The faster it goes, the stronger the magnetic force becomes; eventually it curves the particle back around towards the y-axis. At this point, the charge is moving against the electrical force, so it begins to slow down—the magnetic force then decreases, and the electrical force takes over, bringing the charge to rest at point a. There, the entire process commences anew, carrying over to a point b, and so on.

There being no force in the x-direction, the position of the particle at any time t can be described by vector $(0, y, z)$ and the velocity therefore by $\mathbf{v} = (0, \dot{y}, \dot{z})$, where the dots indicate time derivatives. Thus,

$$\mathbf{v} \times \mathbf{B} = (\dot{y}\hat{\mathbf{y}} + \dot{z}\hat{\mathbf{z}}) \times B\hat{\mathbf{x}} = B\dot{z}\hat{\mathbf{y}} - B\dot{y}\hat{\mathbf{z}}$$

and hence, applying the Newton's second law, we have

$$\mathbf{F} = Q\,(\mathbf{E} + \mathbf{v} \times \mathbf{B}) = Q\,(E\hat{\mathbf{z}} + B\dot{z}\hat{\mathbf{y}} - B\dot{y}\hat{\mathbf{z}}) = m\mathbf{a} = m(\ddot{y}\hat{\mathbf{y}} + \ddot{z}\hat{\mathbf{z}})$$

On comparing the $\hat{\mathbf{y}}$ and $\hat{\mathbf{z}}$ components separately, we get

$$QB\dot{z} = m\ddot{y},\ QE - QB\dot{y} = m\ddot{z}$$

For convenience, let $\omega = (Q/m)B$. Then the equations of motion take the form

$$\ddot{y} = \omega\dot{z},\ \ddot{z} = \omega\,[(E/B) - \dot{y}]$$

Their general solutions are

$$y(t) = C_1 \cos \omega t + C_2 \sin \omega t + (E/B)\,t + C_3$$

$$z(t) = C_2 \cos \omega t - C_1 \sin \omega t + C_4$$

But the particle is initially at the origin, i.e. $t = 0$; therefore, $y(0) = z(0) = 0$ and it started from the rest and hence $\dot{y}(0) = \dot{z}(0) = 0$. These four conditions determine the constants C_1, C_2, C_3 and C_4:

$$C_1 = C_3 = 0 \text{ and } C_2 = -\frac{E}{\omega B} = -C_4,$$

Substitution results in

$$y(t) = \frac{E}{\omega B}(\omega t - \sin \omega t), \quad z(t) = \frac{E}{\omega B}(1 - \cos \omega t)$$

With $R \equiv E/\omega B$

$$y(t) = R(\omega t - \sin \omega t), \quad z(t) = R(1 - \cos \omega t)$$

These two equations can be solved for sine and cosine:

$$\sin \omega t = \omega t - y/R, \quad \cos \omega t = 1 + z/R$$

Using the identity $\sin^2 \omega t + \cos^2 \omega t = 1$, it can be shown

$$(y - R\omega t)^2 + (z - R)^2 = R^2$$

This is the formula for a circle of radius R, whose centre $(0, R\omega t, R)$ travels in the y-direction of a constant speed $v = \omega R = E/B$.

The particle moves as though it were a spot on the rim of the wheel, rolling down the y-axis at speed v. The curve it generates is called *cycloid*. It is shown in Figure 6.4.

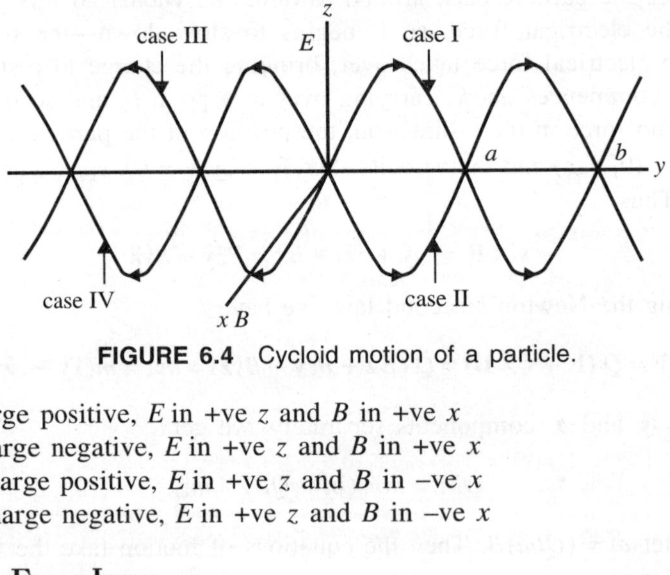

FIGURE 6.4 Cycloid motion of a particle.

Case I: Charge positive, E in +ve z and B in +ve x
Case II: Charge negative, E in +ve z and B in +ve x
Case III: Charge positive, E in +ve z and B in –ve x
Case IV: Charge negative, E in +ve z and B in –ve x

6.4.3 Ampere's Force Law

The Ampere's force law gives the magnetic force that exists between two current-carrying current loops (Figure 6.5). Another version of this law gives the force between the current elements of the different current loops. Both the versions of this law are derivable from the Lorentz force law.

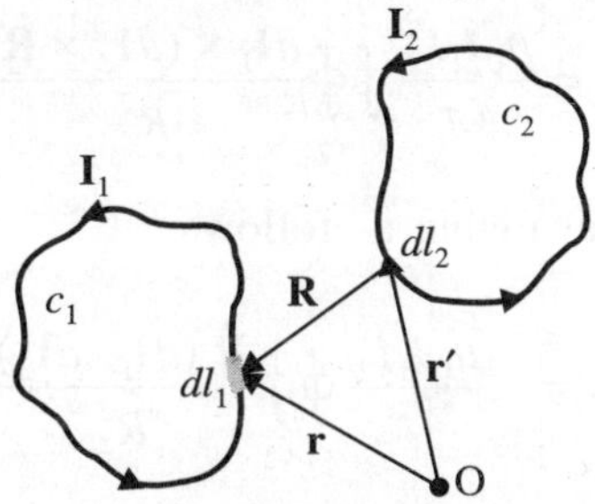

FIGURE 6.5 Depicting Ampere's force law.

For current elements: According to this law, the force between two current elements of different current loops obeys the following:

- The magnitude of the force is proportional to the product of the two currents and also to the product of the lengths of the two current elements.
- The magnitude of the force is inversely proportional to the square of the distance between the current elements.
- For the general case of arbitrary orientations of the current elements $d\mathbf{l}_1$ and $d\mathbf{l}_2$, the forces on them ($d\mathbf{F}_{12}$ and $d\mathbf{F}_{21}$) are not equal and opposite. This is not the violation of the Newton's third law since isolated current elements do not exist.

Mathematically, the force on a current element $I_1 d\mathbf{l}_1$ due to the current element $I_2 d\mathbf{l}_2$, when both are arbitrarily oriented and located in free space, is

$$d\mathbf{F}_{12} = \frac{\mu_0}{4\pi}\frac{I_1 d\mathbf{l}_1 \times (I_2 d\mathbf{l}_2 \times \hat{\mathbf{R}})}{R^2}$$
$$= \frac{\mu_0}{4\pi}\frac{I_1 d\mathbf{l}_1 \times [I_2 d\mathbf{l}_2 \times (\mathbf{r} - \mathbf{r}')]}{|\mathbf{r} - \mathbf{r}'|^3} \tag{6.23}$$

Proof: It can be proved by using the Lorentz force law along with the Biot–Savart law. The force on the current element $\mathbf{I}_1 dl_1$ due to the field $\mathbf{B}_2$ of the current element $\mathbf{I}_2 dl_2$ is

$$d\mathbf{F}_{\text{mag}} = (\mathbf{I}_1 \times \mathbf{B}_2)dl_1$$

The field $\mathbf{B}_2$ of the current element $\mathbf{I}_2 dl_2$ at the current element $\mathbf{I}_1 dl_1$, according to the Biot–Savart law, is

$$\mathbf{B}_2 = \frac{\mu_0}{4\pi}\frac{\mathbf{I}_2 \times \hat{\mathbf{R}}}{R^2} dl_2$$

By combining the above two equations, we can have the Ampere's force law.

For current loops: The law given above for current elements can also be extended to the entire loops. The net force on a current loop '1' due to a current loop '2' when both are arbitrarily oriented and located in free space is

$$\mathbf{F}_{12} = \frac{\mu_0 I_1 I_2}{4\pi} \oint_{c_1} \oint_{c_2} \frac{d\mathbf{l}_1 \times (d\mathbf{l}_2 \times \hat{\mathbf{R}})}{R^2} \tag{6.24}$$

An alternative expression is also available as follows:

$$\mathbf{F}_{12} = -\frac{\mu_0 I_1 I_2}{4\pi} \oint_{c_1} \oint_{c_2} \frac{\hat{\mathbf{R}}\,(d\mathbf{l}_1 \cdot d\mathbf{l}_2)}{R^2} \tag{6.25}$$

The forces on two individual current elements may not necessarily satisfy the Newton's third law but the forces on the loops always obey the Newton's third law.

Lorentz's force law:

$$\mathbf{F} = Q\mathbf{E} + Q(\mathbf{v} \times \mathbf{B}) \text{ for point charges}$$

$$\mathbf{F}_{\text{mag}} = I \int d\mathbf{l} \times \mathbf{B} \text{ for line currents}$$

$$= K \int d\mathbf{a} \times \mathbf{B} \text{ for surface currents}$$

$$= \int \mathbf{J} \times \mathbf{B}\, d\tau \text{ for volume currents}$$

Ampere's force law:

$$d\mathbf{F}_{12} = \frac{\mu_0}{4\pi} \frac{I_1 d\mathbf{l}_1 \times (I_2 d\mathbf{l}_2 \times \hat{\mathbf{R}})}{R^2}$$

$$= \frac{\mu_0}{4\pi} \frac{I_1 d\mathbf{l}_1 \times [I_2 d\mathbf{l}_2 \times (\mathbf{r} - \mathbf{r}')]}{|\mathbf{r} - \mathbf{r}'|^3}$$

$$\mathbf{F}_{12} = \frac{\mu_0 I_1 I_2}{4\pi} \oint_{c_1} \oint_{c_2} \frac{d\mathbf{l}_1 \times (d\mathbf{l}_2 \times \hat{\mathbf{R}})}{R^2}$$

Electric and magnetic forces exhibit a number of important differences:

- Whereas the electric force is always in the direction of the electric field, the magnetic force is always perpendicular to the magnetic field.
- Whereas the electric force acts on a charged particle whether or not moving, the magnetic force acts on it only when it is in motion.
- Whereas the electric force expends energy in displacing a charged particle, the magnetic force does no work when a particle is displaced. Hence, magnetic forces can do no work, the magnetic field can change only the direction of the motion of a charged particle, and it cannot change the particle's speed. Table 6.2 summarizes the differences between electric force and magnetic force.

Table 6.2 Differences between electric force and magnetic force

Electric force	Magnetic force
$\mathbf{F}_{elec} = Q\mathbf{E}$	$\mathbf{F}_{mag} = Q(\mathbf{v} \times \mathbf{B})$
It acts upon charged particles which can be either stationary or moving.	It acts only upon charged particles which are moving.
It is independent of velocity of the charged particle.	Magnitude-wise it is proportional to the speed of the particle and direction-wise it is normal to the velocity of the particle.
It is always parallel to the direction of the field.	It is always perpendicular to the direction of the field.
Electric forces can do work on the particles and change their kinetic energy content.	Magnetic forces do no work on the particles and hence their kinetic energy remains constant.
The total energy remains constant under the influence of electric fields.	The speed of the particle remains constant under the influence of magnetic fields.

6.5 AMPERE'S CIRCUITAL LAW

Ampere's circuital law is one of the fundamental laws of the steady magnetic fields and is analogous to the Gauss law of electrostatics. It states that "the magnetomotive force around a closed path is equal to the net steady current through any surface enclosed by the path." Mathematically, it can be stated as

$$\oint \mathbf{H} \cdot d\mathbf{l} = I_{enc} \tag{6.26}$$

This expression is in integral form and it has a differential counterpart which appears as follows:

$$\nabla \times \mathbf{H} = \mathbf{J}_{enc} \tag{6.27}$$

We can move from one form to the other using the curl theorem of calculus which states that

$$\oint \mathbf{F} \cdot d\mathbf{l} = \int (\nabla \times \mathbf{F}) \cdot d\mathbf{a} \tag{6.28}$$

where **F** is an arbitrary vector. Now, in the case of volume current distributions, the current I_{enc} can be expressed as

$$I_{enc} = \int \mathbf{J}_{enc} \cdot d\mathbf{a} \tag{6.29}$$

and using the Stokes' theorem,

$$\oint \mathbf{H} \cdot d\mathbf{l} = \int (\nabla \times \mathbf{H}) \cdot d\mathbf{a} \tag{6.30}$$

Now the Ampere's law after substitution becomes

$$\oint \mathbf{H} \cdot d\mathbf{l} = \int (\nabla \times \mathbf{H}) \cdot d\mathbf{a} = I_{enc} = \int \mathbf{J}_{enc} \cdot d\mathbf{a} \tag{6.31}$$

Comparing both sides, we get

$$\nabla \times \mathbf{H} = \mathbf{J}_{enc}$$

which is the Ampere's law in differential form.

Note that the Ampere's law in the integral form is valid for all types of current distributions, i.e. line, surface and volume currents but in the differential form as shown above it is valid only for the volume current distribution.

Applications: The Ampere's circuital law in the integral form has extraordinary power of computing fields quickly and easily when the current distribution is symmetrical. It is highly useful in finding the field intensity in the case of infinite straight line currents, infinite plane sheet currents, infinite solenoidal coils and toroidal coils.

Depending upon the symmetry exhibited by the current distributions, a closed path is required to be selected which is called the Amperian loop (path over which integration is performed). It can be either a circle or a rectangle.

The integration of the Ampere's law becomes simpler only if the field is either normal or tangential to the Amperian loop and whenever it is tangential its value must remain constant. Therefore, we have to place the Amperian loop such that the field is either normal or tangential to the loop.

It is also possible to apply the Ampere's law to the combinations of current distributions possessing symmetry, even though the arrangement as a whole is not symmetrical. And then using the superposition principle, the field intensity can be found out.

Limitations:

- It is useful to find the field intensity only when the current distribution exhibits a high level of symmetry.
- The application of the Ampere's law to find the field intensity of the current distribution requires the prior knowledge of the field.

6.6 DIVERGENCE AND CURL

According to the Helmholtz theorem, the divergence and curl of the fields are more than sufficient to describe the fields. They together give the total and complete information regarding the vector fields. As such, divergence and curl of the vector fields are, in general, highly informative parameters. The static magnetic fields exhibit the following important properties:

(a) The magnetic field is non-conservative because the magnetomotive force around a closed path is non-zero. It implies that the magnetic field is rotational; the curl of the magnetic field intensity is non-zero.

$$\oint \mathbf{H} \cdot d\mathbf{l} = I_{\text{enc}} \neq 0 \text{ and } \nabla \times \mathbf{H} = \mathbf{J}_{\text{enc}} \neq 0 \tag{6.32}$$

(b) The magnetic field is solenoidal because the divergence of the magnetic flux density is zero. It implies that the magnetic flux is conserved; the total net flux through any closed surface is zero.

$$\nabla \cdot \mathbf{B} = 0 \text{ and } \oint \mathbf{B} \cdot d\mathbf{a} = 0 \tag{6.33}$$

Proof: The Biot–Savart law for volume currents is

$$\mathbf{H}(\mathbf{r}) = \frac{1}{4\pi}\int \frac{\mathbf{J}(\mathbf{r}') \times \hat{\mathbf{R}}}{R^2}\, d\tau'$$

Here, the field intensity **H** is a function of **r** and **J** is a function of **r′** and the vector joining the source point with the field point is **R** = **r** – **r′**. Note that in the Cartesian system, the tip of **r** is (x, y, z), that of **r′** is (x', y', z'), differential volume of source is $d\tau' = dx'dy'dz'$ and $\mathbf{R} = (x-x')\,\hat{\mathbf{x}} + (y-y')\,\hat{\mathbf{y}} + (z-z')\,\hat{\mathbf{z}}$. Also note that in the above integral relation, the integration is over primed coordinates and divergence and curl are with respect to unprimed quantities.

6.6.1 Divergence

Let us evaluate the divergence of the magnetic field by applying del to both sides of the relation given in the earlier proof.

$$\nabla \cdot \mathbf{H} = \frac{1}{4\pi}\int \nabla \cdot \left(\mathbf{J} \times \frac{\hat{\mathbf{R}}}{R^2}\right) d\tau'$$

Using the vector identity

$$\nabla \cdot (\mathbf{A} \times \mathbf{B}) = \mathbf{B} \cdot (\nabla \times \mathbf{A}) - \mathbf{A} \cdot (\nabla \times \mathbf{B})$$

We expand the expression within the integral as

$$\nabla \cdot \mathbf{H} = \frac{1}{4\pi}\int \left[\frac{\hat{\mathbf{R}}}{R^2} \cdot (\nabla \times \mathbf{J}) - \mathbf{J} \cdot \left(\nabla \times \frac{\hat{\mathbf{R}}}{R^2}\right)\right] d\tau'$$

The first term on the right-hand side of the equation is zero because $\nabla \times \mathbf{J} = 0$ as **J** is a function of primed quantities, independent of unprimed variables. The second term is also zero as we know that

$$\nabla \times \frac{\hat{\mathbf{R}}}{R^2} = 0$$

Therefore, $\nabla \cdot \mathbf{H} = 0$, i.e. steady magnetic field is divergenceless or solenoid field. Also note that it implies

$$\nabla \cdot \mathbf{B} = 0 \text{ and } \oint \mathbf{B} \cdot da = 0$$

$$\nabla \times \mathbf{H} = \mathbf{J} \text{ and } \oint \mathbf{H} \cdot d\mathbf{l} = I_{\text{enc}} \neq 0$$

6.6.2 Curl

Now we evaluate the curl of the magnetic field by applying del to both sides of the relation given in the earlier proof.

$$\nabla \times \mathbf{H} = \frac{1}{4\pi}\int \nabla \times \left(\mathbf{J} \times \frac{\hat{\mathbf{R}}}{R^2}\right) d\tau'$$

Using the vector identity

$$\nabla \times (\mathbf{A} \times \mathbf{B}) = (\mathbf{B} \cdot \nabla)\mathbf{A} - (\mathbf{A} \cdot \nabla)\mathbf{B} + \mathbf{A}(\nabla \cdot \mathbf{B}) - \mathbf{B}(\nabla \cdot \mathbf{A}),$$

we expand the expression within the integral as

$$\nabla \times \mathbf{H} = \frac{1}{4\pi}\int \left[\mathbf{J}\left(\nabla \cdot \frac{\hat{\mathbf{R}}}{R^2}\right) - (\mathbf{J}\cdot\nabla)\frac{\hat{\mathbf{R}}}{R^2}\right] d\tau'$$

after dropping the terms (first and last terms) involving the derivatives of **J**, because **J** does not depend upon unprimed coordinates. Out of the remaining two terms shown above, it can be seen easily that the second term after integration becomes zero. What is ultimately left is the first term which is

$$\nabla \cdot \frac{\hat{\mathbf{R}}}{R^2} = 4\,\pi\delta^3\,(\mathbf{R})$$

Therefore, we have

$$\nabla \times \mathbf{H}(\mathbf{r}) = \frac{1}{4\pi}\int \mathbf{J}\,(\mathbf{r})\,4\pi\delta^3\,(\mathbf{r} - \mathbf{r}')\,d\tau' = \mathbf{J}(\mathbf{r})$$

Therefore, the curl of the magnetic field intensity is current density.

$$\nabla \times \mathbf{H} = \mathbf{J}$$

Also note that it implies

$$\oint \mathbf{H} \cdot d\mathbf{l} = I_{\text{enc}} \neq 0 \text{ and } \nabla \times \mathbf{H} = \mathbf{J}_{\text{enc}} \neq 0$$

6.7 MAGNETIC VECTOR POTENTIAL A

In the case of steady current sources, the magnetic vector potential **A**, measured in webers per metre, is a purely mathematical quantity without any physical significance. If **R** is the vector joining the current element $Id\mathbf{l}'$ and the field point P, then the potential **A** at the field point for line currents is defined as follows:

$$\mathbf{A} = \frac{\mu_0}{4\pi}\int \frac{Id\mathbf{l}'}{R} = \frac{\mu_0}{4\pi}\int \frac{\mathbf{I}d\mathbf{l}'}{R} \tag{6.34}$$

If the position vectors of the source point and field point are $\mathbf{r}'$ and $\mathbf{r}$ respectively, then the magnetic vector potential for line currents is

$$\mathbf{A}(\mathbf{r}) = \frac{\mu_0}{4\pi}\int \frac{I(\mathbf{r}')d\mathbf{l}'}{R} \tag{6.35}$$

Similarly, for surface currents,

$$\mathbf{A}(\mathbf{r}) = \frac{\mu_0}{4\pi}\int \frac{K(\mathbf{r}')d\mathbf{a}'}{R} \tag{6.36}$$

and for volume currents

$$\mathbf{A}(\mathbf{r}) = \frac{\mu_0}{4\pi}\int \frac{\mathbf{J}(\mathbf{r}')d\tau'}{R} \tag{6.37}$$

where $R = |\mathbf{r}' - \mathbf{r}|$

Note that it is a vector quantity always in the direction of current and hence perpendicular to **B**, directly proportional to current element and inversely proportional to distance. It is analogous to the electric scalar potential V but it does not have any physical meaning.

The vector potential **A** is related to the field intensity **H** through

$$\mathbf{A}(\mathbf{r}) = \frac{\mu_0}{4\pi}\int \frac{\mathbf{H}(\mathbf{r}') \times \hat{\mathbf{R}}}{R^2}\, d\tau' \tag{6.38}$$

Proof: The equations $\nabla \cdot \mathbf{B} = 0$ and $\nabla \times \mathbf{B} = \mu_0 \mathbf{J}$ give

$$\mathbf{B}(\mathbf{r}) = \frac{\mu_0}{4\pi}\int \frac{\mathbf{J}(\mathbf{r}') \times \hat{\mathbf{R}}}{R^2}\, d\tau'$$

Now consider the equations $\nabla \cdot \mathbf{A} = 0$ and $\nabla \times \mathbf{A} = \mathbf{B}$ and compare them with the previous two. The comparison must give us

$$\mathbf{A}(\mathbf{r}) = \frac{\mu_0}{4\pi}\int \frac{\mathbf{H}(\mathbf{r}') \times \hat{\mathbf{R}}}{R^2}\, d\tau'$$

> Vector potential:
>
> $$\mathbf{A}(\mathbf{r}) = \frac{\mu_0}{4\pi}\int \frac{I(\mathbf{r}')\, d\mathbf{l}'}{R}$$
>
> $$\mathbf{A}(\mathbf{r}) = \frac{\mu_0}{4\pi}\int \frac{K(\mathbf{r}')\, d\mathbf{a}'}{R}$$
>
> $$\mathbf{A}(\mathbf{r}) = \frac{\mu_0}{4\pi}\int \frac{\mathbf{J}(\mathbf{r}')\, d\tau'}{R}$$
>
> $\nabla \cdot \mathbf{A} = 0$ and $\nabla \times \mathbf{A} = \mathbf{B}$
>
> $$\oint_C \mathbf{A} \cdot d\mathbf{l} = \Phi$$

6.7.1 Divergence and Curl of A

The divergence and curl of the vector potential are given by

$$\nabla \cdot \mathbf{A} = 0 \tag{6.39}$$

and

$$\nabla \times \mathbf{A} = \mathbf{B} \tag{6.40}$$

Proof:

(a) $\nabla \cdot \mathbf{A} = 0$

$$\mathbf{A} = \frac{\mu_0}{4\pi}\int \frac{\mathbf{J}d\tau'}{R}, \quad \text{where } \mathbf{R} = |\mathbf{r} - \mathbf{r}'|.$$

So

$$\mathbf{A}(\mathbf{r}) = \frac{\mu_0}{4\pi}\int \frac{\mathbf{J}(\mathbf{r}')\,d\tau'}{|\mathbf{r} - \mathbf{r}'|}$$

Therefore,

$$\nabla \cdot \mathbf{A} = \frac{\mu_0}{4\pi}\int \nabla \cdot \left(\frac{\mathbf{J}}{R}\right) d\tau'$$

Using $\nabla \cdot (f\mathbf{A}) = f\,(\nabla \cdot \mathbf{A}) + \mathbf{A} \cdot (\nabla f)$, we can have

$$\nabla \cdot \left(\frac{\mathbf{J}}{R}\right) = \mathbf{J} \cdot \nabla\left(\frac{1}{R}\right) + \left(\frac{1}{R}\right)\nabla \cdot \mathbf{J}$$

Here the second term is zero because the source $\mathbf{J}$ is a function of $\mathbf{r}'$ and also

$$\nabla\left(\frac{1}{R}\right) = -\nabla'\left(\frac{1}{R}\right).$$

Therefore,

$$\nabla \cdot \left(\frac{\mathbf{J}}{R}\right) = -\mathbf{J} \cdot \nabla'\left(\frac{1}{R}\right)$$

Similarly

$$\nabla' \cdot \left(\frac{\mathbf{J}}{R}\right) = \mathbf{J} \cdot \nabla'\left(\frac{1}{R}\right) + \left(\frac{1}{R}\right)\nabla' \cdot \mathbf{J}$$

Here also the second term is zero because in megnetostatics $\nabla' \cdot \mathbf{J} = 0$. Therefore,

$$\nabla' \cdot \left(\frac{\mathbf{J}}{R}\right) = \mathbf{J} \cdot \nabla'\left(\frac{1}{R}\right)$$

and hence

$$\nabla \cdot \left(\frac{\mathbf{J}}{R}\right) = -\nabla' \cdot \left(\frac{\mathbf{J}}{R}\right)$$

Now

$$\nabla \cdot \mathbf{A} = \frac{\mu_0}{4\pi}\int \nabla \cdot \left(\frac{\mathbf{J}}{R}\right) d\tau' = -\frac{\mu_0}{4\pi}\int \nabla' \cdot \left(\frac{\mathbf{J}}{R}\right) d\tau'.$$

Using the divergence theorem,

$$\nabla \cdot \mathbf{A} = -\frac{\mu_0}{4\pi}\int \nabla' \cdot \left(\frac{\mathbf{J}}{R}\right) d\tau' = -\frac{\mu_0}{4\pi}\oint \left(\frac{\mathbf{J}}{R}\right) d\mathbf{a}'.$$

where the integral is now over the surface surrounding all the currents. But $\mathbf{J} = 0$ on this surface. So

$$\nabla \cdot \mathbf{A} = 0$$

(b) $\nabla \times \mathbf{A} = \mathbf{B}$

The divergence of the magnetic field is always zero.

$$\nabla \cdot \mathbf{B} = 0$$

It implies the magnetic field is the curl of any vector as the divergence of the curl is always zero. Therefore,

$$\mathbf{B} = \nabla \times \mathbf{A}$$

It can also be proved using the volume current distribution in the following manner:

$$\nabla \times \mathbf{A} = \frac{\mu_0}{4\pi} \int \nabla \times \left(\frac{\mathbf{J}}{R} \right) d\tau'$$

Using $\nabla \times (f\mathbf{A}) = f(\nabla \times \mathbf{A}) - \mathbf{A} \times (\nabla f)$, we have

$$\nabla \times \left(\frac{\mathbf{J}}{R} \right) = \left(\frac{1}{R} \right) \nabla \times \mathbf{J} - \mathbf{J} \times \nabla \left(\frac{1}{R} \right)$$

But $\nabla \times \mathbf{J} = 0$ as the source $\mathbf{J}$ is a function of $\mathbf{r}'$ and

$$\nabla \left(\frac{1}{R} \right) = -\frac{\hat{\mathbf{R}}}{R^2}$$

$$\nabla \times \mathbf{A} = \frac{\mu_0}{4\pi} \int \frac{\mathbf{J} \times \hat{\mathbf{R}}}{R^2} d\tau' = \mathbf{B}$$

Because the magnetic vector potential relates to the magnetic flux density through the curl, the magnetic vector potential $\mathbf{A}$ is at right angles to the magnetic flux density $\mathbf{B}$.

6.7.2 Ampere's Law for Vector Potential

The magnetic flux and vector potential are related through

$$\oint_C \mathbf{A} \cdot d\mathbf{l} = \Phi \tag{6.41}$$

The similarity between the Ampere's law

$$\oint_C \mathbf{H} \cdot d\mathbf{l} = I_{\text{enc}}$$

and the relation

$$\oint_C \mathbf{A} \cdot d\mathbf{l} = \Phi$$

can be exploited to determine $\mathbf{A}$ from Φ in the same way we got $\mathbf{H}$ from I_{enc}.

Proof: $\Phi = \int_s \mathbf{B} \cdot d\mathbf{a} = \int_s \nabla \times \mathbf{A} \cdot d\mathbf{a}$

Using the Stokes' theorem,

$$\int_s \nabla \times \mathbf{A} \cdot d\mathbf{a} = \oint_C \mathbf{A} \cdot d\mathbf{l}$$

Therefore, $\oint_C \mathbf{A} \cdot d\mathbf{l} = \Phi$

6.7.3 Poisson's Equation of Vector Potential

The vector equation

$$\nabla^2 \mathbf{A} = -\mu_0 \mathbf{J} \tag{6.42}$$

is known as Poisson's equation.

Proof:

$$\begin{aligned} \nabla \times (\nabla \times \mathbf{A}) &= \nabla \times \mathbf{B} = \mu_0 \mathbf{J} \\ &= \nabla (\nabla \cdot \mathbf{A}) - \nabla^2 \mathbf{A} \\ &= -\nabla^2 \mathbf{A} \text{ as } \nabla \cdot \mathbf{A} = 0 \end{aligned}$$

Therefore,

$$\nabla^2 \mathbf{A} = -\mu_0 \mathbf{J}$$

6.7.4 Magnetic Scalar Potential

The magnetic scalar potential is defined for situations where the magnetic field intensity is curl free. There are several applications in which the magnetic field exists but there are no current densities involved and hence are curl free. The most obvious are those involving permanent magnets for which:

$$\nabla \times \mathbf{H} = 0 \text{ and hence } \oint \mathbf{H} \cdot d\mathbf{l} = 0$$

For such situations, the magnetic scalar potential ψ can be defined as

$$\mathbf{H} = -\nabla \psi \tag{6.43}$$

The magnetic scalar potential difference can also be defined as

$$\psi_{ba} = \psi_b - \psi_a = -\int_a^b \mathbf{H} \cdot d\mathbf{l} \tag{6.44}$$

The magnetic scalar potential satisfies the Laplace equation.

Observe that the magnetic and electric scalar potentials are defined in exactly the same way and hence their properties are also must be similar. Thus, whenever a magnetic problem is described such that there are no current sources in the region, the properties of the electric scalar potential can be invoked for the magnetic scalar potential.

Even though defined similarly, there exist two fundamental differences between electrostatic potential and scalar magnetic potential. The electrostatic potential is conservative in nature, whereas the scalar magnetic potential is not conservative. The second difference is that the scalar magnetic potential is not a single-valued function of position, whereas the electrostatic potential is a single-valued function of position once zero potential point is assigned.

6.8 TYPES OF MAGNETISM

The origin of magnetism lies in the orbital and spin motions of electrons and how the electrons interact with one another. The magnetic behaviour of the materials can be classified into the following five major groups;

1. Diamagnetism
2. Paramagnetism
3. Ferromagnetism
4. Ferrimagnetism
5. Anti-ferromagnetism

The first two groups of materials exhibit no collective magnetic interactions and are not magnetically ordered, whereas the materials of the next three groups exhibit a long range magnetic order below a certain critical temperature. Ferromagnetic and ferrimagnetic materials are strongly magnetic, whereas the other three are weakly magnetic.

Diamagnetism: These substances are composed of atoms which have no net magnetic moment because all the orbital shells are filled with no unpaired electrons. This phenomenon is usually weak with temperature-independent negative susceptibility. Examples of the materials that exhibit diamagnetism are quartz, calcite and water.

Paramagnetism: In this class of materials, some of the atoms or ions in the materials have a net magnetic moment due to unpaired electrons in partially filled orbitals. One of the most important atoms with unpaired electrons is iron. However, the individual magnetic moments do not interact magnetically and the magnetization is zero when the field is removed. They have temperature-dependant positive susceptibility.

Ferromagnetism: The atomic moments in these materials exhibit very strong interactions produced by very large electronic exchange forces and result in parallel alignment of atomic moments. There are two distinct characteristics of ferromagnetic materials: (1) spontaneous magnetization and (2) the existence of magnetic ordering temperature. The elements Fe, Ni and Co and many of their alloys are examples of ferromagnetic materials. It is due to the magnetic dipoles associated with the spins of unpaired electrons, each dipole points in the same direction as its neighbour, and the alignment occurs in relatively small patches called *domains* and they themselves are randomly oriented. For ferromagnetic materials the susceptance is positive and is approximately 20 to 200 times that of the paramagnetic materials.

Curie temperature and hysteresis are two important properties of ferromagnets. Below certain temperature, called Curie temperature, they are ordered and above it they are disordered. The saturation magnetization goes to zero at this temperature. The ferromagnets can retain a memory of an applied field after its removal. This behaviour is called *hysteresis* and a plot of the variation of magnetization with field intensity is called *hysteresis loop* (Figure 6.6).

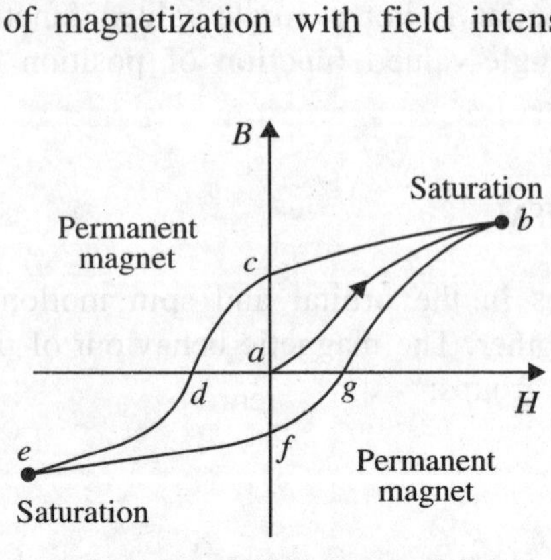

FIGURE 6.6 Hysteresis loop or *B-H* curve of ferromagnetic materials.

Ferrimagnetism: A ferrimagnetic material is one in which the magnetic moments of the atoms on different sub-lattices are opposed and unequal. This happens when the sub-lattices consist of different materials or ions. Ferrimagnetic materials have high resistivity and external field-induced anisotropic properties. Ferrimagnetism properties are similar to those of ferromagnetism in the spontaneous magnetization, Curie temperature, hysteresis and remanence. But they have different magnetic ordering. Ferrimagnetism is exhibited by ferrites and magnetic garnets. The oldest known magnetic substance magnetite is a ferrimagnet. Widely used ferrimagnetic materials are yttrium iron garnet (YIG) and ferrites composed of iron oxides and other elements such as aluminium, nickel, cobalt, manganese and zinc.

Anti-ferromagnetism: A ferromagnetic material is one in which the magnetic moments of the atoms on different sub-lattices are opposed and equal resulting in a net moment of zero.

6.9 MAGNETIZATION OF MATERIALS

When a magnetic field is applied to a material, a net alignment of the magnetic dipoles inside the material occurs resulting in

- Magnetization of the material parallel to **B** (paramagnets) or
- Magnetization of the material opposite to **B** (diamagnets) or
- Retention of substantial magnetization indefinitely after the external field has been removed.

6.9.1 Magnetic Dipole

A circular loop of radius *a* carrying a current *I* is often called magnetic dipole and the product of the area of the loop and the current is known as the magnetic moment. The expression for the field of a magnetic dipole is of some interest to us as it is useful in finding the field of a magnetized object. Now we try to find it via magnetic vector potential **A**.

To find the magnetic vector potential and field intensity at a distant point due to the magnetic dipole, place it over the xy-plane with its centre at the origin. Orient the loop in such a way that the observation point P is in the xz-plane. The current in the loop can be expressed as

$$\begin{aligned}\mathbf{J}(\mathbf{r}') &= \hat{\boldsymbol{\phi}}\, I\delta(\rho' - a)\,\delta(z') \\ &= J_x(\mathbf{r}')\hat{\mathbf{x}} + J_y(\mathbf{r}')\hat{\mathbf{y}} = \sin\phi'\, J(\mathbf{r}')\sin\phi'\hat{\mathbf{x}} + J(\mathbf{r}')\cos\phi'\,\hat{\mathbf{y}}\end{aligned} \tag{6.45}$$

As the direction of the vector potential is the same as that of the current, in the present configuration, the direction of the potential is azimuthal direction. But at the observation point P which is in the xz-plane, the vector potential remains entirely in the y-direction as the x-components get cancelled. Thus,

$$\mathbf{A}(\mathbf{r}) = \hat{\boldsymbol{\phi}}A_\phi = \hat{\mathbf{y}}A_y \tag{6.46}$$

The general expression for the vector potential due to the volume current distribution is

$$\mathbf{A}(\mathbf{r}) = \frac{\mu_0}{4\pi}\int \frac{\mathbf{J}(\mathbf{r}')\,d\tau'}{R} \tag{6.47}$$

If the observation point is far away, then $1/R$ can be expressed as

$$R^{-1} = r^{-1}\left[1 + \frac{\mathbf{r}'\cdot\hat{\mathbf{r}}}{r} + \frac{1}{2}\left\{\frac{3(\mathbf{r}' - \hat{\mathbf{r}})^2}{r^2} - \frac{r'^2}{r^2}\right\} + \cdots\right] \tag{6.48}$$

Substituting this expansion in the expression for the vector potential, we have

$$\mathbf{A}(\mathbf{r}) = \frac{\mu_0}{4\pi r}\int\left[\mathbf{J}(\mathbf{r}')\,d\tau' + \frac{\mathbf{J}(\mathbf{r}')\,\mathbf{r}'\cdot\hat{\mathbf{r}}\,d\tau'}{r} + \frac{\mathbf{J}(\mathbf{r}')}{2}\left\{\frac{3(\mathbf{r}' - \hat{\mathbf{r}})^2}{r^2} - \frac{r'^2}{r^2}\right\}d\tau' + \cdots\right] \tag{6.49}$$

Since the current always flows in closed loops, the first term on integration becomes zero. The contribution by the first term is called monopole moment, by the second term dipole moment and so on. Therefore, the magnetic monopole moment is always zero. Only the second term onwards, they can be non-zero but diminishing very fast. We consider the first non-zero term, i.e. the second term ignoring all others.

$$\mathbf{A}(\mathbf{r}) = \frac{\mu_0}{4\pi r^2}\int[\mathbf{J}(\mathbf{r}')\,\mathbf{r}'\cdot\hat{\mathbf{r}}\,d\tau'] = \hat{\boldsymbol{\phi}}A_\phi = \hat{\mathbf{y}}A_y$$

$$A_\phi = \frac{\mu_0}{4\pi r^2}\int[J_y(\mathbf{r}')\,\mathbf{r}'\cdot\hat{\mathbf{r}}\,d\tau']$$

Now consider

$$\begin{aligned}\mathbf{r}'\cdot\hat{\mathbf{r}} &= (x'\hat{\mathbf{x}} + y'\hat{\mathbf{y}} + z'\hat{\mathbf{z}})\cdot(x\hat{\mathbf{x}} + z\hat{\mathbf{z}}) \\ &= (\rho'\cos\phi'\hat{\mathbf{x}} + y'\rho'\sin\phi'\hat{\mathbf{y}} + z'\hat{\mathbf{z}})\cdot(\sin\theta\hat{\mathbf{x}} + \cos\theta\hat{\mathbf{z}}) \\ &= (\rho'\sin\theta\cos\phi' + z'\cos\theta)\end{aligned}$$

By substituting into the expression for vector potential,

$$A_\phi = \frac{\mu_0 I}{4\pi r^2} \int [\delta(\rho' - a)\,\delta(z')\,(\rho' \sin\theta \cos\phi' + z' \cos\theta)\, d\tau']$$

$$= \frac{\mu_0 I}{4\pi r^2} \int_{-\infty}^{\infty} \int_{0}^{2\pi} \int_{0}^{\infty} [\delta(\rho' - a)\,\delta(z')\,(\rho' \sin\theta \cos\phi' + z' \cos\theta)\, \rho' d\rho' d\phi' dz']$$

$$= \frac{\mu_0 I a^2}{4\pi r^2} \int_0^{2\pi} \sin\theta \cos^2\phi'\, d\phi' = \frac{\mu_0 I a^2 \sin\theta}{4r^2} = \frac{\mu_0 I \pi a^2 \sin\theta}{4\pi r^2}$$

For a current loop carrying uniform current I the magnetic dipole moment is

$$\mathbf{m} \equiv I \int d\mathbf{a} = I\mathbf{a} \tag{6.50}$$

Here $\mathbf{a}$ is the vector area of the loop; if the loop is flat, it is the ordinary area enclosed. In terms of the dipole moment, we have now

$$\mathbf{A}_{\text{dip}}(\mathbf{r}) = \frac{\mu_0}{4\pi} \frac{\mathbf{m} \times \hat{\mathbf{r}}}{r^2} \tag{6.51}$$

There will be two components for the magnetic field intensity at the observation point. Expanding $\mu\mathbf{H} = \nabla \times \mathbf{A}$ in spherical coordinates, we get

$$H_r = -\frac{1}{\mu r \sin\theta} \frac{\partial}{\partial\theta} (\sin\theta A_\phi) = \frac{m \cos\theta}{2\pi r^3} \tag{6.52}$$

$$H_\theta = -\frac{1}{\mu r} \frac{\partial}{\partial r} (r A_\phi) = \frac{I \pi a^2 \sin\theta}{4\pi r^3} = \frac{m \sin\theta}{4\pi r^3} \tag{6.53}$$

Combining Eqs. (6.52) and (6.53), we can express the magnetic field of the dipole as

$$\mathbf{H}_{\text{dip}}(\mathbf{r}) = \frac{m}{4\pi} \frac{1}{r^3} [2\cos\theta\, \hat{\mathbf{r}} + \sin\theta\, \hat{\boldsymbol{\theta}}] \tag{6.54}$$

The expression for the field can also be expressed in another form known as coordinate free form:

$$\mathbf{H}_{\text{dip}}(\mathbf{r}) = \frac{1}{4\pi} \frac{1}{r^3} [3\,(\mathbf{m} \cdot \hat{\mathbf{r}})\,\hat{\mathbf{r}} - \mathbf{m}] \tag{6.55}$$

or

$$\mathbf{B}_{\text{dip}}(\mathbf{r}) = \frac{\mu_0 m}{4\pi} \frac{1}{r^3} [3\,(\mathbf{m} \cdot \hat{\mathbf{r}})\,\hat{\mathbf{r}} - \mathbf{m}] \tag{6.56}$$

Proof: The field due to the dipole is

$$\mathbf{B}_{\text{dip}}(\mathbf{r}) = \mu_0 \mathbf{H}_{\text{dip}} = \frac{\mu_0 m}{4\pi} \frac{1}{r^3} [2\cos\theta\, \hat{\mathbf{r}} + \sin\theta\, \hat{\boldsymbol{\theta}}]$$

A general vector can be resolved into two normal components, so also the dipole moment.

$$\mathbf{m} = (\mathbf{m} \cdot \hat{\mathbf{r}})\, \hat{\mathbf{r}} + (\mathbf{m} \cdot \hat{\boldsymbol{\theta}})\, \hat{\boldsymbol{\theta}} = m \cos \theta \hat{\mathbf{r}} - m \sin \theta \hat{\boldsymbol{\theta}}$$

Then

$$3(\mathbf{m} \cdot \hat{\mathbf{r}})\, \hat{\mathbf{r}} - \mathbf{m} = 3m \cos \theta \hat{\mathbf{r}} - m \cos \theta \hat{\mathbf{r}} + m \sin \theta \hat{\boldsymbol{\theta}} = 2m \cos \theta \hat{\mathbf{r}} + m \sin \theta \hat{\boldsymbol{\theta}}$$

Hence

$$\mathbf{B}_{\text{dip}}(\mathbf{r}) = \frac{\mu_0 m}{4\pi} \frac{1}{r^3} [2 \cos \theta \hat{\mathbf{r}} + \sin \theta \hat{\boldsymbol{\theta}}] = \frac{\mu_0 m}{4\pi} \frac{1}{r^3} [3\,(\mathbf{m} \cdot \hat{\mathbf{r}})\, \hat{\mathbf{r}} - \mathbf{m}]$$

6.9.2 Torque and Force on Planar Loop

A magnetic dipole when placed in a magnetic field experiences a torque. This behaviour is similar to the electric dipole placed in an electric field. Let us try to derive a relation for the torque by considering a rectangular current loop with sides a and b carrying current I placed in a z-directed magnetic field with its centre over the origin. The loop is placed in such a way that its axis makes an angle of θ from the z-axis towards the y-axis.

All the four sides of the loop experience force as they are current-carrying conductors in a magnetic field. The forces on the sloping sides are stretching, equal and opposite resulting in their cancellation. The forces on the horizontal sides are also equal and opposite but they result in a rotating tendency i.e. a torque (Figure 6.7).

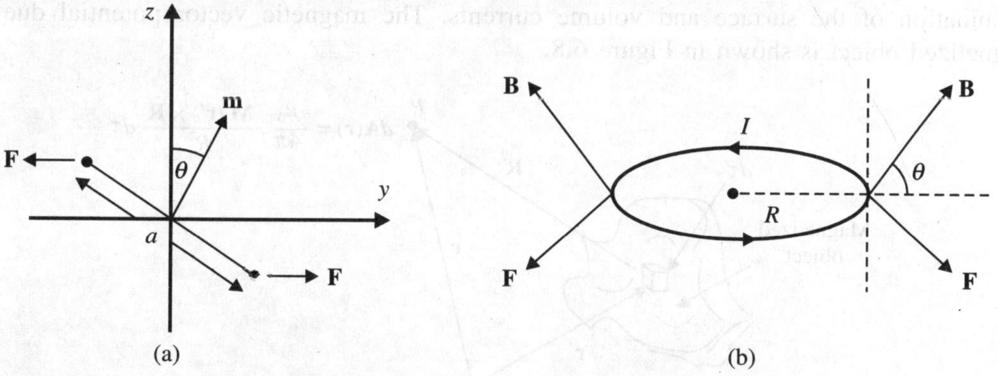

FIGURE 6.7 Torque and force on magnetic dipole.

Now,

$$\mathbf{T} = aF \sin \theta \hat{\mathbf{x}}$$

The magnitude of the force on each of the torque generating sides of length b is

$$F = IbB$$

Therefore the torque becomes

$$\mathbf{T} = aF \sin \theta \hat{\mathbf{x}} = abIB \sin \theta \hat{\mathbf{x}}$$
$$= mB \sin \theta \hat{\mathbf{x}} = \mathbf{m} \times \mathbf{B} \qquad (6.57)$$

Here $m = abI$ is the magnetic dipole moment of the loop.

If the field happens to be a non-uniform one, then in addition to the torque given above, the dipole experiences a net force also. To evaluate the force in non-uniform field, consider a circular loop of radius R over the xy-plane in a magnetic field which is making an angle θ with the horizontal axis. Here the loop experiences a net downward force, which is given by $F = 2\pi IRB \cos\theta$.

It can be expressed as

$$\mathbf{F} = \nabla(\mathbf{m}\cdot\mathbf{B}) \tag{6.58}$$

Therefore, torque on the planar loop in a magnetic field is

$$\mathbf{T} = \mathbf{m}\times\mathbf{B} \tag{6.59}$$

and the Force on the planar loop in a non-uniform magnetic field is

$$\mathbf{F} = \nabla(\mathbf{m}\cdot\mathbf{B}) \tag{6.60}$$

6.9.3 Field of a Magnetized Object

A magnetized object gives rise to magnetic field in its surrounding regions. It can be shown that this field is the same as the sum of fields due to a surface current and a volume current. This relation is useful in the analysis of various situations where steady magnetic fields are involved. Now we try to prove equivalence between the magnetized object and the combination of the surface and volume currents. The magnetic vector potential due to a magnetized object is shown in Figure 6.8.

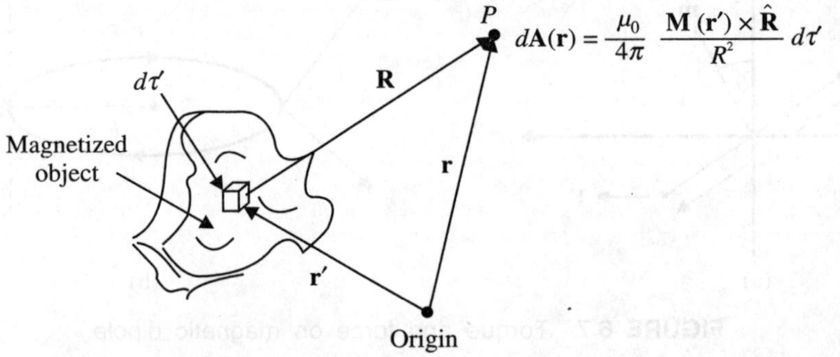

FIGURE 6.8 Magnetic vector potential due to magnetized object.

We know that the vector potential of a single dipole $\mathbf{m}$ is

$$\mathbf{A} = \frac{\mu_0}{4\pi}\frac{\mathbf{m}\times\hat{\mathbf{r}}}{r^2}$$

A quantity used to describe the state of magnetic polarization of a material is magnetization $\mathbf{M}$ which is the magnetic dipole moment per unit volume. To find the potential due to the entire magnetized object, we divide the object into small-sized differential volumes. Each of the differential volumes can be considered as a dipole with a dipole moment equal to $\mathbf{M}d\tau'$. The potential due to each differential volume, therefore, can be written as

$$d\mathbf{A}(\mathbf{r}) = \frac{\mu_0}{4\pi}\frac{\mathbf{M}(\mathbf{r}')\times\hat{\mathbf{R}}}{R^2}d\tau' \tag{6.61}$$

The potential due to the entire magnetized object, therefore, is the sum of the potentials due to various differential volumes of the object which we can get by integrating over the entire volume of the object. Therefore, the vector potential, due to the entire object, is

$$\mathbf{A}(\mathbf{r}) = \int_{\text{vol}} d\mathbf{A} = \frac{\mu_0}{4\pi}\int_{\text{vol}}\frac{\mathbf{M}(\mathbf{r}')\times\hat{\mathbf{R}}}{R^2}d\tau' \tag{6.62}$$

The above expression can be converted into a useful form by using different vector identities. Now replace

$$\frac{\hat{\mathbf{R}}}{R^2} = \nabla'\left(\frac{1}{R}\right)$$

resulting in

$$\mathbf{A}(\mathbf{r}) = \frac{\mu_0}{4\pi}\int_{\text{vol}}\mathbf{M}(\mathbf{r}')\times\nabla'\left(\frac{1}{R}\right)d\tau'$$

Using the vector identity

$$\nabla\times(f\mathbf{F}) = f(\nabla\times\mathbf{F}) - \mathbf{F}\times(\nabla f) \rightarrow \mathbf{F}\times(\nabla' f) = f(\nabla'\times\mathbf{F}) - \nabla'\times(f\mathbf{F})$$

$$\mathbf{A}(\mathbf{r}) = \frac{\mu_0}{4\pi}\int_{\text{vol}}\left(\frac{\nabla'\times\mathbf{M}(\mathbf{r}')}{R} - \nabla'\times\frac{\mathbf{M}(\mathbf{r}')}{R}\right)d\tau' \tag{6.63}$$

The above relation can be put in a more convenient form with the help of the vector identity

$$-\int_{\text{vol}}\nabla'\times\mathbf{F}d\tau' = \oint_{\text{sur}}\mathbf{F}\times d\mathbf{a}'$$

Applying to the second term in the integrand results in

$$\mathbf{A}(\mathbf{r}) = \frac{\mu_0}{4\pi}\int_{\text{vol}}\frac{\nabla'\times\mathbf{M}(\mathbf{r}')}{R}d\tau' + \frac{\mu_0}{4\pi}\oint_{\text{sur}}\frac{\mathbf{M}(\mathbf{r}')\times d\mathbf{a}'}{R}$$

$$= \frac{\mu_0}{4\pi}\int_{\text{vol}}\frac{\mathbf{J}_b}{R}d\tau' + \frac{\mu_0}{4\pi}\int_{\text{sur}}\frac{\mathbf{K}_b d\mathbf{a}'}{R} \tag{6.64}$$

where

$$\mathbf{K}_b = \mathbf{M}\times\hat{\mathbf{n}}$$

and

$$\mathbf{J}_b = \nabla\times\mathbf{M}$$

This relation says that the vector potential and hence also the field of a magnetized object is the same as would be produced by a volume current density $\mathbf{J}_b = \nabla \times \mathbf{M}$ throughout the material plus a surface current $\mathbf{K}_b = \mathbf{M} \times \hat{\mathbf{n}}$ on the boundary.

Magnetic dipole:

$$\mathbf{A}_{\text{dip}}(\mathbf{r}) = \frac{\mu_0}{4\pi} \frac{\mathbf{m} \times \hat{\mathbf{r}}}{r^2}$$

$$\mathbf{B}_{\text{dip}}(\mathbf{r}) = \frac{\mu_0 m}{4\pi} \frac{1}{r^3} [3\,(\mathbf{m} \cdot \hat{\mathbf{r}})\,\hat{\mathbf{r}} - \mathbf{m}]$$

$$\mathbf{T} = \mathbf{m} \times \mathbf{B}$$

$$\mathbf{F} = \nabla\,(\mathbf{m} \cdot \mathbf{B})$$

Currents of magnetized object:

$$\mathbf{K}_b = \mathbf{M} \times \hat{\mathbf{n}}$$

$$\mathbf{J}_b = \nabla \times \mathbf{M}$$

Linear media:

$$\mathbf{J}_b = \chi_m \mathbf{J}_f$$

Magnetic susceptibility:

$$\mu = \mu_0\,(1 + \chi_m)$$

6.9.4 Ampere's Law for Magnetized Materials

The total current of the material can be expressed as $\mathbf{J} = \mathbf{J}_b + \mathbf{J}_f$, where $\mathbf{J}_b$ is bound current, a result of magnetization and $\mathbf{J}_f$ is free current, not a result of magnetization.

The Ampere's law is

$$\frac{1}{\mu_0}(\nabla \times \mathbf{B}) = \mathbf{J} = \mathbf{J}_f + \mathbf{J}_b = \mathbf{J}_f + (\nabla \times \mathbf{M})$$

$$\Rightarrow \qquad \nabla \times \left(\frac{\mathbf{B}}{\mu_0} - \mathbf{M}\right) = \mathbf{J}_f; \quad \text{let } \frac{\mathbf{B}}{\mu_0} - \mathbf{M} = \mathbf{H}$$

$$\Rightarrow \qquad \nabla \times \mathbf{H} = \mathbf{J}_f \quad \Rightarrow \quad \oint \mathbf{H} \cdot d\mathbf{l} = I_{\text{fencl}} \tag{6.65}$$

6.9.5 Linear Magnetic Materials

The volume bound current density in a homogeneous linear material is proportional to the free current density

$$\mathbf{J}_b = \nabla \times \mathbf{M} = \nabla \times \chi_m \mathbf{H} = \chi_m \mathbf{J}_f \tag{6.66}$$

For most substances, the magnetization is proportional to the field $\mathbf{M} = \mu_0 \chi_m \mathbf{H}$, where the proportionality constant χ_m is called the magnetic susceptibility. Now,

$$\mathbf{B} = \mu_0(\mathbf{H} + \mathbf{M}) = \mu_0(1 + \chi_m)\mathbf{H} \tag{6.67}$$

Thus, $\mathbf{B} = \mu\mathbf{H}$, where $\mu = \mu_0\,(1 + \chi_m)$ is called the permeability of the material.

Anisotropic Magnetic Materials

In the case of isotropic magnetic materials, the direction of **B** is the same as the direction of the applied magnetic field **H**. Against this class, there exists another type of linear magnetic materials known as anisotropic magnetic materials. For this type, the induced magnetization vector **M** is not in the direction of **H**. The relationship between **B** and **H** in such type of magnetic materials can be expressed as

$$\begin{bmatrix} B_x \\ B_y \\ B_z \end{bmatrix} = \begin{bmatrix} \mu_{xx} & \mu_{xy} & \mu_{xz} \\ \mu_{yx} & \mu_{yy} & \mu_{yz} \\ \mu_{zx} & \mu_{zy} & \mu_{zz} \end{bmatrix} \begin{bmatrix} H_x \\ H_y \\ H_z \end{bmatrix}$$

Thus, the permeability of an anisotropic magnetic material is not a single constant, instead it is a tensor or 3 × 3 matrix. It indicates that how the three components of **B** depend upon the various components of **H**. Note that both isotropic and anisotropic magnetic materials obey linearity and hence fall in the category of linear magnetic materials.

6.10 MAGNETIC ENERGY

Consider an inductor L connected to a current source and we have to increase the current i flowing through the inductor from zero to a final value I. From the circuit theory, we know that the voltage v across the inductor is given by $v = L\,di/dt$. Power p is equal to the product of v and i and the time integral of power is work or energy. Hence, the total energy expended in building up the current in the inductor is

$$W = \int p\,dt = \int iv\,dt = L\int_0^I i\,dt = \frac{1}{2}LI^2 \tag{6.68}$$

It is possible to put the work or energy in a much nicer way using the field quantities, which can be readily generalized to surface and volume currents. The flux ϕ through the circuit is LI. And it can also be expressed as

$$\phi = \int_s \mathbf{B}\cdot d\mathbf{a} = \int_s (\nabla\times\mathbf{A})\cdot d\mathbf{a} = \oint_c \mathbf{A}\cdot d\mathbf{l} \tag{6.69}$$

where c is the perimeter of the loop and s is any surface bounded by c. Thus,

$$LI = \oint_c \mathbf{A}\cdot d\mathbf{l} \quad \text{and therefore}$$

$$W = \frac{1}{2} I \oint_c \mathbf{A}\cdot d\mathbf{l}$$

or

$$W = \frac{1}{2}\oint_c (\mathbf{A}\cdot\mathbf{I})\,dl \tag{6.70}$$

The expression (6.69) can be generalized to volume currents

$$W = \frac{1}{2}\int (\mathbf{A}\cdot\mathbf{J})\,d\tau \tag{6.71}$$

Using the Ampere's law $\nabla \times \mathbf{H} = \mathbf{J}$, the current $\mathbf{J}$ can be eliminated as

$$W = \frac{1}{2}\int (\mathbf{A} \cdot \nabla \times \mathbf{H})\, d\tau$$

But the vector identity

$$\nabla \cdot (\mathbf{A} \times \mathbf{H}) = \mathbf{H} \cdot (\nabla \times \mathbf{A}) - \mathbf{A} \cdot (\nabla \times \mathbf{H})$$

$$\Rightarrow \quad \mathbf{A} \cdot (\nabla \times \mathbf{H}) = \mathbf{H} \cdot (\nabla \times \mathbf{A}) - \nabla \cdot (\mathbf{A} \times \mathbf{H})$$

Thus,

$$\begin{aligned} W &= \frac{1}{2}\int [\mathbf{H} \cdot (\nabla \times \mathbf{A}) - \nabla \cdot (\mathbf{A} \times \mathbf{H})]\, d\tau \\ &= \frac{\mu_0}{2}\int \mathbf{H} \cdot \mathbf{H}\, d\tau - \frac{1}{2}\int \nabla \cdot (\mathbf{A} \times \mathbf{H})\, d\tau \\ &= \frac{\mu_0}{2}\int \mathbf{H}^2 d\tau - \frac{1}{2}\oint_s (\mathbf{A} \times \mathbf{H}) \cdot d\mathbf{a} \end{aligned} \tag{6.72}$$

where s is the surface bounding the volume v.

The integration in the above expressions needs to be performed over the entire volume occupied by the current. But any region larger than this, even the entire space, will do just as well. If the integration region is infinity, the second integral becomes zero. Thus,

$$W = \frac{\mu_0}{2}\int_{\text{all space}} \mathbf{H}^2 d\tau \tag{6.73}$$

So we can say that the energy stored in the magnetic field is $\mu_0 H^2/2$ per unit volume.

6.11 INDUCTANCE

Inductance is a parameter that represents the energy storage capacity of a coil carrying current. There are two types of inductance: self-inductance L and mutual inductance M.

Self-inductance: For a coil of N turns carrying current I, self-inductance is defined as the ratio of the total flux linkages to the current which they link. Thus,

$$L = \frac{N\Phi}{I} \tag{6.74}$$

where Φ is the flux through any surface having any one turn as perimeter.

Mutual inductance: Consider two coils i and j. Let us suppose a current I_j is the current that is being driven through the coil j. Let us also suppose that a part of the resultant flux ϕ_{if} is linking the coil i. Then the mutual inductance between the coils is defined in the most general way as

$$\begin{aligned} M_{ij} &= \frac{\text{flux linkage through } i\text{th coil due to current in } j\text{th coil}}{\text{current in } j\text{th coil}} \\ &= \frac{N_i \Phi_{ij}}{I_j} \end{aligned} \tag{6.75}$$

The mutual inductance M_{12} between two circuits '1' and '2' then is

$$M_{12} = \frac{N_2 \Phi_{12}}{I_1} \tag{6.76}$$

where N_2 is the number of turns in circuit '2' and Φ_{12} is the flux produced by I_1 which links circuit '2' through any surface having any one turn as perimeter. The self-inductance and mutual inductance are shown in Figure 6.9.

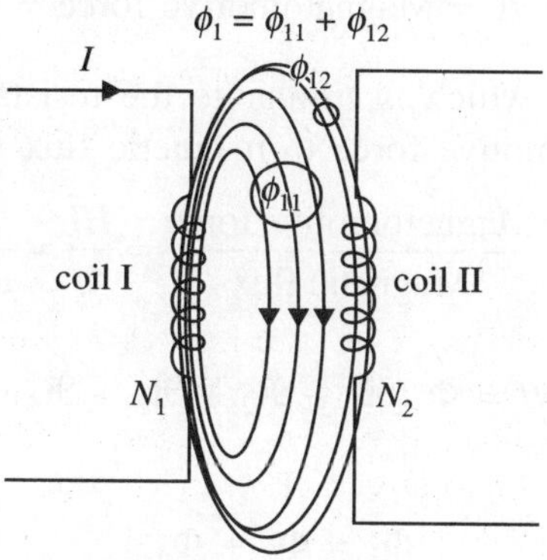

FIGURE 6.9 Self-inductance and mutual inductance.

Neumann formula: The mutual inductance between two filamentary conductor loops is given by

$$M_{12} = \frac{\mu}{4\pi} \oint_{c_1} \oint_{c_2} \frac{d\mathbf{l}_1 \cdot d\mathbf{l}_2}{R} \tag{6.77}$$

Proof: Consider two coils and let us suppose the current through coil '1' is I_1 producing a flux density B_1.

$$\mathbf{B}_1 = \frac{\mu_0 I_1}{4\pi} \oint \frac{d\mathbf{l}_1 \times \hat{\mathbf{R}}}{R^2}$$

Let us suppose a part of the flux produced by coil '1' is passing through coil '2' and let it be ϕ_2 which we can find using

$$\phi_2 = \int \mathbf{B}_1 \cdot d\mathbf{a}_2$$

From the above formula, we can observe that the flux ϕ_2 through the coil '2' is proportional to the current being driven in coil '1'. To continue further,

$$\phi_2 = \int \mathbf{B}_1 \cdot d\mathbf{a}_2 = \int (\nabla \times \mathbf{A}_1) \cdot d\mathbf{a}_2$$

$$= \oint \mathbf{A}_1 \cdot d\mathbf{l}_2 = \frac{\mu_0 I_1}{4\pi} \oint \oint \frac{d\mathbf{l}_1 \cdot d\mathbf{l}_2}{R}$$

Then the ratio of flux to current, the mutual inductance, is

$$M_{21} = \frac{\phi_2}{I_1} = \frac{\mu_0}{4\pi} \oint \oint \frac{d\mathbf{l}_1 \cdot d\mathbf{l}_2}{R}$$

If start with coil '2' carrying current I_2 and compute the mutual inductance M_{12}, then we get the same expression as above for it. Therefore, $M_{12} = M_{21}$.

***Ampere's law for magnetic circuits*:** If we apply Ampere's law to the path shown in the core, then

$$NI = \int_1 \mathbf{H}\cdot d\mathbf{l} + \int_2 \mathbf{H}\cdot d\mathbf{l} + \int_3 \mathbf{H}\cdot d\mathbf{l} + \int_4 \mathbf{H}\cdot d\mathbf{l} \tag{6.78}$$

$$\oint \mathbf{H}\cdot d\mathbf{l} = \text{Magnetomotive force} = NI$$

***Reluctance* $\Re$:** It is a quantity which is similar to the resistance of the circuit theory. It is defined as the ratio of magnetomotive force to magnetic flux in a magnetic circuit. Thus,

$$\Re = \frac{\text{Magnetomotive force}}{\text{Magnetic flux}} = \frac{Hl}{\Phi} = \frac{l}{\mu A} \tag{6.79}$$

For a magnetic circuit,

$$NI = \Phi\,(\Re_1 + \Re_2 + \Re_3 + \Re_4)$$

For a parallel magnetic circuit

$$\Phi_1 = \Phi_2 + \Phi_3$$

Therefore

$$NI = H_1 l_1 + H_2 l_2 = H_1 l_1 + H_3 l_3$$

Cores with air gaps: The apparent area of the rectangular core gap is given by

$$S_a = (a + l_a)\,(b + l_a) \tag{6.80}$$

If the flux flowing is Φ, then the H_a in the core gap can be computed as

$$H_a = \frac{1}{\mu_0}\frac{\Phi}{S_a}$$

Self inductance:

$$L = \frac{N\Phi}{I}$$

Mutual inductance:

$$M_{12} = \frac{N_2\Phi_{12}}{I_1}$$

Neumann formula:

$$M_{12} = \frac{\mu}{4\pi}\oint_{c_1}\oint_{c_2}\frac{d\mathbf{l}_1\cdot d\mathbf{l}_2}{R}$$

Reluctance:

$$\Re = \frac{\text{Magnetomotive force}}{\text{Magnetic flux}} = \frac{Hl}{\Phi} = \frac{l}{\mu A}$$

6.12 BOUNDARY CONDITIONS

The magnetic fields on either side of an interface of magnetic media can be related using the Ampere's law and Gauss law of magnetic fields. The relations are simple and useful to find the fields on one side of the interface when they are known on the other side.

The tangential component of H is discontinuous at the interface of carrying surfacc current. And it is continuous when the two media have finite conductivities. The normal component of B is continuous across the boundary between two adjacent media.

Tangential components: To find the relation between the tangential components, apply the Ampere's circuital law around a small rectangular closed path, as shown in Figure 6.10(a), encompassing the interface in a plane normal to the boundary surface.

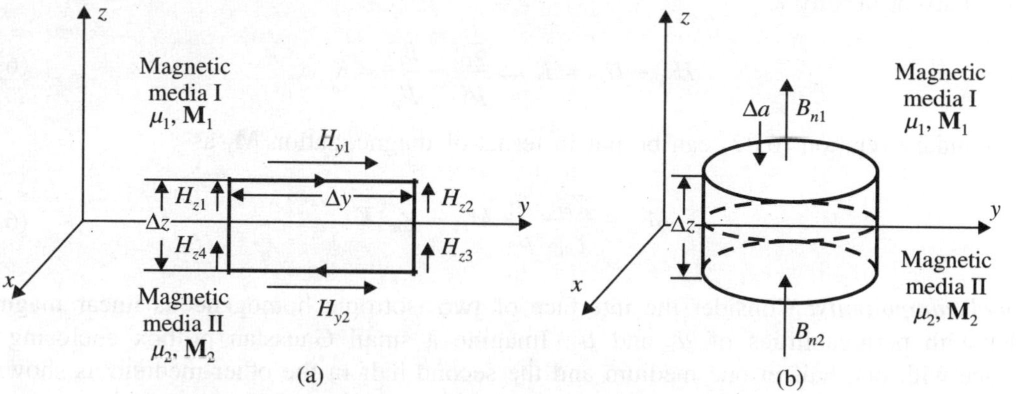

FIGURE 6.10 Boundary conditions—magnetic media interface.

$$\oint \mathbf{H} \cdot d\mathbf{l} = \oint \mathbf{J} \cdot d\mathbf{a}$$

$$\oint \mathbf{H} \cdot d\mathbf{l} = \int_{\text{bottom side}} \mathbf{H} \cdot d\mathbf{l} + \int_{\text{right side}} \mathbf{H} \cdot d\mathbf{l} + \int_{\text{top side}} \mathbf{H} \cdot d\mathbf{l} + \int_{\text{left side}} \mathbf{H} \cdot d\mathbf{l}$$

$$H_{y1}\Delta y - H_{z2}\frac{\Delta z}{2} - H_{z3}\frac{\Delta z}{2} - H_{y2}\Delta y + H_{z4}\frac{\Delta z}{2} + H_{z1}\frac{\Delta z}{2} = J_x \Delta z \Delta y$$

As the width of the loop Δz tends to zero, the entire right-hand side and the terms of the left-hand side with Δz become zero, resulting in

$$H_{y1}\Delta_y - H_{y2}\Delta_y = 0 \text{ or } H_{y1} = H_{y2}$$

As the y-components are tangential to the interface, we have

$$H_{t1} = H_{t2} \tag{6.81}$$

Or the tangential magnetic field is continuous across the interface. If one of the media happens to be a perfect conductor, then there is a possibility of very large surface current and in that case, the right-hand side of the equation, as the loop area width tends to zero, becomes

$$(J_x + \dot{D}_x)\,\Delta z \Delta y = J_x \Delta z\,\Delta y$$
$$H_{y1}\Delta y - H_{y2}\Delta y = J_x \Delta z\,\Delta y$$
$$H_{y1} - H_{y2} = J_x \Delta z = K$$

Therefore, in the case of an interface involving perfect conductor, there is a possibility for the presence of surface currents and consequent discontinuity of tangential magnetic fields.

It results in

$$H_{t1} - H_{t2} = K \tag{6.82}$$

Therefore the tangential magnetic field intensity **H** is discontinuous by an amount equal to the surface current density K.

$$H_{t1} - H_{t2} = K \Rightarrow \frac{B_{t1}}{\mu_1} - \frac{B_{t2}}{\mu_2} = K \tag{6.83}$$

The boundary relation (6.82) can be put in terms of magnetization **M**, as

$$M_{t2} = \frac{\chi_{m1}}{\chi_{m2}} \frac{\mu_1}{\mu_2} M_{t1} - \chi_{m2} K \tag{6.84}$$

Normal components: Consider the interface of two isotropic homogeneous linear magnetic media with permeabilities of μ_1 and μ_2. Imagine a small Gaussian pillbox enclosing the interface with one half in one medium and the second half in the other medium as shown in Figure 6.10(b). Now we apply the Gauss law for magnetic fields. Thus,

$$\oint \mathbf{B} \cdot d\mathbf{a} = 0$$
$$\oint \mathbf{B} \cdot d\mathbf{a} = \oint_{\text{top face}} \mathbf{B} \cdot d\mathbf{a} + \oint_{\text{bottom face}} \mathbf{B} \cdot d\mathbf{a} + \oint_{\text{side face}} \mathbf{B} \cdot d\mathbf{a} = 0$$
$$B_{n1}\Delta a - B_{n2}\Delta a + \phi_{\text{sides}} = 0$$

As the thickness tends to zero, the area of the side wall and the flux through it become zero, resulting in

$$B_{n1} = B_{n2} \tag{6.85}$$

Therefore, we can say that the normal components of the magnetic flux density are continuous across the interface of any two different media. Note that the continuity of the normal components of **B** implies the discontinuous nature of the normal components of **H**. We see that the ratio of intensities is equal to the ratio of permeabilities i.e. μ_1/μ_2.

$$B_{n1} = B_{n2} \Rightarrow \mu_1 H_{n1} = \mu_2 H_{n2} \Rightarrow \frac{H_{n1}}{H_{n2}} = \frac{\mu_2}{\mu_1} \tag{6.86}$$

The relation (6.86) can also be expressed in terms of the magnetization **M** of the linear magnetic media involved as

$$M_{n2} = \chi_{m2} \frac{\mu_1}{\mu_2} H_{n1} = \frac{\chi_{m2}}{\chi_{m1}} \frac{\mu_1}{\mu_2} M_{n1} \qquad (6.87)$$

SOLVED PROBLEMS

1. Find the magnetic field intensity at a distance of ρ from the mid-point of a finite length straight line conductor carrying a current of I.

Solution: We use the result of the previous problem to find the solution. In the present case, $\theta_1 + \theta_2 = 180°$.
Therefore,

$$\mathbf{H} = \frac{I}{4\pi\rho}(\cos\theta_1 - \cos\theta_2)\,\hat{\phi} = \frac{I}{4\pi\rho}(\cos\theta_1 - \cos\overline{180° - \theta_1})\,\hat{\phi}$$

$$= \frac{I}{4\pi\rho}(\cos\theta_1 + \cos\theta_1)\,\hat{\phi} = \frac{I\cos\theta_1}{2\pi\rho}\,\hat{\phi}$$

If the length of the line is $2L$, then the result can be expressed as

$$\mathbf{H} = \frac{I\cos\theta_1}{2\pi\rho}\,\hat{\phi} = \frac{IL}{2\pi\rho\sqrt{L^2 + \rho^2}}\,\hat{\phi}$$

2. Find the magnetic field intensity at a distance of ρ due to an infinite length straight line conductor carrying a current of I.

Solution: The current distribution is cylindrical and so we have to use the cylindrical coordination system. Place the line current over the z-axis (Figure 6.11).

$$\mathbf{H} = \frac{1}{4\pi}\int_{-\infty}^{\infty} \frac{Idz\hat{\mathbf{z}} \times (\rho\hat{\rho} - z\hat{\mathbf{z}})}{(\rho^2 + z^2)^{3/2}} = \frac{1}{4\pi}\int_{-\infty}^{\infty} \frac{Idz\rho\hat{\phi}}{(\rho^2 + z^2)^{3/2}} = \frac{I}{2\pi\rho}\,\hat{\phi}$$

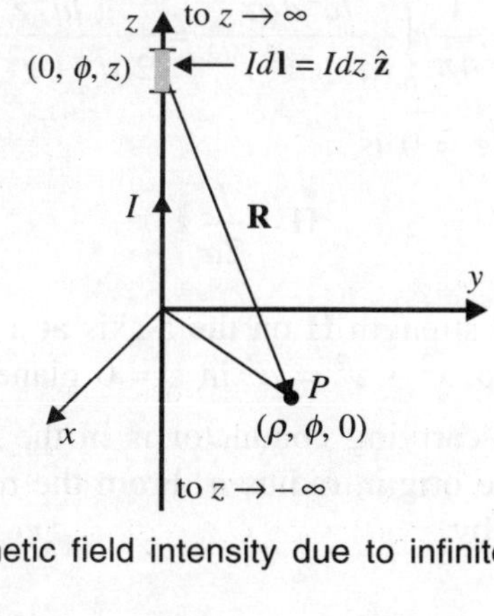

FIGURE 6.11 Magnetic field intensity due to infinite straight line current.

The same result can also be arrived at by using the result of the first problem. In the present case, $\theta_1 = 0°$ and $\theta_2 = 180°$

$$\mathbf{H} = \frac{1}{4\pi\rho}(\cos\theta_1 - \cos\theta_2)\hat{\boldsymbol{\phi}} = \frac{I}{4\pi\rho}(\cos 0° - \cos 180°)\hat{\boldsymbol{\phi}} = \frac{I}{4\pi\rho}(1+1)\hat{\boldsymbol{\phi}} = \frac{I}{2\pi\rho}\hat{\boldsymbol{\phi}}$$

3. Find the magnetic field intensity over the axis of circular loop conductor of radius a carrying a current of I amps at a height of h.

   Solution: The current distribution is cylindrical and so we have to use the cylindrical coordination system. Place the circular line current over the xy-plane with the centre over the origin (Figure 6.12).

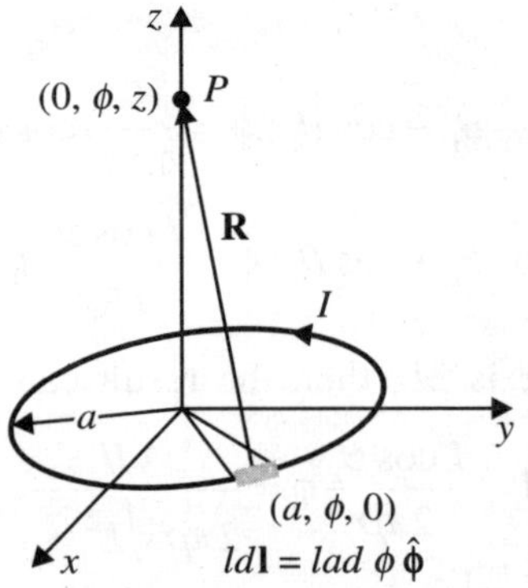

FIGURE 6.12 Magnetic field intensity due to circular current loop.

The field intensity at a height z over the z-axis is

$$\mathbf{H} = \frac{1}{4\pi}\int_0^{\pi}\frac{Iad\,\phi\hat{\boldsymbol{\phi}}\times(-a\hat{\boldsymbol{\rho}} + z\hat{\mathbf{z}})}{(a^2+z^2)^{3/2}} = \frac{1}{4\pi}\int_0^{\pi}\frac{Iad\,\phi\,(z\hat{\boldsymbol{\rho}} + a\hat{\mathbf{z}})}{(a^2+z^2)^{3/2}}$$

The diametrically opposite current elements cancel the radial components. Thus,

$$\mathbf{H} = \frac{1}{4\pi}\int_0^{\pi}\frac{Ia^2 d\phi\hat{\mathbf{z}}}{(a^2+z^2)^{3/2}} = \frac{Ia^2\,\hat{\mathbf{z}}}{2(a^2+z^2)^{3/2}}$$

The field at the origin $z = 0$ is

$$\mathbf{H} = \frac{I}{2a}\hat{\mathbf{z}}$$

4. Find the magnetic field strength **H** on the z-axis at a point $P(0, 0, 4)$ due to a current carrying conductor loop, $x^2 + y^2 = a^2$ in $z = 0$ plane.

   Solution: The current-carrying conductor is in the xy-plane in the form of a circle with the centre over the origin, radius a. From the results of the above problem, the field strength is given by

$$\mathbf{H} = \frac{Ia^2\,\hat{\mathbf{z}}}{2(a^2+z^2)^{3/2}} = \frac{Ia^2\,\hat{\mathbf{z}}}{2(a^2+4^2)^{3/2}}$$

5. Find the magnetic field intensity at the centre of a square loop conductor of side L carrying a current of I amps.

Solution: Place the square current loop in the xy-plane with its centre over the origin and sides parallel to the x- and y-axes. Let us suppose the side of the square loop be L.

$$d\mathbf{H} = \frac{1}{4\pi}\frac{Idx\hat{\mathbf{x}}\times\left[-x\hat{\mathbf{x}}+\left(\frac{L}{2}\right)\hat{\mathbf{y}}\right]}{\left[x^2+\left(\frac{L}{2}\right)^2\right]^{3/2}}$$

$$= \frac{1}{4\pi}\frac{Idx\left(\frac{L}{2}\right)\hat{\mathbf{z}}}{\left[x^2+\left(\frac{L}{2}\right)^2\right]^{3/2}}$$

Total field at the origin is

$$\mathbf{H} = 8\frac{1}{4\pi}\int_0^{L/2}\frac{Idx\left(\frac{L}{2}\right)\hat{\mathbf{z}}}{\left[x^2+\left(\frac{L}{2}\right)^2\right]^{3/2}} = \frac{2\sqrt{2}}{\pi L}\hat{\mathbf{z}}$$

6. Find the magnetic force on a straight conductor of length L carrying current I when placed perpendicularly in a uniform magnetic field B.

Solution: The appropriate formula we have to use is

$$\mathbf{F}_{\text{mag}} = I\int d\mathbf{l}\times\mathbf{B}$$

Given that the conductor is placed normally to the field. It results in the replacement of cross product by ordinary multiplication. It is also given that the field is uniform. The flux density can be brought out of the integral sign. It is given that the conductor is straight with a length L. It results in

$$F = ILB$$

7. A point charge q moving in a region, where there exist electric and magnetic fields **E** and **B**, experience the following three different forces at a point for three different velocities.

$$\mathbf{F}_1 = qE_0(\hat{\mathbf{x}} - 3\hat{\mathbf{y}}) \text{ when the velocity is } \mathbf{v}_1 = v_0\hat{\mathbf{x}}$$

$$\mathbf{F}_2 = qE_0(3\hat{\mathbf{x}} - \hat{\mathbf{y}}) \text{ when the velocity is } \mathbf{v}_2 = v_0\hat{\mathbf{y}}$$

$$\mathbf{F}_3 = qE_0(3\hat{\mathbf{x}} - 5\hat{\mathbf{y}}) \text{ when the velocity is } \mathbf{v}_3 = v_0(2\hat{\mathbf{x}} + \hat{\mathbf{y}})$$

If E_0 and v_0 are constants, find the electric field intensity $\mathbf{E}$ as well as the magnetic field intensity $\mathbf{B}$ at that point.

Solution: Using the Lorentz force equation $\mathbf{F} = Q\mathbf{E} + Q(\mathbf{v} \times \mathbf{B})$, we can write down three vector equations for three different velocities as

$$qE_0(\hat{\mathbf{x}} - 3\hat{\mathbf{y}}) = q\mathbf{E} + q\,(v_0\hat{\mathbf{x}} \times \mathbf{B})$$

$$qE_0(3\hat{\mathbf{x}} - \hat{\mathbf{y}}) = q\mathbf{E} + q\,(v_0\hat{\mathbf{y}} \times \mathbf{B})$$

$$qE_0(3\hat{\mathbf{x}} - 5\hat{\mathbf{y}}) = q\mathbf{E} + q\,[v_0(2\hat{\mathbf{x}} + \hat{\mathbf{y}}) \times \mathbf{B}]$$

By substituting $\mathbf{E} = E_x\hat{\mathbf{x}} + E_y\hat{\mathbf{y}} + E_z\hat{\mathbf{z}}$ and $\mathbf{B} = B_x\hat{\mathbf{x}} + B_y\hat{\mathbf{y}} + B_z\hat{\mathbf{z}}$, we get three scalar equations given by each of the above three vector equations. The total nine equations can be solved to find the six unknowns E_x, E_y, E_z, B_x, B_y and B_z. In fact, six equations are sufficient to find the six unknowns.

Another method, which is simple and easy, is based on the fact that $\mathbf{v}_3 = 2\mathbf{v}_1 + \mathbf{v}_2$. In this method, multiply the first equation by 2, add it to the second equation and subtract the third one from this sum resulting in

$$2q\mathbf{E} + q\mathbf{E} - q\mathbf{E} = 2qE_0(\hat{\mathbf{x}} - 3\hat{\mathbf{y}}) + qE_0(3\hat{\mathbf{x}} - \hat{\mathbf{y}}) - qE_0(3\hat{\mathbf{x}} - 5\hat{\mathbf{y}})$$

$$2\mathbf{E} = E_0(2\hat{\mathbf{x}} - 2\hat{\mathbf{y}}) \quad \Rightarrow \quad \mathbf{E} = E_0(\hat{\mathbf{x}} - \hat{\mathbf{y}})$$

Substituting this result in the first and second equations, we get

$$v_0\hat{\mathbf{x}} \times \mathbf{B} = -2E_0\hat{\mathbf{y}}$$

$$v_0\hat{\mathbf{y}} \times \mathbf{B} = -2E_0\hat{\mathbf{x}}$$

From the above expressions, we can say that the magnetic field is perpendicular to both the x- and y-directions. Let it be $C\hat{\mathbf{z}}$ and substituting this in one of the above two equations, we get

$$C = \frac{2E_0}{v_0}$$

Thus,

$$\mathbf{B} = \frac{2E_0}{v_0}\,\hat{\mathbf{z}}$$

8. Find the magnetic field intensity at a distance of d from a long straight wire carrying a steady current I using the Ampere's circuital law.

Solution: To be able to apply the Ampere's law, we need to have prior information regarding the field. We can have this information by using the Biot–Savart law.

We know from the Biot–Savart law that

- The direction of the field is circumferential $\hat{\phi}$, circling around the wire as indicated by the right-hand rule.
- And by symmetry, the magnitude of the field is constant around the Amperian loop of radius d centred on the wire (see Figure 6.13).

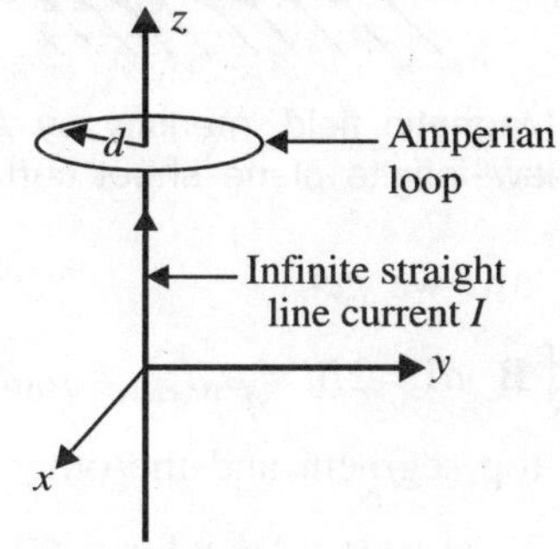

FIGURE 6.13 Magnetic field intensity by Ampere's circuital law-infinite straight line current.

Applying the Ampere's law, we get

$$\oint \mathbf{B} \cdot d\mathbf{l} = B\oint dl = B2\pi d = \mu_0 I_{enc} = \mu_0 I$$

or

$$B = \frac{\mu_0 I}{2\pi d}$$

or

$$\mathbf{B} = \frac{\mu_0 I}{2\pi d}\hat{\phi}$$

9. Find the magnetic field intensity due to an infinite uniform sheet carrying a surface current of $\mathbf{K} = K\hat{\mathbf{x}}$ in the *xy*-plane, using the Ampere's circuital law.

Solution: To be able to apply the Ampere's law, we need to have prior information regarding the field. We can have this information by using the Biot–Savart law.

We can find from the Biot–Savart law that

- The direction of the field is exclusively in $\hat{\mathbf{y}}$ direction and the right-hand rule says it points to the left above the plane and to the right below it.
- And by symmetry, the magnitude of the field is constant around the Amperian loop.

Now consider a rectangular Amperian loop parallel to the *yz*-plane and extending an equal distance above and below the surface (see Figure 6.14).

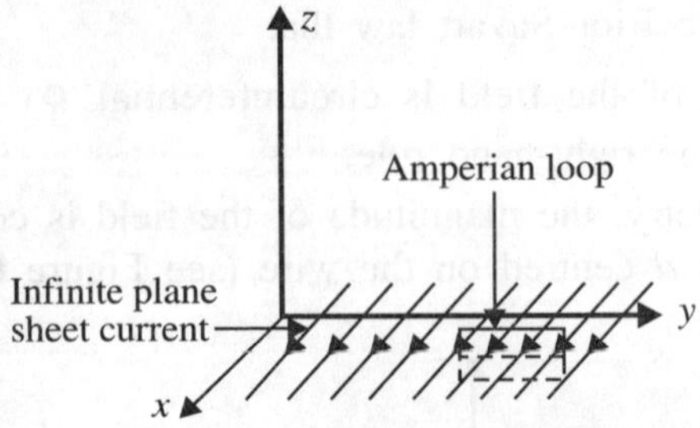

FIGURE 6.14 Magnetic field intensity by Ampere's circuital law–infinite plane sheet current.

Applying the Ampere's law, we get

$$\oint \mathbf{B} \cdot d\mathbf{l} = 2Bl = \mu_0 I_{enc} = \mu_0 kl$$

One **Bl** comes from the top segment and the other from the bottom. So

$$\mathbf{B} = \begin{cases} +(\mu_0/2)k\,\hat{\mathbf{y}} \text{ for } z < 0 \\ -(\mu_0/2)k\,\hat{\mathbf{y}} \text{ for } z > 0 \end{cases}$$

Notice that the field is independent of the distance from the plane.
If we consider the current orientation arbitrary, then

$$\mathbf{H} = \frac{\mathbf{K} \times \hat{\mathbf{n}}}{2}$$

10. Find the magnetic field intensity due to a long coaxial cable of length l with the inner cylinder radius a and the outer cylinder radius b carrying current I, using the Ampere's circuital law.

Solution: To be able to apply the Ampere's law, we need to have prior information regarding the field (see Figure 6.15). We can have this information by using the Biot–Savart law. The direction of the field in the present case is radial.

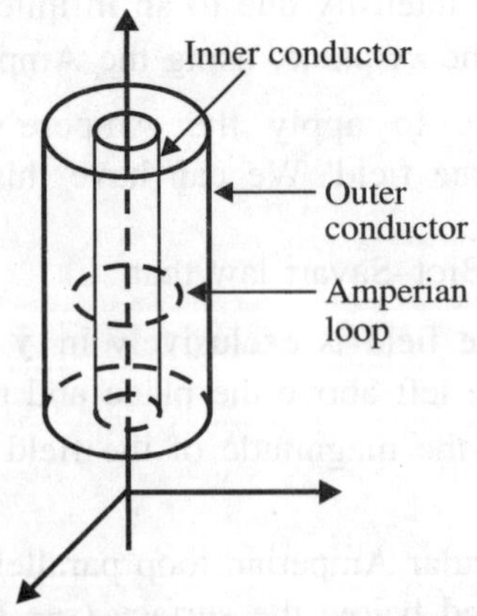

FIGURE 6.15 Magnetic field intensity by Ampere's law—coaxial cable.

By symmetry the magnitude of the field is constant around the Amperian loop of radius ρ centred on the wire.

Applying the Ampere's law, we obtain

$$\oint \mathbf{B} \cdot d\mathbf{l} = B\oint dl = B2\pi\rho = \mu_0 I_{\text{enc}} = \mu_0 I$$

or

$$B = \frac{\mu_0 I}{2\pi\rho}$$

or

$$\mathbf{B} = \frac{\mu_0 I}{2\pi\rho}\,\hat{\boldsymbol{\phi}}$$

11. Find the field intensity of a very long solenoid of n turns per unit length on a cylinder of radius R and carrying steady current I, using the Ampere's circuital law.

Solution: To be able to apply the Ampere's law, we need to have prior information regarding the field. We can have this information by using the Biot–Savart law. It can be shown that

- The field cannot have a radial component.
- The field cannot have a circumferential component.
- The field outside is zero everywhere as it does not depend on the distance from the axis and it goes to zero for large distances from the axis.

So the magnetic field of an infinite, closely wound solenoid runs parallel to the axis (see Figure 6.16).

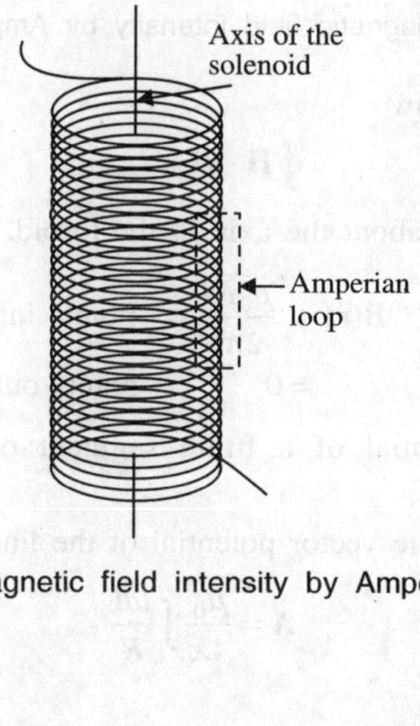

FIGURE 6.16 Magnetic field intensity by Ampere's law–solenoid.

Consider an Amperian loop normal to the current and apply the Ampere's law $\oint \mathbf{H} \cdot d\mathbf{l} = I_{enc}$. It results in

$$\mathbf{B} = \mu_0 nI\,\hat{\mathbf{z}}, \text{ inside the solenoid}$$
$$= 0, \quad \text{outside the solenoid}$$

Notice that the field is uniform inside; in this sense, the solenoid is to magnetostatics what the parallel-plate capacitor is to electrostatics.

12. Find the magnetic field intensity of a toroidal coil consisting of a circular ring or 'donut' around which a long wire is wrapped in N turns and carrying steady current I, using the Ampere's circuital law.

Solution: To be able to apply the Ampere's law, we need to have prior information regarding the field. We can have this information by using the Biot–Savart law.
The cross-sectional shape of the coil is immaterial but it requires to be uniform. It can be shown that by using the Biot–Savart law the magnetic field of the toroid is circumferential at all points, both inside and outside the coil (see Figure 6.17).

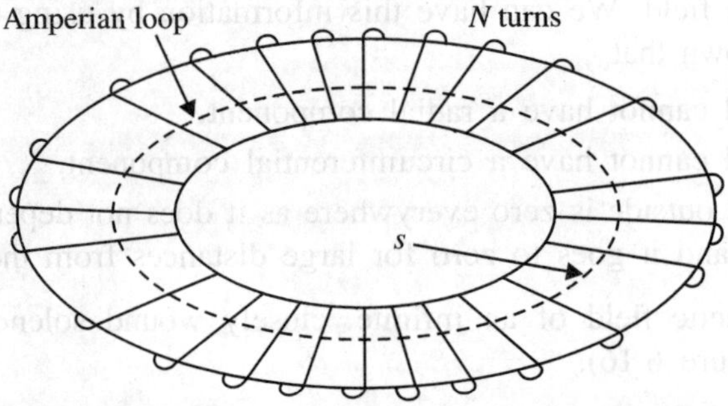

FIGURE 6.17 Magnetic field intensity by Ampere's law—toroid.

Apply the Ampere's law

$$\oint \mathbf{H} \cdot d\mathbf{l} = I_{enc}$$

to a circle of radius s about the axis of the toroid. It results in

$$\mathbf{B(r)} = \frac{\mu_0 NI}{2\pi s}\hat{\boldsymbol{\phi}}, \text{ points inside the coil}$$
$$= 0, \quad \text{points outside the coil}$$

13. Find the vector potential of a finite segment of straight wire carrying steady current I.

Solution: The magnetic vector potential of the line current distribution is given by

$$\mathbf{A} = \frac{\mu_0}{4\pi}\int \frac{I d\mathbf{l}}{R}$$

Arrange the finite length along the z-axis as shown in Figure 6.18 such that the observation point falls over the xy-plane. Let z_1 correspond to the lower end and z_2 correspond to the upper end. Consider a differential length over the conductor at a height of z giving $d\mathbf{l} = dz\hat{\mathbf{z}}$. Now,

$$\mathbf{A} = \frac{\mu_0}{4\pi}\int \frac{Id\mathbf{l}}{R} = \frac{\mu_0 I\hat{\mathbf{z}}}{4\pi}\int_{z_1}^{z_2} \frac{dz}{\sqrt{z^2+R^2}}$$

$$= \frac{\mu_0 I}{4\pi}\hat{\mathbf{z}}\left[\ln\left(z+\sqrt{z^2+R^2}\right)\right]_{z_1}^{z_2} = \frac{\mu_0 I}{4\pi}\hat{\mathbf{z}}\ln\left[\frac{z_2+\sqrt{z_2^2+R^2}}{\left(z_1+\sqrt{z_1^2+R^2}\right)}\right]$$

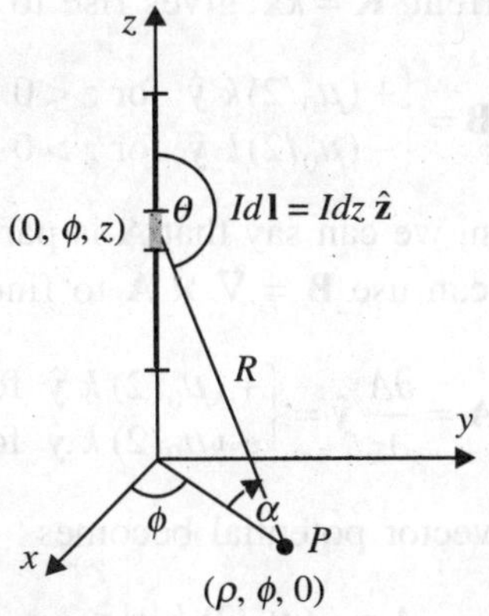

FIGURE 6.18 Vector magnetic potential due to finite length straight conductor.

14. Find the magnetic vector potential at a distance of ρ from a long straight wire carrying a steady current I.

Solution: As the source extends to infinity, we cannot apply the vector potential–source current relation to find the potential. Instead, we use the potential field relation $\mathbf{B} = \nabla \times \mathbf{A}$.

Regarding the nature of the vector potential function, we have observed certain aspects and use them in arriving at the solution. Place the conductor over the z-axis and from the defining relation, we can say that $\mathbf{A}$ is parallel to $\mathbf{I}$. Therefore, in the present case, it can have only the z-component and as the source is exhibiting cylindrical symmetry, the potential cannot be a function of ϕ. And as the source extends to infinity the potential cannot be a function of z. Combining all the aspects, we can say that the potential function consists of only one component and it is a function of only ρ, the distance from the wire. Therefore, in cylindrical coordinates, $\mathbf{A} = A(\rho)\,\hat{\mathbf{z}}$. Now we use $\mathbf{B} = \nabla \times \mathbf{A}$ to find the vector potential $\mathbf{A}$ as follows:

$$\mathbf{B} = \nabla \times \mathbf{A} = -\frac{\partial A}{\partial \rho}\hat{\phi} = \frac{\mu_0 I}{2\pi\rho}\hat{\phi} \qquad \text{(the field of an infinite wire)}$$

$$-\frac{\partial A}{\partial \rho} = \frac{\mu_0 I}{2\pi\rho}$$

$$A(\rho) = \frac{\mu_0 I}{2\pi\rho} \ln\left(\frac{\rho}{a}\right)\hat{\mathbf{z}}$$

where a is anything, including '1'.

15. Find the magnetic vector potential of an infinite uniform surface current $\mathbf{K} = k\hat{\mathbf{x}}$ flowing over the xy-plane.

Solution: Here also as the source extends to infinity, to find the potential function, we use the potential field relation $\mathbf{B} = \nabla \times \mathbf{A}$.
We know the surface current $\mathbf{K} = k\hat{\mathbf{x}}$ gives rise to

$$\mathbf{B} = \begin{cases} +(\mu_0/2)\,k\,\hat{\mathbf{y}} \text{ for } z < 0 \\ -(\mu_0/2)\,k\,\hat{\mathbf{y}} \text{ for } z > 0 \end{cases}$$

From the defining relation, we can say that $\mathbf{A}$ is parallel to $\mathbf{K}$ and depends only on z, so $\mathbf{A} = A(z)\,\hat{\mathbf{x}}$. Now we can use $\mathbf{B} = \nabla \times \mathbf{A}$ to find $\mathbf{A}$ as follows:

$$\mathbf{B} = \nabla \times \mathbf{A} = \frac{\partial A}{\partial z}\hat{\mathbf{y}} = \begin{cases} +(\mu_0/2)\,k\,\hat{\mathbf{y}} \text{ for } z < 0 \\ -(\mu_0/2)\,k\,\hat{\mathbf{y}} \text{ for } z > 0 \end{cases}$$

Therefore, the magnetic vector potential becomes

$$\mathbf{A} = -(\mu_0/2)\,k\,|z|\,\hat{\mathbf{x}}$$

16. Find the magnetic vector potential of a very long solenoid of n turns per unit length on a cylinder of radius R and carrying steady current I.

Solution: We use the relation

$$\oint_C \mathbf{A}\cdot d\mathbf{l} = \Phi = \int_s \mathbf{B}\cdot d\mathbf{a}$$

which is very much similar to the Ampere's circuital law. Considering a circular Amperian loop at radius ρ *inside* and coaxial solenoid, we have

$$\oint_C \mathbf{A}\cdot d\mathbf{l} = A(2\pi\rho) = \int_s \mathbf{B}\cdot d\mathbf{a} = \mu_0 nI\,(\pi\rho^2)$$

$$\mathbf{A} = \frac{\mu_0 nI\rho}{2}\hat{\boldsymbol{\phi}} \text{ for } \rho < R$$

For an Amperian loop *outside* the solenoid, the flux is

$$\int_s \mathbf{B}\cdot d\mathbf{a} = \mu_0 nI\,(\pi R^2)$$

since the field extends up to R only. Thus,

$$\mathbf{A} = \frac{\mu_0 nI}{2}\frac{R^2}{\rho}\hat{\boldsymbol{\phi}} \text{ for } \rho > R$$

17. Find the magnetic vector potential due to spherical shell of radius R_r, carrying a uniform surface charge σ, spinning at angular velocity ω.

Solution: Place the rotating sphere with the angular velocity vector making the angle with z-axis in the xz-plane, with field point over the z-axis. Let us suppose (r', θ', ϕ') be the coordinates of the source point. As the sphere is rotating, its surface charge moves resulting in a surface current density **K** and consequently vector potential which is given by

$$\mathbf{A(r)} = \frac{\mu_0}{4\pi}\int \frac{K\,(\mathbf{r}')\,d\mathbf{a}'}{R}$$

The surface current density **K** depends upon the linear velocity which varies in general from point to point of the spherical surface. Thus,

$$\mathbf{K} = \sigma\mathbf{v}$$

At the source point (r', θ', ϕ'), it is

$$\mathbf{v} = \boldsymbol{\omega} \times \mathbf{r}'$$

The other expressions required to evaluate the integral are

$$R = \sqrt{R_r^2 + r^2 - 2R_r r\cos\theta'}$$

and

$$da' = R_r^2 \sin\theta'\, d\theta'\, d\phi'$$

After substituting the expressions, the integral for vector potential becomes

$$\mathbf{A(r)} = \frac{\mu_0}{4\pi}\int \frac{\sigma\,(\boldsymbol{\omega}\times\mathbf{r}')\,R_r^2 \sin\theta'\, d\theta'\, d\phi'}{\sqrt{R_r^2 + r^2 - 2R_r r\cos\theta'}}$$

Evaluating the above expression results in

$$\mathbf{A(r)} = \begin{cases} \dfrac{\mu_0 R_r \sigma}{3}\,\boldsymbol{\omega}\times\mathbf{r}' & (r \le R_r) \\[2ex] \dfrac{\mu_0 R_r^4 \omega\sigma}{3r^3}\,\boldsymbol{\omega}\times\mathbf{r}' & (r \ge R_r) \end{cases}$$

The above relation for vector potential can be converted to the case of $\boldsymbol{\omega}$ aligned with the z-axis as

$$\mathbf{A}(r,\theta,\phi)=\begin{cases}\dfrac{\mu_0 R_r \omega\sigma}{3} r\sin\theta\,\hat{\boldsymbol{\phi}} & (r\le R_r)\\[2ex] \dfrac{\mu_0 R_r^4 \omega\sigma}{3}\dfrac{\sin\theta}{r^2}\hat{\boldsymbol{\phi}} & (r\ge R_r)\end{cases}$$

From the above expression, we can even evaluate the field inside:

$$\mathbf{B}=\nabla\times\mathbf{A}=\frac{2\mu_0 R_r\omega\sigma}{3}(\cos\theta\hat{\mathbf{r}}-\sin\theta\hat{\boldsymbol{\theta}})=\frac{2\mu_0 R_r\omega\sigma}{3}\hat{\mathbf{z}}$$

We can observe that it is uniform, a quite interesting result.

18. Find the dipole moment of the bookend-shaped loop with sides each of length w and carrying current I.

Solution: The wire could be considered the superposition of two plane square loops one in, let us say, the xz-plane and the other in the xy-plane. Now,

$$\mathbf{m}=Iw^2\hat{\mathbf{y}}+Iw^2\hat{\mathbf{z}}$$

Its magnitude is $\sqrt{2}Iw^2$, and it points along the 45° line $z = y$.

19. Find the magnetic field intensity of a magnetized sphere.

Solution: Choosing the z-axis along the direction of **M**, we have

$$\mathbf{J}_b=\nabla\times\mathbf{M}=0,\ \mathbf{K}_b=\mathbf{M}\times\hat{\mathbf{n}}=M\sin\theta\hat{\boldsymbol{\phi}}$$

Now a rotating spherical shell, of uniform surface charge σ, corresponds to a surface current density

$$\mathbf{K}=\sigma\mathbf{v}=\sigma\omega R_r\sin\theta\hat{\boldsymbol{\phi}}$$

It follows, therefore, that the field of uniformly magnetized sphere is identical to the field of a spinning spherical shell, with the identification of $\sigma\omega R_r\to\mathbf{M}$. Therefore,

$$\mathbf{B}=\frac{2}{3}\mu_0\mathbf{M}\ \text{ inside the sphere,}$$

whereas the field outside is the same as that of a pure dipole with a dipole moment of $\mathbf{m}=\dfrac{4}{3}\pi R_r^3\mathbf{M}$.

20. Find the magnetic field inside an infinite solenoid filled with linear material of susceptibility χ_m.

Solution: Using the Ampere's law, we can find out the magnetic field intensity

$$\mathbf{H}=nI\hat{\mathbf{z}}.$$

Then the magnetic flux density can be computed using

$$\mathbf{B} = \mu_0\,(1 + \chi_m)\,\mathbf{H} = \mu_0\,(1 + \chi_m)\,nI\hat{\mathbf{z}}$$

If the medium is paramagnetic $\chi_m > 0$, the field is slightly enhanced; if it is diamagnetic $\chi_m < 0$, the field is somewhat reduced.

21. Find the magnetic field inside a very long coaxial cable with linear magnetic material of χ_m when current I flows. Also find the same by using the magnetization and bound currents.

Solution: Using the Ampere's law, we can find out the magnetic field intensity

$$\oint \mathbf{H}\cdot d\mathbf{l} = I_{\text{enc}} = I;$$

$$\mathbf{H} = \frac{I}{2\pi\rho}\hat{\boldsymbol{\phi}}$$

The magnetic flux density and magnetization then become

$$\mathbf{B} = \mu_0\,(1 + \chi_m)\,\mathbf{H} = \frac{\mu_0\,(1 + \chi_m)\,I}{2\pi\rho}\hat{\boldsymbol{\phi}}$$

$$\mathbf{M} = \chi_m\mathbf{H} = \frac{\chi_m I}{2\pi\rho}\hat{\boldsymbol{\phi}}$$

Now we try to compute the bound current densities and magnetization as

$$\mathbf{J}_b = \nabla\times\mathbf{M} = \frac{1}{\rho}\frac{\partial}{\partial\rho}\left(\rho\frac{I\chi_m}{2\pi\rho}\right)\hat{\mathbf{z}} = 0$$

$$\mathbf{K}_b = \mathbf{M}\times\hat{\mathbf{n}} = \begin{cases} \dfrac{\chi_m I}{2\pi a}\hat{\mathbf{z}} & \text{at } \rho = a \\ -\dfrac{\chi_m I}{2\pi b}\hat{\mathbf{z}} & \text{at } \rho = b \end{cases}$$

Here a and b are radii of the inner and outer conductors, respectively.
The total current, for the Amperian loop between the cylinders, is

$$I + \frac{\chi_m I}{2\pi a}2\pi a = (1 + \chi_m)\,I$$

$$\oint \mathbf{B}\cdot d\mathbf{l} = \mu_0 I_{\text{enc}} = \mu_0(1 + \chi_m)\,I$$

$$\mathbf{B} = \frac{\mu_0(1 + \chi_m)I}{2\pi\rho}\hat{\boldsymbol{\phi}}$$

22. Find the magnetic field inside a sphere of linear magnetic material placed in uniform magnetic field $\mathbf{B}_0$.

Solution:

Step I: Let us start the solution procedure by moving the initially unmagnetized sphere into the field $\mathbf{B}_0$. The field $\mathbf{B}_0$ magnetizes the sphere which is given by

$$\mathbf{M}_0 = \chi_m \mathbf{H}_0 = \frac{\chi_m}{\mu_0(1+\chi_m)} \mathbf{B}_0 = \frac{k}{\mu_0} \mathbf{B}_0$$

This magnetization sets up a field within the sphere

$$\mathbf{B}_1 = \frac{2}{3} \mu_0 \mathbf{M}_0 = \frac{2}{3} k\mathbf{B}_0$$

Step II: The field $\mathbf{B}_1$ magnetizes the sphere an additional amount:

$$\mathbf{M}_1 = \frac{k}{\mu_0} \mathbf{B}_1$$

This magnetization sets up an additional field within the sphere

$$\mathbf{B}_2 = \frac{2}{3} \mu_0 \mathbf{M}_1 = \frac{2}{3} k\mathbf{B}_1 = \left(\frac{2}{3}k\right)^2 \mathbf{B}_0, \text{ etc.}$$

In a similar manner, we can evaluate the fields $\mathbf{B}_3 + \mathbf{B}_4 + \mathbf{B}_5 + \ldots$. The total resulting field then is

$$\mathbf{B} = \mathbf{B}_0 + \mathbf{B}_1 + \mathbf{B}_2 + \cdots = \left[1 + \frac{2}{3}k + \left(\frac{2}{3}k\right)^2 + \cdots\right]\mathbf{B}_0 = \left(\frac{1}{1 - \frac{2k}{3}}\right)\mathbf{B}_0 = \frac{3(1+\chi_m)}{3+\chi_m} \mathbf{B}_0$$

23. Find the magnetic energy stored in a long coaxial cable of length l with inner cylinder radius a and outer cylinder radius b carrying current I.

Solution: According to the Ampere's law, the field between the cylinders is

$$\mathbf{B} = \frac{\mu_0 I}{2\pi\rho} \hat{\boldsymbol{\phi}}$$

And note that elsewhere the field is zero. Thus, the energy per unit volume is

$$\frac{B^2}{2\mu_0} = \frac{1}{2\mu_0}\left(\frac{\mu_0 I}{2\pi\rho}\right)^2$$

The energy in a cylinder shell of length l of radius ρ carrying thickness $d\rho$ then is

$$\frac{1}{2\mu_0}\left(\frac{\mu_0 I}{2\pi\rho}\right)^2 2\pi l \rho d\rho = \frac{\mu_0 I^2 l}{4\pi} \frac{d\rho}{\rho}$$

Integrating from a to b, we have

$$W = \frac{\mu_0 I^2 l}{4\pi} \ln \frac{b}{a}$$

24. Find the magnetic energy stored in a long solenoid of length l of radius R carrying n turns per unit length and carrying current I.

Solution: The energy can be found using either one of the following:

$$W = \frac{\mu_0}{2} \int_{\text{all space}} H^2 d\tau \text{ or } W = \frac{1}{2} LI^2 \text{ or } W = \frac{1}{2} \oint_c (\mathbf{A} \cdot \mathbf{I})\, dl$$

- The flux density inside the solenoid is $B = \mu_0 nI$ and it is zero outside. Therefore, we need to perform integration only inside the solenoid.

$$W = \frac{1}{2\mu_0} \int_{\text{all space}} B^2 d\tau = \frac{(\mu_0 nI)^2}{2\mu_0} \int d\tau$$

$$= \frac{(\mu_0 nI)^2}{2\mu_0} l\pi R^2 = \frac{\mu_0 \pi R^2 l (nI)^2}{2}$$

- If we have an expression for self-inductance like

$$L = \mu_0 n^2 \pi R^2$$

available, then the energy can be computed using

$$W = \frac{1}{2} LI^2$$

Therefore,

$$W = \frac{1}{2} LI^2 = \frac{1}{2} \mu_0 n^2 I^2 \pi R^2$$

- With the availability of vector potential, it is possible to compute the energy. We know the vector potential over the surface of the solenoid is

$$\mathbf{A} = \frac{\mu_0 nIR}{2} \hat{\phi}$$

Therefore,

$$W_1 = \frac{1}{2} \frac{\mu_0 nI}{2} RI \cdot 2\pi R$$

for one turn. As there are nl turns over the solenoid,

$$W = \frac{\mu_0 \pi R^2 n^2 l I^2}{2}$$

25. Find the magnetic energy stored in the toroidal coil with a rectangular cross section with inner radius a, outer radius b, and height h which carries a total of N turns carrying steady current I.

Solution: In solving this problem, we use the standard relation for energy stored in the magnetic field

$$W = \frac{\mu_0}{2} \int_{\text{all space}} H^2 \, d\tau$$

In this case, we need to integrate only inside the toroid because the field outside is zero. The magnetic field intensity inside the toroidal coil is

$$H = \frac{NI}{2\pi s}$$

Substituting the expression for field intensity in the relation for energy, we get

$$W = \frac{\mu_0}{2} \int_{\text{all space}} H^2 d\tau = \frac{(NI)^2 \mu_0}{2(2\pi)^2} \int_a^b \frac{1}{s^2} shd\phi ds$$

$$= \frac{\mu_0 (NI)^2}{4\pi} h \ln\left(\frac{b}{a}\right)$$

As a by-product, it is possible to find the self-inductance. From

$$W = \frac{1}{2} LI^2$$

$$L = \frac{2W}{I^2} = \frac{\mu_0 N^2 h}{2\pi} \ln\left(\frac{b}{a}\right)$$

26. Find the self-inductance of a long coaxial cable of length l with inner cylinder radius a and outer cylinder radius b carrying current I.

Solution: We have already calculated the magnetic flux density inside the coaxial cable as

$$B = \frac{\mu_0 I}{2\pi\rho} \text{ or } \mathbf{B} = \frac{\mu_0 I}{2\pi\rho} \hat{\phi}$$

Consider a small rectangular loop parallel to the length of the cable and lying entirely in the dielectric. Let us suppose its length is unity. Now we calculate the flux through the loop as follows:

$$\Phi = \int_{\text{loop}} \mathbf{B} \cdot d\mathbf{a} = \int_1 \int_a^b \frac{\mu_0 I}{2\pi\rho} d\,\rho dz = \frac{\mu_0 I}{2\pi} \ln \frac{b}{a}$$

This is the flux linking the current I per unit length of the cable. Then we can calculate the self-inductance from its basic definition.

$$L = \frac{\Phi}{I} = \frac{\mu_0}{2\pi} \ln \frac{b}{a}$$

For a cable of length l, the self-inductance becomes

$$= \frac{\mu_0 l}{2\pi} \ln \frac{b}{a}$$

Another method is with the use of expression for magnetic energy. We have already calculated the magnetic energy stored in the coaxial cable as

$$W = \frac{\mu_0 I^2 l}{4\pi} \ln \frac{b}{a}$$

But it has to be equal to (1/2) LI^2. Therefore,

$$W = \frac{\mu_0 I^2 l}{4\pi} \ln \frac{b}{a} = \frac{1}{2} LI^2 \Rightarrow L = \frac{\mu_0 l}{2\pi} \ln \frac{b}{a}$$

27. Find the self-inductance per unit length of a long solenoid of radius R carrying n turns per unit length.

Solution: The flux density inside the solenoid is

$$B = \mu_0 n I$$

Then the flux through a single turn is

$$\Phi_1 = B\pi R^2 = \mu_0 n I \pi R^2$$

As there are n turns in a unit length, the flux through all the turns in the unit length becomes

$$\Phi = \Phi_1 n = nB\pi R^2 = \mu_0 n^2 I \pi R^2$$

So the self-inductance per unit length is

$$L = \frac{\Phi}{I} = \mu_0 n^2 \pi R^2$$

28. Find the self-inductance of a toroidal coil with rectangular cross section with inner radius a outer radius b, and height h which carries a total of N turns carrying steady current I.

Solution: Apply the Ampere's law $\oint \mathbf{H} \cdot d\mathbf{l} = I_{\text{enc}}$ to a circle of radius s about the axis of the toroid to find the magnetic field inside the coil. Thus,

$$\mathbf{B}(\mathbf{r}) = \frac{\mu_0 N I}{2\pi s} \hat{\boldsymbol{\phi}} \text{ point inside the coil}$$

Note that the flux density is not uniform over the cross section and it is decreasing with radius. Therefore, to get the total flux, simple multiplication with area is adequate and integration is required. The flux through a single turn, then, is

$$\Phi_1 = \int \mathbf{B} \cdot d\mathbf{a} = \frac{\mu_0 NI}{2\pi} h \int_a^b \frac{1}{s}\, ds = \frac{\mu_0 NIh}{2\pi} \ln\left(\frac{b}{a}\right)$$

The total flux is N times the flux through a single turn, so the self-inductance becomes

$$L = \frac{N\Phi_1}{I} = \frac{\mu_0 N^2 h}{2\pi} \ln\left(\frac{b}{a}\right)$$

29. Derive a relation for the mutual inductance between a short solenoid (length l and radius a with n_1 turns per unit length) which lies coaxially within a long solenoid (radius b with n_2 turns per unit length) when current I flows in the short solenoid.

Solution: In solving this problem easily, we use a property of the mutual inductance: it remains the same whether the current is in the long or short solenoid. So let us run the current through the long solenoid and calculate the flux through the inner one. The flux density inside the long solenoid is constant and given by

$$B = \mu_0 n_2 I$$

So the flux through a single loop of the short solenoid is

$$\Phi_1 = B\pi a^2 = \mu_0 n_2 I \pi a^2$$

As there are $n_1 l$ turns in all, so the total flux through the inner solenoid becomes

$$\Phi = \Phi_1 \times n_1 l = \pi a^2 \mu_0 n_1 n_2 Il$$

This is also the flux of a current I in the short solenoid which would put through the long one. Now the mutual inductance is

$$M = \frac{\Phi}{I} = \pi a^2 \mu_0 n_1 n_2 l$$

30. A cast steel parallel magnetic circuit has a coil of 400 turns and $l_1 = 5$ cm, $l_2 = l_3 = 15$ cm. Find the coil current if $\phi_3 = 175$ μWb. The cross-sectional area of each of the side legs is 1.5 cm² and that of the centre leg is equal to double of the side legs (see Figure 6.19).

Solution: From the circuit, $\phi_1 = \phi_2 + \phi_3$ and $\phi_2 = \phi_3$. It implies that $\phi_1 = 2\phi_2 = 2\phi_3$. If A_1 is the cross-sectional area of the side legs, then $A_1 = 2A_2 = 2A_3$ and we get the magnetic flux densities as

$$B_1 = \frac{\phi_1}{A_1}$$

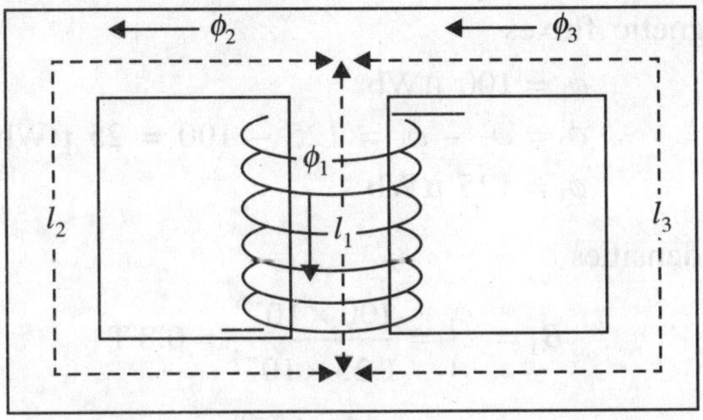

FIGURE 6.19 Magnetic circuit of Problem 30.

$$B_2 = \frac{\phi_2}{A_2} = \frac{\phi_1/2}{A_1/2} = \frac{\phi_1}{A_1}$$

$$B_3 = \frac{\phi_3}{A_3} = \frac{\phi_1/2}{A_1/2} = \frac{\phi_1}{A_1}$$

which are all equal. Their values can be computed as

$$B_1 = B_2 = B_3 = \frac{\phi_1}{A_1} = \frac{0.175\times10^{-3}}{1.5\times10^{-4}} = 1.16\text{ T}$$

From the *B–H* curve, the corresponding value of the field intensity is

$$H = 1040 \text{ A/m}$$

$$F = 1040 \times 0.19 = 197.6 \text{ A}$$

$$I = \frac{F}{N} = \frac{197.6}{400} = 0.494 \text{ A}$$

31. In the parallel magnetic circuit with silicon steel core shown, ϕ_1 = 100 μWb and ϕ_3 = 125 μWb, $l_1 = l_3$ = 20 cm, l_2 = 10 cm, coils have 60 turns each. The cross-sectional area is A = 1.25 cm^2 throughout. Find fluxes, flux densities, field intensities, permeabilities and reluctances of each leg. Also find the currents into each of the coils (see Figure 6.20).

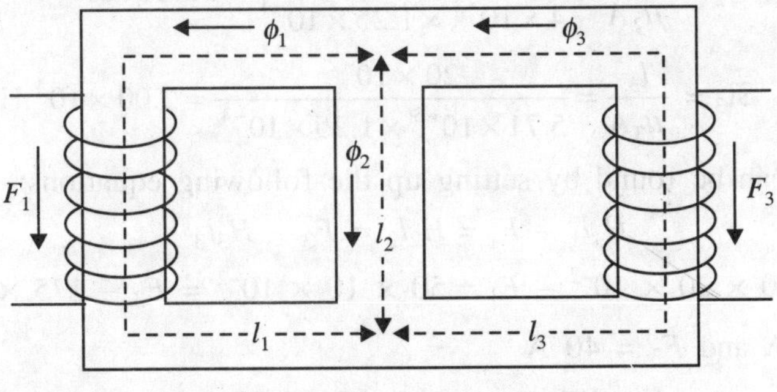

FIGURE 6.20 Parallel magnetic circuit pertaining to Problem 31.

Solution: Magnetic fluxes

$$\phi_1 = 100\ \mu\text{Wb}$$
$$\phi_2 = \phi_3 - \phi_1 = 125 - 100 = 25\ \mu\text{Wb}$$
$$\phi_3 = 125\ \mu\text{Wb}$$

Magnetic flux densities

$$B_1 = \frac{\phi_1}{A} = \frac{100 \times 10^{-6}}{1.25 \times 10^{-4}} = 0.8\ \text{T}$$

$$B_2 = \frac{\phi_2}{A} = \frac{25 \times 10^{-6}}{1.25 \times 10^{-4}} = 0.2\ \text{T}$$

$$B_3 = \frac{\phi_3}{A} = \frac{125 \times 10^{-6}}{1.25 \times 10^{-4}} = 1\ \text{T}$$

From the B–H curve, the field intensity values corresponding to the flux densities are

$$H_1 = 110\ \text{A/m}$$
$$H_2 = 50\ \text{A/m}$$
$$H_3 = 175\ \text{A/m}$$

Permeabilities are

$$\mu_1 = \frac{B_1}{H_1} = \frac{0.8}{110} = 7.27 \times 10^{-3}\ \text{H/m}$$

$$\mu_2 = \frac{B_2}{H_2} = \frac{0.2}{50} = 4 \times 10^{-3}\ \text{H/m}$$

$$\mu_3 = \frac{B_3}{H_3} = \frac{1}{175} = 5.71 \times 10^{-3}\ \text{H/m}$$

Reluctances are

$$\Re_1 = \frac{l_1}{\mu_1 A} = \frac{2 \times 10^{-2}}{7.25 \times 10^{-3} \times 1.25 \times 10^{-4}} = 2.20 \times 10^5\ \text{H}^{-1}$$

$$\Re_2 = \frac{l_2}{\mu_2 A} = \frac{10 \times 10^{-2}}{4 \times 10^{-3} \times 1.25 \times 10^{-4}} = 1 \times 10^5\ \text{H}^{-1}$$

$$\Re_3 = \frac{l_3}{\mu_3 A} = \frac{20 \times 10^{-2}}{5.71 \times 10^{-3} \times 1.25 \times 10^{-4}} = 2.00 \times 10^5\ \text{H}^{-1}$$

Currents can be found by setting up the following equations:

$$H_1 l_1 - F_1 = H_2 l_2 = F_3 - H_3 l_3$$

$$110 \times 20 \times 10^{-2} - F_1 = 50 \times 10 \times 10^{-2} = F_3 - 175 \times 20 \times 10^{-2}$$

$F_1 = 17$ A and $F_3 = 40$ A

$$I_1 = \frac{F_1}{N_1} = \frac{17}{60} = 0.28\ \text{A} \quad \text{and} \quad I_3 = \frac{F_3}{N_3} = \frac{40}{60} = 0.67\ \text{A}$$

EXERCISE SECTION

Essay Type Questions

1. Describe the various current distributions. Derive the equivalence of various current elements.
2. State and explain the Biot–Savart law and derive the expressions for magnetic field intensity due to surface and volume currents.
3. Derive an expression for the magnetic field intensity of a finite length straight line conductor carrying a line current amperes. Make necessary assumptions.
4. What is magnetic flux density? Describe its properties giving units.
5. What is the Lorentz's force law? Differentiate the electric force given by this law from that given by the Coulomb's law. Extend it to various current distributions.
6. State and explain the Ampere's force law for current elements and loops. Does this law obey the Newton's third law?
7. How is vector potential defined in steady magnetic fields? Give relations for vector potential for all the three types of current distributions.
8. Define scalar magnetic potential. Mention its utility and list its salient features.
9. State and explain the Ampere's circuital law for steady currents. Mention its applications and limitations.
10. Derive an expression for the energy density of the steady magnetic fields.
11. Give a classification of magnetic materials. Describe them listing their important feaures.
12. Show that the field of a magnetized body is equivalent to the field given by a volume current and a surface current.
13. Define the terms mutual inductance and self-inductance. What is the Neumann formula for mutual inductance?
14. Write short notes on ,
    (a) linear magnetic media
    (b) reluctance

Numerical Problems

1. There are two line currents, one $10\hat{\mathbf{x}}$ A over the x-axis and the other $-5\hat{\mathbf{y}}$ A over the y-axis. Find the magnetic field intensity at (4, –3, ,2).
2. A conductor filament is from the point $P_1(1, 0, 0)$ to point $P_2(0, 1, 0)$ carrying a current of 10 amps from P_1 to P_2. Find the field intensity at the origin.
3. A current strip of 2 cm wide carries a current of 15 A in x-direction. Find the force on the strip per unit length if the uniform field is $\mathbf{B} = 0.21\,\hat{\mathbf{y}}$.

4. A circular loop of 3 units radius is centred at the origin in $z = 0$ plane and carries a dc current of 10 mA along the ϕ-direction. Find the magnetic flux density at (0, 0, ±4).
5. In a conducting medium, $\mathbf{H} = y^2 z\hat{\mathbf{x}} + 2(x+1)\, yz\hat{\mathbf{y}} - (x+1)\, z^2\hat{\mathbf{z}}$ A/m. Find the current density at (1, 0, –3) and calculate the current passing through $y = 1$ plane, $0 \le x \le 1$ and $0 \le z \le 1$.
6. A vector potential is given as
$$\mathbf{A} = (2x^2 y + yz)\hat{\mathbf{x}} + (2xy^2 - xz)\hat{\mathbf{y}} - (6xyz - 2x^2 y^2)\hat{\mathbf{z}} \text{ Wb/m.}$$
Calculate (a) magnetic flux density, (b) magnetic flux through a loop described by $x < 1$, $0 < y$, $z < 2$ and (c) show that and $\nabla \cdot \mathbf{A} = 0$ and $\nabla \cdot \mathbf{B} = 0$..
7. In a certain material with $\mu = 6.5\mu_0$, the magnetic field intensity is given by $\mathbf{H} = 15\hat{\mathbf{x}} + 20\hat{\mathbf{y}} - 30\hat{\mathbf{z}}$ A/m. Calculate (a) magnetic susceptibility of the material, (b) magnetic flux density, (c) magnetization and (d) magnetic energy density.
8. Region 1, for which $\mu_{r1} = 3$, is defined by $x < 0$ and region 2, $x > 0$, has $\mu_{r2} = 5$ given $\mathbf{H} = 4\hat{\mathbf{x}} + 3\hat{\mathbf{y}} - 6\hat{\mathbf{z}}$ A/m. Determine H_2 for $x > 0$ and the angles that H_1 and H_2 make with the interface.
9. A straight long wire is situated parallel to one side of a square coil. Each side of the coil has a length of 10 cm. the distance between the straight wire and the centre of the coil is 20 cm. Find the mutual inductance of the system. Derive the formula used.

Fill in the Blanks

1. Biot–Savart law is about magnetic field __________.
2. Magnetic field intensity is proportional to the product of current with __________ and inversely proportional to __________.
3. Magnetic field intensity is independent of __________.
4. The proportionality constant in Biot–Savart law is __________.
5. For an infinite straight line current, the direction of the field is __________.
6. The field of an infinite straight line current is __________.
7. The equivalence of differential current elements is __________.
8. The field of a finite straight line current is __________.
9. Amperian circuital law in integral form is __________.
10. Amperian circuital law in differential form is __________.
11. The field inside a long solenoid is __________, whereas outside it is __________.
12. The field inside a long toroid is __________, whereas outside it is __________.
13. The closed integral $\oint_c \mathbf{H} \cdot d\mathbf{l}$ is called __________ of the magnetic field.
14. The unit of magnetic flux density is __________.
15. Permeability of free space is __________.

16. The unit of magnetic flux is __________.
17. Relation between scalar magnetic potential and field intensity is __________.
18. Vector magnetic potential is related to flux density through __________.
19. Expression for vector magnetic potential in terms of current is __________.
20. In paramagnets, the magnetization is __________ to the magnetic field.
21. In diamagnets, the magnetization is __________ to the magnetic field.
22. Retention of magnetism is an important property of __________.
23. In linear magnetic media, bound volume current density is proportional to __________.
24. In linear magnetic media, magnetization is proportional to __________.
25. Expression for the stored energy in magnetic fields is __________.
26. Self-inductance is the ratio of total flux linkages to __________.
27. Mutual inductance M_{ij} is the ratio of flux linkages through the ith coil due to current in the jth coil to __________.
28. Neumann's formula for mutual inductance between two filamentary conductor loops is given by __________.
29. Reluctance is defined as the ratio of magnetomotive force to __________.

Answers

1. intensity
2. differential length, distance square
3. independent
4. $1/4\pi$
5. curling around the conductor
6. $(I/2\pi\rho)\hat{\phi}$
7. $\int_{\text{line}} (\,)\,\mathbf{I}dl \approx \int_{\text{sur}} (\,)\,\mathbf{K}da \approx \int_{\text{vol}} (\,)\,\mathbf{J}d\tau$
8. $(1/4\pi\rho)\,(\sin\alpha_2 - \sin\alpha_1)\,\hat{\phi}$
9. $\oint \mathbf{H}\cdot d\mathbf{l} = I_{\text{enc}}$
10. $\nabla \times \mathbf{H} = \mathbf{J}_{\text{enc}}$
11. uniform, zero
12. non-uniform, zero
13. circulation
14. Wb/m^2 or Tesla
15. $\mu_0 = 4\pi \times 10^{-7}$ H/m
16. Webers
17. $\mathbf{H} = -\nabla\psi\ (\mathbf{J} = 0)$
18. $\nabla \times \mathbf{A} = \mathbf{B}$
19. $\mathbf{A}(\mathbf{r}) = \dfrac{\mu_0}{4\pi}\int \dfrac{I(\mathbf{r}')\,d\mathbf{l}'}{R}$
20. parallel
21. opposite
22. ferromagnetism
23. free volume current density
24. magnetic field intensity
25. $W = \dfrac{\mu_0}{2}\int_{\text{all space}} H^2\,d\tau$
26. linking current

27. current in jth coil

28. $M_{12} = \frac{\mu}{4\pi} \oint_{c_1} \oint_{c_2} \frac{d\mathbf{l}_1 \cdot d\mathbf{l}_2}{R}$

29. magnetic flux

Match the Following

Set I

1. Lorentz force law	(a) $Q(\mathbf{v} \times \mathbf{B})$
2. Line current	(b) $\int \mathbf{J} \times \mathbf{B}\, d\tau$
3. Surface current	(c) $K \int d\mathbf{a} \times \mathbf{B}$
4. Volume current	(d) $I \int d\mathbf{l} \times \mathbf{B}$
5. Point charge	(e) electric plus magnetic

Set II

1. Lorentz force law	(a) $\frac{\mu_0 I_1 I_2}{4\pi} \oint_{c_1} \oint_{c_2} \frac{d\mathbf{l}_1 \times (d\mathbf{l}_2 \times \hat{\mathbf{r}}_{12})}{r^2}$
2. Ampere's force law	(b) force on moving point charge
3. Ampere's circuital law	(c) magnetomotive force
4. Biot–Savart law	(d) $\mu\mathbf{H}$
5. Magnetic flux density	(e) magnetic field intensity

Set III

1. $\nabla \cdot \mathbf{A}$	(a) $\mathbf{J}$
2. $\nabla \times \mathbf{A}$	(b) $-\mu_0 \mathbf{J}$
3. $\oint_c \mathbf{A} \cdot d\mathbf{l}$	(c) $\mathbf{\Phi}$
4. $\nabla^2 \mathbf{A}$	(d) $\mathbf{B}$
5. $\nabla \times \mathbf{H}$	(e) 0

Set IV

1. Torque	(a) $\nabla(\mathbf{m} \cdot \mathbf{B})$
2. Force	(b) $\mathbf{m} \times \mathbf{B}$
3. $\nabla \cdot \mathbf{J}_b$	(c) 0
4. $\mathbf{K}_b$	(d) $\nabla \times \mathbf{M}$
5. $\mathbf{J}_b$	(e) $\mathbf{M} \times \hat{\mathbf{n}}$

Set V

1. Self-inductance	(a) negative susceptibility
2. Mutual inductance	(b) positive susceptibility
3. Neumann's formula	(c) $\frac{\mu}{4\pi}\oint_{c_1}\oint_{c_2}\frac{d\mathbf{l}_1 \cdot d\mathbf{l}_2}{R}$
4. Paramagnetic	(d) flux linkage in one coil due to current in the other coil.
5. Diamagnetic	(e) total flux linkage to the current

Set VI

1. Magnetic force	(a) $\frac{\mu_0}{2}\int_{\text{all space}} H^2 d\tau$
2. Magnetic energy	(b) no work
3. Net bound current	(c) 0
4. Field of solenoid	(d) non-uniform
5. Field of toroid	(e) uniform

Set VII

1. Lorentz force law	(a) $\psi_{ba} = -\int_a^b \mathbf{H} \cdot d\mathbf{l}$
2. Biot–Savart law	(b) $\int \mathbf{J} \times \mathbf{B}\, d\tau$
3. Vector potential	(c) $\nabla \times \mathbf{H} = \mathbf{J}_{\text{enc}}$
4. Scalar potential	(d) $\frac{\mu_0}{4\pi}\int \frac{\mathbf{J}(\mathbf{r}')\, d\tau'}{R}$
5. Ampere's law	(e) $\frac{1}{4\pi}\int \frac{\mathbf{J}(\mathbf{r}') \times \hat{\mathbf{R}}}{R^2}\, d\tau'$

Answers

Set I:	1. (e)	2. (d)	3. (c)	4. (b)	5. (a)
Set II:	1. (b)	2. (a)	3. (c)	4. (e)	5. (d)
Set III:	1. (e)	2. (d)	3. (c)	4. (b)	5. (a)
Set IV:	1. (b)	2. (a)	3. (c)	4. (e)	5. (d)
Set V:	1. (e)	2. (d)	3. (c)	4. (b)	5. (a)
Set VI:	1. (b)	2. (a)	3. (c)	4. (e)	5. (d)
Set VII:	1. (b)	2. (e)	3. (d)	4. (a)	5. (c)

Multiple Choice Questions

1. The line I current in amperes can be

   (a) $\dfrac{dQ}{dt}$ (b) $\int \mathbf{K} dw_\perp$

   (c) $\int \mathbf{J} da_\perp$ (d) all of these

2. The surface current density $\mathbf{K}$ A/m is

   (a) $\dfrac{d\mathbf{I}}{dw_\perp}$ (b) $\dfrac{d\mathbf{I}}{da_\perp}$

   (c) both (a) and (b) (d) none of these

3. The volume current density $\mathbf{J}$ A/m^2 is

   (a) $\dfrac{d\mathbf{I}}{dw_\perp}$ (b) $\dfrac{d\mathbf{I}}{da_\perp}$

   (c) both (a) and (b) (d) none of these

4. The expression which is not a current element is

   (a) $\mathbf{I}dl$ (b) $\mathbf{K}da$

   (c) $\mathbf{J}d\tau$ (d) $\mathbf{I}da$

5. The charge in a current element can be

   (a) $\dfrac{I}{v} dl$ (b) $\dfrac{K}{v} dw$

   (c) $\dfrac{J}{v} da$ (d) all of these

6. The current that obeys the Ohm's law is

   (a) conduction current (b) convection current

   (c) both (a) and (b) (d) none of these

7. The continuity equation for static currents is

   (a) $\nabla \times \mathbf{J} = 0$ (b) $\oint \mathbf{J} \cdot d\mathbf{a} = 0$

   (c) both (a) and (b) (d) none of these

8. According to the Biot–Savart law, the electric field intensity is

   (a) proportional to current element

   (b) inversely proportional to the square of the distance

   (c) normal to the current element and the vector joining the element with field point

   (d) none of these

9. The magnetic flux density $\mathbf{B}$ and field intensity $\mathbf{H}$ are related through

   (a) $\mathbf{B} = \mu\mathbf{H}$ (b) $\mathbf{H} = \mu\mathbf{B}$

   (c) $\mathbf{B} = \varepsilon\mathbf{H}$ (d) none of these

10. The unit of magnetic flux density **B** is
    (a) webers/sq. metre (b) tesla
    (c) both (a) and (b) (d) none of these
11. The magnetic flux ϕ and its density **B** are related through
    (a) $\phi = \int \mathbf{B} \cdot d\mathbf{a}$ (b) . $\phi = \int B d\tau$
    (c) $\phi = \int B dl$ (d) none of these
12. The magnetic flux ϕ in general is
    (a) curls around the current (b) diverges from the current
    (c) both (a) and (b) (d) none of these
13. The permeability of free space in H/m is
    (a) $4\pi \times 10^{-7}$ (b) $10^{-7}/4\pi$
    (c) $10^{-7}/36\pi$ (d) none of these
14. The Lorentz's force law is given for
    (a) moving point charge (b) current carrying conductors
    (c) both (a) and (b) (d) none of these
15. To find the magnetic force on a current-carrying conductor, we need
    (a) Lorentz force law (b) superposition law
    (c) both (a) and (b) (d) none of these
16. The electric force on the moving point charge is given by
    (a) Coulomb's law (b) Lorentz force law
    (c) both (a) and (b) (d) none of these
17. The Ampere's force law is an inclusion of
    (a) Lorentz force law (b) Biot–Savart law
    (c) both (a) and (b) (d) none of these
18. The Ampere's force law for current elements
    (a) obeys the Newton's third law (b) need not obey the Newton's third law
    (c) includes electric force (d) none of these
19. The magnetic force does
    (a) work (b) no work
    (c) work in special circumstances (d) none of these
20. The force on a current-carrying conductor placed normally in magnetic field is
    (a) zero (b) *Blv*
    (c) cannot say (d) none of these
21. The magnetomotive force is
    (a) $\mathbf{E} \cdot d\mathbf{l}$ (b) $\mathbf{H} \cdot d\mathbf{l}$
    (c) both (a) and (b) (d) none of these

22. The Ampere's circuital law is valid only for
    (a) ac (b) dc
    (c) both (a) and (b) (d) none of these
23. The Ampere's circuital law in integral form is
    (a) $\oint \mathbf{H} \cdot d\mathbf{l} = I_{enc}$ (b) $\oint \mathbf{H} \cdot d\mathbf{l} = J_{enc}$
    (c) $\oint \mathbf{H} \cdot d\mathbf{l} = K_{enc}$ (d) none of these
24. The Ampere's circuital law in differential form is
    (a) $\nabla \times \mathbf{H} = \mathbf{I}_{enc}$ (b) $\nabla \times \mathbf{H} = \mathbf{J}_{enc}$
    (c) $\nabla \times \mathbf{H} = \mathbf{K}_{enc}$ (d) none of these
25. The Ampere's circuital law is valid
    (a) always
    (b) only under highly symmetrical conditions
    (c) only under some special conditions
    (d) none of these
26. The Ampere's circuital law is useful to find field intensity **H**
    (a) always
    (b) only under highly symmetrical conditions
    (c) only under some special conditions
    (d) none of these
27. Amperian loop is
    (a) always circular (b) a closed path
    (c) both (a) and (b) (d) none of these
28. A drawback of the application of the Ampere's circuital law is
    (a) prior information of the field is required
    (b) source distribution must be highly symmetrical
    (c) both (a) and (b)
    (d) none of these
29. The chief advantage of the Ampere's circuital law is
    (a) finding field intensity quickly
    (b) finding field intensity easily
    (c) both (a) and (b)
    (d) none of these
30. In paramagnets, the alignment of dipoles is
    (a) opposite to the applied field **B**
    (b) in the direction of the applied field **B**
    (c) perpendicular to the applied field **B**
    (d) none of these

31. In diamagnets, the alignment of dipoles is
    (a) opposite to the applied field **B**
    (b) in the direction of the applied field **B**
    (c) perpendicular to the applied field **B**
    (d) none of these
32. Magnetization retention is a characteristic of
    (a) paramagnets (b) diamagnets
    (c) ferro magnets (d) none of these
33. The field of the magnetized object is equal to
    (a) the field due to volume current density $\mathbf{J}_b = \nabla \times \mathbf{M}$
    (b) the field due to surface current density $\mathbf{K}_b = \mathbf{M} \times \hat{\mathbf{n}}$
    (c) the sum of (a) and (b)
    (d) none of these
34. In the case of the magnetized object, the Ampere's law is
    (a) $\mathbf{J}_b = \nabla \times \mathbf{H}$ (b) $\mathbf{J}_f = \nabla \times \mathbf{H}$
    (c) $\mathbf{J} = \nabla \times \mathbf{H}$ (d) none of these
35. In the case of linear medium, the magnetization **M** is
    (a) proportional to field intensity (b) $\mathrm{M} = \mu_0 \chi_m \mathbf{H}$
    (c) both (a) and (b) (d) none of these
36. The permeability and susceptibility are related through
    (a) $\mu = \mu_0(1 + \chi_m)$ (b) $\mu_r = (1 + \chi_m)$
    (c) both (a) and (b) (d) none of these
37. The self-inductance L is the ratio of
    (a) total flux linkages to current
    (b) current to flux linkages
    (c) magnetomotive force to magnetic flux
    (d) none of these
38. The mutual inductance M is
    (a) the ratio of flux linking the ith coil due to current in the jth coil to current in the jth coil
    (b) purely a geometrical quantity
    (c) both (a) and (b)
    (d) none of these
39. The Neumann's formula is
    (a) the mutual inductance between two filamentary conducting loops
    (b) $M_{12} = \dfrac{\mu}{4\pi} \oint_{c_1} \oint_{c_2} \dfrac{d\mathbf{l}_1 \cdot d\mathbf{l}_2}{r_{21}}$

(c) both (a) and (b)
(d) none of these

40. The reluctance is the ratio of
(a) magnetomotive force to magnetic flux
(b) Equal to $l/\mu A$
(c) both (a) and (b)
(d) none of these

41. The torque on the planar loop in magnetic field is
(a) $\mathbf{T} = \mathbf{m} \times \mathbf{B}$ (b) $\mathbf{T} = \nabla(\mathbf{m} \cdot \mathbf{B})$
(c) both (a) and (b) (d) none of these

42. The force on the planar loop due to non-uniform magnetic field is
(a) $\mathbf{F} = \mathbf{m} \times \mathbf{B}$ (b) $\mathbf{F} = \nabla(\mathbf{m} \cdot \mathbf{B})$
(c) both (a) and (b) (d) none of these

Answers

1. (d)	2. (a)	3. (b)	4. (d)	5. (d)
6. (a)	7. (c)	8. (d)	9. (a)	10. (c)
11. (a)	12. (a)	13. (a)	14. (a)	15. (c)
16. (b)	17. (c)	18. (b)	19. (b)	20. (b)
21. (b)	22. (b)	23. (a)	24. (b)	25. (a)
26. (b)	27. (b)	28. (c)	29. (c)	30. (b)
31. (a)	32. (c)	33. (c)	34. (b)	35. (c)
36. (c)	37. (a)	38. (c)	39. (c)	40. (c)
41. (a)	42. (b)			

UNIT IV

ELECTRODYNAMICS

7

Time-varying Fields and Maxwell's Equations

I honestly think it is better to be a failure at something you love than to be a success at something you hate.

—George Burns

Static fields which are our subject matter till now have several applications, but the most important application of electromagnetic fields is the use of radiation in wireless communications. Radiation is a phenomenon associated with dynamic fields.

We start with time-varying electric fields, which are described by the Faraday's law. Maxwell's correction to the Ampere's law said that the time-varying magnetic field is equivalent to the electric field. Various aspects that led to the correction and ultimately to the enunciation of wave phenomenon in time-varying fields are explained in detail.

Maxwell's equations are relations among the sources and time-varying fields. They form a complete and independent set giving full-fledged information on the electromagnetic fields and waves. They are discussed in their various forms along with their importance.

7.1 ELECTRODYNAMICS

7.1.1 Electromotive Force

To make the current exist in an electric circuit, there must be some agency to push the current carriers around the circuit. The charge carriers drift along the conductor resulting in certain amount of current which depends upon the strength the agency exerts over the carriers and the nature of the material. The driving force around the loop per unit charge, which actually exerts push over the carriers, is called *electromotive force*. In practice, it can be a battery voltage or generator output.

In electric circuits, there are really two forces involved in driving the current: the source $\mathbf{f}_s$, which is ordinarily confined to one portion of the loop and the electrostatic force, which serves to smooth out the flow and communicate the influence of the source to distant parts of the circuit. Thus,

$$\mathbf{f} = \mathbf{f}_s + \mathbf{E} \tag{7.1}$$

The net effect is determined by the line integral of **f** around the circuit which is called electromotive force. Thus,

$$\mathcal{E} \equiv \oint \mathbf{f} \cdot d\mathbf{l} \tag{7.2}$$

We can observe that it is the integral of force per unit charge. It can be considered as the work done per unit charge by the source. It is equal to the potential difference between the terminals of an ideal source. Thus,

$$\mathcal{E} = V \tag{7.3}$$

Battery is one common source of emf. Another source which is most commonly used is the generator.

Motional emf: This kind of emf arises when a wire is moved through a magnetic field. When a rectangular loop of width h is pulled at a velocity v through a normal magnetic field **B**, the emf generated is

$$\mathcal{E} \equiv \oint \mathbf{f}_{\text{mag}} \cdot d\mathbf{l} = vBh \tag{7.4}$$

It can be shown that

$$\mathcal{E} = -\frac{d\Phi}{dt} \tag{7.5}$$

where Φ is the flux of **B** defined by

$$\Phi \equiv \int \mathbf{B} \cdot d\mathbf{a} \tag{7.6}$$

The relation connecting the emf with time rate of change is known as the flux rule for motional emf and it says the emf generated in the loop is equal to the negative rate of change of flux through the loop.

7.1.2 Faraday's Law of Induction

In 1831, Michael Faraday performed three important epoch-making experiments:

- Exp I: He pulled a loop of wire through a magnetic field resulting in the flow of current through the loop.
- Exp II: He moved a magnet moving its field holding the loop still resulting in the current through the loop.
- Exp III: With both the loop and the magnet at rest, he changed the strength of the field by varying the current in the coil resulting in the flow of current in the loop.

In the case of the first experiment, which is an example of motional emf, it is the Lorentz force law at work; the emf is magnetic. To explain the generation of emf in the last two experiments, Faraday assumed that 'a changing magnetic field induces an electric field' and this particular electric field caused the emf. So in the last two cases, the emf is electric. To be specific, Faraday's law of induction states that a time-varying magnetic field produces an emf which may establish a current in a suitable closed circuit. If the circuit is an N turn coil, then

$$\text{emf} = -N\frac{d\phi}{dt} = -N\frac{d}{dt}\int_s \mathbf{B}\cdot d\mathbf{a} \tag{7.7}$$

A non-zero value of $d\phi/dt$ may result due to

- a time-changing flux linking a stationary closed path; the induced emf is then called the 'transformer emf' or V_{tr}
- the relative motion between a steady flux and a closed path; the induced emf is then called the 'motional emf' or V_m
- a combination of the above two.

Transformer emf: The transformer emf comes into being when the coil is stationary and the magnetic field is time-varying. Since the loop is stationary, d/dt operates only on $B(t)$. Hence, the transformer emf becomes

$$V_{tr} = -N\frac{d}{dt}\int_s \mathbf{B}\cdot d\mathbf{a} = -N\int_s \frac{\partial \mathbf{B}}{\partial t}\cdot d\mathbf{a} \tag{7.8}$$

where the ordinary derivative d/dt has been moved inside the integral and changed into a partial derivative $\partial/\partial t$.

Now if we treat the loop as a closed path with contour C, then

$$V_{tr} = \oint_C \mathbf{E}\cdot d\mathbf{l} \tag{7.9}$$

For a loop of one turn, i.e. $N = 1$,

$$\oint_C \mathbf{E}\cdot d\mathbf{l} = -\int_s \frac{\partial \mathbf{B}}{\partial t}\cdot d\mathbf{a} \tag{7.10}$$

which is the integral form of the Faraday's law. In this formula, the direction of the contour C and the direction of the ds are related by the right-hand rule.

Applying the Stoke's theorem to the left-hand side of the equation,

$$\int_s \nabla\times\mathbf{E}\cdot d\mathbf{a} = -\int_s \frac{\partial \mathbf{B}}{\partial t}\cdot d\mathbf{a} \tag{7.11}$$

And in order for the two integrals to be equal, their integrands have to be equal, which gives

$$\nabla\times\mathbf{E} = -\frac{\partial \mathbf{B}}{\partial t} \tag{7.12}$$

which is the differential form of the Faraday's law.

It states that a time-varying magnetic field induces an electric field **E** whose curl is equal to the negative of the time derivative of **B**.

Even though the derivation leading to the Faraday's law started out by considering the field associated with a physical circuit, the equation applies at any point in space, whether or not a physical circuit exists at that point.

Motional emf: It is due to the motion of a conductor in a magnetic field. The force on charge Q located within the conductor, according to the Lorentz's force law, is

$$\mathbf{F} = Q\,(\mathbf{v} \times \mathbf{B}) \tag{7.13}$$

It gives the force on unit charge, i.e. emf, as

$$\frac{\mathbf{F}}{Q} = (\mathbf{v} \times \mathbf{B}) = \mathbf{E}_m \tag{7.14}$$

If the conductor is moving in the time-varying magnetic field, the resultant emf becomes

$$\oint \mathbf{E} \cdot d\mathbf{l} = -\int \frac{\partial \mathbf{B}}{\partial t} \cdot d\mathbf{a} + \oint \mathbf{v} \times \mathbf{B} \cdot d\mathbf{l} \tag{7.15}$$

> The symmetry between the Faraday's induced electric fields
>
> $$\nabla \times \mathbf{E} = -\frac{\partial \mathbf{B}}{\partial t}; \quad \nabla \cdot \mathbf{E} = 0$$
>
> and the Ampere's law of magnetostatic fields
>
> $$\nabla \times \mathbf{H} = \mathbf{J}; \quad \nabla \cdot \mathbf{H} = 0$$
>
> leads to the usage of tricks associated with the Ampere's law
>
> $$\oint \mathbf{H} \cdot d\mathbf{l} = I_{\text{enc}}$$
>
> to find the induced electric field, using
>
> $$\oint \mathbf{E} \cdot d\mathbf{l} = -\frac{d\phi}{dt}$$

7.1.3 Lenz's Law

Lenz's law is a law which is always studied in combination with the Faraday's law of induction. In fact, it gives the information regarding the polarity of the induced voltage by the changing magnetic field. It states that "the induced emf due to the time-varying magnetic field is in such a direction as to produce a current whose flux, if added to the original flux, would reduce the magnitude of the emf."

7.2 EQUATION OF CONTINUITY

Equation of continuity is the only one equation which is considered on par with Maxwell's four equations in field theory. Sometimes it is mentioned as one of the Maxwell's equation sets. Basically, it is a precise mathematical statement of conservation of charge. It states that 'the net current due to flow of charges through a closed surface is equal to the time rate of decrease of the charge within the volume bounded by the surface.'

Consider a closed surface enclosing a volume V containing a net charge of Q. Let us also suppose that the charge is moving out crossing the surface resulting in a current I. This current I must be equal to the rate of change of charge enclosed by the surface

$$I = -\frac{dQ}{dt} \tag{7.16}$$

Here the –ve sign signifies that the flow of charge or current causes the decrease in the enclosed charge. If ρ is the charge density, then the total charge within the volume can be expressed as

$$Q = \int_v \rho d\tau \tag{7.17}$$

Combining Eqs. (7.16) and (7.17), we obtain

$$I = -\frac{dQ}{dt} = -\frac{d}{dt}\int_v \rho d\tau = -\int_v \frac{\partial \rho}{\partial t} d\tau \tag{7.18}$$

If is the current density over the surface s, it can be related to the current through

$$I = \oint_s \mathbf{J} \cdot d\mathbf{a} \tag{7.19}$$

Comparing Eqs. (7.17) and (7.18), we get

$$\oint_s \mathbf{J} \cdot d\mathbf{a} = -\int_v \frac{d\rho}{dt} d\tau \tag{7.20}$$

Equation (7.20) is said to be the equation of continuity in *integral form*. It has another form, the *differential form* which can be derived with the help of the divergence theorem of vector calculus. Using this theorem, the left side of Eq. (7.20) can be expressed as

$$\oint_s \mathbf{J} \cdot d\mathbf{a} = \oint_v \nabla \cdot \mathbf{J} d\tau \tag{7.21}$$

With this replacement, the continuity equation becomes

$$\oint_v \nabla \cdot \mathbf{J} d\tau = -\int_v \frac{\partial \rho}{\partial t} d\tau \tag{7.22}$$

Comparing both sides, we get

$$\nabla \cdot \mathbf{J} = -\frac{\partial \rho}{\partial t} \tag{7.23}$$

This is said to be in the point form or differential form of continuity equation.

$$\oint_s \mathbf{J} \cdot d\mathbf{a} = -\frac{d}{dt}\int_v \rho d\tau$$

$$\nabla \cdot \mathbf{J} = -\frac{\partial \rho}{\partial t}$$

The above two forms of continuity equations are for *time-varying* currents. In the case of static situations, the net steady current passing through the closed surface requires to be zero to conserve the charge. Therefore,

$$\oint_s \mathbf{J} \cdot d\mathbf{a} = 0 \tag{7.24}$$

$$\nabla \cdot \mathbf{J} = 0 \tag{7.25}$$

These two equations are said to be equations of continuity for *steady currents*.

Time-varying form	For steady currents
$\oint_s \mathbf{J} \cdot d\mathbf{a} = -\frac{d}{dt}\int_v \rho d\tau$	$\oint_s \mathbf{J} \cdot d\mathbf{a} = 0$
$\nabla \cdot \mathbf{J} = -\frac{\partial \rho}{\partial t}$	$\nabla \cdot \mathbf{J} = 0$

7.3 MAXWELL'S CORRECTION

The Ampere's circuital law exhibits inconsistency in its results when applied to time-varying fields. To understand the nature of the inconsistency, consider a parallel plate capacitor with a dc source connected. We consider two surfaces s_1 and s_2 and both bounded by the same closed path (Figure 7.1). If we apply the Ampere's law to the above configuration, in steady

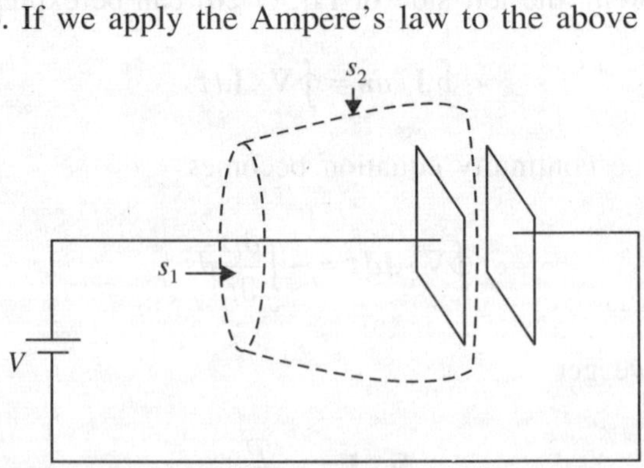

FIGURE 7.1 Inconsistency in Ampere's circuital law.

state or dc conditions, the current crossing the surfaces, either s_1 or s_2, and hence the enclosing path is the same, equal to zero. When the capacitor is charging or discharging, the time-varying conditions prevail and the currents crossing the surfaces are not the same. The current crossing the surface s_1 is non-zero, whereas that crossing s_2 is zero. If we apply the Ampere's law to find the field intensity, it gives different results depending upon the current we pick up leading to the lack of consistency in the result.

We can observe that, when it is dc or time-invariant situation, the Ampere's law works well giving unambiguous results, but when it comes to ac or time-varying currents, its results show inconsistency.

The inconsistency of the Ampere's law can also be observed mathematically by applying divergence to the law. When divergence is applied on both sides, it gives

$$\nabla \cdot \nabla \times \mathbf{H} = \nabla \cdot \mathbf{J} \equiv 0 \tag{7.26}$$

The above relation is identically equal to zero because the div of curl of any vector is always zero.

But according to the equation of continuity, which is fundamental and universal, the divergence of current density, in the case of time varying currents, is non-zero. Thus,

$$\nabla \cdot \mathbf{J} = -\frac{\partial \rho}{\partial t} \neq 0 \tag{7.27}$$

Therefore, the Ampere's law is not consistent with the time-varying form of the continuity equation or time-varying fields even though it is adequate for the static magnetic fields. To eliminate the problem of inconsistency, it is set to modify the Ampere's law. While changing the equation, it should be ensured that the relation that holds for time-varying magnetic fields must converge to the expression to the law for the static case, i.e. $\nabla \times \mathbf{H} = \mathbf{J}$. With this aspect in consideration, let us suppose, for time-varying fields,

$$\nabla \times \mathbf{H} = \mathbf{J} + \mathbf{X}$$

where $\mathbf{X}$ is an unknown vector function to be determined.
Now consider divergence in both sides of this relation.
Now,

$$\nabla \cdot \nabla \times \mathbf{H} = \nabla \cdot \mathbf{J} + \nabla \cdot \mathbf{X}$$

The left-hand side of this equation is always zero. And the first term of the right-hand side, according to the equation of continuity, is

$$\nabla \cdot \mathbf{J} = -\frac{\partial \rho}{\partial t}$$

But the Gauss law says $\nabla \cdot \mathbf{D} = \rho$. With this relation, the equation of continuity becomes

$$\nabla \cdot \mathbf{J} = -\nabla \cdot \frac{\partial \mathbf{D}}{\partial t}$$

resulting in

$$0 = -\nabla \cdot \frac{\partial \mathbf{D}}{\partial t} + \nabla \cdot \mathbf{X}$$

which gives

$$\mathbf{X} = \frac{\partial \mathbf{D}}{\partial t}$$

Hence, in the case of time-varying fields,

$$\nabla \times \mathbf{H} = \mathbf{J} + \frac{\partial \mathbf{D}}{\partial t} \tag{7.28}$$

Note the relations used in the above derivation; both the equation of continuity and the Gauss law are valid for the time-varying fields. Therefore, the resultant expression must also be valid for the time-varying fields.

According to this Maxwell's correction of the Ampere's law, "a changing electric field induces a magnetic field". Maxwell called the term $\partial\mathbf{D}/\partial t$ *displacement current.*

7.4 POYNTING THEOREM

One of the important characteristics of a travelling wave is its ability to carry energy. Without this ability, many of the important applications of electromagnetics could not be realized. Poynting theorem is a statement useful to examine power and energy relations in electromagnetic waves. It is an expression for energy that includes stored electric and magnetic energies and the dissipated energy derived from the Maxwell's equations.

It states that the net power flowing out of a closed surface s enclosing a volume v is equal to the net sum of time rate of decrease in the energy stored, the power dissipated and the power output of the source within volume v. According to this theorem, the vector product $\mathbf{P} = \mathbf{E} \times \mathbf{H}$ at any point is a measure of the rate of energy flow per unit area at that point. Mathematically,

$$\oint_s (\mathbf{E} \times \mathbf{H}) \cdot d\mathbf{a} = -\frac{d}{dt}\int_v \left(\frac{\varepsilon_0 E^2}{2} + \frac{\mu_0 H^2}{2}\right) d\tau - \sigma \int_v E^2 d\tau + \int_v (\mathbf{E} \cdot \mathbf{J}) d\tau \tag{7.29}$$

Proof: This theorem can be proved with the help of Maxwell's curl equations. The Maxwell's curl equations in time-dependent form for electric and magnetic fields are

$$\nabla \times \mathbf{H} = \mathbf{J} + \frac{\partial \mathbf{D}}{\partial t} \qquad \nabla \times \mathbf{E} = -\frac{\partial \mathbf{B}}{\partial t}$$

The dot multiplication of the first equation with $\mathbf{E}$ and of the second equation with $\mathbf{H}$ gives

$$\mathbf{E} \cdot (\nabla \times \mathbf{H}) = \mathbf{E} \cdot \left(\mathbf{J} + \frac{\partial \mathbf{D}}{\partial t}\right) \quad \text{and} \quad \mathbf{H} \cdot (\nabla \times \mathbf{E}) = -\mathbf{H} \cdot \frac{\partial \mathbf{B}}{\partial t}$$

or

$$\mathbf{E} \cdot (\nabla \times \mathbf{H}) = \mathbf{E} \cdot \mathbf{J} + \varepsilon \mathbf{E} \cdot \frac{\partial \mathbf{E}}{\partial t} \quad \text{and} \quad \mathbf{H} \cdot (\nabla \times \mathbf{E}) = -\mu \mathbf{H} \cdot \frac{\partial \mathbf{H}}{\partial t}$$

From these relations, we obtain

$$\mathbf{H} \cdot (\nabla \times \mathbf{E}) - \mathbf{E} \cdot (\nabla \times \mathbf{H}) = -\mu \mathbf{H} \cdot \frac{\partial \mathbf{H}}{\partial t} - \varepsilon \mathbf{E} \cdot \frac{\partial \mathbf{E}}{\partial t} - \mathbf{E} \cdot \mathbf{J}$$

$$\nabla \cdot (\mathbf{E} \times \mathbf{H}) = -\mu \mathbf{H} \cdot \frac{\partial \mathbf{H}}{\partial t} - \varepsilon \mathbf{E} \cdot \frac{\partial \mathbf{E}}{\partial t} - \mathbf{E} \cdot \mathbf{J} = -\mu \frac{\partial H^2}{\partial t} - \varepsilon \frac{\partial E^2}{\partial t} - \mathbf{E} \cdot \mathbf{J}$$

$$-\nabla \cdot (\mathbf{E} \times \mathbf{H}) = \mu \frac{\partial H^2}{\partial t} + \varepsilon \frac{\partial E^2}{\partial t} + \mathbf{E} \cdot \mathbf{J}$$

This is said to be the Poynting theorem in point form. If we consider the energy relations in a volume v, bounded by an area s, then the total power in the volume is obtained by integrating over the volume:

$$-\int_v \nabla \cdot (\mathbf{E} \times \mathbf{H}) d\tau = \int_v \left(\mu \mathbf{H} \cdot \frac{\partial \mathbf{H}}{\partial t} + \varepsilon \mathbf{E} \cdot \frac{\partial \mathbf{E}}{\partial t} \right) d\tau + \int_v \mathbf{E} \cdot \mathbf{J} d\tau$$

This is the Poynting theorem in integral form.
Using the divergence theorem, we have

$$\int_v \nabla \cdot (\mathbf{E} \times \mathbf{H}) d\tau = \oint_s (\mathbf{E} \times \mathbf{H}) \cdot d\mathbf{a}$$

Therefore,

$$\oint_s (\mathbf{E} \times \mathbf{H}) \cdot d\mathbf{a} = -\frac{1}{2} \int_v \left(\mu \frac{\partial H^2}{\partial t} + \varepsilon \frac{\partial E^2}{\partial t} \right) d\tau - \int_v \mathbf{E} \cdot \mathbf{J} d\tau$$

$$= -\frac{d}{dt} \int_v \left(\frac{\mu H^2}{2} + \frac{\varepsilon E^2}{2} \right) d\tau - \int_v \mathbf{E} \cdot \mathbf{J} d\tau$$

The current density **J** in the above expression has two components: one is dissipative $\mathbf{J}_c$ equal to $\sigma \mathbf{E}$ and the second one is source or generated current $\mathbf{J}_s$. Thus,

$$\mathbf{J} = \mathbf{J}_c - \mathbf{J}_s = \sigma \mathbf{E} - \mathbf{J}_s$$

Therefore,

$$-\int_v \mathbf{E} \cdot \mathbf{J} d\tau = \int_v E J_s d\tau - \sigma \int_v E^2 d\tau$$

After incorporating this relation, we have

$$\oint_s (\mathbf{E} \times \mathbf{H}) \cdot da = -\frac{d}{dt} \int_v \left(\frac{\mu H^2}{2} + \frac{\varepsilon E^2}{2} \right) d\tau - \sigma \int_v E^2 d\tau + \int_v E J_s d\tau$$

On the right side, the first term represents the rate of decrease of the stored energy and the second term represents dissipative power, whereas the third one is the power generated within the volume. As the integral on the left side represents the net power flowing out of the enclosed surface, both sides are equal and hence the Poynting theorem is proved.

Importance

- This theorem gives the energy relations of the fields whether static or dynamic in any volume. It also gives the net flow of the power out of the given volume through its surface.
- The poynting vector is the power density on the surface of a volume. The direction of the poynting vector is the direction of the flow of the power.

7.4.1 Complex Poynting Theorem

Note that the above derivation connecting power flow across a closed surface is in time domain. However, most of the applications we come across in the electromagnetic field theory are analysed in phasor domain implying sinusoidal excitation. Thus, it becomes necessary very often to have a relation connecting various power flows across a closed surface in phasor or frequency domain. This required relation can be found in complex poynting theorem. This theorem, apart from providing a definition to the complex poynting vector, gives a connecting relation between real and reactive powers. According to this theorem, the total complex power fed into a volume is equal to the algebraic sum of

- Active power dissipated as heat, plus
- Reactive power proportional to the difference between time-average magnetic and electric energies stored in the volume, plus
- Complex power transmitted across the surface enclosed by the volume.

Mathematically, it can be expressed as

$$\oint_s \frac{1}{2}(\mathbf{E}\times\mathbf{H}^*)\cdot d\mathbf{a} = -\int_v \frac{\sigma}{2}E^2 d\tau - 2j\omega\int_v \frac{1}{2}(w_m - w_e)d\tau - \int_v \frac{1}{2}(\mathbf{E}\cdot\mathbf{J}^*)d\tau \qquad (7.30)$$

Proof: We can derive the above relation by manipulating the Maxwell's equations. Let us consider the Maxwell's curl equations in phasor form for electric and magnetic fields:

$$\nabla\times\mathbf{E} = -j\omega\mu\mathbf{H}$$
$$\nabla\times\mathbf{H} = \mathbf{J} + j\omega\varepsilon\mathbf{E}$$

The first one is the Faraday's law and the second one is the Ampere's law. Now we dot multiply the Faraday's law with $\mathbf{H}^*$ resulting in

$$\mathbf{H}^*\cdot(\nabla\times\mathbf{E}) = -j\omega\mu\mathbf{H}^*\cdot\mathbf{H}$$

Similarly, we dot multiply the conjugate of the Ampere's law with $\mathbf{E}$ giving

$$\mathbf{E}\cdot(\nabla\times\mathbf{H}^*) = (\mathbf{J}^* - j\omega\varepsilon\mathbf{E}^*)$$

The subtraction of the second equation from the first results in

$$\mathbf{H}^*\cdot(\nabla\times\mathbf{E}) - \mathbf{E}\cdot(\nabla\times\mathbf{H}^*) = -\mathbf{E}\cdot(\mathbf{J}^* - j\omega\varepsilon\mathbf{E}^*) - j\omega\mu\mathbf{H}^*\cdot\mathbf{H}$$

Using the vector identity, from vector calculus, the left-hand side of the above relation can be replaced by $\nabla\cdot(\mathbf{E}\times\mathbf{H}^*)$. The right-hand side can also be simplified further.

$$\nabla\cdot(\mathbf{E}\times\mathbf{H}^*) = -\mathbf{E}\cdot(\mathbf{J}^* - j\omega\varepsilon\mathbf{E}^*) - j\omega\mu\mathbf{H}^*\cdot\mathbf{H}$$
$$= -\mathbf{E}\cdot\mathbf{J}^* + j\omega\varepsilon E^2 - j\omega\mu H^2$$

$$= -\mathbf{E} \cdot \mathbf{J}^* + j\omega(\varepsilon E^2 - \mu H^2)$$

$$\nabla \cdot (\mathbf{E} \times \mathbf{H}^*) = -\mathbf{E} \cdot \mathbf{J}^* - j\omega(\mu H^2 - \varepsilon E^2)$$

The current density **J** in the above expression has two components: one is dissipative $\mathbf{J}_c$ equal to $\sigma\mathbf{E}$ and the second one is the source or generated current $\mathbf{J}_s$. Thus,

$$\mathbf{J} = \mathbf{J}_c - \mathbf{J}_s = \sigma\mathbf{E} - \mathbf{J}_s$$

Therefore,

$$-\mathbf{E} \cdot \mathbf{J}^* = \mathbf{E} \cdot \mathbf{J}_s^* - \sigma E^2$$

$$\frac{1}{2}\nabla \cdot (\mathbf{E} \times \mathbf{H}^*) = -\frac{1}{2}\sigma E^2 - \frac{1}{2}j\omega(\mu H^2 - \varepsilon E^2) + \frac{1}{2}\mathbf{E} \cdot \mathbf{J}_s^*$$

Let us suppose $\mathbf{P} = (1/2)(\mathbf{E} \times \mathbf{H}^*)$ and then the integration over a volume and the application of Gauss's theorem to the left-side term give

$$\oint_s \mathbf{P} \cdot d\mathbf{a} = -\int_v \frac{1}{2}\sigma E^2 d\tau - j2\omega \int_v (w_m - w_e) d\tau + \int_v \frac{1}{2}\mathbf{E} \cdot \mathbf{J}_s^* d\tau$$

where

$\frac{1}{2}\sigma E^2$ = Time-averaged dissipative power

$\frac{1}{4}\mu H^2 = w_m$ = Time-averaged magnetic stored energy

$\frac{1}{4}\varepsilon E^2 = w_e$ = Time-averaged electric stored energy

$\frac{1}{2}\mathbf{E} \cdot \mathbf{J}_s^*$ = Complex power output of source into the field

$\frac{1}{2}(\mathbf{E} \times \mathbf{H}^*) = \mathbf{P}$ = Complex Poynting vector

7.5 MAXWELL'S EQUATIONS

The laws of classical electrodynamics were discovered in bits and pieces by Franklin, Coulomb, Ampere, Faraday and others, but the person who completed the job and packaged it all in the compact and consistent form was James Clerk Maxwell.

Maxwell's assumption of changing displacement density was equivalent to an electric current density and as such would produce a magnetic field has had far-reaching effects. The Faraday's law indicates that a changing magnetic field produces electric field.

These two together lead to 'wave equations' predicting the existence of electromagnetic 'wave propagation' even 30 years before Hertz's experimental verification.

The four electromagnetic equations given later by Eq. (7.38) or (7.40) are known as Maxwell's equations because he contributed to their development and established them as a self-consistent set. Out of the four, only the two curl equations are independent. The two divergence equations can be derived from the first two with the aid of the continuity equation.

7.5.1 Phasor Form

The time-varying fields obey the Maxwell's equations irrespective of the type of time variation. The equations get converted into another much simplified form known as phasor form when the time variations are sinusoidal. Our interest in sinusoidal time variations stems from two facts: the first one is that the output waveform of most of the generators is sinusoidal or nearly so and the second one is that the results of the sinusoidal time variations can be extended to other types of the time variations using the linearity property of the equations.

The phasor form is also known as time-harmonic form: the phasor form of the equations assumes linearity of the material properties. When equations are transformed from the time-varying form into the phasor form two changes take place: one is visible, $\partial/\partial t$ gets replaced by $j\omega$, and another is hidden, fields and sources become independent of time. In the phasor form of equations, the fields and sources are functions of position only.

Now we try to derive the equations that are obeyed by fields with sinusoidal time variations. In general, the fields are position dependent as well as time dependent, here, they are sinusoidal. Let us suppose the sinusoidal time-varying electric field is

$$\tilde{\mathbf{E}}(\mathbf{r}, t) = \mathbf{E}(\mathbf{r}) \cos \omega t \tag{7.31}$$

Note that the field is the product of two factors: one gives the space dependence of the field called *phasor* and the second one gives the time dependence. The above relation can be put in a convenient form as

$$\tilde{\mathbf{E}}(\mathbf{r}, t) = \mathbf{E}(\mathbf{r}) \cos \omega t = \text{Re}[\mathbf{E}(\mathbf{r}) e^{j\omega t}] \tag{7.32}$$

We can write similar relations for other field quantities and sources as well and substitute in the Maxwell's time-dependent equations. We consider, to illustrate the procedure, one Maxwell's equation, let us say the Ampere's law. For arbitrary time variations, it is

$$\nabla \times \mathbf{H} = \mathbf{J} + \frac{\partial \mathbf{D}}{\partial t} \tag{7.33}$$

In the case of sinusoidal time variations, it becomes

$$\nabla \times \tilde{\mathbf{H}} = \tilde{\mathbf{J}} + \frac{\partial \tilde{\mathbf{D}}}{\partial t} \tag{7.34}$$

Now substitute time variations of the fields and sources in explicit form as

$$\nabla \times \text{Re}[\mathbf{H}(\mathbf{r})\, e^{j\omega t}] = \text{Re}[\mathbf{J}(\mathbf{r}) e^{j\omega t}] + \frac{\partial}{\partial t} \text{Re}[\mathbf{D}(\mathbf{r}) e^{j\omega t}] \tag{7.35}$$

Then we can interchange the order of taking the real part and differentiation resulting in

$$\text{Re}[\nabla \times \mathbf{H}(\mathbf{r})\, e^{j\omega t}] = \text{Re}[\mathbf{J}(\mathbf{r}) e^{j\omega t}] + \text{Re}\left[\frac{\partial}{\partial t} \mathbf{D}(\mathbf{r}) e^{j\omega t}\right]$$

$$\begin{aligned} \text{Re}[\nabla \times \mathbf{H}(\mathbf{r})\, e^{j\omega t}] &= \text{Re}[\mathbf{J}(\mathbf{r}) e^{j\omega t}] + \text{Re}[j\omega \mathbf{D}(\mathbf{r})\, e^{j\omega t}] \\ &= \text{Re}\left\{[\nabla \times \mathbf{H}(\mathbf{r}) - \mathbf{J}(\mathbf{r}) - j\omega \mathbf{D}(\mathbf{r})] e^{j\omega t}\right\} = 0 \end{aligned} \tag{7.36}$$

As the above relation is valid for all the time t, we can write

$$\nabla \times \mathbf{H}(\mathbf{r}) - \mathbf{J}(\mathbf{r}) - j\omega\mathbf{D}(\mathbf{r}) = 0$$

or

$$\nabla \times \mathbf{H} = \mathbf{J} + j\omega\mathbf{D} \tag{7.37}$$

As the above relation involved the phasor quantities, it is called *phasor equation* and by following the similar procedure other Maxwell's equations can also be converted into phasor form.

Time-dependent to phasor forms: The differential $\partial/\partial t$ gets replaced by $j\omega$ when the time-dependent relation is converted into the phasor form. To convert the phasor form into the time-dependent form, multiply the phasor with $e^{j\omega t}$ and take the real part of the product.

Quasi-static field equations: In the case of slow varying fields, the displacement current can be neglected and it results in the pre-Maxwell system of time-harmonic equations known as the quasi-static field equations. These equations are static line in the sense that the equations satisfy Laplace's or Poisson's equations. One of the advantages of this form is that it extends many of the properties as well as methods used for static fields to time-dependent fields.

Source-free form: The general forms of the Maxwell's equations can be simplified if the sources are not taken into account. Under certain circumstances, the current **J** and charge density ρ_v or both can be removed from the equations resulting in a simpler form of equations known as source-free form. The solutions to the field problems become much simpler provided that the conditions under which these equations apply are satisfied.

7.5.2 Differential Form

Equations in time domain:

$$\left.\begin{aligned} \nabla \times \mathbf{H} &= \mathbf{J} + \frac{\partial \mathbf{D}}{\partial t} \\ \nabla \times \mathbf{E} &= -\frac{\partial \mathbf{B}}{\partial t} \\ \nabla \cdot \mathbf{D} &= \rho \\ \nabla \cdot \mathbf{B} &= 0 \end{aligned}\right\} \tag{7.38}$$

Equations in phasor domain:

$$\left.\begin{aligned} \nabla \times \mathbf{H} &= \mathbf{J} + j\omega\varepsilon\mathbf{E} \\ \nabla \times \mathbf{E} &= -j\omega\mu\mathbf{H} \\ \nabla \cdot \mathbf{D} &= \rho_v \\ \nabla \cdot \mathbf{B} &= 0 \end{aligned}\right\} \tag{7.39}$$

Word form:

- The curl of the magnetic field at a point in a time-varying field is the sum of the conduction and displacement current densities at that point.

- The curl of the electric field at a point in a time-varying field is equal to the negative time rate of the change of the magnetic flux density at that point.
- The divergence of the electric displacement density at a point in a time-varying field is equal to the volume charge density at that point.
- The divergence of the magnetic flux density at a point in a time-varying field is zero.

7.5.3 Integral Form

Equations in time domain:

$$\left.\begin{aligned}\oint \mathbf{H}\cdot d\mathbf{l} &= \int\left(\mathbf{J}+\frac{\partial \mathbf{D}}{\partial t}\right)\cdot d\mathbf{a}\\ \oint \mathbf{E}\cdot d\mathbf{l} &= -\int\frac{\partial \mathbf{B}}{\partial t}\cdot d\mathbf{a}\\ \oint \mathbf{D}\cdot d\mathbf{a} &= Q_{\text{enc}}\\ \oint \mathbf{B}\cdot d\mathbf{a} &= 0\end{aligned}\right\} \tag{7.40}$$

Equations in phasor form:

$$\left.\begin{aligned}\oint \mathbf{H}\cdot d\mathbf{l} &= \int(\mathbf{J}+j\omega\varepsilon\mathbf{E})\cdot d\mathbf{a}\\ \oint \mathbf{E}\cdot d\mathbf{l} &= -\int j\omega\mu\mathbf{H}\cdot d\mathbf{a}\\ \oint \mathbf{D}\cdot d\mathbf{a} &= Q_{\text{enc}}\\ \oint \mathbf{B}\cdot d\mathbf{a} &= 0\end{aligned}\right\} \tag{7.41}$$

Word form:

- The magnetomotive force around a closed path is equal to the conduction current plus the time derivative of electric displacement through any surface.
- The electromotive force around a closed path is equal to the time derivative of the magnetic flux through any surface bounded by the path.
- The net electric displacement through the surface enclosing a volume is equal to the total charge within the volume.
- The net magnetic flux emerging through any closed surface is equal to zero.

Significance

Interdependence of the electric and magnetic fields: The time-varying fields, both electric and magnetic, at a point obey the Maxwell's equations the fact of which is used to compute the fields many times. In the absence of the relations connecting the time-varying fields with their sources, this observation is very significant and useful.

The existence of electromagnetic waves: The Maxwell's equations lead to the development of wave equations. The field intensities of the time-varying fields obey wave equations proving the existence of the wave or energy flow in the time-varying fields.

From the wave equations it can be inferred that the propagation of electromagnetic waves is at a finite speed and it is equal to that of light in free space. Another important implication of the Maxwell's equations is that light itself is an electromagnetic wave.

The electromagnetic phenomenon of any type, i.e. any frequency ranging dc to infinity, any amount of intensity, can be explained, interpreted and understood using the Maxwell's equations.

7.6 BOUNDARY CONDITIONS

Interface is an infinitely thin boundary where two general media are assumed to be in contact with no properties of their own. On each side of the interface, the media are assumed to be linear and have uniform properties. The interface may contain a surface current or a surface charge. The fields on either side of the interface can be related and these relations are mentioned in the name of boundary conditions. These conditions are highly useful in estimating the fields on one side of the interface when they are known on the other side. Also note that the boundary conditions being discussed here are for time-varying fields and these are actually the extended versions of conditions we have for static fields.

The boundary conditions are four in number: two relate to the tangential field components and two relate to the normal components. All the four are derived from the Maxwell's equations. In the case of time-varying fields, they are given below.

Tangential components:

- The tangential component of electric field intensity **E** is continuous across an interface between two general media, regardless of charge densities on the surface.

$$E_{t1} = E_{t2} \tag{7.42}$$

It implies that the tangential component of the electric flux density **D** is discontinuous across the interface. The discontinuity is equal to the ratio of the permittivities of the media.

$$\frac{D_{t1}}{\varepsilon_1} = \frac{D_{t2}}{\varepsilon_2} \quad \Rightarrow \quad \frac{D_{t1}}{D_{t2}} = \frac{\varepsilon_1}{\varepsilon_2} \tag{7.43}$$

Proof: Consider the Faraday's law in integral form and apply it to the small rectangular loop at the interface of two different media. The loop extends half into one medium and the other half into the other medium as shown in Figure 7.2(a). Now,

$$\oint \mathbf{E} \cdot d\mathbf{l} = -\int \dot{\mathbf{B}} \cdot d\mathbf{a}$$

$$\oint \mathbf{H} \cdot d\mathbf{l} = \int_{\text{bottom side}} \mathbf{E} \cdot d\mathbf{l} + \int_{\text{right side}} \mathbf{E} \cdot d\mathbf{l} + \int_{\text{top side}} \mathbf{E} \cdot d\mathbf{l} + \int_{\text{left side}} \mathbf{E} \cdot d\mathbf{l}$$

$$E_{y1}\Delta y - E_{z2}\frac{\Delta z}{2} - E_{z3}\frac{\Delta z}{2} - E_{y2}\Delta y + E_{z4}\frac{\Delta z}{2} - E_{z1}\frac{\Delta z}{2} = \dot{B}_x \Delta z \Delta y$$

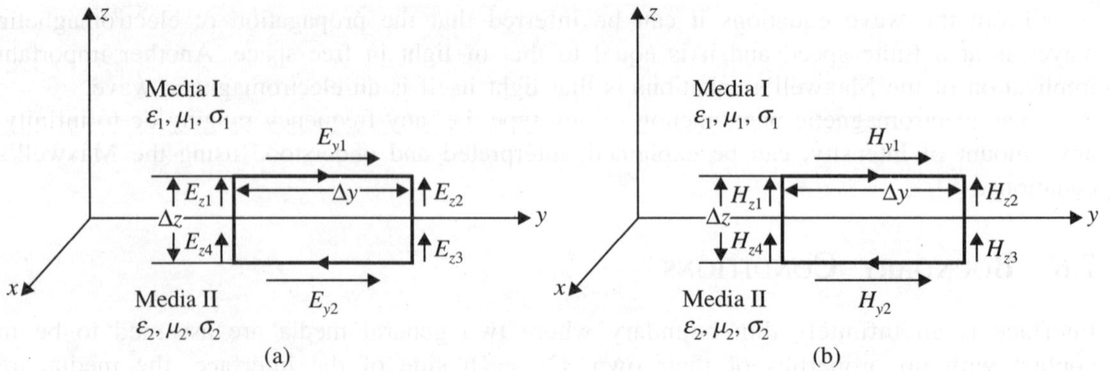

FIGURE 7.2 Boundary conditions—tangential components.

To bring the tangential components closer, we now reduce Δz to zero while maintaining the interface exactly in the middle of the loop. If the magnetic flux density is finite, then the entire right-hand side as well as the terms on the left-hand side with Δz of the above equation become zero, leaving

$$E_{y1}\Delta y - E_{y2}\Delta y = 0 \quad \text{or} \quad E_{y1} = E_{y2}$$

We can observe that the y-components are parallel or tangential to the interface. Therefore, the tangential electric field is the same on both sides of the interface. Thus,

$$E_{t1} = E_{t2}$$

- The tangential component of the magnetic field intensity **H** is continuous across the interface except in the presence of surface current over the interface. The amount of discontinuity is equal to the surface current density. In the absence of the surface current,

$$H_{t1} - H_{t2} = 0 \tag{7.44}$$

and in the presence of the surface current,

$$H_{t1} - H_{t2} = K \tag{7.45}$$

It implies that the tangential component of the magnetic flux density **B** is discontinuous and in the absence of the surface current over the interface, the amount of discontinuity is equal to the ratio of permeabilities of the media and is given by

$$\frac{B_{t1}}{\mu_1} - \frac{B_{t2}}{\mu_2} = 0 \quad \Rightarrow \quad \frac{B_{t1}}{B_{t2}} = \frac{\mu_1}{\mu_2} \tag{7.46}$$

In the presence of the surface current over the interface, the amount of discontinuity in the flux density **B** is given by

$$\frac{B_{t1}}{\mu_1} - \frac{B_{t2}}{\mu_2} = K \tag{7.47}$$

Proof: The procedure we follow here is similar to one we used in case of tangential electric fields. In the present case we have to consider the Faraday's law in integral form and apply it to the small rectangular loop at the interface of two different media. The loop extends half into one medium and the other half into the other medium as shown in Figure 7.2(b).
Now,

$$\oint \mathbf{H}\cdot d\mathbf{l} = \int (\mathbf{J}+\dot{\mathbf{D}})\cdot d\mathbf{a}$$

$$\oint \mathbf{H}\cdot d\mathbf{l} = \int_{\text{bottom side}} \mathbf{H}\cdot d\mathbf{l} + \int_{\text{right side}} \mathbf{H}\cdot d\mathbf{l} + \int_{\text{top side}} \mathbf{H}\cdot d\mathbf{l} + \int_{\text{left side}} \mathbf{H}\cdot d\mathbf{l}$$

$$H_{y1}\Delta y - H_{z2}\frac{\Delta z}{2} - H_{z3}\frac{\Delta z}{2} - H_{y2}\Delta y + H_{z4}\frac{\Delta z}{2} + H_{z1}\frac{\Delta z}{2} = (J_x + \dot{D}_x)\Delta z\Delta y$$

As the width of the loop Δz tends to zero, the entire right-hand side and the terms of the left-hand side with Δz become zero, resulting in

$$H_{y1}\Delta y - H_{y2}\Delta y = 0 \quad \text{or} \quad H_{y1} = H_{y2}$$

As the y-components are tangential to the interface, we have

$$H_{t1} = H_{t2}$$

Or the tangential magnetic field is continuous across the interface. If one of the media happens to be a perfect conductor, then there is a possibility of very large surface current and in that case, the right-hand side of the equation, as the loop area width tends to zero, becomes

$$(J_x + \dot{D}_x)\Delta z\Delta y = J_x\Delta z\Delta y$$

$$H_{y1}\Delta y - H_{y2}\Delta y = J_x\Delta z\Delta y$$

$$H_{y1} - H_{y2} = J_x\Delta z = K$$

Therefore, in the case of an interface involving perfect conductor, there is a possibility for the presence of surface currents and consequent discontinuity of tangential magnetic fields.

Normal components:

- The normal component of the electric flux density **D** is continuous except in the presence of the surface charge density. The discontinuity is equal to the surface charge density present at the interface.

$$D_{n1} - D_{n2} = \sigma \tag{7.48}$$

In the absence of the surface charge over the interface, the amount of discontinuity is

$$D_{n1} - D_{n2} = 0 \tag{7.49}$$

It implies that the normal component of the electric field intensity **E** is discontinuous across the interface and in the presence of surface charge, it is given by

$$\varepsilon_1 E_{n1} - \varepsilon_2 E_{n2} = \sigma \tag{7.50}$$

And in the absence of the surface charge, it is in the ratio of permittivities given by

$$\varepsilon_1 E_{n1} - \varepsilon_2 E_{n2} = 0 \quad \Rightarrow \quad \frac{E_{n1}}{E_{n2}} = \frac{\varepsilon_2}{\varepsilon_1} \tag{7.51}$$

Proof: To derive the above relation, we use the Gauss law in integral form and then apply it to a Gaussian thin pillbox which is placed over the interface of two different media. The box extends half into one medium and the other half into the other medium as shown in Figure 7.3(a).

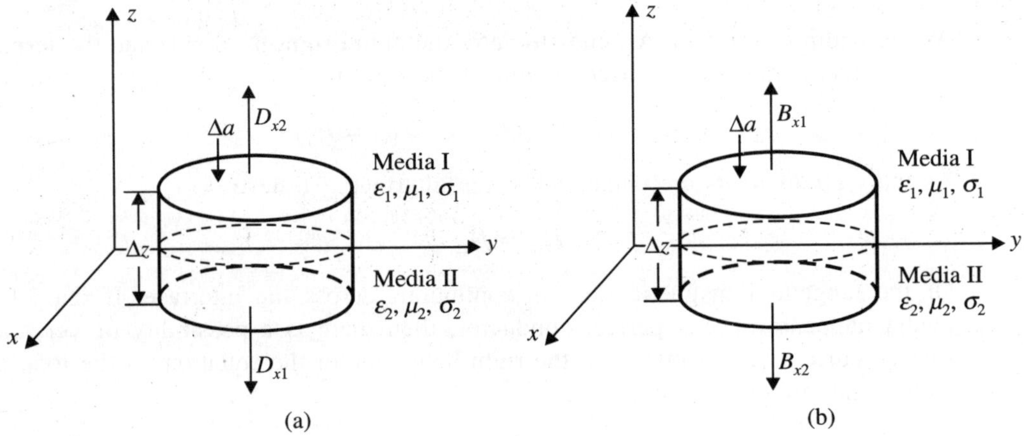

FIGURE 7.3 Boundary conditions—normal components.

Now,

$$\oint \mathbf{D} \cdot d\mathbf{a} = Q_{\text{enc}} = \int \rho d\tau$$

$$\oint \mathbf{D} \cdot d\mathbf{a} = \int_{\text{top face}} \mathbf{D} \cdot d\mathbf{a} + \int_{\text{bottom face}} \mathbf{D} \cdot d\mathbf{a} + \int_{\text{side face}} \mathbf{D} \cdot d\mathbf{a} = Q_{\text{enc}}$$

Initially we consider a charge-free interface $Q_{\text{enc}} = 0$ and applying the above Gauss law to the thin pillbox, we have

$$D_{n1}\Delta a - D_{n2}\Delta a + \psi_{\text{sides}} = 0$$

Here ψ_{sides} represents the flux through the side walls of the pillbox. As the thickness of the pillbox tends to zero, the area of the side wall and consequently the flux through it become zero, resulting in

$$D_{n1} = D_{n2}$$

Therefore, the normal components of the electric displacement density are continuous across the charge-free interface of two different media.

Next we consider the case of an interface with charge. When one of the media is a conductor, then there is a possibility for the charge to exist over the interface. In that case,

$$D_{n1}\Delta a - D_{n2}\Delta a + \psi_{\text{sides}} = \rho\Delta a\Delta z$$

As the thickness tends to zero, the area of the side wall and the flux through it becomes zero, resulting in

$$D_{n1} - D_{n2} = \rho\Delta z = \sigma$$

The above relation shows that the normal components of the electric displacement density are discontinuous across the interface of two different media when the interface carries charge.

It implies that the normal component of the electric field intensity **E** is discontinuous across the interface and in the presence of the surface charge, it is given by

$$\varepsilon_1 E_{n1} - \varepsilon_2 E_{n2} = \sigma$$

And in the absence of the surface charge, it is in the ratio of permittivities given by

$$\varepsilon_1 E_{n1} - \varepsilon_2 E_{n2} = 0 \quad \Rightarrow \quad \frac{E_{n1}}{E_{n2}} = \frac{\varepsilon_2}{\varepsilon_1}$$

- The normal component of the magnetic flux density **B** is continuous across an interface between two general media, regardless of current charge densities on the surface.

$$B_{n1} = B_{n2} \tag{7.52}$$

It implies that the normal component of the magnetic field intensity **H** is discontinuous across the interface. The discontinuity is equal to the ratio between the permeabilities of the materials.

$$\mu_1 H_{n1} = \mu_2 H_{n2} \quad \Rightarrow \quad \frac{H_{n1}}{H_{n2}} = \frac{\mu_2}{\mu_1} \tag{7.53}$$

Proof: We follow a procedure which is similar to the one we applied in the case of normal components of displacement density. Here we use the gauss law of magnetic field and apply it to a pillbox which is placed over the interface of two different media. The box extends half into one medium and the other half into the other medium as shown in Figure 7.3(b).

Now,

$$\oint \mathbf{B}\cdot d\mathbf{a} = 0$$

$$\oint \mathbf{B}\cdot d\mathbf{a} = \int_{\text{top face}} \mathbf{B}\cdot d\mathbf{a} + \int_{\text{bottom face}} \mathbf{B}\cdot d\mathbf{a} + \int_{\text{side face}} \mathbf{B}\cdot d\mathbf{a} = 0$$

$$B_{n1}\Delta a - B_{n2}\Delta a + \varphi_{\text{sides}} = 0$$

As the thickness tends to zero, the area of the side wall and the flux through it become zero, resulting in $B_{n1} = B_{n2}$.

Therefore, we can say that the normal components of the magnetic flux density are continuous across the interface of any two different media.

In general, we can observe that some of the boundary conditions involve surface charge densities, whereas the others involve current densities. However, as a matter of fact, the surface charge can exist only at the interface between lossy dielectrics, i.e. dielectrics having both finite conductivities and finite permittivities or between dielectrics and conductors, whereas the surface current can exist only at the surface of perfect conductors.

The following two types of interfaces are unique and the knowledge of their properties is often useful:

1. Interfaces between perfect or lossless dielectrics
2. Interfaces between a perfect dielectric and a perfect conductor

In the first type of interface, there can be neither current densities nor charge densities at the interface. In the second type of interface, the overriding condition is that of the conductor; that is, all fields in the perfect conductor must be zero.
At PEC–dielectric interface:

$$\left.\begin{aligned} \hat{\mathbf{n}}\times\mathbf{E} &= 0 \\ \hat{\mathbf{n}}\times\mathbf{H} &= \mathbf{K} \\ \hat{\mathbf{n}}\cdot\mathbf{E} &= \frac{\rho_s}{\varepsilon} \\ \hat{\mathbf{n}}\cdot\mathbf{H} &= 0 \end{aligned}\right\} \tag{7.54}$$

At dielectric–dielectric interface:

$$\left.\begin{aligned} \hat{\mathbf{n}}\times(\mathbf{E}_1-\mathbf{E}_2) &= 0 \\ \hat{\mathbf{n}}\times(\mathbf{H}_1-\mathbf{H}_2) &= 0 \\ \hat{\mathbf{n}}\cdot(\varepsilon_1\mathbf{E}_1-\varepsilon_2\mathbf{E}_2) &= 0 \\ \hat{\mathbf{n}}\cdot(\mu_1\mathbf{H}_1-\mu_2\mathbf{H}_2) &= 0 \end{aligned}\right\} \tag{7.55}$$

7.7 MEDIA PROPERTIES

The electromagnetic properties of the medium affect the propagation of waves in a big way. For example, in a good conductor, a wave cannot exist at all and in a perfect dielectric wave, it can travel without any attenuation. The nature of the medium is described in terms of constitutive parameters and constitutive relations which are given as follows:

- Constitutive relations: The electrical permittivity ε, magnetic permeability μ and conductivity σ are called the *constitutive parameters* of material medium. The three relations, namely $\mathbf{D} = \varepsilon\mathbf{E}$, $\mathbf{B} = \mu\mathbf{H}$ and $\mathbf{J} = \varepsilon\mathbf{E}$ are known as the *constitutive relations*.

- Linear medium: A medium is said to be *linear* if its properties ε, μ, σ are independent of the magnitude of the fields **E** and **H**.
- Homogeneous medium: A medium is said to be *homogeneous* if its constitutive parameters ε, μ, σ do not vary from point to point. That is, they remain constant throughout the medium.
- Isotropic medium: A medium is said to be *isotropic* if its constitutive parameters ε, μ, σ are scalars.
- Perfect conductor: It is a medium for which the conductivity σ is infinite. And in perfect conductors, $\mathbf{E} = 0$ regardless of **J**.
- Perfect dielectric: It is a medium for which the conductivity σ is zero. And in perfect dielectrics, $\mathbf{J} = 0$ regardless of **E**.
- Semiconductor: It is a material for which the conductivity falls in between that of conductors and insulators.
- Diamagnetic media: These are the media for which the susceptibility is negative, i.e. $\chi_m < 0$.
- Paramagnetic materials: These are the materials for which the susceptibility is positive, i.e. $\chi_m > 0$.

  The absolute magnitude of χ_m is on the order of 10^{-5} for both the classes of dia- and paramagnetic materials which allows us to ignore χ_m relative to 1. This gives relative permeability $\mu_r \simeq 1$ for both types of materials. Examples are dielectric materials and most of the metals.
- Ferromagnetic materials: These are the media for which the relative permeability $|\mu_r| \gg 1$. Purified iron is an example of this type of material for which the relative permeability is on the order of 2×10^5.

7.8 RETARDED POTENTIALS

By now, it is a well-known fact that the time-varying currents (accelerating charges) produce coupled time-varying electric and magnetic fields. It is always possible to define a scalar potential for an electric field produced by a charge distribution and a vector potential with respect to a magnetic field due to a current distribution. As in time-varying fields both electric and magnetic fields exist, we can define both the potential functions for these fields. The potential functions, thus defined for time-varying fields are called *retarded potentials*. The most significant feature of these functions is the incorporation of time delay or retardation in their structure to account for the finite propagation time of the wave. This is also their distinguishing and differentiating feature from static potential functions.

7.8.1 Definitions

By definition, the vector potential function **A** is given by

$$\mathbf{A}(\mathbf{r}, t) = \frac{\mu}{4\pi} \int \frac{\mathbf{J}\left(\mathbf{r}', t - \dfrac{R}{c}\right)}{R} d\tau' \tag{7.56}$$

The above expression is in time domain. Here **J** represents the source current, the position vectors **r** and **r′** denote the field and source points, respectively and R is the distance between the source point and the field point. Note that the time variable t is retarded by an amount equal to R/c, being the time required for the wave to travel from the source point to the field point. Here c is free space velocity of light equal to 3×10^8 m/s. Also note that **R** is the vector from the source point to the field point and hence $\mathbf{R} = \mathbf{r} - \mathbf{r}'$ and $R = |\mathbf{R}|$.

The differential $d\tau'$ indicates differential volume and the prime over it indicates that the integration is to be performed over the source distribution. The source and observation points are shown in Figure 7.4.

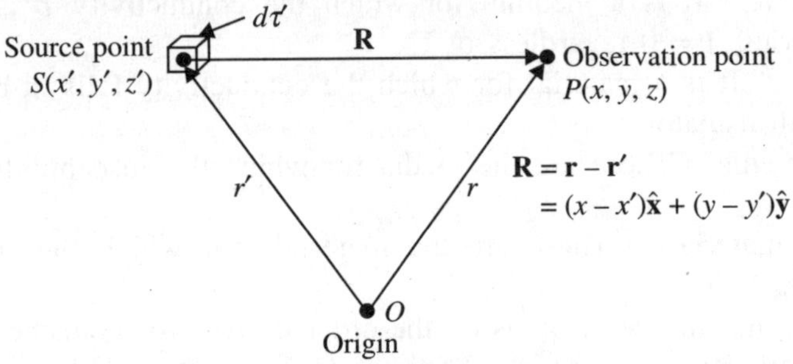

FIGURE 7.4 Illustrating source and observation points.

Similarly a scalar potential function V can also be defined as

$$V(\mathbf{r}, t) = \frac{1}{4\pi\varepsilon} \int \frac{\rho\left(\mathbf{r}', t - \dfrac{R}{c}\right)}{R} d\tau' \tag{7.57}$$

In this expression, the quantity ρ represents the volume charge density of the source and it is related to the volume current density of the source through the equation of continuity

$$\nabla \cdot \mathbf{J} = -\frac{\partial \rho}{\partial t}$$

The above given expressions for potentials, in Eqs. (7.56) and (7.57) when converted into phasor domain, appear as

$$\mathbf{A}(\mathbf{r}) = \frac{\mu}{4\pi} \int \frac{\mathbf{J}(\mathbf{r}')\, e^{-j\beta R}}{R} d\tau' \tag{7.58}$$

and

$$V(\mathbf{r}) = \frac{1}{4\pi\varepsilon} \int \frac{\rho(\mathbf{r}')\, e^{-j\beta R}}{R} d\tau' \tag{7.59}$$

Observe that the time delay R/c of the time domain expressions got converted into a phase delay βR in the phasor domain expressions, here β being the phase-shift constant which is by definition the amount of phase shift per unit length, i.e. $\beta \equiv 2\pi/\lambda$.

If the source distribution happens to be a line current **I** which is the case with wire antennas, then the differential in the above expressions becomes differential length dl' and in the case of surface current **K**, it is the differential area da'.

The relations given in Eqs. (7.56) to (7.59) are in integral form giving potentials directly from the source current or charge distribution. The relations between the potentials and the sources can also be put in differential form which appear as follows:

$$\left.\begin{aligned} \nabla^2\mathbf{A} - \mu\varepsilon\frac{\partial^2\mathbf{A}}{\partial t^2} &= -\mu\mathbf{J} \\ \nabla^2 V - \mu\varepsilon\frac{\partial^2 V}{\partial t^2} &= -\frac{\rho}{\varepsilon} \\ \nabla^2\mathbf{A} + \omega^2\mu\varepsilon\mathbf{A} &= -\mu\mathbf{J} \\ \nabla^2 V + \omega^2\mu\varepsilon V &= -\frac{\rho}{\varepsilon} \end{aligned}\right\} \tag{7.60}$$

Here the first two are in the time domain, whereas the remaining two are in the phasor domain.

In addition to the potential–source relations given above, the potential functions can also be related to the field intensities and the connecting relations appear as follows:

$$\left.\begin{aligned} \mu\mathbf{H} &= \nabla\times\mathbf{A} \\ \mathbf{E} &= -\nabla V - \frac{\partial\mathbf{A}}{\partial t} \\ \mu\mathbf{H} &= \nabla\times\mathbf{A} \\ \mathbf{E} &= -\nabla V - j\omega\mathbf{A} \end{aligned}\right\} \tag{7.61}$$

Here, the first two are in the time domain, whereas the second two are in the phasor domain.

Important relations pertaining to retarded potential with relation to time-varying fields are given hereunder in a box useful for ready reference.

Retarded potentials

- Integral form expressions in time domain:

$$\left.\begin{aligned} A(\mathbf{r}, t) &= \frac{\mu}{4\pi}\int\frac{\mathbf{J}\left(\mathbf{r}', t - \dfrac{R}{c}\right)}{R}d\tau' \\ V(\mathbf{r}, t) &= \frac{1}{4\pi\varepsilon}\int\frac{\rho\left(\mathbf{r}', t - \dfrac{R}{c}\right)}{R}d\tau' \end{aligned}\right\}$$

- Integral form expressions in phasor domain:

$$\left.\begin{aligned} A(\mathbf{r}) &= \frac{\mu}{4\pi}\int\frac{\mathbf{J}(\mathbf{r}')e^{-j\beta R}}{R}d\tau' \\ V(\mathbf{r}) &= \frac{1}{4\pi\varepsilon}\int\frac{\rho(\mathbf{r}')e^{-j\beta R}}{R}d\tau' \end{aligned}\right\}$$

(Contd...)

(*Contd...*)

- Differential form expressions in time domain:

$$\left.\begin{aligned}\nabla^2\mathbf{A}-\mu\varepsilon\frac{\partial^2\mathbf{A}}{\partial t^2}&=-\mu\mathbf{J}\\ \nabla^2 V-\mu\varepsilon\frac{\partial^2 V}{\partial t^2}&=-\frac{\rho}{\varepsilon}\\ \mu\mathbf{H}&=\nabla\times\mathbf{A}\\ \mathbf{E}&=-\nabla V-\frac{\partial\mathbf{A}}{\partial t}\end{aligned}\right\}$$

- Differential form expressions in phasor domain:

$$\left.\begin{aligned}\nabla^2 A+\omega^2\mu\varepsilon\mathbf{A}&=-\mu\mathbf{J}\\ \nabla^2 V+\omega^2\mu\varepsilon V&=-\frac{\rho}{\varepsilon}\\ \mu\mathbf{H}&=\nabla\times\mathbf{A}\\ \mathbf{E}&=-\nabla V-j\omega\mathbf{A}\end{aligned}\right\}$$

In the expressions for potentials in the phasor form, the sources are being convolved with some exponential functions known as Green's functions to get the potentials. To be precise, the functions which convolve with sources, i.e. current and charge distributions, to give the potentials, vector and scalar respectively, are called Green's functions.

7.8.2 Fields in Terms of Potentials

The Maxwell's equations are valid for time-varying fields and hence these equations and any relations that are derived from them hold good for time-varying fields. Now in this section, with the help of the Maxwell's equations, we try to derive the relations for the fields in terms of the retarded potentials.

We start with relation $\mu\mathbf{H} = \nabla \times \mathbf{A}$, which holds good for time-varying fields as it implies $\nabla \cdot \mathbf{B} = 0$, which is one of the Maxwell's equations.

Consider the Maxwell's curl equation

$$\nabla \times \mathbf{E} = -\frac{\partial \mathbf{B}}{\partial t}$$

and substitute the curl of **A** for the magnetic field resulting in

$$\nabla \times \mathbf{E} = -\frac{\partial(\nabla \times \mathbf{A})}{\partial t}$$

leading to

$$\nabla \times \left(\mathbf{E} + \frac{\partial \mathbf{A}}{\partial t}\right) = 0$$

which is a vector equation involving the 'del' operator. The gradient of any scalar field satisfies this equation. We select the solution $-\nabla V$ so that it satisfies even for static fields also. Hence,

$$E = -\nabla V - \frac{\partial \mathbf{A}}{\partial t}$$

7.8.3 Lorentz Gauge Condition

A useful observation that can be made from the above two relations is that these potentials do not uniquely define the fields because they give the same fields for potentials **A**, V, **A′** and V' defined as

$$\mathbf{A}' = \mathbf{A} + \nabla \lambda, \quad V' = V - \frac{\partial \lambda}{\partial t} \tag{7.62}$$

where λ is an arbitrary scalar function. So without affecting the physical quantities **E** and **H**, variations can be made in **A** and V. These changes in potentials are called *gauge transformations* which are exploited to adjust the divergence of **A** with a view to simplifying the complex equations.

One such transformation is *Coulomb gauge* transformation, which sets the divergence of vector potential to zero, i.e. $\nabla \cdot \mathbf{A} = 0$.

The main advantage of Coulomb gauge is that it gives a scalar potential that is particularly simple to calculate. The serious drawback of this gauge is that the scalar potential V does not exhibit retardation; therefore, it does not obey the special theory of relativity. Apart from a causal appearance of V, the vector potential **A** is particularly difficult to calculate. But this gauge gives **E** which obeys the special theory of relativity.

Another one which is widely used is *Lorentz gauge* transformation, i.e.

$$\nabla \cdot \mathbf{A} = -\mu\varepsilon \frac{\partial V}{\partial t} \tag{7.63}$$

Its importance lies in the fact that it treats both **A** and V on equal footing leading to similar equations for them in terms of d'Alembertian

$$\square^2 = \nabla^2 - \mu\varepsilon \frac{\partial^2}{\partial t^2} \tag{7.64}$$

resulting in similar and simpler solutions.

7.8.4 Potentials in Terms of Sources

Time-varying potential functions **A** and V of the time-varying fields can be related through *Lorentz gauge condition for potentials*, i.e.

$$\left.\begin{aligned} \nabla \cdot \mathbf{A} &= -\mu\varepsilon \frac{\partial V}{\partial t} \\ \nabla \cdot \mathbf{A} &= -j\omega\mu\varepsilon V \end{aligned}\right\} \tag{7.65}$$

or in phasor form

Consider the Maxwell's equation

$$\nabla \cdot \mathbf{E} = \frac{\rho}{\varepsilon}$$

Replacing the field with the potential it becomes

$$\nabla \cdot \left[-\nabla V - \frac{\partial \mathbf{A}}{\partial t} \right] = \frac{\rho}{\varepsilon}$$

which can be expanded as

$$\nabla^2 V + \frac{\partial (\nabla \cdot \mathbf{A})}{\partial t} = -\frac{\rho}{\varepsilon}$$

Using the Lorentz gauge to replace the divergence of vector potential, we get

$$\nabla^2 V - \mu\varepsilon \frac{\partial^2 V}{\partial t^2} = -\frac{\rho}{\varepsilon}$$

and in phasor form, it becomes

$$\nabla^2 V + \omega^2 \mu\varepsilon V = -\frac{\rho}{\varepsilon}$$

Note that it is a relation connecting scalar potential with its source distribution These equations can be solved to get

$$V(\mathbf{r}, t) = \frac{1}{4\pi\varepsilon} \int \frac{\rho\left(\mathbf{r}', t - \frac{R}{c}\right)}{R} d\tau'$$

and

$$V(\mathbf{r}) = \frac{1}{4\pi\varepsilon} \int \frac{\rho(\mathbf{r}')e^{-j\beta R}}{R} d\tau'$$

respectively. Now consider another Maxwell's curl equation

$$\nabla \times \mathbf{H} = \mathbf{J} + \frac{\partial \mathbf{D}}{\partial t}$$

Replacing the field with potential, becomes

$$\nabla \times \frac{1}{\mu} (\nabla \times \mathbf{A}) = \mathbf{J} + \varepsilon \frac{\partial}{\partial t}\left(-\nabla V - \frac{\partial \mathbf{A}}{\partial t} \right)$$

$$\nabla \times (\nabla \times \mathbf{A}) = \mu \mathbf{J} + \mu\varepsilon \frac{\partial}{\partial t}\left(-\nabla V - \frac{\partial \mathbf{A}}{\partial t} \right)$$

$$\nabla\nabla \cdot \mathbf{A} - \nabla^2 \mathbf{A} = \mu \mathbf{J} - \mu\varepsilon \nabla \frac{\partial V}{\partial t} - \mu\varepsilon \frac{\partial^2 \mathbf{A}}{\partial t^2}$$

Using the Lorentz gauge condition

$$\nabla\nabla\cdot\mathbf{A} - \nabla^2\mathbf{A} = \mu\mathbf{J} + \nabla\nabla\cdot\mathbf{A} - \mu\varepsilon\frac{\partial^2\mathbf{A}}{\partial t^2}$$

$$\nabla^2\mathbf{A} - \mu\varepsilon\frac{\partial^2\mathbf{A}}{\partial t^2} = -\mu\mathbf{J}$$

and in the phasor form, it becomes

$$\nabla^2\mathbf{A} + \omega^2\mu\varepsilon\mathbf{A} = -\mu\mathbf{J}$$

which is a relation connecting vector potential with source. These equations can be solved to get

$$\mathbf{A}(\mathbf{r},\, t) = \frac{\mu}{4\pi}\int\frac{\mathbf{J}\left(\mathbf{r}', t - \frac{R}{c}\right)}{R}\,d\tau'$$

and

$$\mathbf{A}(\mathbf{r}) = \frac{\mu}{4\pi}\int\frac{J(\mathbf{r}')e^{-j\beta R}}{R}\,d\tau'$$

respectively. It is also possible to express both the fields, electric and magnetic, in terms of one single function known as Hertzian vector usually indicated by **Z**. It is defined in terms of the vector potential as

$$\mathbf{Z} = \frac{1}{\mu\varepsilon}\int\mathbf{A}\,dt \tag{7.66}$$

In terms of this vector, we can express the electric and magnetic fields as

$$\mathbf{E} = \nabla\nabla\cdot\mathbf{Z} - \mu\varepsilon\ddot{\mathbf{Z}} \tag{7.67}$$

and

$$\mathbf{H} = \varepsilon\nabla\times\dot{\mathbf{Z}} \tag{7.68}$$

Proof: From the definition,

$$\mathbf{Z} = \frac{1}{\mu\varepsilon}\int\mathbf{A}dt \quad\Rightarrow\quad \mathbf{A} = \mu\varepsilon\dot{\mathbf{Z}}$$

and as

$$\mathbf{H} = \frac{1}{\mu}\nabla\times\mathbf{A}$$

It gives

$$\mathbf{H} = \frac{1}{\mu}\nabla\times\mu\varepsilon\dot{\mathbf{Z}} = \varepsilon\nabla\times\dot{\mathbf{Z}}$$

Using the gauge condition

$$\nabla\cdot\mathbf{A} = -\mu\varepsilon\dot{V}$$

we express the scalar potential as

$$V = -\frac{1}{\mu\varepsilon}\nabla \cdot \int \mathbf{A} dt = -\nabla \cdot \mathbf{Z}$$

Now substituting

$$V = -\nabla \cdot \mathbf{Z}$$

and

$$\nabla \cdot \dot{\mathbf{A}} = -\mu\varepsilon\ddot{V}$$

in

$$\mathbf{E} = -\nabla V - \frac{\partial \mathbf{A}}{\partial t}$$

we get

$$\mathbf{E} = -\nabla\nabla \cdot \mathbf{Z} - \mu\varepsilon\ddot{\mathbf{Z}}$$

Significance of Retarded Potentials

Many a time we come across problems wherein it is required to find the field intensity due to an arbitrary current distribution. In such cases, the fields can be computed, at least theoretically, by either of the two methods. The first method is a single-step method in which fields are calculated from the given source distributions, but it requires relations connecting the source directly with the fields. The second method involves a two-step procedure. In this method, the potentials are calculated first from the source distribution and then the fields are computed from the potentials. The second method is usually resorted to when the relations directly connecting the source and fields are not available or when there is computational difficulty in evaluating the fields due to the first method.

In the case of time-varying fields, relations do not exist that connect the source distributions with the fields directly. Because of this reason, the second method is usually resorted to which involves a two-step procedure, first calculating the retarded potentials and then finding the fields from the available potentials.

The wire antennas are the earliest antennas used by mankind and in the analysis of these antennas, the retarded potentials are widely used.

Maxwell's Equations in Time-dependent Form		
	Differential form	**Integral form**
Ampere's law with modification	$\nabla \times \mathbf{H} = \mathbf{J} + \frac{\partial \mathbf{D}}{\partial t}$	$\oint \mathbf{H} \cdot d\mathbf{l} = \int \left(\mathbf{J} + \frac{\partial \mathbf{D}}{\partial t}\right) \cdot d\mathbf{a}$
Faraday's law	$\nabla \times \mathbf{E} = -\frac{\partial \mathbf{B}}{\partial t}$	$\oint \mathbf{E} \cdot d\mathbf{l} = -\int \frac{\partial \mathbf{B}}{\partial t} \cdot d\mathbf{a}$
Gauss law	$\nabla \cdot \mathbf{D} = \rho$	$\oint \mathbf{D} \cdot d\mathbf{a} = Q_{\text{enc}}$
No name	$\nabla \cdot \mathbf{B} = 0$	$\oint \mathbf{B} \cdot d\mathbf{a} = 0$

Maxwell's Equations in Phasor Form		
	Differential form	**Integral form**
Ampere's law with modification	$\nabla \times \mathbf{H} = \mathbf{J} + j\omega\varepsilon\mathbf{E}$	$\oint \mathbf{H} \cdot d\mathbf{l} = \int (\mathbf{J} + j\omega\varepsilon\mathbf{E}) \cdot d\mathbf{a}$
Faraday's law	$\nabla \times \mathbf{E} = -j\omega\mu\mathbf{H}$	$\oint \mathbf{E} \cdot d\mathbf{l} = -\int j\omega\mu\mathbf{H} \cdot d\mathbf{a}$
Gauss law	$\nabla \cdot \mathbf{D} = \rho_v$	$\oint \mathbf{D} \cdot d\mathbf{a} = Q_{enc}$
No name	$\nabla \cdot \mathbf{B} = 0$	$\oint \mathbf{B} \cdot d\mathbf{a} = 0$

SOLVED PROBLEMS

1. Find the voltage induced in a loop of area A in the xy-plane which started rotating at $t = 0$ with an angular velocity ω about the x-axis in a time-varying magnetic field $\mathbf{B} = B(t)\hat{\mathbf{z}}$. Find the solution using both the forms of Faraday's law.

Solution: It can found in two different ways:

(a) Using differential form of the Faraday's law:

Induced voltage $v = -N\dfrac{d\Phi}{dt}$ and the flux $\Phi = \int \mathbf{B} \cdot d\mathbf{a}$

Given that $N = 1$ and $\mathbf{B} = B(t)\hat{\mathbf{z}}$

As the loop is rotating, the flux through it varies from a maximum $\Phi = AB(t)$ when it is in the xy-plane to a minimum of zero when the loop is normal in the xy-plane in a sinusoidal manner. Hence,

$$\Phi = \int \mathbf{B} \cdot d\mathbf{a} = AB(t) \cos \omega t$$

Therefore,

$$v = -N\frac{d\Phi}{dt} = -A\frac{dB(t)}{dt} \cos \omega t + AB(t)\, \omega \sin \omega t$$

(b) Using the integral form of the Faraday's law:

Induced voltage is $\oint \mathbf{E} \cdot d\mathbf{l} = -\int \dfrac{\partial \mathbf{B}}{\partial t} \cdot d\mathbf{a} + \int \mathbf{v} \times \mathbf{B} \cdot d\mathbf{l}$

The first part is the transformer emf and the second term represents the generator emf. Let us first calculate the transformer emf v_t.

$$v_t = -\int \frac{\partial \mathbf{B}}{\partial t} \cdot d\mathbf{a} = -\int \frac{\partial \mathbf{B}(t)}{\partial t} \hat{\mathbf{z}} \cdot da\hat{\mathbf{a}} = -\frac{\partial B(t)}{\partial t} A\hat{\mathbf{z}} \cdot \hat{\mathbf{a}}$$

$$= -\frac{\partial B(t)}{\partial t} A \cos \omega t \quad \text{as} \quad \hat{\mathbf{z}} \cdot \hat{\mathbf{a}} = \cos \omega t$$

In the above expression, $\hat{\mathbf{a}}$ is normal to the loop. Next move to the generator emf v_g.

$$v_g = \oint \mathbf{v} \times \mathbf{B} \cdot d\mathbf{l}$$

In this expression,

$$\mathbf{v} = r\omega\hat{\mathbf{a}} = \frac{y}{\cos\omega t}\omega\hat{\mathbf{a}}$$

therefore,

$$\mathbf{v} \times \mathbf{B} = \frac{y}{\cos\omega t}\omega\hat{\mathbf{a}} \times B(t)\hat{\mathbf{z}} = \frac{-\omega B(t)\sin\omega t}{\cos\omega t}y\hat{\mathbf{x}}$$

$$v_g = \frac{-\omega B(t)\sin\omega t}{\cos\omega t}\oint y\hat{\mathbf{x}} \cdot d\mathbf{l}$$

To evaluate this integral we use the relation $\nabla \times y\hat{\mathbf{x}} = -\hat{\mathbf{z}}$.

$$v_g = \frac{-\omega B(t)\sin\omega t}{\cos\omega t}\oint y\hat{\mathbf{x}} \cdot d\mathbf{l} = \frac{-\omega B(t)\sin\omega t}{\cos\omega t}\int_s \nabla \times y\hat{\mathbf{x}} \cdot d\mathbf{a}$$

$$= \frac{-\omega B(t)\sin\omega t}{\cos\omega t}\int_s -\hat{\mathbf{z}} \cdot da\hat{\mathbf{a}} = \frac{\omega B(t)\sin\omega t}{\cos\omega t}A\cos\omega t$$

$$= A\omega B(t)\sin\omega t$$

$$v = v_t + v_g = -\frac{\partial B(t)}{\partial t}A\cos\omega t + A\omega B(t)\sin\omega t$$

2. A rigid vertical rectangular loop in the yz-plane with width b and height a of a metallic wire is falling under the influence of gravity in the presence of a magnetic field $\mathbf{B} = B_0(1 + z)\,\hat{\mathbf{x}}$, where B_0 and k are constants. Show that the emf induced around the closed path of the loop is B_0kabv, where v is the downward velocity of the loop.

Solution: Using the integral form of the Faraday's law

$$\oint \mathbf{E} \cdot d\mathbf{l} = -\frac{d}{dt}\int \mathbf{B} \cdot d\mathbf{a}$$

$$\int \mathbf{B} \cdot d\mathbf{a} = \int_{z=z}^{z=z+a}\int_{y=y}^{y=y+b} B_0(1+kz)\hat{\mathbf{x}} \cdot dydz\hat{\mathbf{x}}$$

$$= \int_{z=z}^{z=z+a}\int_{y=y}^{y=y+b} B_0(1+kz)dydz = B_0b\left[z + \frac{kz^2}{2}\right]_z^{z+a} = B_0ba + \frac{B_0bk}{2}(2az + a^2)$$

$$\oint \mathbf{E} \cdot d\mathbf{l} = -\frac{d}{dt}\int \mathbf{B} \cdot d\mathbf{a}$$

$$= -\frac{d}{dt}\left[B_0ba + \frac{B_0bk}{2}(2az + a^2)\right] = -B_0abk\frac{dz}{dt} = B_0abkv$$

Therefore, the induced emf around the loop is B_0akv, where $v = -dz/dt$ is the downward velocity of the loop.

3. Show that the displacement current through the capacitor is equal to the conduction current I when the voltage source $v = V_m \sin \omega t$ is connected to a capacitance C.

Solution: Using circuit theory principles, the conduction current is

$$i_c = C\frac{dv}{dt} = \frac{\varepsilon A}{d} V_m \omega \cos \omega t$$

The displacement current exists only within the capacitor. As a first step to evaluate it, let us evaluate the electric displacement density.

$$\mathbf{D} = \varepsilon \mathbf{E} = \varepsilon \frac{V_m \sin \omega t}{d}$$

$$i_d = A\mathbf{J}_d = A\frac{\partial \mathbf{D}}{\partial t} = A\varepsilon \mathbf{E} = \frac{\varepsilon A}{d} V_m \omega \cos \omega t$$

We can observe their equality.

4. A plane wave travelling in free space has an average Poynting vector of 5 W/m^2. Find the average energy density.

Solution: Assuming phasor values are given, the average Poynting vector is

$$P = \frac{E^2}{2\eta} = \frac{E^2}{2 \times 120\,\pi} = 5 \text{ W/m}^2$$

It gives the electric field intensity V/m.
The magnetic field intensity then becomes

$$\frac{E}{H} = 120\,\pi \quad \Rightarrow \quad H = \frac{E}{120\,\pi} = \frac{\sqrt{2 \times 5 \times 120\,\pi}}{120\,\pi}$$

$$= \sqrt{\frac{2 \times 5}{120\,\pi}} = 0.163 \text{ A/m}$$

The average energy density is

$$W_{\text{ave}} = \frac{\varepsilon E^2}{2} + \frac{\mu H^2}{2}$$

$$= \frac{10^{-9} \times 5 \times 120\pi}{2 \times 36\pi} + \frac{4\pi \times 10^{-7} \times 5}{2 \times 120\pi}$$

$$= 0.167 \times 10^{-7} + 0.167 \times 10^{-7} = 0.334 \times 10^{-7} \text{ J/m}^2$$

5. Let us suppose that the magnetic flux density in medium '1' , free space, is given by $B_1 = 0.8\hat{\mathbf{x}} + 1.2\hat{\mathbf{y}} + 0.4\hat{\mathbf{z}}$ T. Find the magnetic field intensity $\mathbf{H}_2$ in medium '2' whose relative permeability $\mu_r = 15$ at the current-free interface

(a) x = constant and
(b) y = constant and the angles made by $\mathbf{B}_1$ and $\mathbf{B}_2$ and with the interfaces.

Solution:

(a) x = constant

Firstly we write down the flux density from the field intensity using the relation $\mathbf{B} = \mu\mathbf{H}$. Then we estimate the y- and z-components of $\mathbf{H}_2$ by using the fact that the tangential components of the magnetic field intensity must be the same across the interface.

$$\mathbf{B}_1 = 0.8\hat{\mathbf{x}} + 1.2\hat{\mathbf{y}} + 0.4\hat{\mathbf{z}}$$

$$\mathbf{H}_1 = \frac{1}{\mu_0}(0.8\hat{\mathbf{x}} + 1.2\hat{\mathbf{y}} + 0.4\hat{\mathbf{z}})$$

$$\mathbf{H}_2 = \frac{1}{\mu_0}(\mu_0 H_x\hat{\mathbf{x}} + 1.2\hat{\mathbf{y}} + 0.4\hat{\mathbf{z}})$$

$$\mathbf{B}_2 = \mu_2\mathbf{H}_2 = \mu_{r2}(\mu_0 H_x\hat{\mathbf{x}} + 1.2\hat{\mathbf{y}} + 0.4\hat{\mathbf{z}})$$

Now, we use the fact of continuity of normal components of flux density.

$$\mu_2 H_x = 0.8 \quad \Rightarrow \quad 15\mu_0 H_x = 0.8 \quad \Rightarrow \quad H_x = \frac{0.8}{15\mu_0}$$

Therefore,

$$\mathbf{H}_2 = \frac{1}{\mu_0}(\mu_0 H_x\hat{\mathbf{x}} + 1.2\hat{\mathbf{y}} + 0.4\hat{\mathbf{z}})$$

$$= \frac{1}{\mu_0}\left[\mu_0\left(\frac{0.8}{15\mu_0}\right)\hat{\mathbf{x}} + 1.2\hat{\mathbf{y}} + 0.4\hat{\mathbf{z}}\right]$$

$$= \frac{1}{\mu_0}(0.053\hat{\mathbf{x}} + 1.2\hat{\mathbf{y}} + 0.4\hat{\mathbf{z}})$$

Angles: Note that the direction of the interface x = constant is along the x-axis. If θ_1 is the angle between the field and the interface, then the angle between the field and the x-axis becomes $(90° - \theta_1)$. Therefore,

$$\mathbf{B}_1 \cdot \hat{\mathbf{x}} = |\mathbf{B}_1| \cos(90° - \theta_1)$$

$$0.8 = 1.49 \sin\theta_1 \quad \Rightarrow \quad \theta_1 = \sin^{-1}\left(\frac{0.8}{1.49}\right) = 32.47°$$

$$\mathbf{B}_2 \cdot \hat{\mathbf{x}} = |\mathbf{B}_2| \cos(90° - \theta_1)$$

$$0.053 = 1.49 \sin\theta_2$$

$$0.053 = 1.49 \sin\theta_2 \quad \Rightarrow \quad \theta_1 = \sin^{-1}\left(\frac{0.053}{1.49}\right) = 2.03°$$

(b) y = constant

We follow the procedure same as in the previous part of the problem.

$$\mathbf{B}_1 = 0.8\hat{\mathbf{x}} + 1.2\hat{\mathbf{y}} + 0.4\hat{\mathbf{z}}$$

$$\mathbf{H}_1 = \frac{1}{\mu_0}(0.8\hat{\mathbf{x}} + 1.2\hat{\mathbf{y}} + 0.4\hat{\mathbf{z}})$$

$$\mathbf{H}_2 = \frac{1}{\mu_0}(0.8\hat{\mathbf{x}} + \mu_0 H_y\hat{\mathbf{y}} + 0.4\hat{\mathbf{z}})$$

$$\mathbf{B}_2 = \mu_2\mathbf{H}_2 = \mu_{r2}(0.8\hat{\mathbf{x}} + \mu_0 H_y\hat{\mathbf{y}} + 0.4\hat{\mathbf{z}})$$

Now we use the fact of continuity of normal components of flux density.

$$\mu_2 H_y = 1.2 \rightarrow 15\mu_0 H_y = 1.2 \quad \Rightarrow \quad H_y = \frac{1.2}{15\mu_0}$$

Therefore,

$$\mathbf{H}_2 = \frac{1}{\mu_0}(0.8\hat{\mathbf{x}} + \mu_0 H_y\hat{\mathbf{y}} + 0.4\hat{\mathbf{z}})$$

$$= \frac{1}{\mu_0}\left[0.8\hat{\mathbf{x}} + \mu_0\left(\frac{1.2}{15\mu_0}\right)\hat{\mathbf{y}} + 0.4\hat{\mathbf{z}}\right]$$

$$= \frac{1}{\mu_0}(0.8\hat{\mathbf{x}} + 0.08\hat{\mathbf{y}} + 0.4\hat{\mathbf{z}})$$

Angles:

$$\mathbf{B}_1 \cdot \hat{\mathbf{y}} = |\mathbf{B}_1| \cos(90° - \theta_1)$$

$$1.2 = 1.49 \sin\theta_1 \rightarrow \theta_1 = \sin^{-1}\left(\frac{1.2}{1.49}\right) = 53.6°$$

$$\mathbf{B}_2 \cdot \hat{\mathbf{y}} = |\mathbf{B}_2| \cos(90° - \theta_1)$$

$$0.08 = 1.49 \sin\theta_2 \rightarrow \theta_2 = \sin^{-1}\left(\frac{0.08}{1.49}\right) = 3.07°$$

6. A current I flows from a point charge $Q(t)$ at the origin to infinity along a straight wire lying over the positive half of the z-axis. Consider a circular path in the plane normal to the wire and centred over the z-axis. Find the magnetomotive force $\oint \mathbf{H} \cdot d\mathbf{l}$ around the circular path (see Figure 7.5).

Solution: As the problem involves both the time-varying charge and current, we have to use the Ampere's circuital law with Maxwell's correction.

$$\oint \mathbf{H} \cdot d\mathbf{l} = \int \mathbf{J} \cdot d\mathbf{a} + \frac{d}{dt}\int \mathbf{D} \cdot d\mathbf{a}$$

Consider a spherical surface radius r_0, bounded by the closed curve under reference with its centre at the origin and apply the Ampere's circuital law.

$$\int \mathbf{J} \cdot d\mathbf{a} = I$$

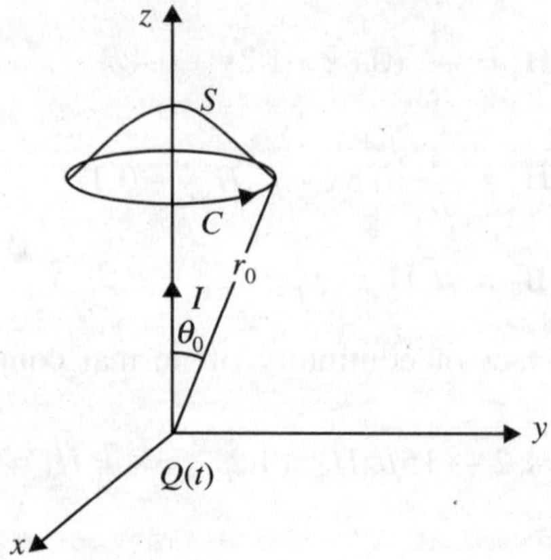

FIGURE 7.5 Configuration pertaining to Problem 6.

$$\int \mathbf{D}\cdot d\mathbf{a} = \int_{\theta=0}^{\theta_0}\int_{\phi=0}^{2\pi} \frac{Q}{4\pi r_0^2}\hat{\mathbf{r}}\cdot r_0^2 \sin\theta \, d\theta d\phi \hat{\mathbf{r}} = \frac{Q}{2}(1-\cos\theta_0)$$

$$\frac{d}{dt}\int \mathbf{D}\cdot d\mathbf{a} = \frac{dQ}{dt}\frac{(1-\cos\theta_0)}{2}$$

But when we apply the law of conservation of charge to the entire spherical surface, we get

$$I + \frac{dQ}{dt} = 0 \quad \Rightarrow \quad \frac{dQ}{dt} = -I$$

Therefore,

$$\frac{d}{dt}\int \mathbf{D}\cdot d\mathbf{a} = \frac{dQ}{dt}\frac{(1-\cos\theta_0)}{2} = -I\frac{(1-\cos\theta_0)}{2}$$

Finally

$$\oint \mathbf{H}\cdot d\mathbf{l} = \int \mathbf{J}\cdot d\mathbf{a} + \frac{d}{dt}\int \mathbf{D}\cdot d\mathbf{a} = I - I\frac{(1-\cos\theta_0)}{2}$$
$$= I\frac{(1+\cos\theta_0)}{2}$$

7. The electric field in a region is given by $\mathbf{E} = y\hat{\mathbf{x}} - x\hat{\mathbf{y}}$. Find the emf, i.e. $\oint \mathbf{E}\cdot d\mathbf{l}$, around a clockwise *xy*-plane closed path (see Figure 7.6) of segments given by

(a) $y = 0.848x + \sqrt{2}(x \to 0, -0.5)$, $x = -1(y \to 0.5, 1)$, $y = \frac{1}{x}(x \to -1, -0.5)$,

$y = 2(x \to -0.5, 0)$, $x = -1(y \to 0.5, 1)$

(b) $y = \frac{x}{2}(x \to 0, -1)$,

$y = 0.848x + \sqrt{2}(x \to -0.5, 0)$, $x^2 + y^2 = 2(x \to 0, 1) y = x^2(x \to 1, 0)$

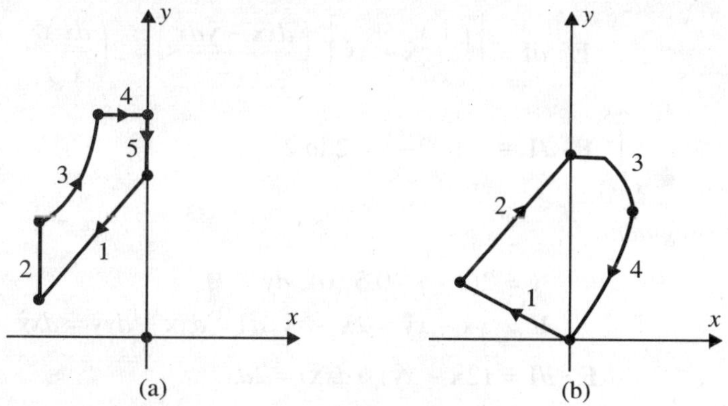

FIGURE 7.6 Configurations pertaining to Problem 7.

Solution:

(a) To find the emf over the entire closed path, we integrate the field over each segment and then sum them up.

First segment:

$$y = 0.848x + \sqrt{2}(x \to 0, -0.5)$$
$$dy = 0.848\,dx$$
$$\mathbf{E} = y\hat{\mathbf{x}} - x\hat{\mathbf{y}} = \left(0.848x + \sqrt{2}\right)\hat{\mathbf{x}} - x\hat{\mathbf{y}}$$
$$d\mathbf{l} = dx\hat{\mathbf{x}} + dy\hat{\mathbf{y}} = dx\hat{\mathbf{x}} + 0.848dx\hat{\mathbf{y}}$$
$$\mathbf{E} \cdot d\mathbf{l} = \left[\left(0.848x + \sqrt{2}\right)\hat{\mathbf{x}} - x\hat{\mathbf{y}}\right] \cdot (dx\hat{\mathbf{x}} + 0.848dx\hat{\mathbf{y}}) = \sqrt{2}dx$$
$$\int_{x=0}^{-0.5} \mathbf{E} \cdot d\mathbf{l} = \int_{x=0}^{-0.5} \sqrt{2}dx = \sqrt{2}x\Big|_{0}^{-0.5} = -0.5\sqrt{2}$$

Second segment:

$$x = -1(y \to 0.5,\ 1),\ dx = 0$$
$$\mathbf{E} = y\hat{\mathbf{x}} - x\hat{\mathbf{y}} = y\hat{\mathbf{x}} + \hat{\mathbf{y}},\ d\mathbf{l} = dx\hat{\mathbf{x}} + dy\hat{\mathbf{y}} = dy\hat{\mathbf{y}}$$
$$\mathbf{E} \cdot d\mathbf{l} = (y\hat{\mathbf{x}} + \hat{\mathbf{y}}) \cdot (dy\hat{\mathbf{y}}) = dy$$
$$\int_{y=1/2}^{1} \mathbf{E} \cdot d\mathbf{l} = \int_{y=1/2}^{1} dy = \frac{1}{2}$$

Third segment:

$$y = \frac{1}{x}(x \to -1, -0.5),\ dy = -\frac{dx}{x^2}$$
$$\mathbf{E} = y\hat{\mathbf{x}} - x\hat{\mathbf{y}} = \left(\frac{1}{x}\right)\hat{\mathbf{x}} - x\hat{\mathbf{y}},\ d\mathbf{l} = dx\hat{\mathbf{x}} + dy\hat{\mathbf{y}} = dx\hat{\mathbf{x}} - \frac{\hat{\mathbf{y}}dx}{x^2}$$

$$\mathbf{E}\cdot d\mathbf{l} = \left[\left(\frac{1}{x}\right)\hat{\mathbf{x}} - x\hat{\mathbf{y}}\right]\cdot\left(\frac{dx\hat{\mathbf{x}} - \hat{\mathbf{y}}dx}{x^2}\right) = 2\left(\frac{dx}{x}\right)$$

$$\int_{x=-1}^{-0.5} \mathbf{E}\cdot d\mathbf{l} = \int_{x=-1}^{-0.5} 2\frac{dx}{x} = 2\ln 2$$

Fourth segment:

$$y = 2(x \to -0.5,\ 0),\ dy = 0$$

$$\mathbf{E} = y\hat{\mathbf{x}} - x\hat{\mathbf{y}} = 2\hat{\mathbf{x}} - x\hat{\mathbf{y}},\ d\mathbf{l} = dx\hat{\mathbf{x}} + dy\hat{\mathbf{y}} = dx\hat{\mathbf{x}}$$

$$\mathbf{E}\cdot d\mathbf{l} = (2\hat{\mathbf{x}} - x\hat{\mathbf{y}})\cdot(dx\hat{\mathbf{x}}) = 2dx$$

$$\int_{x=-0.5}^{\sqrt{2}} \mathbf{E}\cdot d\mathbf{l} = \int_{x=-0.5}^{0} 2dx = 1$$

Fifth segment:

$$x = -1\left(y \to 2, \sqrt{2}\right), dx = 0$$

$$\mathbf{E} = y\hat{\mathbf{x}} - x\hat{\mathbf{y}} = y\hat{\mathbf{x}},\ d\mathbf{l} = dx\hat{\mathbf{x}} + dy\hat{\mathbf{y}} = dy\hat{\mathbf{y}}$$

$$\mathbf{E}\cdot d\mathbf{l} = (y\hat{\mathbf{x}})\cdot(dy\hat{\mathbf{y}}) = 0$$

$$\int_{y=2}^{\sqrt{2}} \mathbf{E}\cdot d\mathbf{l} = 0$$

We obtain the emf around the closed path by summing up contributions from all the segments as

$$\oint \mathbf{E}\cdot d\mathbf{l} = -0.5\sqrt{2} + \frac{1}{2} + 2\ln 2 + 1 + 0 = 2.179$$

(b) To find the emf over the entire closed path, we integrate the field over each segment and then sum them up.
First segment:

$$y = \frac{x}{2}(x \to 0, -1), dy = \frac{dx}{2}$$

$$\mathbf{E} = y\hat{\mathbf{x}} - x\hat{\mathbf{y}} = \left(\frac{1}{2}\right)x\hat{\mathbf{x}} - x\hat{\mathbf{y}},\ d\mathbf{l} = dx\hat{\mathbf{x}} + dy\hat{\mathbf{y}} = dx\hat{\mathbf{x}} + \left(\frac{dx}{2}\right)\hat{\mathbf{y}}$$

$$\mathbf{E}\cdot d\mathbf{l} = \left[\left(\frac{1}{2}\right)x\hat{\mathbf{x}} - x\hat{\mathbf{y}}\right]\cdot\left[dx\hat{\mathbf{x}} + \left(\frac{dx}{2}\right)\hat{\mathbf{y}}\right] = 0$$

Second segment:

$$y = 0.848 + \sqrt{2}(x \to -0.5, 0)$$

$$dy = 0.848\ dx$$

$$\mathbf{E} = y\hat{\mathbf{x}} - x\hat{\mathbf{y}} = (0.848x + \sqrt{2})\hat{\mathbf{x}} - x\hat{\mathbf{y}}$$

$$d\mathbf{l} = dx\hat{\mathbf{x}} - dy\hat{\mathbf{y}} = dx\hat{\mathbf{x}} + 0.848dx\hat{\mathbf{y}}$$

$$\mathbf{E}\cdot d\mathbf{l} = \left[\left(0.848x + \sqrt{2}\right)\hat{\mathbf{x}} - x\hat{\mathbf{y}}\right]\cdot(dx\hat{\mathbf{x}} + 0.848dx\hat{\mathbf{y}})$$

$$= \sqrt{2}dx$$

$$\int_{x=-0.5}^{0} \mathbf{E}\cdot d\mathbf{l} = \int_{x=-0.5}^{0} \sqrt{2}dx = \sqrt{2}x\Big|_{-0.5}^{0} = 0.5\sqrt{2}$$

Third segment:

$$x^2 + y^2 = 2(x \to 0,\ 1)\ xdx + ydy = 0$$

$$\mathbf{E} = y\hat{\mathbf{x}} - x\hat{\mathbf{y}} = \sqrt{2 - x^2}\,\hat{\mathbf{x}} - x\hat{\mathbf{y}},\ d\mathbf{l} = dx\hat{\mathbf{x}} + dy\hat{\mathbf{y}} = dx\hat{\mathbf{x}} - \left(\frac{x}{\sqrt{2 - x^2}}\right)dx\hat{\mathbf{y}}$$

$$\mathbf{E}\cdot d\mathbf{l} = \left(\sqrt{2 - x^2}\,\hat{\mathbf{x}} - x\hat{\mathbf{y}}\right)\cdot\left[dx\hat{\mathbf{x}} - \left(\frac{x}{\sqrt{2 - x^2}}\right)dx\hat{\mathbf{y}}\right]$$

$$= dx\left(\sqrt{2 - x^2} + \frac{x^2}{\sqrt{2 - x^2}}\right) = \frac{2dx}{\sqrt{2 - x^2}}$$

$$\int_{x=0}^{1} \mathbf{E}\cdot d\mathbf{l} = \int_{x=0}^{1} \frac{2dx}{\sqrt{2 - x^2}} = \frac{\pi}{2}$$

Fourth segment:

$$y = x^2(x \to 1,\ 0),\ dy = 2xdx$$

$$\mathbf{E} = y\hat{\mathbf{x}} - x\hat{\mathbf{y}} = x^2\hat{\mathbf{x}} - x\hat{\mathbf{y}},\ d\mathbf{l} = dx\hat{\mathbf{x}} + 2xdx\hat{\mathbf{y}} = dx(\hat{\mathbf{x}} + 2x\hat{\mathbf{y}})$$

$$\mathbf{E}\cdot d\mathbf{l} = (x^2\hat{\mathbf{x}} - x\hat{\mathbf{y}})\cdot[dx(\hat{\mathbf{x}} + 2x\hat{\mathbf{y}})]$$

$$= (x^2dx - 2x^2dx) = -x^2dx$$

$$\int_{x=1}^{0} \mathbf{E}\cdot d\mathbf{l} = \int_{x=1}^{0} -x^2dx = \frac{1}{3}$$

We obtain the emf around the closed path by summing up contributions from all the segments as

$$\oint \mathbf{E}\cdot d\mathbf{l} = 0 + 0.5\sqrt{2} + \frac{\pi}{2} + \frac{1}{3} = 2.611$$

8. Using the gauss law, Ohm's law and the equation of continuity, show that the charge density ρ that existed within a conductor decreases to $1/e$ times its initial value in a time ε/σ. What is the value of this time for copper?

Solution: Continuity equation is

$$\nabla \cdot \mathbf{J} = -\frac{\partial \rho}{\partial t}$$

Using the constitutive relation $\mathbf{J} = \varepsilon \boldsymbol{E}$, the continuity equation becomes

$$\frac{\partial \rho}{\partial t} = -\nabla \cdot \mathbf{J} = -\nabla \cdot \sigma \mathbf{E} = -\sigma \nabla \cdot \mathbf{E}$$

Using the Gauss law $\nabla \cdot \mathbf{E} = \rho/\varepsilon$, the continuity equation becomes

$$\frac{\partial \rho}{\partial t} = -\sigma \nabla \cdot \mathbf{E} = -\frac{\sigma}{\varepsilon}\rho$$

The solution of this first order differential equation is

$$\rho = \rho_0 e^{-(\sigma/\varepsilon)t}$$

From the equation we can observe that the charge density within the conductor decreases to $1/e$ times its initial value in a time ε/σ. In the case of copper,

$$\frac{\varepsilon}{\sigma} = 1.53 \times 10^{-19}\,\text{s}$$

9. A non-uniform charge distribution given by $\rho(x, y, z) = \rho_0 x$ C/m^3, where ρ_0 is a constant, exists within a volume defined by $x > 0$, $y > 0$, $x + y < 1$ and $0 < z < 1$. Find the displacement flux emanating from the closed surface bounding the volume. Also find the uniform charge density with which the given volume has to be filled to give the same amount of displacement flux.

Solution: We use the Gauss law to evaluate the electric displacement out of the closed surface.

$$\oint \mathbf{D} \cdot d\mathbf{a} = Q_{\text{enc}} = \int \rho d\tau$$

$$= \int_{x=0}^{1} \int_{y=0}^{1-x} \int_{z=0}^{1} \rho_0 x dx dy dz = \frac{\rho_0}{6}\,\text{C}$$

Note that the given volume is a triangular cube with volume 1/2. Let the required unifrom charge density be ρ_0'. Then

$$\frac{\rho_0'}{2} = \frac{\rho_0}{6} \rightarrow \rho_0' = \frac{\rho_0}{3}$$

10. Convert the Maxwell's equations for mmf and emf from integral form to differential form.

Solution: Maxwell's relations for mmf and emf are

$$\oint \mathbf{H} \cdot d\mathbf{l} = \int \left(\mathbf{J} + \frac{\partial \mathbf{D}}{\partial t} \right) \cdot d\mathbf{a}$$

$$\oint \mathbf{E} \cdot d\mathbf{l} = -\int \frac{\partial \mathbf{B}}{\partial t} \cdot d\mathbf{a}$$

We use the Stokes' theorem of vector calculus to convert the closed line integrals that appear in the above relations into surface integrals.

$$\oint \mathbf{H} \cdot d\mathbf{l} = \int \nabla \times \mathbf{H} \cdot d\mathbf{a}$$

$$\oint \mathbf{E} \cdot d\mathbf{l} = \int \nabla \times \mathbf{E} \cdot d\mathbf{a}$$

Substitution of the above surface integrals in place of the line integrals in Maxwell's relations results in

$$\int \nabla \times \mathbf{H} \cdot d\mathbf{a} = \int \left(\mathbf{J} + \frac{\partial \mathbf{D}}{\partial t} \right) \cdot d\mathbf{a}$$

$$\int \nabla \times \mathbf{E} \cdot d\mathbf{a} = -\int \frac{\partial \mathbf{B}}{\partial t} \cdot d\mathbf{a}$$

Comparing both sides of the equations results in

$$\nabla \times \mathbf{H} = \mathbf{J} + \frac{\partial \mathbf{D}}{\partial t}$$

$$\nabla \times \mathbf{E} = -\frac{\partial \mathbf{B}}{\partial t}$$

Which are the required Maxwell's curl relations.

11. Convert the Maxwell's flux relations from integral form to differential form.

Solution: Maxwell's relations for electric displacement and magnetic flux are

$$\oint \mathbf{D} \cdot d\mathbf{a} = Q_{\text{enc}}$$

$$\oint \mathbf{B} \cdot d\mathbf{a} = 0$$

We use the divergence theorem of vector calculus to convert the closed surface integrals that appear in the above relations into the volume integrals.

$$\oint \mathbf{D} \cdot d\mathbf{a} = \oint \nabla \cdot \mathbf{D} d\tau$$

$$\oint \mathbf{B} \cdot d\mathbf{a} = \oint \nabla \cdot \mathbf{B} d\tau$$

Substitution of the above surface integrals in place of the line integrals in Maxwell's relations results in

$$\oint \nabla \cdot \mathbf{D} d\tau = Q_{\text{enc}}$$

$$\oint \nabla \cdot \mathbf{B} d\tau = 0$$

Note that the charge enclosed Q_{enc} can be expressed as a volume integral as shown below.

$$Q_{\text{enc}} = \int \rho d\tau$$

With the substitution the above two equations become

$$\int \nabla \cdot \mathbf{D} d\tau = Q_{\text{enc}} = \int \rho d\tau$$

$$\int \nabla \cdot \mathbf{B} d\tau = 0$$

Comparing both sides of the equations results in

$$\nabla \cdot \mathbf{D} = \rho$$
$$\nabla \cdot \mathbf{B} = 0$$

Which are the required Maxwell's divergence relations.

EXERCISE SECTION

Essay Type Questions

1. What is the Faraday's law of induction? Explain the significance of the terms 'transformer emf' and 'generator emf'.
2. What is the continuity equation? Derive it from the basics and describe all its forms.
3. State the Maxwell's correction to the Ampere's law for the steady currents and derive this correction from the equation of continuity.
4. Differentiate the Poynting theorem from the complex Poynting theorem. Give a proof for the complex Poynting theorem.
5. What is the significance of the Maxwell's equations? Mention them in their various forms.
6. What is the phasor form of the Maxwell's equations? How can they be derived from the time domain equations? Explain.
7. State the boundary conditions at the interface of two different general media and derive them from the Maxwell's equations.
8. Express the time-varying fields in terms of retarded potential functions. Derive these relations from the Maxwell's equations.
9. Derive the expressions for retarded potentials in terms of source distribution.

Numerical Problems

1. In a medium of $\varepsilon = 5\varepsilon_0$, $\mu = 2.5\mu_0$, $s = 0.2$ mΩ/m and $E = 20$ μV/m, find the conduction current density. If this current density exists in a cylindrical rod of 2 cm diameter, evaluate the current that can flow through the rod.

2. A copper wire carries a conduction current of 1 A. Determine the displacement current in the wire at 100 MHz. Take $\varepsilon = \varepsilon_0$ and conductivity 5.8×10^7 ℧/m.
3. In a material for which $\sigma = 5.0(\Omega\text{m})^{-1}$ and $\varepsilon_r = 1$, the electric field intensity is $E = 250 \sin 10^{10} t$ V/m. Find the conduction and displacement current densities and the frequency at which they have equal magnitudes.
4. In free space, $\mathbf{D} = D_m \sin(\omega t + bz)\,\hat{\mathbf{x}}$. Determine $\mathbf{B}$ and the displacement current density.
5. In a perfect dielectric medium, the EM wave has a maximum value of E of 10 V/m with $\mu_r = 1$ and $\varepsilon_r = 4$. Find the velocity of the wave, peak Poynting vector, average Poynting vector, impedance of the medium and peak value of the magnetic field.

Fill in the Blanks

1. Equation of continuity signifies __________ conservation.
2. Faraday's law gives __________ force.
3. Ampere's law gives __________ force.
4. Guass law gives electric __________.
5. Divergence-less of magnetic field indicates non-existence of magnetic __________.
6. The phasors are the functions of __________ and independent of time.
7. To get the phasor form equations, $\partial/\partial t$ is replaced by __________.
8. Faraday's law says changing magnetic field induces __________ field.
9. Maxwell's correction says changing electric field is equivalent to __________ field.
10. Relative motion between field and coil results in __________.
11. Time-varying magnetic flux results in __________.
12. Lenz's law gives the __________ of induced emf.
13. Tangential __________ field is always continuous across the interface.
14. Tangential __________ field is continuous across the surface current-free interface.
15. Normal __________ flux is always continuous across the interface.
16. Normal __________ displacement is continuous across the surface charge-free interface.
17. Point form is another name of __________ form Maxwell's equations.
18. Point form equations relate fields at a __________.
19. Integral form equations relate fields of different __________.
20. Phasor form equations are __________ independent.

Answers

1. charge
2. electromotive
3. magnetomotive
4. displacement
5. monopoles
6. position

7. $j\omega$
8. electric
9. magnetic
10. motional emf
11. transformer emf
12. polarity
13. electric
14. magnetic
15. magnetic
16. electric
17. differential
18. point
19. locations
20. time

Match the Following

Set I

1. Faraday's law	(a) changing electric field is magnetic field
2. Transformer emf	(b) polarity of induced emf
3. Generator emf	(c) relative motion between field and coil
4. Lenz's law	(d) time-varying magnetic flux
5. Maxwell's correction	(e) emf $= -N\dfrac{d\phi}{dt}$

Set II

1. Faraday's law	(a) $\oint \mathbf{E}\cdot d\mathbf{l} = -\int \dfrac{\partial \mathbf{B}}{\partial t}\cdot d\mathbf{a}$
2. Transformer emf	(b) $\oint \mathbf{E}\cdot d\mathbf{l} = -\dfrac{d}{dt}\int \mathbf{B}\cdot d\mathbf{a}$
3. Generator emf	(c) $\oint \mathbf{E}\cdot d\mathbf{l} = \oint \mathbf{v}\times\mathbf{B}\cdot d\mathbf{l}$
4. Lenz's law	(d) displacement current
5. Maxwell's correction	(e) induced current opposes the change of flux

Set III

1. Ampere's law	(a) $\nabla\cdot\mathbf{J} = -\dfrac{\partial\rho}{\partial t}$
2. Faraday's law	(b) $\nabla\cdot\mathbf{B} = 0$
3. Gauss law	(c) $\nabla\cdot\mathbf{D} = \rho$
4. No monopole	(d) $\nabla\times\mathbf{E} = -\dfrac{\partial\mathbf{B}}{\partial t}$
5. Continuity equation	(e) $\nabla\times\mathbf{H} = \mathbf{J} + \dfrac{\partial\mathbf{D}}{\partial t}$

Set IV

1. Curl of magnetic field	(a) Faraday's law
2. Mmf	(b) Ampere's law
3. Electric displacement	(c) Gauss law
4. Divergence-less magnetic field	(d) continuity equation
5. Charge conservation	(e) no monopole

Set V

1. Ampere's law	(a) $\oint_s \mathbf{J}\cdot d\mathbf{a} = -\dfrac{d}{dt}\int_v \rho d\tau$
2. Faraday's law	(b) $\oint \mathbf{B}\cdot d\mathbf{a} = 0$
3. Gauss law	(c) $\oint \mathbf{D}\cdot d\mathbf{a} = Q_{\text{enc}}$
4. No monopole	(d) $\oint \mathbf{E}\cdot d\mathbf{l} = -\int \dfrac{\partial \mathbf{B}}{\partial t}\cdot d\mathbf{a}$
5. Continuity equation	(e) $\oint \mathbf{H}\cdot d\mathbf{l} = \int\left(\mathbf{J} + \dfrac{\partial \mathbf{D}}{\partial t}\right)\cdot d\mathbf{a}$

Set VI

1. Emf	(a) Faraday's law
2. Curl of magnetic field	(b) Ampere's law
3. Divergence of electric field	(c) Gauss law
4. Divergence of magnetic field	(d) continuity equation
5. Divergence of current density	(e) no monopole

Set VII

1. Coulomb gauge	(a) time delay
2. Lorentz gauge	(b) does not obey relativity
3. Retarded potentials	(c) transverse nature wave
4. Radiation	(d) obeys relativity
5. Retardation	(e) time-varying potentials

Answers

Set I:	1. (e)	2. (d)	3. (c)	4. (b)	5. (a)
Set II:	1. (b)	2. (a)	3. (c)	4. (e)	5. (d)
Set III:	1. (e)	2. (d)	3. (c)	4. (b)	5. (a)

Set IV:	1. (b)	2. (a)	3. (c)	4. (e)	5. (d)
Set V:	1. (e)	2. (d)	3. (c)	4. (b)	5. (a)
Set VI:	1. (b)	2. (a)	3. (c)	4. (e)	5. (d)
Set VII:	1. (b)	2. (e)	3. (d)	4. (a)	5. (c)

Multiple Choice Questions

1. The changing magnetic field induces electric field is the law given by
   (a) Maxwell's correction (b) Faraday's law
   (c) both (a) and (b) (d) none of these
2. The changing electric field is equivalent to magnetic field is given by
   (a) Maxwell's correction (b) Faraday's law
   (c) both (a) and (b) (d) none of these
3. The motional emf is due to
   (a) changing magnetic field
   (b) relative motion between electric field and coil
   (c) both (a) and (b)
   (d) none of these
4. The transformer emf is due to
   (a) changing magnetic field
   (b) relative motion between electric field and coil
   (c) both (a) and (b)
   (d) none of these
5. The Lenz's law is about
   (a) amount of induced emf (b) direction of induced emf
   (c) both (a) and (b) (d) none of these
6. The motional emf is given by
   (a) $\oint \mathbf{E} \cdot d\mathbf{l} = \oint \mathbf{v} \times \mathbf{B} \cdot d\mathbf{l}$ (b) $\oint \mathbf{E} \cdot d\mathbf{l} = -\int \frac{\partial \mathbf{B}}{\partial t} \cdot d\mathbf{a}$
   (c) both (a) and (b) (d) none of these
7. The transformer emf is given by
   (a) $\oint \mathbf{E} \cdot d\mathbf{l} = \oint \mathbf{v} \times \mathbf{B} \cdot d\mathbf{l}$ (b) $\oint \mathbf{E} \cdot d\mathbf{l} = -\int \frac{\partial \mathbf{B}}{\partial t} \cdot d\mathbf{a}$
   (c) both (a) and (b) (d) none of these
8. The Maxwell's correction is to
   (a) Gauss law (b) Ampere's law
   (c) Faraday's law (d) none of these

9. The displacement current density $\mathbf{J}_D$ is
   (a) $\dfrac{\partial \mathbf{D}}{\partial t}$ (b) $\dfrac{\partial \mathbf{B}}{\partial t}$
   (c) $\dfrac{\partial \mathbf{E}}{\partial t}$ (d) none of these
10. The conduction current density $\mathbf{J}_C$ is
   (a) $\dfrac{\partial \mathbf{D}}{\partial t}$ (b) $\dfrac{\partial \mathbf{B}}{\partial t}$
   (c) $\dfrac{\partial \mathbf{E}}{\partial t}$ (d) $\sigma \mathbf{E}$
11. The electric field due to changing magnetic field is
   (a) conservative (b) non-conservative
   (c) both (a) and (b) (d) none of these
12. Transformer and generator work based on
   (a) Faraday's law (b) Ampere's law
   (c) both (a) and (b) (d) none of these
13. The Poynting vector is
   (a) $\mathbf{E} \times \mathbf{H}$ (b) power flow per unit area
   (c) both (a) and (b) (d) none of these
14. The complex Poynting vector is
   (a) $\dfrac{1}{2}\mathbf{E} \times \mathbf{H}^*$ (b) $\dfrac{1}{2}\mathbf{E} \times \mathbf{H}$
   (c) $\mathbf{E} \times \mathbf{H}^*$ (d) none of these
15. The power loss in plane conductor is
   (a) $R_s K_{\text{eff}}^2$ (b) $R_s J_{\text{eff}}^2$
   (c) $R_s I_{\text{eff}}^2$ (d) none of these
16. The Maxwell's equations are
   (a) div and curl of **E** and **H** fields
   (b) coupled equations of **E** and **H** fields
   (c) only curl equations which are independent
   (d) all are correct
17. The differential form of Maxwell's equations give
   (a) relation between fields at a point
   (b) relation between fields at different points
   (c) both (a) and (b)
   (d) none of these

18. The integral form of Maxwell's equations give
    (a) relation between fields at a point
    (b) relation between fields at different points
    (c) both (a) and (b)
    (d) none of these
19. The transformation between differential and integral forms requires
    (a) Stokes theorem (b) Gauss theorem
    (c) both (a) and (b) (d) none of these
20. The phasor form results in
    (a) replacement of $\partial/\partial t$ with $j\omega$
    (b) fields and sources becoming independent of time
    (c) both (a) and (b)
    (d) none of these
21. The equation of continuity is
    (a) $\nabla \cdot \mathbf{J} = -\dfrac{\partial \rho}{\partial t}$
    (b) charge conservation principle
    (c) both (a) and (b)
    (d) none of these
22. The tangential **E** is continuous across the interface between
    (a) two conductors (b) two dielectrics
    (c) a conductor and a dielectric (d) any two general media
23. The tangential **D** is discontinuous across the interface equal to the ratio of
    (a) permittivities of media (b) permeabilities of media
    (c) conductivities of media (d) none of these
24. The tangential **H** is discontinuous across the interface equal to the ratio of
    (a) permittivities of media (b) surface current density
    (c) conductivities of media (d) none of these
25. The tangential **B** is in the absence of currents, discontinuous across the interface equal to the ratio of
    (a) permittivities of media (b) permeabilities of media
    (c) conductivities of media (d) none of these
26. Surface charge density can exist only at the interface of
    (a) lossy dielectrics (b) a dielectric and a conductor
    (c) both (a) and (b) (d) none of these
27. Surface current density can exist only at the interface of
    (a) perfect conductor (b) a dielectric and a conductor
    (c) both (a) and (b) (d) none of these

28. At the interface of perfect dielectrics, there can exist
    (a) surface current (b) surface charge
    (c) both (a) and (b) (d) none of these
29. At the interface of perfect conductor and perfect dielectric, there can exist
    (a) surface current (b) surface charge
    (c) both (a) and (b) (d) none of these
30. In the absence of surface charge, the normal **E** across the interface is in the ratio of
    (a) permittivities of media (b) permeabilities of media
    (c) conductivities of media (d) none of these
31. The normal **D** across the interface is discontinuous by an amount equal to
    (a) surface charge density (b) surface current density
    (c) ratio of permittivities of media (d) none of these
32. The normal **H** is continuous across the interface of
    (a) two conductors (b) two dielectrics
    (c) a conductor and a dielectric (d) any two general media
33. The normal **B** across the interface is discontinuous by an amount equal to the ratio of
    (a) permittivities of media (b) permeabilities of media
    (c) conductivities of media (d) none of these

Answers

1. (b)	2. (a)	3. (d)	4. (a)	5. (b)
6. (a)	7. (b)	8. (b)	9. (a)	10. (d)
11. (b)	12. (a)	13. (c)	14. (a)	15. (a)
16. (d)	17. (a)	18. (b)	19. (c)	20. (c)
21. (c)	22. (d)	23. (a)	24. (b)	25. (b)
26. (c)	27. (a)	28. (d)	29. (c)	30. (a)
31. (a)	32. (d)	33. (b)		

UNIT V

WAVE THEORY

8 Electromagnetic Wave Theory

I've failed over and over and over again in my life and that is why I succeed.

—Michael Jordan

The ubiquitous wireless systems depend upon the electromagnetic radiation phenomenon to transmit signals from one point to another. Radiation is a phenomenon that is associated with the time-varying fields. It was Maxwell who first predicted the existence of radiation in the time-varying fields. Later, Hertz developed the necessary experimental tools to prove the existence of radiation. In this chapter, we discuss the theoretical background to the radiation mechanism.

We derive wave equations for different types of media and observe the dependence of the behaviour of the wave on the properties of the medium. Different configurations of the wave and their properties, particularly the uniform plane wave, are discussed.

8.1 Wave Equations

The existence of wave or radiation in time-varying fields was first predicted by deriving wave equations from the Maxwell's equations which are obeyed by time-varying fields. The experimental proof has come much later. Now we try to derive wave equations and hence the possibility of EM wave in different media using the Maxwell's equations. First, we consider a perfect dielectric, i.e. for which the conductivity is zero and then we consider another medium for which the conductivity is a finite non-zero.

Perfect dielectric medium: In the present derivation, we use the Maxwell's equations in time-varying form. Each dot over the vector denotes partial time differentiation once. Consider the Faraday's law and then apply curl on both sides,

$$\nabla \times \nabla \times \mathbf{E} = -\mu \nabla \times \dot{\mathbf{H}}$$

Replace the curl of the magnetic field on the right-side with the electric field using the Ampere's law with the Maxwell's correction in free space. Now,

$$\nabla \times \mathbf{H} = \varepsilon \dot{\mathbf{E}}$$

So,

$$\nabla \times \nabla \times \mathbf{E} = -\mu\varepsilon \ddot{\mathbf{E}}$$

The left side can be expanded as

$$\nabla \times \nabla \times \mathbf{E} = \nabla\nabla \cdot \mathbf{E} - \nabla^2 \mathbf{E}$$

Therefore, it results in

$$\nabla\nabla \cdot \mathbf{E} - \nabla^2 \mathbf{E} = -\mu\varepsilon \ddot{\mathbf{E}}$$

According to the Gauss law, at any point in the charge-free medium like free space,

$$\nabla \cdot \mathbf{E} = \frac{1}{\varepsilon} \nabla \cdot \mathbf{D} = 0$$

Substituting this relation in the above equation, we obtain

$$\nabla^2 \mathbf{E} = \mu\varepsilon \ddot{\mathbf{E}}$$

Similarly, with the aid of the Maxwell's equations it is possible to derive the following equation:

$$\nabla^2 \mathbf{H} = \mu\varepsilon \ddot{\mathbf{H}}$$

These are wave equations involving time-varying **E** and **H**. Therefore, wave and energy flow must exist in the time-varying fields. In the phasor form, these two equations become

$$\nabla^2 \mathbf{E} = -\omega^2 \mu\varepsilon \mathbf{E}$$
$$\nabla^2 \mathbf{H} = -\omega^2 \mu\varepsilon \mathbf{H}$$

The above two equations are called vector *Helmholtz equations.*

Thus, the wave equation set for free space is

$$\left.\begin{aligned} \nabla^2 \mathbf{E} &= \mu\varepsilon \ddot{\mathbf{E}} \\ \nabla^2 \mathbf{H} &= \mu\varepsilon \ddot{\mathbf{H}} \\ \nabla^2 \mathbf{E} &= -\omega^2 \mu\varepsilon \mathbf{E} \\ \nabla^2 \mathbf{H} &= -\omega^2 \mu\varepsilon \mathbf{H} \end{aligned}\right\} \tag{8.1}$$

Conducting media: Note here that we are not considering a perfect conducting medium: in perfect conducting medium no fields can exist and no wave can travel. The present medium is one for which the conductivity is finite and not infinite.

Consider the Faraday's law in time domain and then apply curl on both sides,

$$\nabla \times \nabla \times \mathbf{E} = -\mu \nabla \times \dot{\mathbf{H}}$$

But in free space, according to the Ampere's law with the Maxwell's correction

$$\nabla \times \mathbf{H} = \mathbf{J} + \varepsilon \dot{\mathbf{E}} = \sigma \mathbf{E} + \varepsilon \dot{\mathbf{E}}$$

as

$$\mathbf{J} = \sigma\mathbf{E}$$

according to the Ohm's law. Therefore,

$$\nabla \times \nabla \times \mathbf{E} = -\mu\sigma\dot{\mathbf{E}} - \mu\varepsilon\ddot{\mathbf{E}}$$

Expanding the left side,

$$\nabla \times \nabla \times \mathbf{E} = \nabla\nabla \cdot \mathbf{E} - \nabla^2\mathbf{E}$$

So

$$\nabla\nabla \cdot \mathbf{E} - \nabla^2 \mathbf{E} = -\mu\sigma\dot{\mathbf{E}} - \mu\varepsilon\ddot{\mathbf{E}}$$

According to the Gauss law, as the medium is free space without any charge,

$$\nabla \cdot \mathbf{E} = \frac{1}{\varepsilon}\nabla \cdot \mathbf{D} = 0$$

Hence,

$$\nabla^2\mathbf{E} = \mu\sigma\dot{\mathbf{E}} + \mu\varepsilon\ddot{\mathbf{E}}$$

This is the wave equation in **E** and similarly the wave equation for **H**

$$\nabla^2\mathbf{H} = \mu\sigma\dot{\mathbf{H}} + \mu\varepsilon\ddot{\mathbf{H}}$$

can be derived using the Maxwell's equation.
In phasor form, these two equations become

$$\begin{aligned}\nabla^2\mathbf{E} &= (j\omega\mu\sigma - \omega^2\mu\varepsilon)\mathbf{E}\\ &= j\omega\mu(\sigma + j\omega\varepsilon)\mathbf{E} = \gamma^2\mathbf{E}\\ \nabla^2\mathbf{H} &= (j\omega\mu\sigma - \omega^2\mu\varepsilon)\mathbf{H}\\ &= j\omega\mu(\sigma + j\omega\varepsilon)\mathbf{H} = \gamma^2\mathbf{H}\end{aligned}$$

Thus, the wave equation set for conducting the medium is

$$\left.\begin{aligned}\nabla^2\mathbf{E} &= \mu\sigma\dot{\mathbf{E}} + \mu\varepsilon\ddot{\mathbf{E}}\\ \nabla^2\mathbf{H} &= \mu\sigma\dot{\mathbf{H}} + \mu\varepsilon\ddot{\mathbf{H}}\\ \nabla^2\mathbf{E} &= \gamma^2\mathbf{E}\\ \nabla^2\mathbf{H} &= \gamma^2\mathbf{H}\end{aligned}\right\} \tag{8.2}$$

8.1.1 Propagation Constant γ

The constant γ is known as *propagation constant* of the medium and in general a complex quantity having both the real and imaginary parts. The real part is called *attenuation constant* σ and the imaginary part is called *phase shift constant β*.

$$\gamma = \alpha + j\beta = \sqrt{j\omega\mu(\sigma + j\omega\varepsilon)} \tag{8.3}$$

$$\alpha = \omega\sqrt{\frac{\mu\varepsilon}{2}\left(\sqrt{1+\frac{\sigma^2}{\omega^2\varepsilon^2}}-1\right)} \tag{8.4}$$

$$\beta = \omega\sqrt{\frac{\mu\varepsilon}{2}\left(\sqrt{1+\frac{\sigma^2}{\omega^2\varepsilon^2}}+1\right)} \tag{8.5}$$

Proof: Both the propagation constant and its square are complex quantities.

$$\gamma = \alpha + j\beta = \sqrt{j\omega\mu(\sigma + j\omega\varepsilon)}$$
$$\gamma^2 = \alpha^2 - \beta^2 + 2j\alpha\beta = j\omega\mu(\sigma + j\omega\varepsilon)$$

Comparing the real and imaginary parts of both sides, we obtain

$$\alpha^2 - \beta^2 = -\omega^2\mu\varepsilon$$
$$2\alpha\beta = \omega\mu\sigma$$

Finding the value of β from the second relation, we get

$$\beta = \frac{\omega\mu\sigma}{2\alpha}$$

and substituting it in the first one, we have

$$\alpha^2 - \left(\frac{\omega\mu\sigma}{2\alpha}\right)^2 = -\omega^2\mu\varepsilon$$

$$\alpha^4 + \alpha^2\omega^2\mu\varepsilon - \left(\frac{\omega\mu\sigma}{2}\right)^2 = 0$$

Solving this quadratic equation in α^2, we get

$$\alpha^2 = \frac{1}{2}\left[-\omega^2\mu\varepsilon \pm \sqrt{(\omega^2\mu\varepsilon)^2 + 4\left(\frac{\omega\mu\sigma}{2}\right)^2}\right]$$
$$= \frac{\omega^2\mu\varepsilon}{2}\left[-1 \pm \sqrt{1+\left(\frac{\sigma}{\omega\varepsilon}\right)^2}\right]$$

Here we have to consider positive sign because the free space is a passive medium and it cannot have negative attenuation. Thus,

$$\alpha^2 = \frac{\omega^2\mu\varepsilon}{2}\left[\sqrt{1+\left(\frac{\sigma}{\omega\varepsilon}\right)^2} - 1\right]$$

It results in

$$\alpha = \sqrt{\frac{\omega^2\mu\varepsilon}{2}\left[\sqrt{1+\left(\frac{\sigma}{\omega\varepsilon}\right)^2} - 1\right]}$$

$$= \omega\sqrt{\frac{\mu\varepsilon}{2}\left[\sqrt{1+\left(\frac{\sigma}{\omega\varepsilon}\right)^2}-1\right]}$$

In a similar manner, it can be proved very easily that

$$\beta = \omega\sqrt{\frac{\mu\varepsilon}{2}\left(\sqrt{1+\frac{\sigma^2}{\omega^2\varepsilon^2}}+1\right)}$$

For good dielectrics

$$\alpha = \omega\sqrt{\frac{\mu\varepsilon}{2}\left(\sqrt{1+\frac{\sigma^2}{\omega^2\varepsilon^2}}-1\right)}$$

$$\cong \omega\sqrt{\frac{\mu\varepsilon}{2}\left(1+\frac{\sigma^2}{2\omega^2\varepsilon^2}-1\right)} = \frac{\sigma}{2}\sqrt{\frac{\mu}{\varepsilon}} \tag{8.6}$$

$$\beta = \omega\sqrt{\frac{\mu\varepsilon}{2}\left(\sqrt{1+\frac{\sigma^2}{\omega^2\varepsilon^2}}+1\right)}$$

$$\cong \omega\sqrt{\frac{\mu\varepsilon}{2}\left(1+\frac{\sigma^2}{2\omega^2\varepsilon^2}+1\right)}$$

$$= \omega\sqrt{\mu\varepsilon}\left(1+\frac{\sigma^2}{8\omega^2\varepsilon^2}\right) \tag{8.7}$$

For good conductors

$$\gamma = \sqrt{j\omega\mu(\sigma+j\omega\varepsilon)}$$

$$= \sqrt{j\omega\mu\sigma\left(1+\frac{j\omega\varepsilon}{\sigma}\right)}$$

$$\cong \sqrt{j\omega\mu\sigma} = \sqrt{\omega\mu\sigma}\,\angle 45° \tag{8.8}$$

Therefore,

$$\alpha = \beta = \sqrt{\frac{\omega\mu\sigma}{2}}$$

8.2 ELECTROMAGNETIC WAVES

The term *wave* refers to a phenomenon in which repetition of some activity occurs with respect to space as well as time coordinates. Formally it can be defined as a physical phenomenon that occurs at one place at a given time is reproduced at other places at later times, the time delay being proportional to the space separation from the first location. Then the group of phenomena constitutes the wave.

8.2.1 Classification

Wherever there exists the time-varying field, there exists the wave and the converse is also true. The electromagnetic waves can be classified into following four categories:

- *Transverse electromagnetic (TEM) wave:* In this type of wave, also known as *principal* wave, both the electric vector **E** and the magnetic vector **H** are entirely normal to the direction of the propagation of the wave. In addition, the electric vector **E**, the magnetic vector **H** and the direction of propagation, all the three vectors, form a right-handed vector system. The electromagnetic energy travels as TEM wave in free space and over parallel wire transmission line. The coaxial lines can also hold this type of wave.

  The phase velocity and the group velocity are the same for TEM wave. Neither one depends upon the frequency. So the TEM wave is a non-dispersive wave.

  The TEM wave can be either a plane wave or a cylindrical or spherical wave. A plane wave is characterized by a disturbance that at a given point in time has uniform properties across an infinite plane perpendicular to the direction of wave propagation, and similarly for cylindrical and spherical waves the disturbances are uniform across the cylindrical and spherical surfaces. All the three are three-dimensional waves as they propagate through a volume and its disturbance may be a function of all three space variables (Figure 8.1).

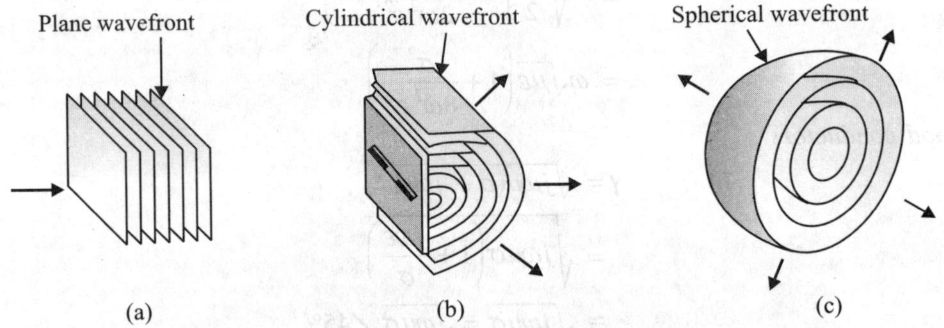

FIGURE 8.1 Different types of waves: (a) plane waves, (b) cylindrical waves and (c) spherical waves.

- *Transverse electric (TE) wave:* In this wave, the electric vector is entirely normal to the direction of propagation and hence it has no component in the direction of propagation. The magnetic vector has both the normal and parallel components.
- *Transverse magnetic (TM) wave:* In this wave, the magnetic vector is entirely normal to the direction of propagation and hence it has no component in the direction of propagation. The electric vector has both the normal and parallel components.
- *Hybrid or mixed wave:* It is a linear combination of TE and TM waves. In this wave, both the electric and magnetic vectors possess both the components, normal and parallel, to the direction of propagation of the wave.

Non-TEM waves, i.e. TE, TM and TE + TM waves, exist in hollow-pipe waveguides. The phase velocity differs from the group velocity in the case of non-TEM waves. And both depend upon the frequency. So the non-TEM waves are always dispersive in nature.

The coaxial line, in addition to the TEM wave, can carry higher-order forms of TM and TE waves with components of electric or magnetic field in the direction of the line axis. However, for the usual coaxial lines, the dimensions are small enough that the lines are operating at frequencies far below cut-off for these modes.

8.2.2 Uniform Plane Waves

We frequently encounter uniform plane waves (UPWs) in practice, particularly at the receiving antenna. Let us first define the term. A plane wave is a wave whose equi-phase surfaces are planes. A plane wave has no electric- or magnetic field components along its direction of propagation. A uniform plane wave on the other hand can be defined as a wave in which the electric and magnetic field intensities are directed in fixed directions in space and are constant in magnitude and phase on planes perpendicular to the direction of propagation. It is characterized by electric and magnetic fields that have uniform properties at all points across an infinite plane.

For a field to be constant in amplitude and phase on infinite planes, the source must also be infinite in extent. In this sense, a plane wave cannot be generated in practice. However, in many practical situations, like signal from the far away TV transmitter or signal of a satellite at its earth station, wave can be approximated to a plane wave. Now we try to find out the implications of the basic defining properties of a uniform plane wave.

Property 1. There exists no component along the direction of propagation of the wave.

It implies that both the electric vector and the magnetic vector must be entirely normal to the direction of propagation.

Proof: By definition, UPW is one for which the amplitude and phase are constant over the planes which are normal to the direction of propagation. Therefore, **E** and **H** must be constant over the normal planes and as a consequence, their derivatives with respect to the coordinates normal to the propagation direction must be zero.

Consider the UPW travelling in the z^+-direction. Its derivatives with respect to x and y must be zero, i.e.,

$$\frac{\partial \mathbf{E}}{\partial x} = \frac{\partial \mathbf{E}}{\partial y} = \frac{\partial \mathbf{H}}{\partial x} = \frac{\partial \mathbf{H}}{\partial y} = 0$$

The wave equation in this case becomes

$$\nabla^2 \mathbf{E} = \frac{\partial^2 \mathbf{E}}{\partial x^2} + \frac{\partial^2 \mathbf{E}}{\partial y^2} + \frac{\partial^2 \mathbf{E}}{\partial z^2} = \frac{\partial^2 \mathbf{E}}{\partial z^2} = \mu\varepsilon \frac{\partial^2 \mathbf{E}}{\partial t^2}$$

$$\frac{\partial^2 E_x}{\partial z^2}\hat{\mathbf{x}} + \frac{\partial^2 E_y}{\partial z^2}\hat{\mathbf{y}} + \frac{\partial^2 E_z}{\partial z^2}\hat{\mathbf{z}} = \mu\varepsilon\left(\frac{\partial^2 E_x}{\partial t^2}\hat{\mathbf{x}} + \frac{\partial^2 E_y}{\partial t^2}\hat{\mathbf{y}} + \frac{\partial^2 E_z}{\partial t^2}\hat{\mathbf{z}}\right)$$

Comparing the z-components,

$$\frac{\partial^2 E_z}{\partial z^2} = \mu\varepsilon \frac{\partial^2 E_z}{\partial t^2} \tag{8.9}$$

Now consider the divergence of electric field in free space.

$$\varepsilon\nabla \cdot \mathbf{E} = \varepsilon\left(\frac{\partial E_x}{\partial x} + \frac{\partial E_y}{\partial y} + \frac{\partial E_z}{\partial z}\right) = \rho$$

As it is UPW in the z^+-direction, the derivatives with respect to x and y become zero and it is in free space, so the charge density $\rho = 0$. Incorporating these two aspects, we have

$$\frac{\partial E_z}{\partial z} = 0 \tag{8.10}$$

Substituting Eq. (8.10) in Eq. (8.9), we have

$$\frac{\partial^2 E_z}{\partial t^2} = 0 \tag{8.11}$$

This differential equation says that the field component E_z must be either zero or constant or increasing uniformly with time. Neither one of the last two can be part of wave motion. Therefore, a UPW progressing in the z-direction cannot have a z-component of **E**. A similar analysis can show that **H** cannot have a component along the z-direction.

It follows, therefore, that the uniform plane waves are transverse in nature and have components of fields only in directions normal to the directions of propagation.

Property 2. The ratio of **E** to **H** for a UPW in free space is given by $\mathbf{E}/\mathbf{H} = \sqrt{\mu/\varepsilon}$.

Proof: In the case of a UPW travelling in the z-direction, the derivatives of **E** and **H** with respect to and x and y must be zero as well as their components along the z-direction. Inserting these two conditions and expanding one of the Maxwell's curl equations

$$\nabla \times \mathbf{H} = \varepsilon\frac{\partial E}{\partial t}$$

we have

$$-\frac{\partial H_y}{\partial z}\hat{\mathbf{x}} + \frac{\partial H_x}{\partial z}\hat{\mathbf{y}} = \varepsilon\left(\frac{\partial E_x}{\partial t}\hat{\mathbf{x}} + \frac{\partial E_y}{\partial t}\hat{\mathbf{y}}\right)$$

Comparing the y-components, we get

$$\frac{\partial H_x}{\partial z} = \varepsilon\frac{\partial E_y}{\partial t}$$

Now let us suppose $E_y = f_1(z - v_0 t)$ with $v_0 = 1/\sqrt{\mu\varepsilon}$, speed of the wave. Substituting in the above expression,

$$\frac{\partial H_x}{\partial z} = \varepsilon \frac{\partial E_y}{\partial t} = \varepsilon(-v_0)\frac{\partial f_1(z - v_0 t)}{\partial (z - v_0 t)}$$

$$H_x = -v_0 \varepsilon \int \frac{\partial f_1(z - v_0 t)}{\partial (z - v_0 t)} dz = -v_0 \varepsilon \int \frac{\partial f_1(z - v_0 t)}{\partial z} \frac{\partial z}{\partial (z - v_0 t)} dz$$

$$= -v_0 \varepsilon \int \frac{\partial f_1(z - v_0 t)}{\partial z} dz = -v_0 \varepsilon f_1 + C$$

Here C is the integration constant which has to be ignored as it cannot represent wave motion. Therefore,

$$\frac{f_1}{H_x} = \frac{E_y}{H_x} = -\frac{1}{v_0 \varepsilon} = \frac{-\sqrt{\mu\varepsilon}}{\varepsilon} = -\sqrt{\frac{\mu}{\varepsilon}}$$

Similarly it can be shown that

$$\frac{E_x}{H_y} = \sqrt{\frac{\mu}{\varepsilon}} \tag{8.12}$$

The ratio of E to H, where E and H are total electric and magnetic field intensities is known as intrinsic impedance of the medium. Its symbol is η. As

$$E = \sqrt{E_x^2 + E_y^2}$$

and

$$H = \sqrt{H_x^2 + H_y^2}$$

the intrinsic impedance of the medium becomes

$$\eta = \frac{E}{H} = \frac{\sqrt{E_x^2 + E_y^2}}{\sqrt{H_x^2 + H_y^2}} = \sqrt{\frac{\mu}{\varepsilon}}$$

which can be proved as shown below.

Let us suppose $\frac{E_x}{H_y} = \sqrt{\frac{\mu}{\varepsilon}} = k \Rightarrow E_x^2 = kH_y^2$ and $H_x^2 = kE_y^2$

Substituting in the expression for intrinsic impedance, we get

$$\eta = \frac{E}{H} = \frac{\sqrt{E_x^2 + E_y^2}}{\sqrt{H_x^2 + H_y^2}} = \frac{\sqrt{kH_y^2 + E_y^2}}{\sqrt{\frac{E_y^2}{k} + H_y^2}}$$

$$= k\frac{\sqrt{kH_y^2 + E_y^2}}{\sqrt{E_y^2 + kH_y^2}} = k = \sqrt{\frac{\mu}{\varepsilon}}$$

In the case of free space, the intrinsic impedance becomes

$$\eta_v = \sqrt{\frac{\mu_0}{\varepsilon_0}} = 377\ \Omega \tag{8.13}$$

The intrinsic impedance of a medium in general is a complex quantity and is given by

$$\boldsymbol{\eta} = \sqrt{\frac{j\omega\mu}{\sigma + j\omega\varepsilon}}$$

In the present case, as the free space is a perfect dielectric, the conductivity is zero, resulting in a pure real value. Thus,

$$\boldsymbol{\eta} = |\boldsymbol{\eta}| = \eta = \sqrt{\frac{\mu}{\varepsilon}}$$

Property 3. If the medium happens to be of non-zero conductivity, then it can be shown that its intrinsic impedance is

$$\frac{\mathbf{E}}{\mathbf{H}} = \sqrt{\frac{j\omega\mu}{\sigma + j\omega\varepsilon}} \tag{8.14}$$

Proof: Consider the phasor form of the Ampere's law and then

$$\nabla \times \mathbf{H} = \mathbf{J} + j\omega\varepsilon\mathbf{E} = \sigma\mathbf{E} + j\omega\varepsilon\mathbf{E} = (\sigma + j\omega\varepsilon)\mathbf{E}$$

Expand the curl and equate the x-components. It results in

$$\frac{\partial H_z}{\partial y} - \frac{\partial H_y}{\partial z} = (\sigma + j\omega\varepsilon)E_x$$

The first term on the left side is zero because for the z^+-travelling wave, the derivatives of **E** and **H** with respect to x and y must be zero. And

$$\frac{\partial H_y}{\partial z} = -\gamma H_y = -\sqrt{j\omega\mu(\sigma + j\omega\varepsilon)}H_y$$

Therefore,

$$\sqrt{j\omega\mu(\sigma + j\omega\varepsilon)}H_y = (\sigma + j\omega\varepsilon)E_x$$

$$\frac{\sqrt{j\omega\mu(\sigma + j\omega\varepsilon)}}{(\sigma + j\omega\varepsilon)} = \frac{E_x}{H_y}$$

$$\frac{E_x}{H_y} = \sqrt{\frac{j\omega\mu}{(\sigma + j\omega\varepsilon)}}$$

Therefore, the intrinsic impedance of a medium in general is a complex quantity and is given by

$$\boldsymbol{\eta} = \sqrt{\frac{j\omega\mu}{(\sigma + j\omega\varepsilon)}}$$

Property 4. The relative orientation of **E** and **H** is normal to each other. The direction of the wave propagation is indicated by the vector $\mathbf{E} \times \mathbf{H}$.

Proof: The relative orientation of **E** and **H** is determined by taking their dot product.

$$\mathbf{E} \cdot \mathbf{H} = E_x H_x + E_y H_y = \eta H_y H_x - \eta H_y H_x = 0$$

Therefore, in a uniform plane wave, **E** and **H** are perpendicular to each other. Additional information regarding the orientation of vectors and direction of propagation can be obtained by considering their cross product.

$$\mathbf{E} \times \mathbf{H} = \hat{\mathbf{z}}(\mu H_x^2 + \mu H_y^2) = \hat{\mathbf{z}}\eta H^2$$

As the wave under consideration is z-travelling one, we can conclude that electric field vector crossed magnetic field vector gives the direction of propagation of the wave. Hence, we can state that the field vectors **E** and **H** are perpendicular to the direction of propagation of the wave in addition to being perpendicular to each other. In fact, the vectors **E**, **H** and the direction of propagation of the wave form the RH vector system. When we rotate the four fingers of the right hand from the direction of **E** towards the direction of **H**, the thumb will point in the direction of the wave travel.

Property 5. Expressions for fields in the uniform plane wave in the phasor form are: for the electric vector of a wave travelling in the z^+-direction, it is $\mathbf{E}(z) = \mathbf{E}_0 e^{-j\beta z}$; for a wave travelling in the z-direction, it is $\mathbf{E}(z) = \mathbf{E}_0 e^{j\beta z}$. In a similar manner expressions for other field components and other directions can be written. In the case of arbitrary direction travelling wave, we have to use direction cosines.

$$\mathbf{E}(\mathbf{r}) = \mathbf{E}_0 e^{-j\beta \hat{\mathbf{n}} \cdot \mathbf{r}} = \mathbf{E}_0 e^{-j\beta(x \cos A + y \cos B + z \cos C)}$$

Proof: The wave equation in electric field **E** is

$$\nabla^2 \mathbf{E} = \mu\varepsilon \ddot{\mathbf{E}}$$

In the case of z^+-travelling uniform plane wave, it becomes

$$\frac{\partial^2 E_x}{\partial z^2} = \mu\varepsilon \frac{\partial^2 E_x}{\partial t^2}$$

$$\frac{\partial^2 E_y}{\partial z^2} = \mu\varepsilon \frac{\partial^2 E_y}{\partial t^2}$$

Both are partial differential equations and are in the same form. Therefore, their solutions also remain in the same form. If we consider the first equation, its general solution, from differential calculus, is

$$E_x = f_1(z - v_0 t) + f_2(z + v_0 t) \tag{8.14a}$$

Here $v_0 = 1/\sqrt{\mu\varepsilon}$, f_1 and f_2 are two functions, which need not be the same. Examples for these functions are $A \sin \beta(z - v_0 t)$, $B \cos \beta(z + v_0 t)$, $Ce^{k(z - v_0 t)}$ $\sqrt{(z - v_0 t)}$, etc.

For a wave travelling in the z^+-direction, the expression is $\mathbf{E}(z) = \mathbf{E}_0 e^{-j\beta z}$ and for a wave travelling in the z-direction, it is $\mathbf{E}(z) = \mathbf{E}_0 e^{j\beta z}$.

The most important aspect of all these expressions is that they represent wave phenomenon. The first term in the right-hand side of the Eq. (8.14a) represents wave motion in the z^+-direction and the second term represents wave travelling in the z^+-direction.

Similar will be the expression for the field component E_y because the form of governing differential equation is the same. The magnetic field components also possess expressions of the same form.

Direction cosines: Sometimes it is necessary to write the expression for a plane wave that is travelling in some arbitrary direction with respect to a set of fixed axes. This is most conveniently done in terms of direction cosines of the normal to the plane of the wave.

The equation of a plane wave travelling in the direction $\hat{\mathbf{n}}$, normal to the planes of constant phase, can be written as

$$\mathbf{E}(\mathbf{r}) = \mathbf{E}_0 e^{-j\beta \hat{\mathbf{n}}\cdot r} = \mathbf{E}_0 e^{-j\beta(x\cos A + y\cos B + z\cos C)} \tag{8.15}$$

Here, A, B and C are the angles the ray or wave normal makes with the x, y and z axes respectively. cos A, cos B and cos C are called direction cosines of the wave.

Salient features of UPWs

1. No component along the direction of propagation of the wave.
2. The ratio of **E** to **H** in UPW, i.e. its intrinsic impedance, is

$$\frac{\mathbf{E}}{\mathbf{H}} = \sqrt{\frac{j\omega\mu}{\sigma + j\omega\varepsilon}}$$

3. The relative orientation of **E** and **H** is normal to each other.
4. The vectors **E**, **H** and the direction of propagation form a right-handed vector system.
5. The velocity of the wave propagation in unbounded medium is $1/\sqrt{\mu\varepsilon}$.
6. The power flow in UPW per unit area is E^2/η.
7. Expression for fields in uniform plane wave

$E_x = f_1(z - v_0 t) + f_2(z + v_0 t)$:	Time domain
$\mathbf{E}(z) = \mathbf{E}_0 e^{-j\beta z}$:	Phasor domain
$\mathbf{E}(r) = \mathbf{E}_0 e^{-j\beta\hat{\mathbf{n}}\cdot r} = \mathbf{E}_0 e^{-j\beta(x\cos A + y\cos B + z\cos C)}$:	Direction cosines

8.2.3 Skin Depth and Surface Impedance

A wave falling over the surface of a medium tries to penetrate into the medium. Depending upon the parameters, particularly conductivity, the wave is able to enter into the medium only up to certain depth before it gets weakened completely. Skin depth is a parameter which describes the depth up to which the wave is allowed to enter into the medium.

Another phenomenon that takes place on the fall of the wave over the surface of a medium is flow of currents. If the medium is conductive, the fields are mainly confined to the surface driving surface currents. Surface impedance is a parameter useful to describe the surface currents that flow over the surface of a conductor upon the incidence of wave.

Skin Depth

Skin depth is also known as depth of penetration. When the medium is conductive, the wave gets attenuated as it progresses into the medium. Skin depth or depth of penetration δ is defined as the depth in which the wave gets attenuated to $1/e$ or approximately 37 per cent of its original value.

As the amplitude of the wave decreases exponentially with depth $\alpha\delta = 1$,

$$\delta = \frac{1}{\alpha} = \frac{1}{\omega\sqrt{\frac{\mu\varepsilon}{2}\left(\sqrt{1+\frac{\sigma^2}{\omega^2\varepsilon^2}}-1\right)}} \approx \sqrt{\frac{2}{\omega\mu\sigma}} \tag{8.16}$$

for a good conductor. Skin depth of a conductor is shown in Figure 8.2.

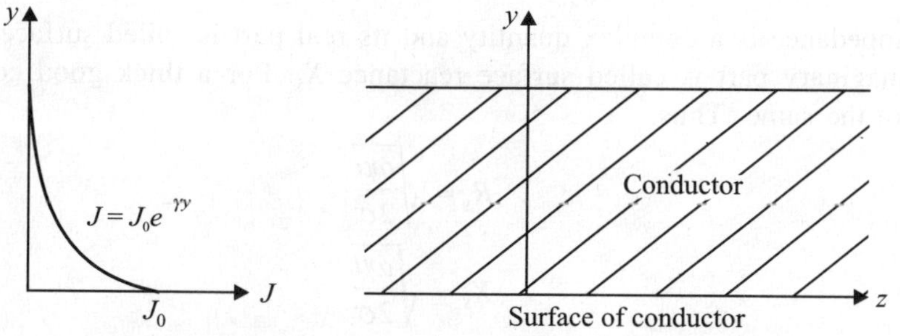

FIGURE 8.2 Illustrating skin depth of a conductor.

Surface Impedance

At high frequencies, when the wave falls over the conductor, the current flows and this current is confined almost entirely to a very thin sheet at the surface of the conductor. The amount of the surface current depends upon the impedance of the surface.

Surface impedance is defined as the ratio of tangential electric field strength at the surface of the conductor to the surface current density that flows as a result of the incident wave. Thus,

$$Z_s = \frac{E_{\tan}}{J_s} \tag{8.17}$$

If the conductor is a flat plate with its surface at $y = 0$ plane, then the current distribution in the y-direction will be $\mathbf{J} = \mathbf{J}_0 e^{-\gamma y}$, where $\mathbf{J}_0$ is the current density at the surface which is related to the tangential electric field through $\mathbf{J}_0 = \sigma\mathbf{E}_{\tan}$.

Assuming the thickness of the conductor plate to be larger than the depth of penetration so that no reflection from the back surface of the conductor occurs, the surface current density becomes

$$\mathbf{J}_s = \int_0^{\infty} \mathbf{J}dy = \mathbf{J}_0 \int_0^{\infty} e^{-\gamma y} dy = \frac{\mathbf{J}_0}{\gamma} \tag{8.18}$$

Therefore, the surface impedance

$$\mathbf{Z}_s = \frac{\mathbf{E}_{\tan}}{\mathbf{J}_s} = \frac{\mathbf{J}_0 \gamma}{\sigma \mathbf{J}_0} = \frac{\gamma}{\sigma}$$

For conducting medium

$$\gamma = \sqrt{j\omega\mu(\sigma + j\omega\mu)} \approx \sqrt{j\omega\mu\sigma}$$

Hence, the surface impedance becomes

$$\mathbf{Z}_s = \frac{\gamma}{\sigma} = \sqrt{\frac{j\omega\mu}{\sigma}} = \sqrt{\frac{\omega\mu}{\sigma}} \angle 45^\circ \tag{8.19}$$

The surface impedance is a complex quantity and its real part is called surface resistance R_s, whereas its imaginary part is called surface reactance X_s. For a thick good conductor, their magnitudes are the same. Thus,

$$R_s \approx \sqrt{\frac{\omega\mu}{2\sigma}} \tag{8.20}$$

$$X_s \approx \sqrt{\frac{\omega\mu}{2\sigma}} \tag{8.21}$$

Observations: The surface impedance is equal to the intrinsic impedance η for the conducting medium. It is also equal to the characteristic impedance of the thick plane conductor. It can be shown that it is also equal to the input impedance of the thick plane conductor when viewed as a transmission line conducting energy into the interior of the metal.

The real part of the surface impedance is surface resistance, with units of ohms. It is the same as the high frequency skin effect resistance per unit length of a flat conductor of unit width. The surface resistance R_s can be related to the depth of penetration or skin depth δ in a conductor. The relation is given by $R_s = 1/\sigma\delta$.

The surface resistance of a flat conductor at any frequency is equal to the dc resistance of a thickness δ of the same conductor. This means that the conductor having a thickness very much greater than δ and having exponential current distribution throughout its depth has the same resistance as would a thickness δ of the conductor with the current distributed uniformly throughout this thickness. It can be shown that the power loss per unit area of the plane conductor is $J_{s\,\text{eff}}^2 R_s$.

8.2.4 Power Loss in Plane Conductor

To find the power loss in the conductor, we need to evaluate the normal component of the Poynting vector at the surface of the conductor. The average or real power flow per unit area normal to the surface is

$$P_n = \frac{1}{2}\mathrm{Re}(\mathbf{E}_{\tan} \times \mathbf{H}_{\tan}^*) \tag{8.22}$$

In good conductor, tangential components of electric and magnetic fields are normal to each other and the electric field leads the magnetic field by 45° in time phase. Thus,

$$P_n = \frac{1}{2}|\mathbf{E}_{\tan}||\mathbf{H}_{\tan}|\cos 45° = \frac{1}{2\sqrt{2}}|\mathbf{E}_{\tan}||\mathbf{H}_{\tan}|$$

$$= \frac{1}{2\sqrt{2}}|\eta_m||\mathbf{H}_{\tan}|^2 = \frac{1}{2\sqrt{2}}\frac{|\mathbf{E}_{\tan}|^2}{|\eta_{\tan}|} \tag{8.23}$$

Here η_m is intrinsic impedance of the conductor and it is given by

$$\eta_m = \sqrt{\frac{\omega\mu_m}{\sigma_m}}\angle 45° \tag{8.24}$$

It is related to the fields through

$$\frac{|\mathbf{E}_{\tan}|}{|\mathbf{H}_{\tan}|} = |\eta_m| \tag{8.25}$$

For good conductors, it is equal to the surface impedance, i.e. $\eta_m = \mathbf{Z}_s$. It results in

$$P_n = \frac{1}{2\sqrt{2}}|\mathbf{Z}_s||\mathbf{H}_{\tan}|^2 = \frac{1}{2\sqrt{2}}\frac{|\mathbf{E}_{\tan}|^2}{|\mathbf{Z}_s|} \tag{8.26}$$

In the case of good conductors, the tangential magnetic field is equal to the surface current, i.e. $|\mathbf{H}_{\tan}| = |\mathbf{K}|$. In the above expression, the fields are peak values which can be converted into rms values. Incorporating these two changes, we have

$$P_n = \frac{1}{\sqrt{2}}\frac{|\mathbf{E}_{\tan}|^2}{|\mathbf{Z}_s|} = \frac{1}{\sqrt{2}}|\mathbf{Z}_s||\mathbf{K}|^2 = R_s\,|\mathbf{K}|^2 \tag{8.27}$$

8.3 POLARIZATION OF WAVES

Basically, polarization refers to the time-varying behaviour of orientation of the electric vector **E** in a uniform plane wave at some fixed point. For a radiated wave, it is defined as that property of an electromagnetic wave which describes the time-varying direction and relative magnitude of the electric field vector. Specifically the figure is traced as a function of time by the extremity of the vector at a fixed location in space and the sense in which it is traced is observed along the direction of propagation.

The importance of polarization lies in the fact that to be able to have a satisfactory wireless communication link, the polarizations of the transmitting antenna or wave must be matched to that of the receiving antenna. If the match is not there, the received signal strength falls correspondingly and in the case of total mismatch, the signal cannot be received at all whatever may be the strength of the wave at the location receiving antenna.

Several types of polarizations are there: linear, elliptical and circular. In the linear type, we have again two important types, one is horizontal type and another is vertical type. Out of the three major types, elliptical polarization is considered the most general one as the other types, are derivable from this polarization.

Polarization is a ***time-domain*** specification of the uniform ***plane wave*** about the ***orientation*** of ***electric vector***.

Linear polarization: If the direction of the electric vector **E** at a point in space in the uniform plane wave remains the same with respect to time, then the wave is said to be *linearly* polarized. Note that in this case, the tip of the **E** vector traces a straight line with time. If the **E** vector is horizontal, then the wave is said to be horizontally polarized and if it is vertical, then the wave is called vertically polarized. In addition to horizontal and vertical polarizations, a linearly polarized wave can have its **E** vector arbitrarily oriented making some angle with the horizontal.

Mathematical expressions: The electric field of a uniform plane wave travelling in the positive z-direction can be expressed in time-varying form as

$$\tilde{\mathbf{E}}(z,t) = \text{Re}(\mathbf{E}_0 e^{-j\beta z} e^{j\omega t}) \tag{8.28}$$

Here $\mathbf{E}_0$ is a complex vector with an arbitrary orientation in the xy-plane and it must be normal to the z-direction as it is of the wave travelling in that direction.

In the case of the x-polarized wave, $\mathbf{E}_0 = \hat{\mathbf{x}}E$ and the corresponding expression of the time-varying electric field becomes

$$\tilde{\mathbf{E}}(z,t) = \text{Re}(\hat{\mathbf{x}}Ee^{-j\beta z} e^{j\omega t}) = \hat{\mathbf{x}}E\cos(\omega t - \beta z) \tag{8.29}$$

In the case of the y-polarized wave, $\mathbf{E}_0 = \hat{\mathbf{y}}E$ and the corresponding expression of the time-varying electric field becomes

$$\tilde{\mathbf{E}}(z,t) = \text{Re}(\hat{\mathbf{y}}Ee^{-j\beta z} e^{j\omega t}) = \hat{\mathbf{y}}E\cos(\omega t - \beta z) \tag{8.30}$$

Two linear polarized waves, which are spatially normal and timewise in phase, with amplitudes E_1 and E_2 are added, resulting in another linear polarized wave. To prove this, let us consider two waves: one polarized along the x-direction and the other along the y-direction both in phase. When combined,

$$\begin{aligned}\tilde{\mathbf{E}}(z,t) &= \hat{\mathbf{x}}E_1\cos(\omega t - \beta z) + \hat{\mathbf{y}}E_2\cos(\omega t - \beta z)\\ &= (\hat{\mathbf{x}}E_1 + \hat{\mathbf{y}}E_2)\cos(\omega t - \beta z)\end{aligned} \tag{8.31}$$

This is an expression for a linear polarized wave; the angle the electric vector makes with the x-axis is given by $\tan^{-1}(E_1/E_2)$. The electric vector in linear polarization is shown in Figure 8.3.

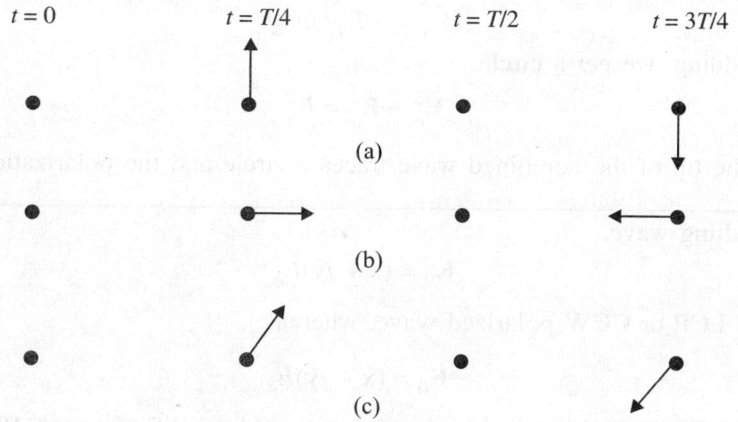

FIGURE 8.3 Electric vector in linear polarization: (a) vertical polarization, (b) horizontal polarization and (c) polarization at an angle to horizontal.

Circular polarization: If the tip of the **E** vector of a travelling plane wave at a point in space with respect to time traces a circle, then the wave is said to be *circularly* polarized as shown in Figures 8.4(a) and (b). It can be shown that the sum of the two equal normal linear polarizations in phase quadrature results in circular polarization.

Proof: To give circular polarization when combined, the equal electric vectors of the two linear polarizations must be spatially perpendicular to each other and timing-wise, they must have a 90° phase difference. Now we try to deduce the result mathematically.

Consider two waves, both linear polarized, travelling along the positive z-direction. Let us suppose that the electric vectors of the two waves are normal to each other, one let us say along the x-direction and the other along the y-direction with amplitudes same and equal to E_a. Let us also suppose that one of the components, let us say the y-component leads the x-component by 90°. Now the expression for the amplitude of electric field $\mathbf{E}_0$ of the combined wave becomes

$$\mathbf{E}_0 = (\hat{\mathbf{x}} + j\hat{\mathbf{y}})\, E_a$$

And the expression in time-varying form for the electric field of the combined wave is

$$\tilde{\mathbf{E}}(z, t) = \mathrm{Re}(\mathbf{E}_0 e^{-j\beta z} e^{j\omega t})$$

Now we show that the tip of the electric vector of the combined wave traces a circle with time resulting in circular polarization. Consider the time-varying field at some point in space let us say at $z = 0$, then the above equation becomes

$$\begin{aligned}\tilde{\mathbf{E}}(0, t) &= \mathrm{Re}[(\hat{\mathbf{x}} + j\hat{\mathbf{y}})\, E_a e^0 e^{j\omega t}] \\ &= \mathrm{Re}[(\hat{\mathbf{x}} \cos \omega t - \hat{\mathbf{y}} \sin \omega t)\, E_a]\end{aligned}$$

In component form

$$\tilde{\mathbf{E}}_x = E_a \cos \omega t$$

and

$$\tilde{\mathbf{E}}_y = - E_a \sin \omega t$$

Squaring and adding, we get a circle.

$$\tilde{\mathbf{E}}_x^2 + \tilde{\mathbf{E}}_y^2 = E_a^2 \tag{8.32}$$

It proves that the tip of the combined wave traces a circle and the polarization is circular.

> For a z-travelling wave,
>
> $$\mathbf{E}_0 = (\hat{\mathbf{x}} + j\hat{\mathbf{y}})E_a$$
>
> gives rise to LCP or CCW polarized wave, whereas
>
> $$\mathbf{E}_0 = (\hat{\mathbf{x}} - j\hat{\mathbf{y}})E_a$$
>
> gives rise to RCP or CW polarized wave because a phase difference of 180° in the form of negative sign is included.
>
> Similarly,
>
> $$\mathbf{E}_0 = (A\hat{\mathbf{x}} + jB\hat{\mathbf{y}})$$
>
> gives rise to LEP or CCW polarized wave, whereas
>
> $$\mathbf{E}_0 = (A\hat{\mathbf{x}} - jB\hat{\mathbf{y}})$$
>
> gives rise to REP or CW polarized wave.

Elliptical polarization: If the tip of the **E** vector of a travelling plane wave at a point in space traces an ellipse with respect to time, then the wave is said to be *elliptically* polarized as shown in Figures 8.4(c) and (d). It is considered the most general form of polarization, because all the other types can be derived from this one.

It can be shown that the sum of two normal linear polarizations with a phase difference and/or unequal amplitudes results in elliptical polarization.

Proof: To result in an elliptically polarized wave when combined, the electric vectors of the two linear polarizations must be spatially perpendicular to each other and timing-wise, they must not be in phase and/or the electric vectors are unequal. Now we try to deduce the result mathematically.

Consider two waves, both linear polarized, travelling along the positive z-direction. Let us suppose that the electric vectors of the two waves are normal to each other, one let us say along the x-direction and the other along the y-direction with amplitudes let us say E_x and E_y. Let us also suppose that one of the components let us say the y-component leads the x-component by 90°. Now the expression for the amplitude of the electric field $\mathbf{E}_0$ of the combined wave becomes

$$\mathbf{E}_0 = (\hat{\mathbf{x}}E_x + j\hat{\mathbf{y}}E_y)$$

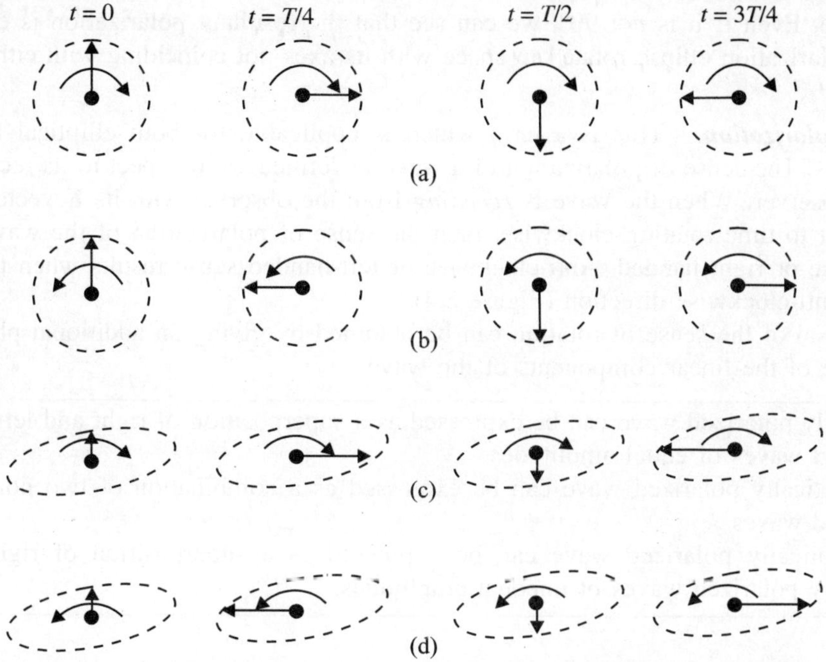

FIGURE 8.4 Rotation of electric vector in circular [(a) and (b)] and elliptical [(c) and (d)] polarizations. Note (a) and (c) denote CW rotation and [(b) and (d)] denote ACW rotation.

And the expression in time-varying form for the electric field of the combined wave is

$$\tilde{\mathbf{E}}(z, t) = \text{Re}(\mathbf{E}_0 e^{-j\beta z} e^{j\omega t})$$

Now we show that the tip of the electric vector of the combined wave traces an ellipse with time resulting in elliptical polarization. Consider the time-varying field at some point in space let us say at $z = 0$; then the above equation becomes

$$\begin{aligned}\tilde{\mathbf{E}}(0, t) &= \text{Re}[(\hat{\mathbf{x}} + j\hat{\mathbf{y}})\, E_a e^0 e^{j\omega t}]\\ &= \text{Re}[\hat{\mathbf{x}}E_x \cos \omega t - \hat{\mathbf{y}}E_y \sin \omega t]\end{aligned}$$

In component form

$$\tilde{\mathbf{E}}_x = E_x \cos \omega t$$

and

$$\tilde{\mathbf{E}}_y = - E_y \sin \omega t$$

Squaring and adding, we get the equation of an ellipse as

$$\frac{\tilde{E}_x^2}{E_x^2} + \frac{\tilde{E}_y^2}{E_y^2} = 1 \tag{8.33}$$

It proves that the tip of the combined wave traces an ellipse and the polarization is elliptical. In the above derivation, we considered a phase difference of 90° for the sake of mathematical convenience. Even if it is not 90°, we can see that the resultant polarization is elliptical but with the polarization ellipse rotated in space with its axes not coinciding with either the x- or the y-axis.

Sense of polarization: This is a term which is applicable for both elliptical and circular polarizations. The sense of polarization of a wave is defined with respect to its receding travel from the observer. When the wave is *receding* from the observer with its **E** vector at a point with respect to time rotating clockwise, then the sense of polarization of the wave is said to be clockwise or right handed. Anti-clockwise or left-handed sense results when the **E** vector rotates in anti-clockwise direction (Figure 8.4).

Reversal of the sense of rotation can be obtained by giving an additional phase shift of 180° to one of the linear components of the wave.

- A linearly polarized wave can be expressed as a superposition of right and left circularly polarized waves of equal amplitudes.
- An elliptically polarized wave can be expressed as a combination of two normal linear polarized waves.
- An elliptically polarized wave can be expressed as a superposition of right and left circularly polarized waves of unequal amplitudes.

8.3.1 Specification of Polarization

The nature of the polarization can be determined completely from the properties of the polarization ellipse. One important parameter of the ellipse is the axial ratio and it is defined as

$$\text{Axial ratio}\left(\frac{b}{a}\right) = \frac{\text{semi-minor axis of the ellipse}}{\text{semi-major axis of the ellipse}} \tag{8.34}$$

Actually it is this factor that determines the shape of the polarization ellipse. An axial ratio of zero or infinity means that the wave is a linearly polarized one. And if it is one, then the wave is circularly polarized. Otherwise, the wave is an elliptically polarized one.

Another parameter of polarization specification is its orientation angle. The orientation of the ellipse is specified in general by the angle the major axis of the ellipse makes with the horizontal axis of the coordinate system. Sense of polarization is also an important polarization specification. Note that for LH polarization, the axial ratio is positive and for RH polarization, it is negative. The polarization ellipse showing the various angles in shown in Figure 8.5.

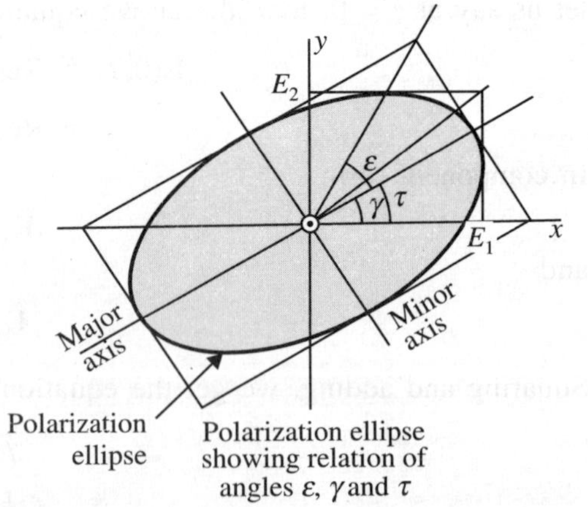

FIGURE 8.5 Polarization ellipse displaying various angles of importance.

Representations for polarization states: There are two types of representations, one is wave polarization representation ($\pm AR$, τ) which uses the axial ratio AR which is +ve for LH and –ve for RH polarizations, tilt angle τ which is the angle the major axis makes with the x-axis, and the other one is electrical quantities representation (E_{y0}/E_{x0}, ϕ). It uses the ratio E_{y0}/E_{x0}, the angle between them is ϕ. If $\varepsilon = \cot^{-1}(\pm AR)$ and $\gamma = \tan^{-1}(E_{y0}/E_{x0})$, then the state can be represented either by the pair (ε, τ) or by the pair (γ, ϕ).

AR varies from 1 for a circle to ∞ for a line, ε varies from $-\pi/4$ to $\pi/4$ and τ varies from 0 to π. ϕ varies from $-\pi/2$ to $\pi/2$ and γ from 0 to π.

8.3.2 Polarization of Antennas

Polarization of the antenna is defined in its transmitting mode. It is, in a given direction, defined as 'polarization of the wave transmitted by the antenna'.

Polarization pattern represents polarization characteristics of the wave the antenna emits and it is defined as 'the spatial distribution of the polarization of a field vector radiated by the antenna taken over its radiation sphere'. At each point over the sphere, the polarization is usually resolved into a pair of orthogonal components along the θ and ϕ directions. These components are called co-polarization and cross polarization.

Primary polarization and cross polarization: An antenna may have two orthogonal linear polarizations. The desired polarization is defined as primary or co-polarization and the other polarization is called cross polarization. But by proper design of the antenna and/or proper selection of the coordinate system, the antenna can be made to have only one polarization, either θ or ϕ.

As an example, we consider the polarization of the dish antenna. When the incident wave is y-polarized, the wave reflected by the dish antenna may contain the x-polarized component in addition to the y-polarized component. The y-component is called principal polarization and the x-component is called cross polarization.

Polarization mismatch occurs due to the mismatch between the polarization of the receiving antenna and the polarization of the incoming wave. Due to this mismatch, the receiving antenna cannot extract maximum amount of power from the incoming wave.

Poincare sphere provides a compact graphical representation of all types of polarizations (Figure 8.6). It is useful to find how close two polarization states are or how much interaction takes place between two states of polarization. Its salient features are given as follows:

- The equator $\varepsilon = 0$, $AR = \infty$ represents linear polarization. The longitude point $\tau = 0$ represents horizontal polorization and $\tau = 90°$ vertical polarization.
- The north pole represents left circular polarization and the south pole right circular polarization.

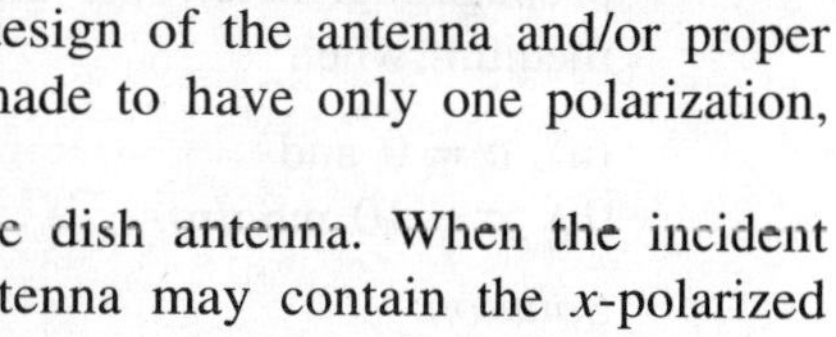

FIGURE 8.6 Poincare sphere.

- The northern hemisphere represents left-handed polarization and the southern hemisphere represents right-handed polarization.
- An arbitrary point on the sphere is denoted by $(2\varepsilon, 2\tau)$. The *xy*-plane or horizontal plane represents $\varepsilon = 0$ and the *xz*-plane or vertical plane represents $\tau = 0$.

Matched states: When the two states of polarization fall on the same point over the Poincare sphere, then they are said to be matched states of polarization. And they can interact maximum.

Orthogonal states: When the two states of polarization fall on radially opposite points on the Poincare sphere, then they are said to be orthogonal states. And no interaction is possible between them.

The power transfer efficiency between the two states represented by M and M' on the Poincare sphere is given by

$$\eta_{\text{pol}} = \cos^2\left\{\frac{\angle MOM'}{2}\right\} \tag{8.35}$$

SOLVED PROBLEMS

1. If $\varepsilon_r = 1$ and $\mu_r = 1$ for the medium in which a wave with frequency $f = 0.3$ GHz is propagating, determine the propagation constant and the intrinsic impedance of the medium when

   (a) $\sigma = 0$ and

   (b) $\sigma = 10$ mho/m.

   Solution:

   (a) Conductivity $\sigma = 0$ signifies that the medium is lossless. For a lossless medium, we have already seen that the propagation constant is pure imaginary. It is given by

   $$\gamma = \sqrt{j\omega\mu(\sigma + j\omega\varepsilon)} = j\omega\sqrt{\mu_0\varepsilon_0}$$

   $$= j2\pi\times 0.3\times 10^9\sqrt{4\pi\times 10^{-7}\times\frac{10^{-9}}{36\pi}} = j2\pi$$

   In the case of lossless medium, the intrinsic impedance is pure real. It is given by

   $$\eta = \sqrt{\frac{j\omega\mu}{\sigma + j\omega\varepsilon}}$$

   $$= \sqrt{\frac{\mu_0}{\varepsilon_0}} = \sqrt{4\pi\times 10^{-7}\times\frac{36\pi}{10^{-9}}} = 120\pi\ \Omega = \sqrt{\frac{\mu_0}{\varepsilon_0}}$$

   (b) In the second case, the conductivity is non-zero given by $\sigma = 10$ mho/m. This is a case of lossy medium and in this case, the propagation constant and the intrinsic impedance are complex.

$$\gamma = \sqrt{j\omega\mu_0(\sigma + j\omega\varepsilon_0)}$$

$$= \sqrt{j2\pi \times 0.3 \times 10^9 \times 4\pi \times 10^{-7}\left[10 + j2\pi \times 0.3 \times 10^9 \times \left(\frac{10^{-9}}{36\pi}\right)\right]}$$

$$= j10\pi\sqrt{24 - j0.04}$$

$$\eta = \sqrt{\frac{j\omega\mu}{\sigma + j\omega\varepsilon}} = \sqrt{\frac{j(2\pi \times 0.3 \times 10^9)\,4\pi \times 10^{-7}}{10 + j(2\pi \times 0.3 \times 10^9)\left(\frac{10^{-9}}{36\pi}\right)}}$$

$$= 2\pi\sqrt{\frac{j60}{10 + j0.0166}}$$

2. Write expressions for travelling wave functions for the following cases:

(a) time variation at $x = 0$ is $10t \cos 20t$ and propagating in the positive x-direction with a velocity of 2 m/s and

(b) distance variation at $t = 0$ in the manner $e^{-|z|} \sin 10z$ and propagating in the positive z-direction with a velocity of 5 m/s.

Solution:

(a) As the wave is propagating in the positive x-direction with a velocity of 2 m/s, the wave function must be of the form

$$f(x, t) = f\left(t - \frac{x}{v_p}\right) = f\left(t - \frac{x}{2}\right)$$

It is given that the time variation at $x = 0$ is $10t \cos 20t$, i.e.

$$f(0, t) = f(t) = 10t \cos 20t$$

Then the required function is

$$f(x, t) = [f(t)]_{t \to t - x/2} = 10\left(t - \frac{x}{2}\right)\cos 20\left(t - \frac{x}{2}\right)$$

$$= 10t - 5x \cos (20t - 10x)$$

(b) As the wave is propagating in the positive z-direction with a velocity of 5 m/s, the wave function must be of the form

$$f(z, t) = f\left(t - \frac{z}{v_p}\right) = f\left(t - \frac{z}{5}\right) = F(z - 5t)$$

It is given that the time variation at $t = 0$ is $e^{-|z|} \sin 10z$, i.e.

$$f(z, 0) = F(z) = e^{-|z|} \sin 10z$$

Then the required function is

$$f(z, t) = [f(t)]_{z \to z-5t} = e^{-|z-5t|} \sin 10(z - 5t)$$
$$= 10t - 5x \cos(20t - 10x)$$

3. The magnetic field intensity associated with a uniform plane wave propagating in free space is given by

$$\mathbf{H} = H_0\left[\cos(9\pi \times 10^8 t + 3\pi y)\hat{\mathbf{x}} + \sin(9\pi \times 10^8 t + 3\pi y)\hat{\mathbf{z}}\right] \text{ A/m}$$

Find the frequency, wavelength, the direction of propagation and the associated electric field intensity.

Solution: From the expression of the field intensity, the angular frequency is

$$\omega = 9\pi \times 10^8$$

Therefore, the frequency becomes

$$f = \frac{\omega}{2\pi} = \frac{9\pi \times 10^8}{2\pi} = 450 \text{ MHz}$$

From the expression of field intensity, the phase shift constant is

$$\beta = 3\pi$$

Therefore, the wavelength becomes

$$\lambda = \frac{2\pi}{\beta} = \frac{2\pi}{3\pi} = \frac{2}{3} \text{ m}$$

The wavelength can also be found from the relation

$$\lambda f = c = 3 \times 10^8 \text{ m/s}$$

By observing the argument ($9\pi \times 10^8 t + 3\pi y$), we can say that the direction of propagation is the negative y-direction. To write down the expression for the electric field, we have to keep two facts in mind: ratio of amplitudes of electric to magnetic fields is $\eta_0 = 120\pi$ and the direction of propagation is in the direction of the cross product of the electric and magnetic fields.

Now let

$$\mathbf{H} = \mathbf{H}_1 + \mathbf{H}_2 = H_0[\cos(9\pi \times 10^8 t + 3\pi y)\hat{\mathbf{x}}] + H_0[\sin(9\pi \times 10^8 t + 3\pi y)\hat{\mathbf{z}}]$$

Keeping the above two points in mind, write down the expressions for the corresponding electric fields due to the above two component magnetic fields.

For the first term, the direction of propagation is negative y and the direction of magnetic field is positive x, therefore, the direction of electric field must be negative z. And the amplitude of the electric field must be $\eta_0 = 120\pi$ times that of the magnetic field. Therefore,

$$\mathbf{E}_1 = -\,\eta_0 H_0[\cos(9\pi\times10^8 t + 3\pi y)]\hat{\mathbf{z}}$$

Similarly we can write down the expression for the second term as

$$\mathbf{E}_2 = \eta_0 H_0[\sin(9\pi\times10^8 t + 3\pi y)]\hat{\mathbf{x}}$$

Therefore,

$$\mathbf{E} = \mathbf{E}_1 + \mathbf{E}_2 = -\,\eta_0 H_0[\cos(9\pi\times10^8 t + 3\pi y)]\hat{\mathbf{z}} + \eta_0 H_0\,[\sin(9\pi\times10^8 t + 3\pi y)]\hat{\mathbf{z}}$$

4. The electric field intensity associated with a plane wave travelling in a perfect dielectric medium is given by $E_x(z,t) = 10\cos(2\pi 10^7 t - 0.1\pi z)$ V/m. Calculate the velocity of propagation and write down the expression for the magnetic field intensity associated with the wave if $\mu = \mu_0$.

Solution: From the given expression of the field, $\omega = 2\pi\times10^7$ rad/s and $\beta = 0.1\pi$. But we know that the velocity of the wave v is

$$v = \frac{\omega}{\beta} = \frac{2\pi\times10^7}{0.1\pi} = 2\times10^8 \text{ m/s}$$

To write down the expression for the electric field, we have to keep two facts in mind: the ratio of amplitudes of electric to magnetic fields is $\eta_0 = 120\pi$ and the direction of propagation is in the direction of the cross product of the electric and magnetic fields.

From the argument of the given expression of the field, we can say that the wave is travelling in the positive z-direction. The electric field is in the positive x-direction. Therefore, the magnetic field must be in the positive y-direction. The ratio of amplitudes is η_0. Incorporating the two aspects, the expression for the magnetic field intensity becomes

$$H_x(z,t) = \left(\frac{10}{120\pi}\right)\cos(2\pi 10^7 t - 0.1\pi z)\text{ A/m}$$

5. In a non-magnetic medium, the wave has an electric vector given by $\mathbf{E} = 50\cos(10^9 t - 8x)\hat{\mathbf{y}} + 40\sin(10^9 t - 8x)\hat{\mathbf{z}}$. Find the dielectric constant ε_r and the corresponding magnetic field vector $\mathbf{H}$.

Solution: From the given expression of the field, $\omega = 10^9$ rad/s and $\beta = 8$. But we know that the velocity of the wave v is

$$v = \frac{\omega}{\beta} = \frac{10^9}{8} = \frac{1}{\sqrt{\mu\varepsilon}} = \frac{1}{\sqrt{\mu_0\varepsilon_0\mu_r\varepsilon_r}} = \frac{3\times10^8}{\sqrt{\mu_r\varepsilon_r}}$$

Given that the material is non-magnetic, i.e. $\mu_r = 1$. Therefore,

$$\frac{10^9}{8} = \frac{3\times10^8}{\sqrt{\varepsilon_r}} \quad\Rightarrow\quad \sqrt{\varepsilon_r} = \frac{3\times10^8\times8}{10^9} = 2.4$$

And the dielectric constant therefore is

$$\varepsilon_r = 2.4^2$$

For the given data, in the present case,

- The wave is travelling in the positive x-direction.
- The intrinsic impedance is $120\pi/2.4$.
- The magnetic field corresponding to the first term must be in the z-direction and that of the second term must be in the negative y-direction.
  Therefore, the magnetic field must be

$$\mathbf{H} = \left(\frac{2.4 \times 50}{120\pi}\right)\cos\,(10^9 t - 8x)\hat{\mathbf{z}} - \left(\frac{2.4 \times 40}{120\pi}\right)\sin\,(10^9 t - 8x)\hat{\mathbf{y}}$$

6. The electric vector of a uniform plane wave propagating in a perfect dielectric medium with $\varepsilon = 9\varepsilon_0$ and $\mu = \mu_0$ is

$$\mathbf{E} = 10\left(-\hat{\mathbf{x}} - 2\sqrt{2}\hat{\mathbf{y}} + \sqrt{3}\hat{\mathbf{z}}\right)\cos\left[16\pi \times 10^6 t - 0.04\pi\left(\sqrt{3}x - 2y - 3z\right)\right]$$

Find

(a) the frequency of the wave,
(b) the direction of propagation,
(c) the wavelength along the direction of propagation,
(d) the wavelengths along the x- and y-directions and
(e) the apparent velocities along the x-, y- and z-directions.

Solution:

(a) Frequency of the wave
In the expression for field intensity, the factor that multiplies the time is the angular frequency.

$$\omega = 16\pi \times 10^6 \text{ rad/s}$$

$$f = \frac{\omega}{2\pi} = \frac{16\pi \times 10^6}{2\pi} = 8\text{MHz}$$

(b) Direction of propagation
We can observe from the second part of the argument of the field expression that the wave propagation has components in all the three directions. The unit vector of the direction is

$$\hat{\boldsymbol{\beta}} = \frac{\sqrt{3}\hat{\mathbf{x}} - 2\hat{\mathbf{y}} - 3\hat{\mathbf{z}}}{\sqrt{3+4\,9}} = \frac{1}{4}\left(\sqrt{3}\hat{\mathbf{x}} - 2\hat{\mathbf{y}} - 3\hat{\mathbf{z}}\right)$$

(c) Wavelength along the direction of propagation

$$\lambda = \frac{c}{f} = \frac{3 \times 10^8}{8 \times 10^6} = 37.5 \text{ m}$$

(d) Wavelengths along the x-, y- and z-directions
The wavelength along the x-direction is

$$\lambda_x = \frac{\lambda}{\cos\theta}$$

Here θ is the angle between the direction of propagation and the x-axis.

$$\cos\theta = |\hat{\beta}\cdot\hat{\mathbf{x}}| = \frac{\sqrt{3}}{4}$$

Therefore,

$$\lambda_x = \frac{\lambda}{\cos\theta} = 37.5\frac{4}{\sqrt{3}} = 86.60\,\text{m}$$

Similarly, the wavelength along the y-direction is

$$\lambda_y = \frac{\lambda}{\cos\theta} = 37.5\frac{4}{2} = 75\,\text{m}$$

And along the z-direction, it is

$$\lambda_z = \frac{\lambda}{\cos\theta} = 37.5\frac{4}{3} = 50\,\text{m}$$

(e) Apparent velocities along the x-, y- and z-directions
These are given by

$$v_x = f\lambda_x = 8\times10^6\times37.5\frac{4}{\sqrt{3}} = 6.93\times10^8\ \text{m/s}$$

$$v_y = f\lambda_y = 8\times10^6\times37.5\frac{4}{2} = 6\times10^8\ \text{m/s}$$

$$v_z = f\lambda_z = 8\times10^6\times37.5\frac{4}{3} = 4\times10^8\ \text{m/s}$$

7. Find the expressions for electric and magnetic vectors of the incident, reflected and refracted waves shown in Figure 8.7 using direction cosines.

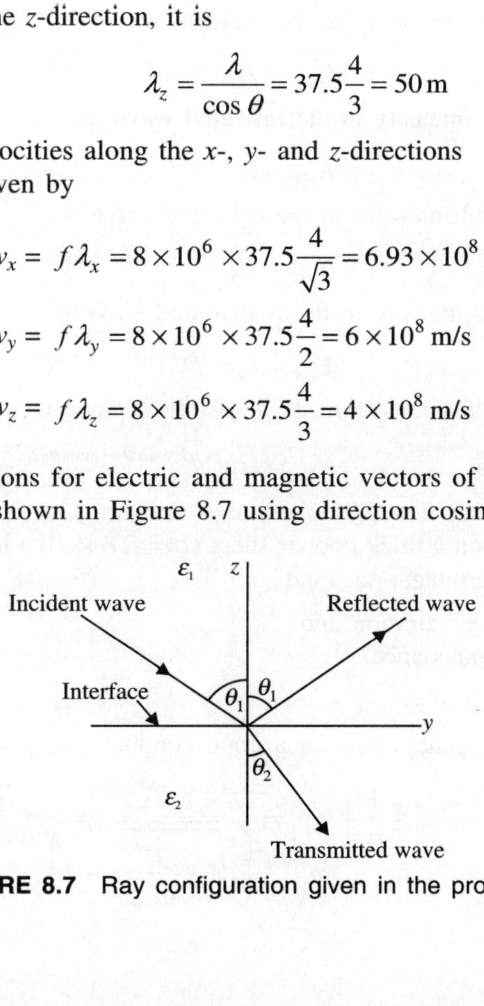

FIGURE 8.7 Ray configuration given in the problem.

Solution: In this configuration, the angles the incident ray makes with the *x*-, *y*- and *z*-axes are 90°, 90° – θ_1 and 180° – θ_1, respectively, whereas the corresponding angles of the reflected ray are 90°, –90° + θ_1 and θ_1, respectively. In the case of refracted wave, these angles become 90°, 90° – θ_2 and 180° – θ_2, respectively. The resulting direction cosines in *x*-, *y*- and *z*-directions respectively are 0, sin θ_1 and –cos θ_1 in the case of incident wave, 0, –sin θ_1 and cos θ_1 in the case of reflected wave and 0, sin θ_2 and –cos θ_2 in the case of refracted ray. Here θ_1 represents the angle of incidence as well as reflection and θ_2 represents the angle of refraction.

We know that the equation of a plane wave travelling in the direction $\hat{\mathbf{n}}$, normal to the planes of constant phase, can be written as

$$\mathbf{E}(\mathbf{r}) = \mathbf{E}_0 e^{-j\beta\hat{\mathbf{n}}\cdot r} = \mathbf{E}_0 e^{-j\beta(x\cos A + y\cos B + z\cos C)}$$

Here *A*, *B* and *C* are the angles the ray or wave normal makes with the *x*-, *y*- and *z*-axes, respectively. cos *A*, cos *B* and cos *C* are called direction cosines of the wave. The electric field intensity in the incident wave is

$$E_{\text{inc}} = E_i e^{-j\beta(y\sin\theta_1 - z\cos\theta_1)}$$

The magnetic field intensity in the incident wave is

$$H_{\text{inc}} = H_i e^{-j\beta(y\sin\theta_1 - z\cos\theta_1)}$$

The electric field intensity in the reflected wave is

$$E_{\text{ref}} = E_r e^{-j\beta(y\sin\theta_1 + z\cos\theta_1)}$$

The magnetic field intensity in the reflected wave is

$$H_{\text{ref}} = H_r e^{-j\beta(y\sin\theta_1 + z\cos\theta_1)}$$

The electric field intensity in the transmitted wave is

$$E_{\text{tra}} = E_t e^{-j\beta(y\sin\theta_2 - z\cos\theta_2)}$$

The magnetic field intensity in the transmitted wave is

$$H_{\text{tra}} = H_t e^{-j\beta(y\sin\theta_2 - z\cos\theta_2)}$$

8. A plane wave with E = 2.0 V/m and a frequency of 300 MHz is moving in free space impinging on a thick copper sheet ($\sigma = 5.8 \times 10^7$) located perpendicular to the direction of the propagation. Find

(a) the depth of penetration and
(b) the surface impedance.

Solution:

(a) The depth of penetration for a good conductor is given by

$$\delta = \frac{1}{\alpha} = \frac{1}{\omega\sqrt{\dfrac{\mu\varepsilon}{2}\left(\sqrt{1+\dfrac{\sigma^2}{\omega^2\varepsilon^2}}-1\right)}} \approx \sqrt{\frac{2}{\omega\mu\sigma}}$$

Substituting

$$\omega = 2\pi \times 300 \times 10^6 \text{ rad/s}$$
$$\sigma = 5.8 \times 10^7$$

The depth of penetration is

$$\delta \approx \sqrt{\frac{2}{2\pi \times 300 \times 10^6 \times 5.8 \times 10^7 \times 4\pi \times 10^{-7}}} = 0.038 \times 10^{-4} \text{ m}$$

(b) The surface impedance

$$Z_s = \frac{\gamma}{\sigma} = \sqrt{\frac{j\omega\mu}{\sigma}} = \sqrt{\frac{\omega\mu}{\sigma}} \angle 45^\circ$$

$$\approx \sqrt{\frac{2\pi \times 300 \times 10^6 \times 4\pi \times 10^{-7}}{5.8 \times 10^7}} \angle 45^\circ = 63.9 \times 10^{-4} \angle 45^\circ \ \Omega$$

9. A wave has two linearly polarized electric fields at a point given by $\mathbf{E}_1 = (2\hat{\mathbf{x}} - c\hat{\mathbf{y}} + c\hat{\mathbf{z}}) \cos \omega t$ and $\mathbf{E}_2 = (c\hat{\mathbf{x}} + c\hat{\mathbf{y}} + 2\hat{\mathbf{z}}) \sin \omega t$.

(a) Find the polarization of ($\mathbf{E}_1 + \mathbf{E}_2$) for $c = 1$.

(b) Also find the values of c for which it is circular polarization and linear polarization.

Solution:

(a) For $c = 1$

$$\mathbf{E}_1 = (2\hat{\mathbf{x}} - \hat{\mathbf{y}} + \hat{\mathbf{z}}) \cos \omega t \text{ and}$$
$$\mathbf{E}_2 = (\hat{\mathbf{x}} - \hat{\mathbf{y}} + 2\hat{\mathbf{z}}) \sin \omega t$$

We can observe from the expressions that the fields are in time phase quadrature having equal magnitude, i.e. $|\mathbf{E}_1| = |\mathbf{E}_2| = \sqrt{6}$. The angle between them is not 90°, i.e.

$$\mathbf{E}_1 \cdot \mathbf{E}_2 = 2 - 1 + 2 \neq 0$$

By considering all the three points, we can say that the wave is elliptically polarized.

(b) For the wave to be circularly polarized, the three conditions to be satisfied are:

(i) the linearly polarized fields must be in time phase quadrature,

(ii) both the fields must have the same magnitude and

(iii) they must be normal to each other spatially. In the given wave, the first two conditions are satisfied for any value of k. To satisfy the third condition,

$$\mathbf{E}_1 \cdot \mathbf{E}_2 = 2c - c^2 + 2c$$
$$= 4c - c^2 = c(4 - c) = 0$$

It gives two values for c. One is 0 and the other is 4. As the given fields are in time phase quadrature, for the wave to be linearly polarized, the fields must be parallel to each other, i.e. $\mathbf{E}_1 \times \mathbf{E}_2 = 0$ giving $c = -2$.

10. Consider a general expression for the field of an elliptically polarized wave as

$$\tilde{\mathbf{E}} = E_1 \cos \omega t\, \hat{\mathbf{x}} + E_2 \cos(\omega t - \phi)\, \hat{\mathbf{y}}$$

and show that at the ends of the major and minor axes

$$\tan 2\omega t = \frac{E_2^2 \sin 2\phi}{E_1^2 + E_2^2 \cos 2\phi}$$

Also explain how to obtain the axial ratio and tilt angle.

Solution: The above expression in general represents an elliptically polarized wave and $d\mathbf{E}/dt$ gives tangential to the ellipse. Since at the ends of the major and minor axes the tangents to the ellipse are normal to the axes, it follows that at those points on the ellipse

$$\tilde{\mathbf{E}} \cdot \frac{d\tilde{\mathbf{E}}}{dt} = 0$$

If $\tilde{\mathbf{E}} = E_1 \cos \omega t \hat{\mathbf{x}} + E_2 \cos (\omega t - \phi) \hat{\mathbf{y}}$, then

$$\frac{d\tilde{\mathbf{E}}}{dt} = -\,\omega E_1 \sin \omega t \hat{\mathbf{x}} - \omega E_2 \sin (\omega t - \phi)\, \hat{\mathbf{y}}$$

At the ends of the major and minor axes

$$-\omega E_1^2 \cos \omega t \sin \omega t - \omega E_2^2 \cos(\omega t - \phi) \sin(\omega t - \phi) = 0$$

$$E_1^2 \sin 2\omega t + E_2^2 \sin 2(\omega t - \phi) = 0$$

$$E_1^2 \sin 2\omega t + E_2^2 (\sin 2\omega t \cos 2\phi - \cos 2\omega t \sin 2\phi) = 0$$

$$(E_1^2 + E_2^2 \cos 2\phi) - E_2^2 \sin 2\phi \cos 2\omega t = 0$$

leading to

$$\tan 2\omega t = \frac{E_2^2 \sin 2\phi}{E_1^2 + E_2^2 \cos 2\phi}$$

By solving the above equation, we can find the angle ωt at the tips of the axes, and substituting these values in the expression of field, we can arrive at the major and minor axes and also their orientation. Their amplitude ratio gives the axial ratio and their orientation gives the tilt angle.

11. The electric field intensity of a uniform plane wave in free space is given by

$$\mathbf{E} = E_0 \left[\cos(\omega t - \beta z)\, \hat{\mathbf{x}} + \sqrt{2} \cos\left(\omega t - \beta z - \frac{\pi}{3} \right) \hat{\mathbf{y}} \right] \text{V/m}$$

Find

(a) the polarization of the wave

(b) the axial ratio, i.e. the ratio of the semi-major axis to the semi-minor axis
(c) the tilt angle, i.e. the angle that the major axis makes with the x-axis.

Solution:

(a) $$\mathbf{E} = E_0\left[\cos(\omega t - \beta z)\hat{\mathbf{x}} + \sqrt{2}\cos\left(\omega t - \beta z - \frac{\pi}{3}\right)\hat{\mathbf{y}}\right]$$

The field has two unequal components in normal directions, in the x- and y-directions, indicating two unequal linear normal polarizations with a phase difference of 60° in between them. Therefore, the wave is an elliptically polarized one.

Now at $(\omega t - \beta z) = 0$, the field is $\mathbf{E} = E_0\left(\hat{\mathbf{x}} + \sqrt{2}\cos\pi/3\hat{\mathbf{y}}\right)$, at $(\omega t - \beta z) = \pi/2$, the field is $\mathbf{E} = \sqrt{2}E_0\cos\pi/6\hat{\mathbf{y}}$ and $\left(\hat{\mathbf{x}} + \sqrt{2}\cos\pi/3\hat{\mathbf{y}}\right)\times\cos\pi/6\hat{\mathbf{y}}$ is in the positive z-direction, which is the same as the direction of propagation of the wave. Therefore, it is a right elliptically polarized wave.

(b) The general expression for the field of an elliptically polarized wave is $\mathbf{E} = E_1\cos\omega t\hat{\mathbf{x}} + E_2\cos(\omega t - \phi)\hat{\mathbf{y}}$ and at the ends of the major and minor axes

$$\tan 2\omega t = \frac{E_2^2\sin 2\phi}{E_1^2 + E_2^2\cos 2\phi}$$

For the given problem, $E_1 = E_0$, $E_2 = \sqrt{2}E_0$ and $\phi = 60°$. The values of ωt corresponding to the major and minor axes are

$$\tan 2\omega t = \frac{\sin 120°}{1 + 2\cos 120°} = \frac{\sqrt{3}}{1-1} = \infty$$

$$2\omega t = 90°,\ 270° \Rightarrow \quad \omega t = 45°,\ 135°$$

Now

$$\begin{aligned}\mathbf{E}|_{\omega t=45°} &= E_0\left[\cos 45°\,\hat{\mathbf{x}} + \sqrt{2}\cos\left(45° - \frac{\pi}{3}\right)\hat{\mathbf{y}}\right]\\ &= E_0[-0.707\hat{\mathbf{x}} + 1.366\hat{\mathbf{y}}]\\ &= 1.5382\,E_0 \text{ at } 62.63° \text{ from the } x\text{-axis}\end{aligned}$$

Similarly,

$$\begin{aligned}\mathbf{E}|_{\omega t=135°} &= E_0\left[\cos 135°\,\hat{\mathbf{x}} + \sqrt{2}\cos\left(135° - \frac{\pi}{3}\right)\hat{\mathbf{y}}\right]\\ &= E_0[-0.707\hat{\mathbf{x}} + 0.366\hat{\mathbf{y}}]\\ &= 0.7962\,E_0 \text{ at } 152.63° \text{ from the } y\text{-axis}\end{aligned}$$

Therefore, the axial ratio $= \dfrac{1.5382\,E_0}{0.7962\,E_0} = 1.932$

The tilt angle is at 62.63° from the x-axis.

12. For each of the following fields of uniform plane waves, determine the polarization and express it as a superposition of the right and left circularly polarized fields:

(a) $\mathbf{E} = E_0(2\hat{\mathbf{x}} - 3\hat{\mathbf{y}})\cos(\omega t - \beta z)$

(b) $\mathbf{E} = 5E_0\hat{\mathbf{x}}\cos(\omega t - \beta z) - E_0\hat{\mathbf{y}}\sin(\omega t - \beta z)$

Solution:

(a) $\mathbf{E} = E_0\,(2\hat{\mathbf{x}} - 3\hat{\mathbf{y}})\cos(\omega t - \beta z) = 2E_0\hat{\mathbf{x}}\cos(\omega t - \beta z) - 3E_0\hat{\mathbf{y}}\cos(\omega t - \beta z)$

The field consists of two normal linear polarizations in the x- and y-directions with a time phase difference of 180° (due to the negative sign of the second term) in between them. The combination, therefore, gives rise to linear polarization.

The given field can be expressed as

$$\begin{aligned}\mathbf{E} &= E_0(2\hat{\mathbf{x}} - 3\hat{\mathbf{y}})\cos(\omega t - \beta z)\\ &= \left(\frac{E_0}{2}\right)(2\hat{\mathbf{x}} - 3\hat{\mathbf{y}})\cos(\omega t - \beta z) + \left(\frac{E_0}{2}\right)(3\hat{\mathbf{x}} + 2\hat{\mathbf{y}})\sin(\omega t - \beta z)\\ &\quad + \left(\frac{E_0}{2}\right)(2\hat{\mathbf{x}} - 3\hat{\mathbf{y}})\cos(\omega t - \beta z) - \left(\frac{E_0}{2}\right)(3\hat{\mathbf{x}} + 2\hat{\mathbf{y}})\sin(\omega t - \beta z)\\ &= \mathbf{E}_1 + \mathbf{E}_2\end{aligned}$$

Consider $\mathbf{E}_1$ consisting of the first two terms: They are equal in amplitude, direction-wise perpendicular to each other because $(2\hat{\mathbf{x}} - 3\hat{\mathbf{y}})\cdot(3\hat{\mathbf{x}} + 2\hat{\mathbf{y}}) = 0$ and they differ 90° in time phase. Therefore, their combination gives rise to circular polarization.

Now at $(\omega t - \beta z) = 0$, the field is $\mathbf{E}_1 = (E_0/2)(2\hat{\mathbf{x}} - 3\hat{\mathbf{y}})$, and at $(\omega t - \beta z) = \pi/2$, the field is $\mathbf{E}_1 = (E_0/2)(3\hat{\mathbf{x}} + 2\hat{\mathbf{y}})$ and $(2\hat{\mathbf{x}} - 3\hat{\mathbf{y}})\times(3\hat{\mathbf{x}} + 2\hat{\mathbf{y}})$ is in the positive z-direction. Therefore, it is wave propagating in the positive z-direction with right circularly polarization.

Similar analysis can be done for the second part and it can be shown that it is also a circularly polarized wave with left-handed sense.

(b) $\mathbf{E} = 5E_0\,\hat{\mathbf{x}}\cos(\omega t - \beta z) - E_0\hat{\mathbf{y}}\sin(\omega t - \beta z)$

This expression represents a combination of two linearly polarized fields, in x- and y-directions, so spatially normal to each other, time phase-wise differing by 90° but amplitude-wise differing from each other. Naturally the resulting wave must be an elliptically polarized one.

Now at $(\omega t - \beta z) = 0$, the field is $\mathbf{E} = 5E_0\hat{\mathbf{x}}$, at $(\omega t - \beta z) = \pi/2$, the field is $\mathbf{E} = -E_0\hat{\mathbf{y}}$ and $\hat{\mathbf{x}}\times(-\hat{\mathbf{y}})$ is in the negative z-direction, which is opposite to the direction of propagation of the wave. Therefore, it is a left elliptically polarized wave.

Now,

$$\begin{aligned}\mathbf{E} &= 5E_0\hat{\mathbf{x}}\cos(\omega t - \beta z) - E_0\hat{\mathbf{y}}\sin(\omega t - \beta z)\\ &= 2E_0\hat{\mathbf{x}}\cos(\omega t - \beta z) + 2E_0\hat{\mathbf{y}}\sin(\omega t - \beta z)\\ &\quad +3E_0\hat{\mathbf{x}}\cos(\omega t - \beta z) - 3E_0\hat{\mathbf{y}}\sin(\omega t - \beta z)\\ &= \mathbf{E}_1 + \mathbf{E}_2\end{aligned}$$

We can observe that the first two components combine to give the right circular polarization and the second two components the left circular polarization.

EXERCISE SECTION

Essay Type Questions

1. Derive the wave equations from the Maxwell's equations for free space propagation.
2. Derive the wave equations for conducting a medium from the Maxwell's equations.
3. Explain propagation constant, attenuation constant and phase shift constant.
4. Derive the expressions for propagation constant, attenuation constant and phase shift constant for a perfect dielectric and a good conductor.
5. What is uniform plane wave? Show that the field in the uniform plane wave is independent of two dimensions.
6. Draw the expressions for field intensity from the wave equations when the wave is a uniform plane wave.
7. Evaluate the expressions for the velocity of wave, electric to magnetic fields ratio and orientation of field vectors with respect to the direction of propagation.
8. What are direction cosines? How can the fields for uniform plane waves be expressed using direction cosines?
9. What is polarization? What is the significance of polarization in wireless communications?
10. Define and explain the terms linear, elliptical and circular polarizations.

Numerical Problems

1. An EM wave is propagating through a medium having $\mu_r = 5$ and $\varepsilon_r = 10$. Determine the velocity of propagation, intrinsic impedance of free space and of medium and wavelength in free space and medium, if the frequency is 1 GHz.
2. In a medium, $\mathbf{E} = 16e^{-x/20}\sin(2\times 10^8 t - 2x)\hat{\mathbf{z}}$ V/m. Find the direction of propagation, propagation constant, wavelength, speed of the wave and skin depth.
3. A travelling wave has two linearly polarized components $E_x = 2\cos\omega t$ and $E_y = 3\cos(\omega t + \pi/2)$.
   (a) What is the axial ratio?
   (b) What is the tilt angle of the major axis of the polarization ellipse?
   (c) What is the sense of rotation?

Fill in the Blanks

1. The wave equation for the electric field in free space is __________.
2. The wave equation for the magnetic field in free space is __________.
3. The wave equation for the electric field in conductive media is __________.
4. The wave equation for the magnetic field in conductive media is __________.
5. The velocity of wave in free space is __________.
6. The propagation constant for a general medium is __________.
7. The propagation constant for free space is __________.
8. The real part of the propagation constant is __________.
9. The imaginary part of the propagation constant is __________.
10. For a good dielectric, the attenuation constant is __________.
11. For a good dielectric, the phase shift constant is __________.
12. For a good conductor, the attenuation constant is __________.
13. For a good conductor, the phase shift constant is __________.
14. In TEM wave, the electric and magnetic vectors are normal to the direction of __________.
15. An example of TEM wave is __________.
16. In uniform plane waves, the equiphase surfaces are parallel __________.
17. In a free space UPW, the intrinsic impedance is __________ and power density is __________.
18. The expression for the electric field of $+z$ travelling UPW is __________.
19. The electric vector remains in the same __________ in linear polarization.
20. The electric vector traces a circle with __________ in circular polarization.
21. The electric vector traces an ellipse with __________ in elliptical polarization.
22. Th sense of polarization is defined with respect to __________ wave.
23. Two equal normal linear in-phase polarizations combine to give __________.
24. The depth of penetration is __________.
25. The surface impedance is __________.

Answers

1. $\nabla^2\mathbf{E} = \mu\varepsilon\ddot{\mathbf{E}}$
2. $\nabla^2\mathbf{H} = \mu\varepsilon\ddot{\mathbf{H}}$
3. $\nabla^2\mathbf{E} = \gamma^2\mathbf{E}$
4. $\nabla^2\mathbf{H} = \gamma^2\mathbf{H}$
5. $c = \dfrac{1}{\sqrt{\mu_0\varepsilon_0}}$
6. $\gamma = \sqrt{j\omega\mu(\sigma + j\omega\varepsilon)}$
7. $j\omega\sqrt{\mu\varepsilon}$
8. $\alpha = \omega\sqrt{\dfrac{\mu\varepsilon}{2}\left(\sqrt{1+\dfrac{\sigma^2}{\omega^2\varepsilon^2}}-1\right)}$

9. $\beta = \omega\sqrt{\frac{\mu\varepsilon}{2}\left(\sqrt{1+\frac{\sigma^2}{\omega^2\varepsilon^2}}+1\right)}$

10. $\alpha = \frac{\sigma}{2}\sqrt{\frac{\mu}{\varepsilon}}$

11. $\beta = \omega\sqrt{\mu\varepsilon}\left(1+\frac{\sigma^2}{8\omega^2\varepsilon^2}\right)$

12. $\alpha = \sqrt{\frac{\omega\mu\sigma}{2}}$

13. $\beta = \sqrt{\frac{\omega\mu\sigma}{2}}$

14. propagation

15. uniform plane waves

16. planes

17. $\sqrt{\frac{\mu_0}{\varepsilon_0}}, \frac{E^2}{\eta}$

18. $E = E_0 e^{-j\beta b}$

19. direction

20. time

21. time

22. receding

23. circular

24. $\delta = \sqrt{\frac{2}{\omega\mu\sigma}}$

25. $Z_s = \frac{\gamma}{\sigma}$

Match the Following

Set I

1. Wave equation in free space — (a) $\omega\sqrt{\frac{\mu\varepsilon}{2}\left(\sqrt{1+\frac{\sigma^2}{\omega^2\varepsilon^2}}+1\right)}$
2. Wave equation in conducting medium — (b) $\omega\sqrt{\frac{\mu\varepsilon}{2}\left(\sqrt{1+\frac{\sigma^2}{\omega^2\varepsilon^2}}-1\right)}$
3. Propagation constant — (c) $\gamma = \sqrt{j\omega\mu(\sigma + j\omega\varepsilon)}$
4. Attenuation constant — (d) $\nabla^2\mathbf{E} = \gamma^2\mathbf{E}$
5. Phase shift constant — (e) $\nabla^2\mathbf{E} = \mu\varepsilon\ddot{\mathbf{E}}$

Set II

1. Depth of penetration — (a) $\frac{\gamma}{\sigma}$
2. Surface impedance — (b) $\sqrt{\frac{2}{\omega\mu\sigma}}$
3. Complex quantity — (c) propagation constant

4. Attenuation constant — (d) imaginary part of propagation constant
5. Phase shift constant — (e) real part of propagation constant

Set III

1. Velocity of wave — (a) orientation of electric vector
2. Intrinsic impedance — (b) $c = \dfrac{1}{\sqrt{\mu_0 \varepsilon_0}}$
3. Power density — (c) $\dfrac{E^2}{\eta}$
4. Field in UPW — (d) $E = E_0 e^{-j\beta b}$
5. Polarization — (e) $\sqrt{\dfrac{j\omega\mu}{(\sigma + j\omega\varepsilon)}}$

Set IV

1. Current density — (a) $\varepsilon\mathbf{E}$
2. Displacement density — (b) $\sigma\mathbf{E}$
3. Flux density — (c) $\mu\mathbf{H}$
4. Isotropic medium — (d) ε and μ are the same at all points
5. Homogeneous medium — (e) ε and μ scalar constants

Answers

Set I:	1. (e)	2. (d)	3. (c)	4. (b)	5. (a)
Set II:	1. (b)	2. (a)	3. (c)	4. (e)	5. (d)
Set III:	1. (e)	2. (d)	3. (c)	4. (b)	5. (a)
Set IV:	1. (b)	2. (a)	3. (c)	4. (e)	5. (d)

Multiple Choice Questions

1. TEM wave is
   (a) a plane wave (b) a cylindrical wave
   (c) a spherical wave (d) all of these
2. In a uniform plane wave (UPW), **E** and **H** are
   (a) parallel (b) perpendicular
   (c) opposite (d) none of these
3. In a UPW, the directions of propagation of wave, **E** and **H** form
   (a) an LH vector system (b) an RH vector system
   (c) a non-orthogonal system (d) none of these

4. In a UPW while travelling, the ratio of E to H is

(a) $\sqrt{\dfrac{\mu}{\varepsilon}}$ (b) $\sqrt{\dfrac{\varepsilon}{\mu}}$

(c) $\sqrt{\dfrac{1}{\mu\varepsilon}}$ (d) none of these

5. The speed of a UPW in space is

(a) $\sqrt{\dfrac{\mu}{\varepsilon}}$ (b) $\sqrt{\dfrac{\varepsilon}{\mu}}$

(c) $\sqrt{\dfrac{1}{\mu\varepsilon}}$ (d) none of these

6. In space, the wave is always

(a) plane (b) TEM
(c) both (a) and (b) (d) none of these

7. The intrinsic impedance of free space is

(a) 127 Ω (b) 177 Ω
(c) 120π Ω (d) none of these

8. In a UPW, **E** and **H** are independent of

(a) one dimension (b) two dimensions
(c) three dimensions (d) none of these

9. In a UPW, the direction of power flow is

(a) normal to **E** and **H** (b) parallel to **E** and **H**
(c) normal to **E** and parallel to **H** (d) none of these

10. In UPW, the power density is given by

(a) $\dfrac{E^2}{\eta}$ (b) ηH^2
(c) EH (d) all of these

11. The propagation constant in general is

(a) complex (b) real
(c) imaginary (d) none of these

12. For a good conductor

(a) α and β are the same (b) $\alpha = \beta = \sqrt{\dfrac{\omega\mu\sigma}{2}}$
(c) both (a) and (b) (d) none of these

13. The depths of penetration δ and α are related as

(a) $\alpha\delta = 1$ (b) $\alpha = \delta$
(c) $\delta = \alpha^2$ (d) none of these

14. The depths of penetration δ and α are related as

(a) $\dfrac{\gamma}{\sigma}$ (b) $\dfrac{\sigma}{\gamma}$

(c) $\gamma\sigma$ (d) none of these

15. If the **E** vector is entirely transverse to the direction of propagation, then the wave is

(a) TE wave (b) H wave

(c) both (a) and (b) (d) none of these

16. If the **H** vector is entirely transverse to the direction of propagation, then the wave is

(a) TM wave (b) E wave

(c) both (a) and (b) (d) none of these

17. The direction of an electric field remains constant w.r.t. time in

(a) linear (b) circular

(c) elliptical (d) all of these

18. The sum of two elliptical polarizations can give rise to

(a) linear (b) circular

(c) elliptical (d) all of these

19. The sum of two normal, linear, in-phase polarizations gives rise to

(a) linear (b) circular

(c) elliptical (d) all of these

20. In linear polarization, w.r.t. time the magnitude and direction of **E**

(a) do not change, change respectively

(b) change, do not change respectively

(c) change, change respectively

(d) do not change, do not change respectively

21. In elliptical polarization, w.r.t. time the magnitude and direction of **E**

(a) do not change, change respectively

(b) change, do not change respectively

(c) change, change respectively

(d) do not change, do not change respectively

22. In circular polarization, w.r.t. time the magnitude and direction of **E**

(a) do not change, change respectively

(b) change, do not change respectively

(c) change, change respectively

(d) do not change, do not change respectively

23. The sense of polarization is defined w.r.t. to wave when it is

(a) approaching (b) receding

(c) passing by (d) none of these

24. Horizontal and vertical polarizations are, respectively,
    (a) linear, linear (b) linear, circular
    (c) circular, linear (d) circular, circular
25. To give rise to elliptical polarization, the two linear polarizations must be
    (a) at phase quadrature (b) normal to each other spatially
    (c) same amplitude (d) none of these
26. To alter the sense of rotation, the phase added to the normal components of the electric vector is
    (a) 90° (b) 270°
    (c) 180° (d) none of these
27. For the RH circular polarization, the axial ratio is
    (a) infinity (b) –1
    (c) zero (d) none of these

Answers

1. (d)	2. (b)	3. (b)	4. (a)	5. (c)
6. (b)	7. (c)	8. (b)	9. (a)	10. (d)
11. (a)	12. (c)	13. (a)	14. (a)	15. (c)
16. (c)	17. (a)	18. (d)	19. (b)	20. (b)
21. (c)	22. (a)	23. (b)	24. (a)	25. (b)
26. (c)	27. (b)			

9

Transmission and Reflection

If at first you don't succeed, try, try again. Then quit.
There's no point in being a damn fool about it
—W.C. Fields

When a wave falls over the interface of two different media, in general, it partly gets reflected and the remaining part gets refracted. The proportion of the reflected and transmitted waves, we will see soon, depends upon the relative values of permittivities and permeabilities of the media.

To begin with, various fundamental laws describing wave reflection and refraction are explained. Then the situation where wave falls over a conductor from free space is thoroughly analysed. The reflection and refraction that take place at the interface of two dielectric media are analysed and expressions for reflection coefficient and transmission coefficient are derived. This chapter is concluded with a discussion on transmission and reflection in lossy media.

9.1 INTRODUCTION

In the ensuing analysis we assume the interfacing media as uniform. The **E** and **H**, denoting electric and magnetic field intensities, respectively, are phasors and hence complex quantities representing the amplitudes of the waves. We consider the cases where the media involved are lossless as well as lossy. The wave amplitudes remain the same at any point in the medium if it is lossless like perfect conductor and perfect dielectric. Therefore, the **E** and **H** also remain constant while the wave is travelling in a lossless medium.

The wave on reflection and refraction obeys the Snell's laws provided the media are uniform. The Snell's law of reflection says that the angle of incidence is equal to the angle of reflection and the incident ray, reflected ray and normal to the interface, all, always remain in one plane. The Snell's law of refraction states that the ratio of sine of the angle of incidence to sine of the angle of refraction is equal to the ratio of the velocity of the wave in incidence medium to the velocity of the wave in refracting medium. And the incident ray, refracted ray and normal to the interface always remain in one plane. Mathematically,

$$\frac{\sin\theta_i}{\sin\theta_r} = \frac{v_i}{v_r} = \sqrt{\frac{(\mu\varepsilon)_r}{(\mu\varepsilon)_i}} \tag{9.1}$$

If the permeabilities of the media are the same, then

$$\frac{\sin\theta_i}{\sin\theta_r} = \sqrt{\frac{\varepsilon_r}{\varepsilon_i}} \tag{9.2}$$

In the analysis, it is considered that the incident wave with its electric vector is either normal or parallel to the plane of incidence, i.e. the plane containing the incident ray and the normal to the surface. It is because a wave with an arbitrary polarization can be decomposed into two waves, one with electric vector normal and another with electric vector parallel to the incident plane. As a consequence, it is possible to extend the results obtained with the electric vector normal and parallel to the wave with the electric vector in the arbitrary direction.

We consider the incidence of a uniform plane wave and if its orientation is such that the electric vector is normal to the plane of incidence, then the wave is called *perpendicularly* or *horizontally polarized wave*. Note that, in this case, the electric vector remains parallel to the reflecting surface and the magnetic vector is parallel to the plane of incidence. If the electric vector of the wave is parallel to the plane of incidence, then it is called *parallel polarized wave* or *vertical polarized wave* and in this case, the magnetic vector remains parallel to the reflecting surface.

In the analysis, we consider two different uniform media, a plane interface surface and the incidence of a uniform plane wave over the interface. We locate the interface over the *xy* or $z = 0$ plane and the plane of incidence over the *yz*-plane resulting in the normal to the surface along the *z*-axis.

9.2 Perfect Conductor

In this section, we consider the reflection of the plane wave from the plane surface of a perfect conductor. The wave is assumed to have incident on the conductor surface while travelling in free space or perfect dielectric. Another aspect we need to consider here is that the wave cannot enter into a perfect conductor resulting in no refracted wave. We analyse both the cases of normal incidence and oblique incidence.

When a wave falls over the plane surface of a perfect conductor, reflection, obeying Snell's law takes place. In front of the surface there exist two traveling waves; one incident and the other reflected. Due to the interference of these two waves, standing wave pattern comes into being. In the case of normal incidence, we will see soon, the standing wave pattern exists along the direction of the incidence and in the case of oblique incidence, along with a standing wave pattern in the direction of normal to the surface, a travelling wave also exists parallel to the surface.

9.2.1 Normal Incidence

In the case of normal incidence, the reflected wave also is in the normal direction but with the direction of propagation opposite to that of the incident wave. Now we have two waves travelling in opposite directions but over the same path normal to the reflecting surface. As a result, interference takes place between these two waves forming a standing wave pattern in a direction normal to the surface.

Now we try to find out the fields in front of the conductor surface. According to one of the boundary conditions, the tangential component of electric field is zero at the surface of the conductor. In the present case, as the wave's incidence and reflection both are normal, the electric field vector (magnetic field vectors also) of both the waves are tangential to the surface. And the total electric field strength at the surface is the sum of the fields due to incident and reflected waves. Therefore, if E_i is the field of the incident wave and E_r is that of the reflected wave, then

$$E_i + E_r = 0$$

or

$$E_i = -E_r \tag{9.3}$$

The above mathematical relation says that the tangential component of the electric field on reflection at the surface of a perfect conductor undergoes a phase change of 180°.

On the other hand, the magnetic vector gets reflected without any phase change, because to have the reversal of the direction propagation with reversal of the phase of the electric vector, the magnetic vector must maintain the same direction. If both the electric and magnetic vectors change a phase of 180°, then the direction of propagation does not change. Hence, at the surface of the perfect conductor, the tangential component of the magnetic vector gets reflected without any phase change, i.e. $H_r = H_i$.

> *The tangential component of an electric vector gets reflected with phase reversal whereas that of the magnetic vector gets reflected without phase reversal, at the surface of a perfect conductor.*

As the standing wave is a combination of incident and reflected waves, the field strength in the standing wave is the sum of the fields of incident and reflected waves.

$$\begin{aligned} E_T &= E_i e^{j\beta z} + E_r e^{-j\beta z} \\ &= E_i e^{j\beta z} - E_i e^{-j\beta z} \\ &= 2jE_i \sin \beta z \end{aligned} \tag{9.4}$$

The above phasor expression for total field can be converted into time domain resulting in

$$\tilde{E}_T(z, t) = 2E_i \sin \beta z \sin \omega t \tag{9.5}$$

Similarly, the total magnetic field strength can be computed as

$$\begin{aligned} H_T &= H_i e^{j\beta z} + H_r e^{-j\beta z} \\ &= H_i e^{j\beta z} + H_i e^{-j\beta z} \\ &= 2H_i \cos \beta z \end{aligned} \tag{9.6}$$

The total magnetic field strength given by the above expression can be converted into time domain resulting in

$$\tilde{H}_T(z, t) = 2H_i \cos \beta z \sin \omega t \tag{9.7}$$

We can observe that in the standing wave, the total field strengths $\tilde{E}_T$ and $\tilde{H}_T$ are 90° apart in time-phase. Also there exists a surface current of J_s A/s = $H_T(x = 0)$ over the surface of the reflecting conductor. The standing wave pattern of an electric field and a magnetic field is shown in Figure 9.1.

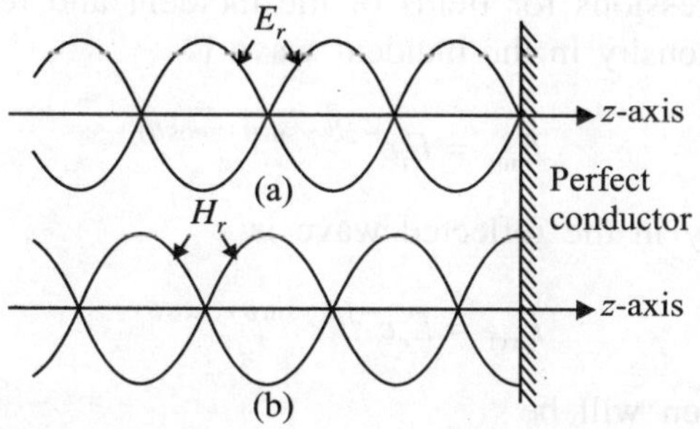

FIGURE 9.1 Standing wave pattern (a) electric field and, (b) magnetic field.

Perfect conductor—standing wave pattern	
$\tilde{E}_T(z, t) = 2E_i \sin \beta z \sin \omega t$	Normal incidence
$E_T = 2jE_i \sin \beta_z z e^{-j\beta_y y}$	Oblique incidence: ⊥ polarization
$H_T = 2H_i \cos \beta_z z e^{-j\beta_y y}$	Oblique incidence: || polarization
$E_{T,z} = 2E_i \sin \theta \cos (\beta_z z) e^{-j\beta_y y}$	
$E_{T,y} = 2jE_i \cos \theta \sin (\beta_z z) e^{-j\beta_y y}$	

9.2.2 Oblique Incidence

Perpendicular polarization: In the case of oblique incidence, i.e. when the angle of incidence is in between 0 and 90°, we need to consider the situations separately according to the polarization. It can be shown that in the case of perpendicular polarization (Figure 9.2), the total electric field strength is

$$E_T = 2jE_i \sin \beta_z z e^{-j\beta_y y} \tag{9.8a}$$

where $\beta_z = \beta \cos \theta$ and $\beta_y = \beta \sin \theta$.

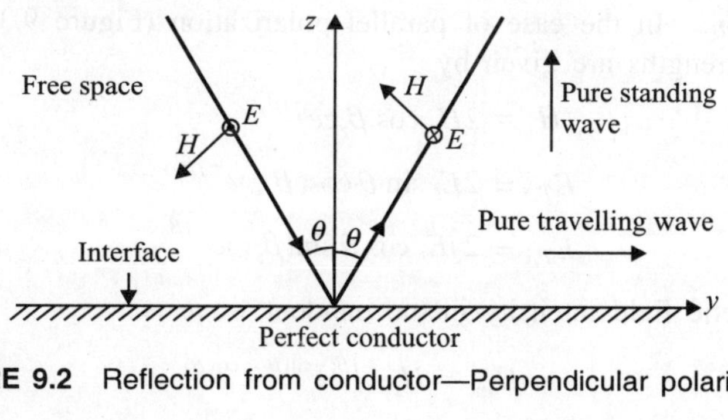

FIGURE 9.2 Reflection from conductor—Perpendicular polarization.

Proof: Let us suppose a uniform plane wave, a perpendicularly polarized one, is falling over the interface with an incident angle of θ. And according the Snell's law, it gets reflected with the same angle. No transmission into the conducting medium takes place because the wave cannot penetrate into a perfect conductor. Using direction cosines, it is possible to write the expressions for fields of the incident and reflected waves.

The electric field intensity in the incident wave is

$$E_{\text{inc}} = E_i e^{-j\beta(y\sin\theta - z\cos\theta)}$$

The electric field intensity in the reflected wave is

$$E_{\text{ref}} = E_r e^{-j\beta(y\sin\theta + z\cos\theta)}$$

The total electric field then will be

$$E_T = E_{\text{inc}} + E_{\text{ref}} = E_i e^{-j\beta(y\sin\theta - z\cos\theta)} + E_r e^{-j\beta(y\sin\theta + z\cos\theta)}$$

The magnitudes of the incident and reflected waves are the same because they are in loss-free regions. As the wave is perpendicularly polarized, the electric vector is tangential to the reflecting surface and hence it undergoes a phase reversal on reflection. Therefore, $E_r = -E$ Incorporating this, total field then becomes

$$E_T = E_i\left[e^{-j\beta(y\sin\theta - z\cos\theta)} - e^{-j\beta(y\sin\theta + z\cos\theta)}\right]$$

$$= E_i e^{-j\beta y\sin\theta}(e^{j\beta z\cos\theta} - e^{-j\beta z\cos\theta}) = E_i e^{-j\beta y\sin\theta}[2j\sin(\beta z\cos\theta)]$$

Thus,

$$E_T = 2jE_i \sin\beta_z z e^{-j\beta_y y}$$

where $\beta_z = \beta\cos\theta$ and $\beta_y = \beta\sin\theta$.

It can be converted into a time-dependent form by multiplying with $e^{j\omega t}$ and then taking the real part of the product.

$$\tilde{E}_T = \text{Re}(2jE_i \sin\beta_z z e^{-j\beta_y y} e^{j\omega t})$$

$$= -2E_i \sin\beta_z z \sin(\omega t - \beta_y y)$$

Parallel polarization: In the case of parallel polarization (Figure 9.3), the total magnetic and electric field strengths are given by

$$H_T = 2H_i \cos\beta_z z e^{-j\beta_y y} \tag{9.8b}$$

$$E_{T,z} = 2E_i \sin\theta\cos(\beta_z z)e^{-j\beta_y y} \tag{9.8c}$$

$$E_{T,y} = 2jE_i \cos\theta\sin(\beta_z z)e^{-j\beta_y y} \tag{9.8d}$$

Proof: The magnetic field intensity in the incident wave is

$$H_{\text{inc}} = H_i e^{-j\beta(y\sin\theta - z\cos\theta)}$$

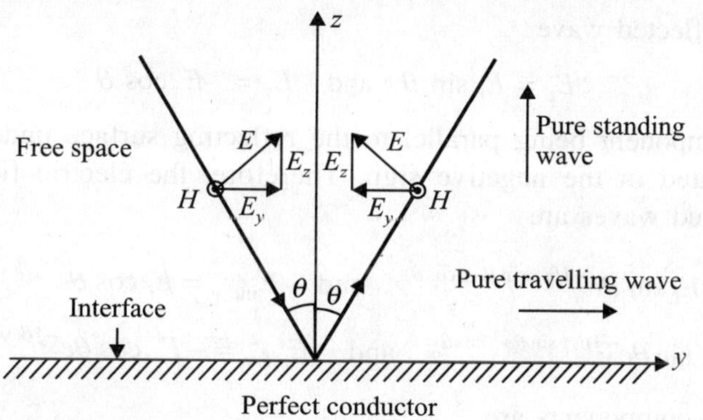

FIGURE 9.3 Reflection from conductor—Parallel polarization.

The electric field intensity in the reflected wave is

$$H_{\text{ref}} = H_r e^{-j\beta(y\sin\theta + z\cos\theta)}$$

The total electric field then is

$$H_T = H_{\text{inc}} + H_{\text{ref}} = H_i e^{-j\beta(y\sin\theta - z\cos\theta)} + H_r e^{-j\beta(y\sin\theta + z\cos\theta)}$$

The magnitudes of the incident and reflected waves are the same because their interactions are with regions, free space and perfect conductor, both are loss-free. As the wave is parallel polarized, the magnetic vector is tangential to the reflecting surface and hence it undergoes no phase change on reflection. Therefore, $H_r = H_i$. Incorporating this, the total field then becomes

$$\begin{aligned} H_T &= H_i\left[e^{-j\beta(y\sin\theta - z\cos\theta)} - e^{-j\beta(y\sin\theta + z\cos\theta)}\right] \\ &= H_i e^{-j\beta y\sin\theta}\left[e^{j\beta z\cos\theta} + e^{-j\beta z\cos\theta}\right] = H_i e^{-j\beta y\sin\theta}\left[2\cos(\beta z\cos\theta)\right] \end{aligned}$$

Therefore,

$$H_T = 2H_i \cos\beta_z z e^{-j\beta_y y}, \text{ where } \beta_z = \beta\cos\theta \text{ and } \beta_y = \beta\sin\theta$$

It can be converted into a time-dependent form by multiplying with $e^{j\omega t}$ and then taking the real part of the product.

$$\begin{aligned} \tilde{H}_T &= \text{Re}(2H_i \cos\beta_z z e^{-j\beta_y y} e^{j\omega t}) \\ &= 2H_i \cos\beta_z z \cos(\omega t - \beta_y y) \end{aligned}$$

This is the expression for the total magnetic field intensity. Most of the times, it is the electric field intensity that is of some interest to us and we try to find it out. To find the total electric field intensity, it is required to decompose the electric vector into two components, one parallel to the reflecting surface and the other normal to the surface in both the incident and reflected waves.

In the incident wave, the magnitudes of the electric field are

$$E_z = E_i \sin\theta \text{ and } E_y = E_i \cos\theta$$

Similarly for the reflected wave

$$E_z = E_r \sin\theta \quad \text{and} \quad E_y = -E_r \cos\theta$$

Note that the y-component being parallel to the reflecting surface undergoes phase reversal and it is incorporated in the negative sign. Therefore, the electric field intensities in the incident and reflected waves are

$$E_{\text{inc},z} = E_i \sin\theta e^{-j\beta(y\sin\theta - z\cos\theta)} \quad \text{and} \quad E_{\text{inc},y} = E_i \cos\theta e^{-j\beta(y\sin\theta - z\cos\theta)}$$

$$E_{\text{ref},z} = E_r \sin\theta e^{-j\beta(y\sin\theta + z\cos\theta)} \quad \text{and} \quad E_{\text{ref},y} = -E_r \cos\theta e^{-j\beta(y\sin\theta + z\cos\theta)}$$

Now the total field components are

$$E_{T,z} = E_{\text{inc},z} + E_{\text{ref},z} = E_i \sin\theta e^{-j\beta(y\sin\theta - z\cos\theta)} + E_r \sin\theta e^{-j\beta(y\sin\theta + z\cos\theta)}$$

$$E_{T,y} = E_{\text{inc},y} + E_{\text{ref},y} = E_i \cos\theta e^{-j\beta(y\sin\theta - z\cos\theta)} - E_r \cos\theta e^{-j\beta(y\sin\theta + z\cos\theta)}$$

The magnitudes of the incident and reflected waves are the same because they are in loss-free regions. Therefore, $E_r = E_i$. Incorporating this change

$$E_{T,z} = E_i \sin\theta e^{-j\beta y\sin\theta}\left(e^{j\beta z\cos\theta} + e^{-j\beta z\cos\theta}\right) = 2E_i \sin\theta \cos(\beta_z z)e^{-j\beta_y y}$$

$$E_{T,y} = E_i \cos\theta e^{-j\beta y\sin\theta}\left(e^{j\beta z\cos\theta} - e^{-j\beta z\cos\theta}\right) = 2jE_i \cos\theta \sin(\beta_z z)e^{-j\beta_y y}$$

In both the cases, there exist standing wave distribution along the z-axis, i.e. $\cos\beta_z z$ or $\sin\beta_z z$ and travelling wave along the y-direction, i.e. $e^{-j\beta_y y}$

9.3 PERFECT DIELECTRIC

In this section, we consider the reflection and refraction of a uniform plane wave at the interface of two different dielectric media. The amount of reflection and refraction depends upon the magnitude of discontinuity in the parameters of the media. We try to estimate two important parameters, the reflection coefficient and transmission coefficient, that represent the amount of reflection and refraction that has taken place at the interface. We derive expressions for these parameters in terms of the media characteristics.

In the following analysis, E_i, E_r and E_t denote the amplitudes of the electric field of the incident, reflected and transmitted waves; H_i, H_r and H_t are corresponding magnetic fields; and η_1 and η_2 are intrinsic impedances of the media. The intrinsic impedance of a medium in general is a complex quantity and is given by

$$\eta = \sqrt{\frac{j\omega\mu}{\sigma + j\omega\varepsilon}} \tag{9.9}$$

In the present case, as both the media are perfect dielectrics, the conductivity is zero, resulting in a pure real value, i.e.

$$\eta = |\eta| = \eta = \sqrt{\frac{\mu}{\varepsilon}} \tag{9.10}$$

9.3.1 Normal Incidence

In the case of wave incidence on a perfect dielectric, either normally or obliquely, a part of the incident wave gets reflected and the remaining goes into the dielectric as the transmitted wave. Boundary conditions dictate that, across the interface, the tangential components of the electric field must remain continuous. The tangential magnetic field must also be continuous as we are dealing with an interface devoid of any surface charge. Mathematically, we can put the conditions on fields as

$$E_i + E_r = E_t \tag{9.11}$$

and

$$H_i + H_r = H_t \tag{9.12}$$

In addition

$$E_i = \eta_1 H_i \tag{9.13}$$

$$E_r = -\eta_1 H_r \tag{9.14}$$

and

$$E_t = \eta_2 H_t \tag{9.15}$$

Now we try to evaluate the reflection and transmission coefficients which are of some importance in evaluating the amount of discontinuity at the interface. By definition, the ratio of reflected electric field to incident field is called the *reflection coefficient* and it is usually represented by the symbol Γ. The ratio of transmitted electric field to incident field is called the *transmission coefficient* and it is usually represented by the letter T.

Consider

$$H_i + H_r = H_t = \frac{E_t}{\eta_2}$$

$$\frac{E_i}{\eta_1} - \frac{E_r}{\eta_1} = \frac{E_i + E_r}{\eta_2} = \frac{E_i}{\eta_2} + \frac{E_r}{\eta_2}$$

$$\frac{E_i}{\eta_1} - \frac{E_i}{\eta_2} = \frac{E_r}{\eta_1} + \frac{E_r}{\eta_2} \quad \Rightarrow \quad E_i \frac{\eta_2 - \eta_1}{\eta_1 \eta_2} = E_r \frac{\eta_1 + \eta_2}{\eta_1 \eta_2}$$

Therefore, the reflection coefficient is

$$\frac{E_r}{E_i} = \Gamma = \frac{\eta_2 - \eta_1}{\eta_1 + \eta_2} \tag{9.16}$$

The transmission coefficient is

$$T = \frac{E_t}{E_i} = \frac{E_i + E_r}{E_i} = 1 + \frac{E_r}{E_i} = \frac{2\eta_2}{\eta_1 + \eta_2} \tag{9.17}$$

The permeabilities of known dielectrics do not differ appreciably from that of free space. In such a case, the reflection coefficient becomes

$$\Gamma = \frac{\eta_2 - \eta_1}{\eta_1 + \eta_2} = \frac{\sqrt{\dfrac{\mu_2}{\varepsilon_2}} - \sqrt{\dfrac{\mu_1}{\varepsilon_1}}}{\sqrt{\dfrac{\mu_2}{\varepsilon_2}} + \sqrt{\dfrac{\mu_1}{\varepsilon_1}}}$$

$$= \frac{\sqrt{\dfrac{\mu}{\varepsilon_2}} - \sqrt{\dfrac{\mu}{\varepsilon_1}}}{\sqrt{\dfrac{\mu}{\varepsilon_2}} + \sqrt{\dfrac{\mu}{\varepsilon_1}}} = \frac{\sqrt{\varepsilon_1} - \sqrt{\varepsilon_2}}{\sqrt{\varepsilon_1} + \sqrt{\varepsilon_2}} \tag{9.18}$$

Similarly, the transmission coefficient becomes

$$T = \frac{E_t}{E_i} = \frac{2\sqrt{\varepsilon_1}}{\sqrt{\varepsilon_1} + \sqrt{\varepsilon_2}} \tag{9.19}$$

Note that in the case of perfect dielectrics, the reflection and transmission coefficients and the intrinsic impedances are pure real.

9.3.2 Matched Condition

When the intrinsic impedances of the two media are the same, i.e. $\eta_1 = \eta_2$, the reflection coefficient becomes zero. It implies that the incident wave is getting transmitted entirely. This is the situation which is very often called *matched condition*. Note that the matching does not require the media to have all their parameters same; only the intrinsic impedances are required to be the same.

With further analysis, it can also be shown that the reflections from the surface of a medium can be eliminated by interspersing another medium of appropriate thickness and dielectric constant, provided the incidence is normal and all the three media are non-magnetic perfect dielectrics.

The thickness of the medium used for matching should be of quarter wavelength and its intrinsic impedance should be the geometric mean of the intrinsic impedances of the media.

Derivation:
Let us suppose there are two non-magnetic dielectric media '1' and '2' of large thickness, and for matching purpose another non-magnetic dielectric medium '3' with finite thickness d is introduced in between them (Figure 9.4). The media are having intrinsic impedances of η_1, η_2 and η_3. Let us suppose the wave incidence is normal and as a result the electric fields in the media are E_1, E_2 and E_3 and the magnetic fields are H_1, H_2 and H_3 in the media '1', '2' and '3', respectively.

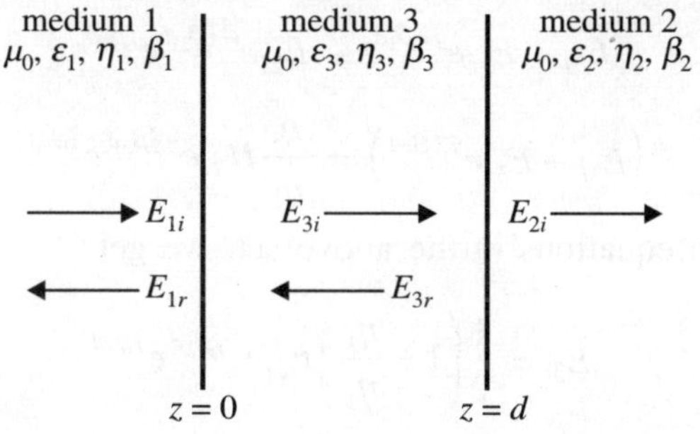

FIGURE 9.4 Matching two different non-magnetic dielectric media.

The fields in the media '1' and '3' consist of two components, incident and reflected, whereas in medium '2' only one component, incident wave, exists. They are given by

$$E_1 = E_{1i}e^{-j\beta_1 z} + E_{1r}e^{j\beta_1 z}$$

$$E_3 = E_{3i}e^{-j\beta_3 z} + E_{3r}e^{j\beta_3 z}$$

$$E_2 = E_{2i}e^{-j\beta_2 z}$$

$$H_1 = \frac{1}{\eta_1}\left(E_{1i}e^{-j\beta_1 z} - E_{1r}e^{j\beta_1 z}\right)$$

$$H_3 = \frac{1}{\eta_3}\left(E_{3i}e^{-j\beta_3 z} - E_{3r}e^{j\beta_3 z}\right)$$

$$H_2 = \frac{1}{\eta_2}H_{2i}e^{-j\beta_2 z}$$

As the incidence is normal, the electric and magnetic field vectors are parallel to the interface and the boundary conditions require them to be continuous across the interface. Two equations each at the interfaces result in

$$E_{1i} + E_{1r} = E_{3i} + E_{3r}$$

$$\frac{1}{\eta_1}(E_{1i} - E_{1r}) = \frac{1}{\eta_3}(E_{3i} - E_{3r})$$

$$E_{3i}e^{-j\beta_3 d} + E_{3r}e^{j\beta_3 d} = E_{2i}e^{-j\beta_2 d}$$

$$\frac{1}{\eta_3}\left(E_{3i}e^{-j\beta_3 d} - E_{3r}e^{j\beta_3 d}\right) = \frac{1}{\eta_2}H_{2i}e^{-j\beta_2 d}$$

The above four equations can be rearranged as

$$E_{1i} + E_{1r} = E_{3i} + E_{3r}$$

$$(E_{1i} - E_{1r}) = \frac{\eta_1}{\eta_3}(E_{3i} - E_{3r})$$

$$E_{3i} + E_{3r}e^{j2\beta_3 d} = E_{2i}e^{-j\beta_2 d}e^{j\beta_3 d}$$

$$\left(E_{3i} - E_{3r}e^{j2\beta_3 d}\right) = \frac{\eta_3}{\eta_2}H_{2i}e^{-j\beta_2 d}e^{j\beta_3 d}$$

Manipulating the last two equations in the above set, we get

$$E_{3i} = \frac{1}{2}\left(1+\frac{\eta_3}{\eta_2}\right)E_{2i}e^{-j\beta_2 d}e^{j\beta_3 d}$$

$$E_{3r} = \frac{1}{2}\left(1+\frac{\eta_3}{\eta_2}\right)E_{2i}e^{-j\beta_2 d}e^{-j\beta_3 d}$$

Manipulating the first two equations, i.e. subtracting the second equation from the first one, we obtain

$$E_{1r} = \frac{1}{2}(E_{3i} + E_{3r}) - \frac{\eta_1}{2\eta_3}(E_{3i} - E_{3r})$$

$$= \frac{E_{3i}}{2}\left(1-\frac{\eta_1}{\eta_3}\right) + \frac{E_{3r}}{2}\left(1+\frac{\eta_1}{\eta_3}\right)$$

Substituting the values of E_{3i} and E_{3r}, derived above, we get

$$E_{1r} = \frac{1}{4}\left(1+\frac{\eta_3}{\eta_2}\right)\left(1-\frac{\eta_1}{\eta_3}\right)E_{2i}e^{-j\beta_2 d}e^{j\beta_3 d} + \frac{1}{4}\left(1+\frac{\eta_3}{\eta_2}\right)\left(1+\frac{\eta_1}{\eta_3}\right)E_{2i}e^{-j\beta_2 d}e^{-j\beta_3 d}$$

To make it zero,

$$\frac{1}{4}\left(1+\frac{\eta_3}{\eta_2}\right)\left(1-\frac{\eta_1}{\eta_3}\right)e^{j\beta_3 d} + \frac{1}{4}\left(1+\frac{\eta_3}{\eta_2}\right)\left(1+\frac{\eta_1}{\eta_3}\right)e^{-j\beta_3 d} = 0$$

$$(\eta_3 + \eta_2)(\eta_3 - \eta_1) + (\eta_1 + \eta_3)(\eta_2 - \eta_3)e^{-j2\beta_3 d} = 0$$

The above equation gets satisfied only when

$$(\eta_3 + \eta_2)(\eta_3 - \eta_1) = (\eta_1 + \eta_3)(\eta_2 - \eta_3) \text{ and } e^{-j2\beta_3 d} = -1$$

$$\eta_3^2 + \eta_2\eta_3 - \eta_2\eta_1 - \eta_3\eta_1 = \eta_2\eta_1 - \eta_3\eta_1 + \eta_2\eta_3 - \eta_3^2 2 \text{ and } 2\beta_3 d = \pi$$

$$\eta_3^2 = \eta_2\eta_1$$

and

$$\frac{4\pi}{\lambda_3}d = \pi$$

Ultimately, we get

$$\frac{\mu_0}{\varepsilon_3} = \sqrt{\frac{\mu_0}{\varepsilon_1}\frac{\mu_0}{\varepsilon_2}} \tag{9.20}$$

and

$$d = \frac{\lambda_3}{4} \tag{9.21}$$

9.3.3 Oblique Incidence

Here we consider the oblique incidence of wave on a dielectric–dielectric interface (Figure 9.5). We try to estimate the reflection and transmission coefficients for both types of polarizations.

According to the Poynting theorem, the power per unit area in the incident wave is E_i^2/η_1 as the wave is the uniform plane wave. But this power spreads over the reflecting surface whose area increases with the angle of incidence, resulting in decrease of power per unit surface area with incidence angle.

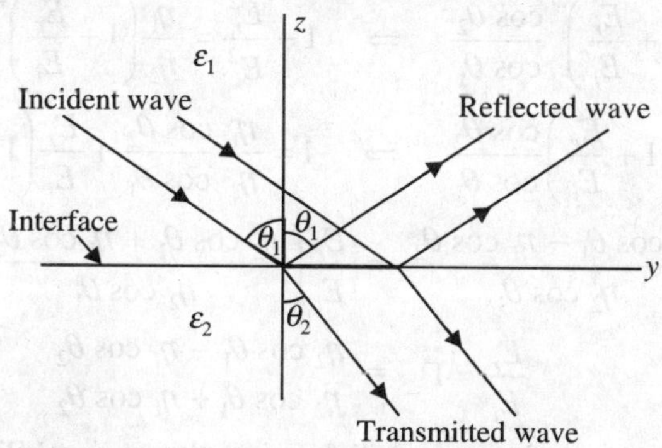

FIGURE 9.5 Reflection and transmission at dielectric–dielectric interface.

It can be shown that the power incident on a unit area of reflecting surface is $(E_i^2/\eta_1)\cos\theta_1$. With similar reasoning, we can show that powers per unit area of reflected surface and transmitted surface are $(E_r^2/\eta_1)\cos\theta_1$ and $(E_t^2/\eta_2)\cos\theta_2$, respectively.

The energy conservation requires that the power in the incident wave striking the unit area of the interface must be the sum of the powers in the reflected wave and transmitted wave emanating from that area. Therefore,

$$\frac{1}{\eta_1}E_i^2\cos\theta = \frac{1}{\eta_1}E_r^2\cos\theta_1 + \frac{1}{\eta_2}E_t^2\cos\theta_2 \tag{9.22}$$

$$\frac{E_r^2}{E_i^2} = 1 - \frac{\eta_1}{\eta_2}\frac{E_t^2}{E_i^2}\frac{\cos\theta_2}{\cos\theta_1} \tag{9.23}$$

If the permeabilities of the media are the same, then, we have seen earlier that

$$\frac{\eta_1}{\eta_2} = \sqrt{\frac{\varepsilon_2}{\varepsilon_1}} \tag{9.24}$$

Incorporation of Eq. (9.24) in Eq. (9.23) results in

$$\frac{E_r^2}{E_i^2} = 1 - \sqrt{\frac{\varepsilon_2}{\varepsilon_1}} \frac{E_t^2}{E_i^2} \frac{\cos\theta_2}{\cos\theta_1} \tag{9.25}$$

We are going to use Eq. (9.25) in the analysis of both the polarizations.

Perpendicular polarization: In this case, the electric vectors of the incident, reflected and transmitted waves are parallel or tangential to the reflecting surface and hence, according to the boundary condition, the electric field must be continuous across the interface. Mathematically, this condition is expressed as

$$E_i + E_r = E_t$$

$$\frac{E_t}{E_i} = 1 + \frac{E_r}{E_i}$$

$$\frac{E_r^2}{E_i^2} = 1 - \frac{\eta_1}{\eta_2}\left(1 + \frac{E_r}{E_i}\right)^2 \frac{\cos\theta_2}{\cos\theta_1} \quad \Rightarrow \quad 1 - \frac{E_r^2}{E_i^2} = \frac{\eta_1}{\eta_2}\left(1 + \frac{E_r}{E_i}\right)^2 \frac{\cos\theta_2}{\cos\theta_1}$$

$$1 - \frac{E_r}{E_i} = \frac{\eta_1}{\eta_2}\left(1 + \frac{E_r}{E_i}\right)\frac{\cos\theta_2}{\cos\theta_1} \quad \Rightarrow \quad 1 = \frac{\eta_1}{\eta_2}\frac{\cos\theta_2}{\cos\theta_1} + \frac{E_r}{E_i}\left(1 + \frac{\eta_1}{\eta_2}\frac{\cos\theta_2}{\cos\theta_1}\right)$$

$$\frac{\eta_2\cos\theta_1 - \eta_1\cos\theta_2}{\eta_2\cos\theta_1} = \frac{E_r}{E_i}\left(\frac{\eta_2\cos\theta_1 + \eta_1\cos\theta_2}{\eta_2\cos\theta_1}\right)$$

$$\frac{E_r}{E_i} = \Gamma_\perp = \frac{\eta_2\cos\theta_1 - \eta_1\cos\theta_2}{\eta_2\cos\theta_1 + \eta_1\cos\theta_2} \tag{9.26}$$

As was mentioned in the case of practical dielectrics, the permeabilities do not change much and we can safely consider that $\mu_1 = \mu_2 = \mu$. And it results in $\eta_1 = \sqrt{\mu/\varepsilon_1}$ and $\eta_2 = \sqrt{\mu/\varepsilon_2}$. Incorporating these relations

$$\Gamma_\perp = \frac{\sqrt{\varepsilon_1}\cos\theta_1 - \sqrt{\varepsilon_2}\cos\theta_2}{\sqrt{\varepsilon_1}\cos\theta_1 + \sqrt{\varepsilon_2}\cos\theta_2} = \frac{\cos\theta_1 - \sqrt{\frac{\varepsilon_2}{\varepsilon_1}}\cos\theta_2}{\cos\theta_1 + \sqrt{\frac{\varepsilon_2}{\varepsilon_1}}\cos\theta_2} \tag{9.27}$$

The reflection coefficient can also be expressed in terms of only the incident angle using the Snell's law.

$$\frac{\sin\theta_1}{\sin\theta_2} = \sqrt{\frac{\varepsilon_2}{\varepsilon_1}} \quad \Rightarrow \quad \sin\theta_2 = \sin\theta_1\sqrt{\frac{\varepsilon_1}{\varepsilon_2}} \quad \Rightarrow \quad \cos\theta_2 = \sqrt{1 - \sin^2\theta_2} = \sqrt{\frac{(\varepsilon_2 - \varepsilon_1\sin^2\theta_1)}{\varepsilon_2}}$$

$$\Gamma_\perp = \frac{\cos\theta_1 - \sqrt{\frac{\varepsilon_2}{\varepsilon_1}}\sqrt{\frac{(\varepsilon_2 - \varepsilon_1\sin^2\theta_1)}{\varepsilon_2}}}{\cos\theta_1 + \sqrt{\frac{\varepsilon_2}{\varepsilon_1}}\sqrt{\frac{(\varepsilon_2 - \varepsilon_1\sin^2\theta_1)}{\varepsilon_2}}} = \frac{\cos\theta_1 - \sqrt{\frac{(\varepsilon_2 - \varepsilon_1\sin^2\theta_1)}{\varepsilon_1}}}{\cos\theta_1 + \sqrt{\frac{(\varepsilon_2 - \varepsilon_1\sin^2\theta_1)}{\varepsilon_1}}}$$

$$= \frac{\cos\theta_1 - \sqrt{\dfrac{(\varepsilon_2 - \varepsilon_1 \sin^2\theta_1)}{\varepsilon_1}}}{\cos\theta_1 + \sqrt{\dfrac{(\varepsilon_2 - \varepsilon_1 \sin^2\theta_1)}{\varepsilon_1}}} = \frac{\cos\theta_1 - \sqrt{\left(\dfrac{\varepsilon_2}{\varepsilon_1} - \sin^2\theta_1\right)}}{\cos\theta_1 + \sqrt{\left(\dfrac{\varepsilon_2}{\varepsilon_1} - \sin^2\theta_1\right)}}$$

The above expression for reflection coefficient can be put in another compact form shown below which is known as Fresnel relation.

$$\Gamma_\perp = \frac{\sin(\theta_2 - \theta_1)}{\sin(\theta_2 + \theta_1)} \tag{9.28}$$

Proof: According to the Snell's law,

$$\sqrt{\varepsilon_1}\sin\theta_1 = \sqrt{\varepsilon_2}\sin\theta_2$$

$$\Gamma_\perp = \frac{\sqrt{\varepsilon_1}\cos\theta_1 - \sqrt{\varepsilon_2}\cos\theta_2}{\sqrt{\varepsilon_1}\cos\theta_1 + \sqrt{\varepsilon_2}\cos\theta_2}$$

$$= \frac{\sin\theta_2\cos\theta_1 - \cos\theta_2\sin\theta_1}{\sin\theta_2\cos\theta_1 + \cos\theta_2\sin\theta_1} = \frac{\sin(\theta_2 - \theta_1)}{\sin(\theta_2 + \theta_1)}$$

In the Fresnel equations shown above, the relation for reflection is being expressed only in terms of geometric quantities, i.e. angle. However, material parameters are implicitly involved.

Perfect dielectric—reflection coefficient

$\Gamma = \dfrac{\sqrt{\varepsilon_1} - \sqrt{\varepsilon_2}}{\sqrt{\varepsilon_1} + \sqrt{\varepsilon_2}}$	Normal incidence
$\Gamma_\perp = \dfrac{\sqrt{\varepsilon_1}\cos\theta_1 - \sqrt{\varepsilon_2}\cos\theta_2}{\sqrt{\varepsilon_1}\cos\theta_1 + \sqrt{\varepsilon_2}\cos\theta_2}$	Oblique incidence: $\perp$ polarization
$\Gamma_\parallel = \dfrac{\sqrt{\varepsilon_2}\cos\theta_1 - \sqrt{\varepsilon_1}\cos\theta_2}{\sqrt{\varepsilon_2}\cos\theta_1 + \sqrt{\varepsilon_1}\cos\theta_2}$	Oblique incidence: $\parallel$ polarization

Parallel polarization: In the case of parallel polarization, the electric vector components, which are parallel to the reflecting surface, require to be continuous across the interface. This condition transforms to

$$E_i\cos\theta_1 - E_r\cos\theta_1 = E_t\cos\theta_2$$

$$(E_i - E_r)\cos\theta_1 = E_t\cos\theta_2$$

$$\frac{E_t}{E_i} = \left(1 - \frac{E_r}{E_i}\right)\frac{\cos\theta_1}{\cos\theta_2}$$

$$\frac{E_r^2}{E_i^2} = 1 - \frac{\eta_1}{\eta_2}\left(1 - \frac{E_r}{E_i}\right)^2 \frac{\cos\theta_1}{\cos\theta_2} \quad\Rightarrow\quad 1 - \frac{E_r^2}{E_i^2} = \frac{\eta_1}{\eta_2}\left(1 - \frac{E_r}{E_i}\right)^2 \frac{\cos\theta_1}{\cos\theta_2}$$

$$1 + \frac{E_r}{E_i} = \frac{\eta_1}{\eta_2}\left(1 - \frac{E_r}{E_i}\right)\frac{\cos\theta_1}{\cos\theta_2} \quad\Rightarrow\quad 1 = \frac{\eta_1}{\eta_2}\frac{\cos\theta_1}{\cos\theta_2} - \frac{E_r}{E_i}\left(1 + \frac{\eta_1}{\eta_2}\frac{\cos\theta_1}{\cos\theta_2}\right)$$

$$\frac{\eta_2\cos\theta_2 - \eta_1\cos\theta_1}{\eta_2\cos\theta_2} = -\frac{E_r}{E_i}\left(\frac{\eta_2\cos\theta_2 + \eta_1\cos\theta_1}{\eta_2\cos\theta_2}\right)$$

$$\frac{E_r}{E_i} = \Gamma_{\parallel} = \frac{\eta_1\cos\theta_1 - \eta_2\cos\theta_2}{\eta_1\cos\theta_1 + \eta_2\cos\theta_2} \tag{9.29}$$

In the case of practical dielectrics, the permeabilities do not change much and we can safely consider that $\mu_1 = \mu_2 = \mu$. And it results in $\eta_1 = \sqrt{\mu/\varepsilon_1}$ and $\eta_2 = \sqrt{\mu/\varepsilon_2}$. Incorporating these relations

$$\Gamma_{\parallel} = \frac{\sqrt{\varepsilon_2}\cos\theta_1 - \sqrt{\varepsilon_1}\cos\theta_2}{\sqrt{\varepsilon_2}\cos\theta_1 + \sqrt{\varepsilon_1}\cos\theta_2} = \frac{\cos\theta_1 - \sqrt{\dfrac{\varepsilon_1}{\varepsilon_2}}\cos\theta_2}{\cos\theta_1 + \sqrt{\dfrac{\varepsilon_1}{\varepsilon_2}}\cos\theta_2} \tag{9.30}$$

Using the Snell's law it is possible to express the reflection coefficient in terms of incidence angle as

$$\Gamma_{\parallel} = \frac{\left(\dfrac{\varepsilon_2}{\varepsilon_1}\right)\cos\theta_1 - \sqrt{\left(\dfrac{\varepsilon_2}{\varepsilon_1}\right) - \sin^2\theta_1}}{\left(\dfrac{\varepsilon_2}{\varepsilon_1}\right)\cos\theta_1 + \sqrt{\left(\dfrac{\varepsilon_2}{\varepsilon_1}\right) - \sin^2\theta_1}} \tag{9.31}$$

It is also possible to express the reflection coefficient only in terms of the angles as

$$\Gamma_{\parallel} = \frac{\tan(\theta_1 - \theta_2)}{\tan(\theta_1 + \theta_2)} \tag{9.32}$$

It is often called Fresnel relation for reflection coefficient. It can be derived as follows: Using the Snell's law $\sqrt{\varepsilon_1}\sin\theta_1 = \sqrt{\varepsilon_2}\sin\theta_2$,

$$\Gamma_{\parallel} = \frac{\sqrt{\varepsilon_2}\cos\theta_1 - \sqrt{\varepsilon_1}\cos\theta_2}{\sqrt{\varepsilon_2}\cos\theta_1 + \sqrt{\varepsilon_1}\cos\theta_2}$$

$$= \frac{\sin\theta_1\cos\theta_1 - \sin\theta_2\cos\theta_2}{\sin\theta_1\cos\theta_1 + \sin\theta_2\cos\theta_2}$$

$$= \frac{\sin 2\theta_1 - \sin 2\theta_2}{\sin 2\theta_1 + \sin 2\theta_2} = \frac{\sin(\theta_1 - \theta_2)\cos(\theta_1 + \theta_2)}{\sin(\theta_1 + \theta_2)\cos(\theta_1 - \theta_2)} = \frac{\tan(\theta_1 - \theta_2)}{\tan(\theta_1 + \theta_2)}$$

9.3.4 Grazing Angle

Sometimes it is useful to have expressions for reflection coefficients in terms of grazing angle which is the angle ψ between the incident ray and the medium boundary. It is related to the incidence angle θ_1 through $\theta_1 + \psi = 90°$ or $\sin \psi = \cos \theta_1$. And with the help of the Snell's law, we can easily express the refraction angle in terms of the grazing angle.

$$\frac{\sin \theta_1}{\sin \theta_2} = \sqrt{\frac{\varepsilon_2}{\varepsilon_1}}$$

$$\Rightarrow \frac{\cos \psi}{\sin \theta_2} = \sqrt{\frac{\varepsilon_2}{\varepsilon_1}}$$

$$\Rightarrow \sin \theta_2 = \cos \psi \sqrt{\frac{\varepsilon_1}{\varepsilon_2}}$$

$$\Rightarrow \sin \theta_2 = \sqrt{1 - \left(\frac{\varepsilon_1}{\varepsilon_2}\right) \cos^2 \psi}$$

By substituting the above relation in the expression for the reflection coefficient for horizontal polarization,

$$\frac{E_r}{E_i} = \Gamma_\perp = \frac{\eta_2 \cos \theta_1 - \eta_1 \cos \theta_2}{\eta_2 \cos \theta_1 + \eta_1 \cos \theta_2}$$

$$= \frac{\eta_2 \sin \psi - \eta_1 \sqrt{1 - \left(\frac{\varepsilon_1}{\varepsilon_2}\right) \cos^2 \psi}}{\eta_2 \sin \psi + \eta_1 \sqrt{1 - \left(\frac{\varepsilon_1}{\varepsilon_2}\right) \cos^2 \psi}} \tag{9.33}$$

Similarly the reflection coefficient for the vertical polarization becomes

$$\frac{E_r}{E_i} = \Gamma_\parallel = \frac{\eta_1 \cos \theta_1 - \eta_2 \cos \theta_2}{\eta_1 \cos \theta_1 + \eta_2 \cos \theta_2}$$

$$= \frac{\eta_1 \sin \psi - \eta_2 \sqrt{1 - \left(\frac{\varepsilon_1}{\varepsilon_2}\right) \cos^2 \psi}}{\eta_1 \sin \psi + \eta_2 \sqrt{1 - \left(\frac{\varepsilon_1}{\varepsilon_2}\right) \cos^2 \psi}} \tag{9.34}$$

9.4 OBSERVATIONS

Certain points are worth mentioning regarding the reflection and refraction at the interface of two different dielectric media.

(a) Normal incidence: In the case of normal incidence, i.e. $\theta_1 = 0$, of uniform plane wave

- The reflection coefficients and transmission coefficients are the same for both the cases of polarizations and are given by

$$\frac{E_r}{E_i} = \frac{\eta_2 - \eta_1}{\eta_1 + \eta_2} \tag{9.35}$$

$$\frac{E_t}{E_i} = \frac{2\eta_2}{\eta_1 + \eta_2} \tag{9.36}$$

- There is no bending of the wave, i.e. the angle of refraction is zero ($\theta_2 = 0$).

(b) Total internal reflection: We can show that the wave gets reflected completely when the angle of incidence is more than or equal to $\theta_1 = \sin^{-1}\sqrt{\varepsilon_2/\varepsilon_1}$.

- In the case of total reflection, there exists no refracted wave and the angle of refraction $\theta_2 = 90°$. Therefore,

$$\cos\theta_2 = \sqrt{1 - \sin^2\theta_2} = \cos 90° = 0$$

$$\sqrt{1 - \sin^2\theta_2} = 0$$

$$\sqrt{1 - \frac{\mu_1\varepsilon_1}{\mu_2\varepsilon_2}\sin^2\theta_1} = 0$$

$$\sin\theta_1 = \sqrt{\frac{\mu_2\varepsilon_2}{\mu_1\varepsilon_1}}$$

Usually $\mu_1 = \mu_2$, resulting in

$$\theta_1 = \sin^{-1}\sqrt{\frac{\varepsilon_2}{\varepsilon_1}} \tag{9.37}$$

The angle given by the above expression is called *critical angle* and for total reflection the incidence angle must be equal to or more than this value.

$$\theta_c = \sin^{-1}\sqrt{\frac{\varepsilon_2}{\varepsilon_1}} \tag{9.38}$$

If the angle of incidence is more than θ_c, the reflection coefficients become complex but the total internal reflection continues to occur.

- It can take place only when the medium '1' is denser than the medium '2' and θ_1 is large enough to satisfy

$$\theta_1 > \sin^{-1}\sqrt{\frac{\varepsilon_2}{\varepsilon_1}}$$

- During total internal reflection, the reflection coefficients are $E_r/E_i = 1$ for perpendicular polarization and $E_r/E_i = -1$ for parallel polarization.
- The total internal reflection does not imply that there is no field in the medium '2'. In fact, a field does exist in the less dense medium. This field has a phase progression along the boundary but decreases exponentially away from it, so it is a non-uniform plane wave.
- This is the phenomenon based on the light signal travelling through fibre-optic cable.

(c) Brewster angle: The reflection coefficient becomes zero, $E_r/E_i = 0$, i.e. with parallel polarization when the angle of incidence is $\theta_1 = \tan^{-1}\sqrt{\varepsilon_2/\varepsilon_1}$.
The reflection coefficient in the case of parallel polarization is

$$\frac{E_r}{E_i} = \frac{\eta_2 \cos\theta_2 - \eta_1 \cos\theta_1}{\eta_2 \cos\theta_2 + \eta_1 \cos\theta_1} = 0$$

Solving this equation,

$$\eta_2 \cos\theta_2 = \eta_1 \cos\theta_2 \quad \Rightarrow \quad \eta_2\sqrt{\left(1 - \frac{\mu_1\varepsilon_1}{\mu_2\varepsilon_2}\sin^2\theta_1\right)} = \eta_1\sqrt{(1-\sin^2\theta_1)}$$

$$\sin^2\theta_1 = \frac{\eta_2^2 - \eta_1^2}{\eta_2^2\left(\frac{\mu_1\varepsilon_1}{\mu_2\varepsilon_2}\right) - \eta_1^2}$$

$$= \varepsilon_2 \frac{\left(\frac{\mu_2}{\mu_1}\right)\varepsilon_1 - \varepsilon_2}{\varepsilon_2^2 - \varepsilon_1^2}$$

If we assume $\mu_1 = \mu_2$ and $\varepsilon_2 \neq \varepsilon_1$, then

$$\sin^2\theta_1 = \varepsilon_2 \frac{\varepsilon_1 - \varepsilon_2}{\varepsilon_2^2 - \varepsilon_1^2} = \frac{\varepsilon_2}{\varepsilon_1 + \varepsilon_2}$$

It then gives

$$\cos^2\theta_1 = 1 - \sin^2\theta_1 = \frac{\varepsilon_1}{\varepsilon_1 + \varepsilon_2}$$

And $\tan\theta_1 = \sqrt{\dfrac{\varepsilon_2}{\varepsilon_1}}$

Thus, there exists an incidence angle for which the reflection coefficient is zero implying complete transmission of the wave. This angle is called **Brewster angle** which can be formally defined as the angle of incidence at which there is no reflected wave when the incident wave is parallel or vertically polarized. It is given by

$$\theta_1 = \tan^{-1}\sqrt{\frac{\varepsilon_2}{\varepsilon_1}} \tag{9.39}$$

(d) The reflection coefficient can never be zero in the case of perpendicular polarization.

The items (c) and (d) explain why a perpendicularly plane polarized wave gets reflected when an elliptically polarized wave incidents upon the surface at the Brewster angle.

9.5 LOSSY MEDIA

Lossy medium is one for which the conductivity is a finite non-zero i.e. $\sigma \neq 0$. For the convenience of analysis the lossy media can be divided into three categories: good conductor, poor conductor and lossy dielectric. We consider the plane wave propagation in each of them one by one.

The analysis of the wave reflection and transmission in lossy media is similar to that we adopted in the case of perfect dielectrics and conductors but with one major change: when the medium is lossy, the parameters, such as propagation constant, transmission and reflection coefficients, intrinsic impedance, etc., change from real to complex.

9.5.1 Good Conductor

A good conductor is one for which the conductivity is very high. Expressed mathematically, a good conductor satisfies the criterion $\sigma \gg \omega\varepsilon$. For this type of media we have already derived an expression for propagation constant

$$\gamma = \sqrt{j\omega\mu(\sigma + j\omega\varepsilon)}$$

$$= \sqrt{j\omega\mu\sigma\left(1 + \frac{j\omega\varepsilon}{\sigma}\right)}$$

$$\cong \sqrt{j\omega\mu\sigma} = \sqrt{\omega\mu\sigma}\angle 45°$$

Therefore,

$$\alpha = \beta = \sqrt{\frac{\omega\mu\sigma}{2}}$$

The intrinsic impedance of good conductors then is

$$\eta = \sqrt{\frac{j\omega\mu}{\sigma + j\omega\varepsilon}} \approx \sqrt{\frac{j\omega\mu}{\sigma}} = \sqrt{\frac{\omega\mu}{\sigma}}\angle 45°$$

By substituting the above relation in the expression for the reflection coefficient for horizontal polarization, we obtain

$$\frac{E_r}{E_i} = \Gamma_\perp = \frac{\eta_2 \cos\theta_1 - \eta_1 \cos\theta_2}{\eta_2 \cos\theta_1 + \eta_1 \cos\theta_2}$$

$$= \frac{\eta_2 \sin\psi - \eta_1 \sqrt{1 - \left(\frac{\varepsilon_1}{\varepsilon_2}\right)\cos^2\psi}}{\eta_2 \sin\psi + \eta_1 \sqrt{1 - \left(\frac{\varepsilon_1}{\varepsilon_2}\right)\cos^2\psi}} \tag{9.40}$$

Similarly the reflection coefficient for vertical polarization becomes

$$\frac{E_r}{E_i} = \Gamma_\parallel = \frac{\eta_1 \cos\theta_1 - \eta_2 \cos\theta_2}{\eta_1 \cos\theta_1 + \eta_2 \cos\theta_2}$$

$$= \frac{\eta_1 \sin\psi - \eta_2 \sqrt{1 - \left(\frac{\varepsilon_1}{\varepsilon_2}\right)\cos^2\psi}}{\eta_1 \sin\psi + \eta_2 \sqrt{1 - \left(\frac{\varepsilon_1}{\varepsilon_2}\right)\cos^2\psi}} \tag{9.41}$$

Now the reflection coefficients are no more real and they have become complex.

9.5.2 Poor Conductor

Some conducting materials with low conductivity can be considered as neither good conductors nor good dielectrics. An example is sea water. In general, the propagation constant and intrinsic impedance of a poor conductor are given by

$$\gamma = \sqrt{j\omega\mu(\sigma + j\omega\varepsilon)} \tag{9.42}$$

$$\eta = \sqrt{\frac{j\omega\mu}{\sigma + j\omega\varepsilon}} \tag{9.43}$$

These are the actual defining relations without any approximations. For computing the reflection and transmission coefficients, we have to consider the above expressions.

9.5.3 Lossy Dielectrics

In practice, it is not possible to achieve either a zero conductivity dielectric or an infinite conductivity conductor. If the conductivity of the dielectric is considerable and cannot be neglected, in that case the dielectric is called *lossy dielectric*. In this type of media, the electric and magnetic fields do not remain in time phase. It can be shown in lossy dielectrics, that the attenuation constant and phase shift constant are

$$\alpha = \omega\sqrt{\frac{\mu\varepsilon}{2}\left(\sqrt{1 + \frac{\sigma^2}{\omega^2\varepsilon^2}} - 1\right)}$$

$$\cong \omega\sqrt{\frac{\mu\varepsilon}{2}\left(1+\frac{\sigma^2}{2\omega^2\varepsilon^2}-1\right)} = \frac{\sigma}{2}\sqrt{\frac{\mu}{\varepsilon}}$$

$$\beta = \omega\sqrt{\frac{\mu\varepsilon}{2}\left(\sqrt{1+\frac{\sigma^2}{\omega^2\varepsilon^2}}+1\right)}$$

$$\cong \omega\sqrt{\frac{\mu\varepsilon}{2}\left(1+\frac{\sigma^2}{2\omega^2\varepsilon^2}+1\right)}$$

$$= \omega\sqrt{\mu\varepsilon}\left(1+\frac{\sigma^2}{8\omega^2\varepsilon^2}\right)$$

By substituting the above relations in the expression for the reflection coefficient for horizontal polarization, we get

$$\frac{E_r}{E_i} = \Gamma_\perp = \frac{\eta_2 \cos\theta_1 - \eta_1 \cos\theta_2}{\eta_2 \cos\theta_1 + \eta_1 \cos\theta_2}$$

$$= \frac{\sin\psi - \sqrt{(\varepsilon_r - jx) - \cos^2\psi}}{\sin\psi + \sqrt{(\varepsilon_r - jx) - \cos^2\psi}} \tag{9.44}$$

Similarly the reflection coefficient for vertical polarization becomes

$$\frac{E_r}{E_i} = \Gamma_\parallel = \frac{\sqrt{\left(\frac{\varepsilon_2}{\varepsilon_1}\right) - \sin^2\theta_1} - \left(\frac{\varepsilon_2}{\varepsilon_1}\right)\cos\theta_1}{\sqrt{\left(\frac{\varepsilon_2}{\varepsilon_1}\right) - \sin^2\theta_1} + \left(\frac{\varepsilon_2}{\varepsilon_1}\right)\cos\theta_1} \tag{9.45}$$

The reflection coefficient becomes zero, i.e. the complete transmission of the wave takes place when the angle of incidence is

$$\theta_1 = \tan^{-1}\sqrt{\frac{\varepsilon_2}{\varepsilon_1}}$$

This angle, we know well, is called the Brewster angle or polarizing angle. Furthermore, if

$$\varepsilon_2 = \varepsilon\left(1 - j\frac{\sigma}{\omega\varepsilon}\right)$$

and $\varepsilon_1 = \varepsilon_0$, i.e. for the case of wave falling on the dielectric from air, the reflection coefficient becomes

$$\frac{E_r}{E_i} = \Gamma_\parallel = \frac{(\varepsilon_r - jx)\sin\psi - \sqrt{(\varepsilon_r - jx) - \cos^2\psi}}{(\varepsilon_r - jx)\sin\psi + \sqrt{(\varepsilon_r - jx) - \cos^2\psi}} \tag{9.46}$$

where $\varepsilon_r = \varepsilon/\varepsilon_0$, $x = \sigma/\omega\varepsilon_0$ and $\psi = 90^\circ - \theta_1$ called pseudo Brewster angle.

For total reflection, we set $\theta_2 = 90°$ and the angle of incidence becomes

$$\theta_c = \sin^{-1}\sqrt{\frac{\varepsilon_2}{\varepsilon_1}}$$

which is the critical angle for total internal reflection. A wave incident on the boundary at an angle equal to or greater than the critical angle will be totally reflected. And there is the real critical angle only when $\varepsilon_1 > \varepsilon_2$. Hence, the total reflection takes place only if the wave propagates from a dense medium to a rarer medium.

SOLVED PROBLEMS

1. Show that, starting from boundary conditions, the reflection coefficient at the interface of two dielectrics for oblique incidence of a perpendicular polarized wave is given by $\Gamma_\perp = \sin(\theta_2 - \theta_1)/\sin(\theta_2 + \theta_1)$.

Solution: One of the boundary conditions requires continuity of tangential components of magnetic field across any interface with no surface currents. In the present case, we are dealing with interface of two dielectric media, where there is no chance for the surface currents to exist. With symbols carrying their usual meanings, therefore,

$$-H_i \cos\theta_1 + H_r \cos\theta_1 = -H_t \cos\theta_2$$

$$\left(1 - \frac{H_r}{H_i}\right) = \frac{H_t}{H_i}\frac{\cos\theta_2}{\cos\theta_1} \quad \Rightarrow \quad \frac{H_t}{H_i} = \left(1 - \frac{H_r}{H_i}\right)\frac{\cos\theta_1}{\cos\theta_2}$$

Another boundary condition requires that the normal components of magnetic flux density be continuous across the boundary. If the media are of the same permeability, as in the present case, the normal magnetic field components also require to be the same across the boundary. Therefore,

$$H_i \sin\theta_1 + H_r \sin\theta_1 = H_t \sin\theta_2$$

$$\left(1 + \frac{H_r}{H_i}\right) = \frac{H_t}{H_i}\frac{\sin\theta_2}{\sin\theta_1} = \left(1 - \frac{H_r}{H_i}\right)\frac{\cos\theta_1}{\cos\theta_2}\frac{\sin\theta_2}{\sin\theta_1}$$

$$\frac{H_r}{H_i}\left(1 + \frac{\cos\theta_1}{\cos\theta_2}\frac{\sin\theta_2}{\sin\theta_1}\right) = \left(\frac{\cos\theta_1}{\cos\theta_2}\frac{\sin\theta_2}{\sin\theta_1} - 1\right)$$

$$\frac{H_r}{H_i}\frac{(\sin\theta_1\cos\theta_2 + \cos\theta_1\sin\theta_2)}{(\sin\theta_1\cos\theta_2)} = \frac{(\cos\theta_1\sin\theta_2 - \sin\theta_1\cos\theta_2)}{(\sin\theta_1\cos\theta_2)}$$

$$\Gamma_\perp = \frac{\sin(\theta_2 - \theta_1)}{\sin(\theta_2 + \theta_1)}$$

2. Show that the reflection coefficient at the interface of two dielectrics for oblique incidence of a parallel polarized wave is given by

$$\Gamma_{\|} = \frac{\tan(\theta_1 - \theta_2)}{\tan(\theta_1 + \theta_2)}$$

Solution: The tangential components of electric field are continuous across the boundary. Thus,

$$E_i \cos\theta_1 - E_r \cos\theta_1 = E_t \cos\theta_2$$

$$\left(1 + \frac{E_r}{E_i}\right) = \frac{E_t \cos\theta_2}{E_i \cos\theta_1} \quad \Rightarrow \quad \frac{E_t}{E_i} = \frac{\cos\theta_1}{\cos\theta_2}\left(1 + \frac{E_r}{E_i}\right)$$

The normal components of the electric displacement density are continuous across the boundary in the absence of surface charge over the interface which is the present case. Thus,

$$\varepsilon_1 E_i \sin\theta_1 + \varepsilon_1 E_r \sin\theta_1 = \varepsilon_2 E_t \sin\theta_2$$

Here we assumed both media of the same permeability which is the case usually prevalent. Now,

$$1 + \frac{E_r}{E_i} = \frac{\varepsilon_2}{\varepsilon_1} \frac{E_t \sin\theta_2}{E_i \sin\theta_1}$$

Using the Snell's law

$$\frac{\sin^2\theta_1}{\sin^2\theta_2} = \frac{\varepsilon_2}{\varepsilon_1}$$

we can have

$$1 + \frac{E_r}{E_i} = \frac{\sin^2\theta_1}{\sin^2\theta_2} \frac{E_t \sin\theta_2}{E_i \sin\theta_1} = \frac{\sin\theta_1}{\sin\theta_2} \frac{E_t}{E_i}$$

$$= \frac{\sin\theta_1}{\sin\theta_2} \frac{\cos\theta_1}{\cos\theta_2}\left(1 - \frac{E_r}{E_i}\right)$$

$$= \frac{\sin 2\theta_1}{\sin 2\theta_2}\left(1 - \frac{E_r}{E_i}\right)$$

Bringing all the terms involving the reflection coefficient to the left side, we have

$$\frac{E_r}{E_i}\left(1 + \frac{\sin 2\theta_1}{\sin 2\theta_2}\right) = \frac{\sin 2\theta_1}{\sin 2\theta_2} - 1$$

$$\frac{E_r}{E_i}(\sin 2\theta_1 + \sin 2\theta_2) = (\sin 2\theta_1 - \sin 2\theta_2)$$

$$\Gamma_{\|} = \frac{(\sin 2\theta_1 - \sin 2\theta_2)}{(\sin 2\theta_1 + \sin 2\theta_2)} = \frac{2\sin(\theta_1 - \theta_2)\cos(\theta_1 + \theta_2)}{2\sin(\theta_1 + \theta_2)\cos(\theta_1 - \theta_2)} = \frac{\tan(\theta_1 - \theta_2)}{\tan(\theta_1 + \theta_2)}$$

3. Prove that the reflection coefficient can never be zero in the case of perpendicular polarization.

Solution: To prove the point, we equate the expression for reflection coefficient in the case of perpendicular polarization to zero and then show that this equation has no solution.

$$\frac{E_r}{E_i} = \frac{\eta_2 \cos\theta_1 - \eta_1 \cos\theta_2}{\eta_2 \cos\theta_1 + \eta_1 \cos\theta_2} = 0$$

$$\eta_2 \cos\theta_1 = \eta_1 \cos\theta_2 = \eta_1 \sqrt{\left(1 - \frac{\mu_1\varepsilon_1}{\mu_2\varepsilon_2}\sin^2\theta_1\right)}$$

$$\sin^2\theta_1 = \frac{\eta_2^2 - \eta_1^2}{\eta_2^2 - \eta_1^2\left(\frac{\mu_1\varepsilon_1}{\mu_2\varepsilon_2}\right)} = \mu_2 \frac{\mu_2 - \mu_1\left(\frac{\varepsilon_2}{\varepsilon_1}\right)}{\mu_2^2 - \mu_1^2}$$

If we consider a practical situation where the transmission is in between two dielectrics, then $\mu_1 = \mu_2$ and $\varepsilon_2 \neq \varepsilon_1$, resulting in no solution to the above equation. Therefore, a perpendicularly polarized incident wave can never be totally transmitted into another medium and its reflection coefficient can never be zero.

4. A uniform plane wave of frequency 2.45×10^9 Hz is falling on the surface $z = 0$ of a perfect dielectric with $\varepsilon_r = 4$ from free space. Find the values of the permittivity and thickness of the medium to be placed in between so that no reflections occur at $z = 0$.

Solution: For the reflection-less wave transmission, the dielectric medium that is placed should have intrinsic impedance which is the geometric mean of the impedances of the media on either side.

$$\eta^2 = \eta_1\eta_2 = \sqrt{\frac{\mu_0}{\varepsilon_1}\frac{\mu_0}{\varepsilon_2}} = (120\pi)^2\sqrt{\frac{1}{\varepsilon_{r1}\varepsilon_{r2}}} = (120\pi)^2\sqrt{\frac{1}{\varepsilon_{r2}}}$$

We have incorporated the given information that $\varepsilon_{r1} = 1$. Now let us suppose the relative permittivity of the dielectric to be inserted is ε_{mr}. Then

$$\eta = \sqrt{\frac{\mu_0}{\varepsilon_{mr}\varepsilon_0}} = 120\pi\sqrt{\frac{1}{\varepsilon_{mr}}}$$

Therefore,

$$\frac{(120\pi)^2}{\varepsilon_{mr}} = (120\pi)^2\sqrt{\frac{1}{\varepsilon_{r2}}} \rightarrow \varepsilon_{mr} = (\varepsilon_{r2})^2 = (4)^2 = 16$$

The thickness of the dielectric should be of quarter wavelength. Thus,

$$d = \frac{\lambda_m}{4} = \frac{v}{4f} = \frac{1}{4f\sqrt{\mu\varepsilon}} = \frac{1}{4f\sqrt{\mu_0\varepsilon_0\varepsilon_{mr}}}$$

$$= \frac{3\times10^8}{4\times2.45\times10^9\times4} = 0.0076 \text{ m}$$

5. A plane wave is incident from the region '1' which is free space over to a material medium which is characterized by $\sigma = 2.17$ S/m, $\mu = 47\varepsilon_0$ and $\mu = \mu_0$. The electric field of the wave is given by

$$\mathbf{E}_i = E_0 \cos(5.5\pi\times10^9 t - 10^{-3}\pi z)\,\hat{\mathbf{x}}$$

Find the fields in the reflected as well as transmitted wave.

Solution: The medium '2' is lossy and hence the transmitted wave undergoes attenuation. We need attenuation constant, which is the real part of the propagation constant. The reflected wave is the product of incident wave with reflection coefficient. To compute the reflection coefficient, we need intrinsic impedance of the medium. From the given values of the parameters, we try to compute the propagation constant and intrinsic impedance of the dielectric.

$$\gamma = \sqrt{j\omega\mu(1 + j\omega\varepsilon)}$$

$$= j\omega\sqrt{\mu\varepsilon}\sqrt{\left(1+\frac{1}{j\omega\varepsilon}\right)} = \frac{j\omega\sqrt{\varepsilon_r}}{3\times10^8}\sqrt{\left(1+\frac{1}{j\omega\varepsilon_0\varepsilon_r}\right)}$$

$$= \frac{j2\pi\times2.45\times10^9\sqrt{47}}{3\times10^8}\sqrt{\left(1+\frac{2.17\times36\pi}{j2\pi\times2.45\times10^9\times10^{-9}\times47}\right)}$$

$$= 58.85 + j35.67 = 361.5\angle 80.6°$$

$$\eta = \sqrt{\frac{j\omega\mu}{(\sigma + j\omega\varepsilon)}} = \sqrt{\frac{\mu_0}{\varepsilon_0}}\sqrt{\frac{1}{\varepsilon_r}}\sqrt{\frac{1}{\left[1+\left(\frac{\sigma}{j\omega\varepsilon}\right)\right]}}$$

$$= 120\pi\sqrt{\frac{1}{47}}\sqrt{\frac{1}{\left[1+\left(\frac{2.17\times36\pi}{j2\pi\times2.45\times10^9\times10^{-9}\times47}\right)\right]}}$$

$$= 53.5\angle 9.4°\Omega$$

Now we compute reflection coefficient and transmission coefficient.

$$\Gamma = \frac{\eta_2 - \eta_1}{\eta_2 + \eta_1}$$

$$\tau = 1 + \Gamma = \frac{2\eta_2}{\eta_1 + \eta_2}$$

In the above formulae, $\eta_1 = 377\ \Omega$ pertains to free space and η_2 to the medium and is equal to $53.5\angle 9.4^\circ\ \Omega$. Therefore,

$$\Gamma = \frac{\eta_2 - \eta_1}{\eta_2 + \eta_1} = \frac{53.5\angle 9.4^\circ - 377}{53.5\angle 9.4^\circ + 377} = 0.754\angle 177.3^\circ$$

$$\tau = 1 + \Gamma = \frac{2\eta_2}{\eta_1 + \eta_2} = \frac{2 \times 53.5\angle 9.4^\circ}{53.5\angle 9.4^\circ + 377} = 0.249\angle 8.235^\circ$$

Given that

$$\mathbf{E}_i = E_0 \cos(5.5\pi \times 10^9 t - 10^{-3}\pi z)\,\hat{\mathbf{x}}$$

In phasor form it is

$$\mathbf{E}_i(z) = \hat{\mathbf{x}} E_0 \angle -10^{-3}\pi z$$

Then the fields in the reflected wave become

$$\mathbf{E}_r(z) = \Gamma \mathbf{E}_i(z) = \hat{\mathbf{x}} |\Gamma| E_0 (\angle -10^{-3}\pi z + \angle\Gamma)$$

In time domain it becomes

$$\begin{aligned}\mathbf{E}_r &= |\Gamma| E_0 \cos(5.5\pi \times 10^9 t - 10^{-3}\pi z + \angle\Gamma)\,\hat{\mathbf{x}} \\ &= 0.754\, E_0 \cos(5.5\pi \times 10^9 t - 10^{-3}\pi z + \angle 177.3^\circ)\,\hat{\mathbf{x}}\end{aligned}$$

$$\mathbf{H}_r(z) = -\frac{\mathbf{E}_r(z)}{\eta_1}$$

In time domain

$$\begin{aligned}\mathbf{H}_r &= -\left(\frac{|\Gamma|}{377}\right) E_0 \cos(5.5\pi \times 10^9 t - 10^{-3}\pi z - \angle\eta_1 + \angle\Gamma)\,\hat{\mathbf{x}} \\ &= -\left(\frac{0.754}{377}\right) E_0 \cos(5.5\pi \times 10^9 t - 10^{-3}\pi z - 0 + \angle 177.3^\circ)\,\hat{\mathbf{x}}\end{aligned}$$

$$\mathbf{E}_t(z) = \tau \mathbf{E}_i(z)\, e^{-\gamma z}$$

In time domain

$$\begin{aligned}\mathbf{E}_t &= |\tau| E_0 e^{-\alpha z} \cos(5.5\pi \times 10^9 t - 10^{-3}\pi z - \beta z + \angle\tau)\,\hat{\mathbf{x}} \\ &= 0.249 E_0 e^{-\alpha z} \cos(5.5\pi \times 10^9 t - 10^{-3}\pi z - \beta z + \angle 8.235^\circ)\,\hat{\mathbf{x}}\end{aligned}$$

$$\mathbf{H}_t(z) = \frac{\mathbf{E}_t(z)}{\eta_2}$$

In time domain

$$\begin{aligned}\mathbf{H}_t &= \left(\frac{|\tau|}{|\eta_2|}\right) E_0 e^{-\alpha z} \cos(5.5\pi \times 10^9 t - 10^{-3}\pi z - \beta z - \angle\eta_2 + \angle\tau)\,\hat{\mathbf{x}} \\ &= \left(\frac{0.249}{53.5}\right) E_0 e^{-\alpha z} \cos(5.5\pi \times 10^9 t - 10^{-3}\pi z - \beta z - \angle 9.4^\circ + \angle 8.235^\circ)\,\hat{\mathbf{x}}\end{aligned}$$

6. The electric vector of a uniform plane wave propagating in free space is

$$\mathbf{E}_i = \frac{E_0}{\sqrt{2}}(\hat{\mathbf{x}} - \hat{\mathbf{z}}) \cos\left[8\pi \times 10^9 t - \frac{20}{\sqrt{2}}\pi(x+z)\right]$$

incident on a dielectric interface over the *yz*-plane with $\varepsilon = 1.5\,\varepsilon_0$ and $\mu = \mu_0$. Find the expressions for electric fields of the reflected and transmitted waves.

Solution: From the second part of the argument of field expression, we can observe that the propagation direction is entirely in the *x*-*z* plane. From the expression of the amplitude, we can also note that the electric vector is entirely in the *x*-*z* plane. Therefore, we can conclude that the wave given is a parallel polarized one.

Once again from the second part of the argument, we can observe that the unit vector along the propagation direction is $(\hat{\mathbf{x}} + \hat{\mathbf{z}})/\sqrt{2}$, which makes an angle of 45° with the *z*- as well as the *x*-axes. Therefore, the angle of incidence is 45°. According to the Snell's law, it is also the angle of reflection. And the angle of refraction can be computed as

$$\sin\theta_2 = \sqrt{\frac{\varepsilon_0}{1.5\varepsilon_0}}\sin 45^\circ = \frac{1}{\sqrt{3}}$$

$$\theta_2 = 35.26^\circ$$

As we need the reflection and transmission coefficients to write down the field expressions, now we find their values.

$$\Gamma_{\parallel} = \frac{\sqrt{\varepsilon_2}\cos\theta_1 - \sqrt{\varepsilon_1}\cos\theta_2}{\sqrt{\varepsilon_2}\cos\theta_1 + \sqrt{\varepsilon_1}\cos\theta_2}$$

$$= \frac{\sqrt{1.5\varepsilon_0}\cos 45^\circ - \sqrt{\varepsilon_0}\cos 35.26^\circ}{\sqrt{1.5\varepsilon_0}\cos 45^\circ + \sqrt{\varepsilon_0}\cos 35.26^\circ}$$

$$= \frac{\sqrt{1.5}\times 0.707 - 0.8165}{\sqrt{1.5}\times 0.707 + 0.8165} = \frac{0.049}{1.682} = 0.03$$

From the wave directions,

$$\tau_{\parallel} = \frac{E_t}{E_i} = \left(1 - \frac{E_r}{E_i}\right)\frac{\cos\theta_1}{\cos\theta_2}$$

$$= (1 - 0.03)\frac{\cos 45^\circ}{\cos 35.26^\circ} = 0.97\frac{0.707}{0.8165} = 0.84$$

Now to write down the expressions for fields, we need some more quantities. The unit vector along the incident wave direction is $(\hat{\mathbf{x}} + \hat{\mathbf{z}})/\sqrt{2}$. Therefore, for the reflected wave it must be $(-\hat{\mathbf{x}} + \hat{\mathbf{z}})/\sqrt{2}$ and for the refracted wave it is

$$(\hat{\mathbf{x}} \cos 35.26° + \hat{\mathbf{z}} \cos 35.26°) = (0.8165\,\hat{\mathbf{x}} + 0.577\,\hat{\mathbf{z}})$$

The value of the phase shift constant for the incident wave as well as the reflected wave is 20π and for the transmitted wave it is $20\pi\sqrt{1.5}$.

With the available information, now we can write down the expressions for fields as

$$\mathbf{E}_r = 0.03\frac{E_0}{\sqrt{2}}(\hat{\mathbf{x}} + \hat{\mathbf{z}}) \cos\left[8\pi \times 10^9 t - \frac{20}{\sqrt{2}}\pi(-x + z)\right]$$

$$\mathbf{E}_t = 0.84\frac{E_0}{\sqrt{2}}(\hat{\mathbf{x}} - \hat{\mathbf{z}}) \cos\left[8\pi \times 10^9 t - \frac{20\sqrt{1.5}}{\sqrt{2}}\pi(0.8165\,\hat{\mathbf{x}} + 0.577\,\hat{\mathbf{z}})\right]$$

7. A plane wave with $E = 2.0$ V/m and a frequency of 300 MHz is moving in free space impinging on a thick copper sheet ($\sigma = 5.8 \times 10^7$) located perpendicular to the direction of the propagation. Find E and H at the plane surface.

Solution: The tangential component of the electric vector gets reflected with phase reversal, whereas that of the magnetic vector gets reflected without phase reversal at the surface of a perfect conductor. As a result, on normal incidence, at the surface of the conductor the amplitude of the electric field becomes zero and that of the magnetic field becomes twice that of the incident wave.

In the incident travelling wave, it is given $E = 2.0$ V/m. Then

$$H = \frac{E}{120\pi} = \frac{2.0}{120\pi} = \frac{1}{60\pi} \text{ A/m}$$

The angular frequency $\omega = 2\pi f = 2\pi \times 300 \times 10^6$ rad/s

In front of the surface, the standing wave pattern comes into being with a node for the electric field and an anti-node for the magnetic field.

At the surface, $E = 0$ and $H = 2/60\pi = 1/30\pi$ A/m.

EXERCISE SECTION

Essay Type Questions

1. Analyse the reflection of a plane wave from a plane surface of a perfect conductor for
   (a) normal incidence and
   (b) oblique incidence.
2. Find the reflection coefficients of a plane wave when the reflection is from a plane surface of a perfect dielectric for
   (a) normal incidence and
   (b) oblique incidence.
3. Explain the following phenomena:
   (a) Brewster angle and
   (b) total internal reflection

Numerical Problems

1. An EM wave in a dielectric medium 1(ε_1, μ_0) impinges obliquely on a boundary plane with another dielectric medium 2(ε_2, μ_0). Let θ_i and θ_t denote the incident and refraction angles respectively and show that for parallel polarization, reflection coefficient is $(\sin 2\theta_i - \sin 2\theta_t)/(\sin 2\theta_i + \sin 2\theta_t)$ and transmission coefficient is $(4 \sin \theta_t \cos \theta_i)/(\sin 2\theta_i + \sin 2\theta_t)$.
2. A uniform plane wave at 1 MHz travels in air in a direction that makes 30° with the x-axis, 60° with the y-axis and 90° with the z-axis. It has a z-directed electric field of magnitude 5 V/m. Express the electric and magnetic fields in vector form.
3. A y-polarized uniform plane wave with fields (E_i, H_i) and a frequency of 100 MHz propagates in air in the +x-direction and impinges normally on a perfectly conducting plane at $x = 0$. Assuming the amplitude of E_i to be 6 mV/m, write the phasor and instantaneous expressions for
   (a) E_i and H_i of the incident wave,
   (b) E_r and H_r of the reflected wave and
   (c) E_T and H_T of the total wave in air, and
   (d) determine the location nearest to the conducting plane, where E_T and H_T are zero.

Fill in the Blanks

1. Snell's law of refraction is given by ________.
2. Plane of polarization contains the incident ray and is ________ to the interface.
3. For perpendicular polarization, the **E** is ________ to the plane of polarization.
4. For perpendicular polarization, the **H** is ________ to the reflecting surface.
5. For parallel polarization, the **E** is ________ to the plane of polarization.
6. Perpendicular polarization is known as horizontal polarization, whereas parallel polarization is known as ________ polarization.
7. The tangential electric vector gets reflected with a phase shift of ________ at the surface of a perfect conductor.
8. The tangential magnetic vector gets reflected with a phase shift of ________ at the surface of a perfect conductor.
9. Normal to the perfect conductor reflecting surface, there exists a ________ wave pattern.
10. Along the perfect conductor reflecting surface, there exists a ________ wave.
11. Normal to the perfect conductor reflecting surface, the phase shift constant is ________.
12. Along the perfect conductor reflecting surface, the phase shift constant is ________.
13. In the case of normal incidence over the surface of a perfect dielectric, the reflection coefficient is ________.
14. In the case of normal incidence over the surface of a perfect dielectric, the transmission coefficient is ________.

15. In the case of oblique incidence/⊥ polarization over the surface of a perfect dielectric, the reflection coefficient is ________.
16. In case of oblique incidence/polarization over the surface of a perfect dielectric, the transmission coefficient is ________.
17. Brewster's angle is given by ________.
18. Brewster's angle exists only for ________ polarization
19. For total internal reflection to occur, the incidence medium should be a ________ reflecting medium.
20. Critical angle is given by ________.

Answers

1. $\dfrac{\sin\theta_i}{\sin\theta_r} = \dfrac{v_i}{v_r} = \sqrt{\dfrac{(\mu\varepsilon)_r}{(\mu\varepsilon)_i}}$
2. normal
3. perpendicular
4. not parallel
5. parallel
6. vertical
7. 180°
8. zero
9. standing wave
10. travelling
11. $\beta \sin\theta$
12. $\beta \cos\theta$
13. $\dfrac{\sqrt{\varepsilon_1} - \sqrt{\varepsilon_2}}{\sqrt{\varepsilon_1} + \sqrt{\varepsilon_2}}$
14. $\dfrac{2\sqrt{\varepsilon_1}}{\sqrt{\varepsilon_1} + \sqrt{\varepsilon_2}}$
15. $\dfrac{\sqrt{\varepsilon_1}\cos\theta_1 - \sqrt{\varepsilon_2}\cos\theta_2}{\sqrt{\varepsilon_1}\cos\theta_1 + \sqrt{\varepsilon_2}\cos\theta_2}$
16. $\dfrac{\sqrt{\varepsilon_2}\cos\theta_1 - \sqrt{\varepsilon_1}\cos\theta_2}{\sqrt{\varepsilon_2}\cos\theta_1 + \sqrt{\varepsilon_1}\cos\theta_2}$
17. $\tan^{-1}\sqrt{\dfrac{\varepsilon_2}{\varepsilon_1}}$
18. parallel
19. thicker
20. $\sin^{-1}\sqrt{\dfrac{\varepsilon_2}{\varepsilon_1}}$

Match the Following

Set I

1. Perpendicular polarization	(a) $\dfrac{\sin\theta_i}{\sin\theta_r} = \dfrac{v_i}{v_r} = \sqrt{\dfrac{(\mu\varepsilon)_r}{(\mu\varepsilon)_i}}$
2. Parallel polarization	(b) travelling wave
3. Normal to surface	(c) standing wave pattern
4. Parallel to surface	(d) **E** vector ∥ to plane of polarization
5. Snell's law	(e) **E** vector ⊥ to plane of polarization

Set II

1. Tangential **E** vector	(a) zero
2. Tangential **H** vector	(b) 180°
3. Normal to surface	(c) standing wave pattern
4. Parallel to surface	(d) vertical polarization
5. Parallel polarization	(e) travelling wave

Set III

1. Normal incidence	(a) $\tan^{-1}\sqrt{\dfrac{\varepsilon_2}{\varepsilon_1}}$
2. Critical angle	(b) $\dfrac{\sqrt{\varepsilon_2}\cos\theta_1 - \sqrt{\varepsilon_1}\cos\theta_2}{\sqrt{\varepsilon_2}\cos\theta_1 + \sqrt{\varepsilon_1}\cos\theta_2}$
3. Perpendicular polarization	(c) $\dfrac{\sqrt{\varepsilon_1}\cos\theta_1 - \sqrt{\varepsilon_2}\cos\theta_2}{\sqrt{\varepsilon_1}\cos\theta_1 + \sqrt{\varepsilon_2}\cos\theta_2}$
4. Parallel polarization	(d) $\sin^{-1}\sqrt{\dfrac{\varepsilon_2}{\varepsilon_1}}$
5. Brewster angle	(e) $\dfrac{\sqrt{\varepsilon_1} - \sqrt{\varepsilon_2}}{\sqrt{\varepsilon_1} + \sqrt{\varepsilon_2}}$

Answers

Set I:	1. (e)	2. (d)	3. (c)	4. (b)	5. (a)
Set II:	1. (b)	2. (a)	3. (c)	4. (e)	5. (d)
Set III:	1. (e)	2. (d)	3. (c)	4. (b)	5. (a)

Multiple Choice Questions

1. Reflection by a perfect conductor on a normal incidence results in
   (a) standing waves (b) phase reversal of **E** field
   (c) no phase reversal of **H** field (d) all of these
2. The plane of incidence is the plane containing
   (a) incident ray (b) normal to surface.
   (c) both (a) and (b) (d) none of these
3. The **E** vector of the wave is parallel to the boundary surface in
   (a) perpendicular polarization (b) horizontal polarization
   (c) both (a) and (b) (d) none of these

4. The **H** vector of the wave is perpendicular to the plane of incidence in
   (a) parallel polarization (b) vertical polarization
   (c) both (a) and (b) (d) none of these
5. Reflection by a perfect conductor on an oblique incidence results in
   (a) standing wave along normal to surface with wavelength $\lambda/\cos\theta$
   (b) traveling wave along parallel to surface with wavelength $\lambda/\sin\theta$
   (c) both (a) and (b) (d) none of these
6. Reflection by perfect dielectric on normal incidence results in
   (a) reflection coefficient $\dfrac{E_r}{E_i} = \dfrac{\sqrt{\varepsilon_1} - \sqrt{\varepsilon_2}}{\sqrt{\varepsilon_1} + \sqrt{\varepsilon_2}}$
   (b) transmission coefficient $\dfrac{E_t}{E_i} = \dfrac{2\sqrt{\varepsilon_1}}{\sqrt{\varepsilon_1} + \sqrt{\varepsilon_2}}$
   (c) $\dfrac{H_t}{H_i} = \dfrac{\eta_1}{\eta_2}\dfrac{E_t}{E_i}$
   (d) all of these
7. Reflection by a perfect dielectric on an oblique incidence results in
   (a) $\dfrac{E_r}{E_i} = \dfrac{\cos\theta_1 - \sqrt{\left(\dfrac{\varepsilon_2}{\varepsilon_1}\right) - \sin^2\theta_1}}{\cos\theta_1 + \sqrt{\left(\dfrac{\varepsilon_2}{\varepsilon_1}\right) - \sin^2\theta_1}} = \dfrac{\sqrt{\varepsilon_1}\cos\theta_1 - \sqrt{\varepsilon_2}\cos\theta_1}{\sqrt{\varepsilon_1}\cos\theta_1 + \sqrt{\varepsilon_2}\cos\theta_2}$; $\perp$ polarization
   (b) $\dfrac{E_r}{E_i} = \dfrac{\left(\dfrac{\varepsilon_2}{\varepsilon_1}\right)\cos\theta_1 - \sqrt{\left(\dfrac{\varepsilon_2}{\varepsilon_1}\right) - \sin^2\theta_1}}{\left(\dfrac{\varepsilon_2}{\varepsilon_1}\right)\cos\theta_1 + \sqrt{\left(\dfrac{\varepsilon_2}{\varepsilon_1}\right) - \sin^2\theta_1}} = \dfrac{\sqrt{\varepsilon_2}\cos\theta_1 - \sqrt{\varepsilon_1}\cos\theta_1}{\sqrt{\varepsilon_2}\cos\theta_1 + \sqrt{\varepsilon_1}\cos\theta_2}$; $\parallel$ polarization
   (c) both (a) and (b)
   (d) none of these
8. The Brewster angle is characterized by
   (a) no reflected wave
   (b) incidence angle $\theta_1 = \tan^{-1}\sqrt{\dfrac{\varepsilon_2}{\varepsilon_1}}$
   (c) the occurrence only for $\parallel$ polarized wave
   (d) all of these

Answers

1. (d)	2. (c)	3. (c)	4. (c)	5. (c)
6. (d)	7. (c)	8. (d)		

10 Guided Waves and Waveguides

In order to succeed you must fail, so that you know what not to do the next time.

—Anthony J. D'Angelo

Wave propagation, in general, can be divided into guided and unguided wave propagations. Wave in free space is an example of the unguided wave propagation. Waves over transmission lines and waves through waveguides fall in the guided wave propagation category.

In this chapter, the wave propagation between two infinite parallel plates is analysed. The results of the analysis are going to be used extensively in transmission line theory and waveguide propagation. Then hollow-pipe waveguides are taken up for analysis. The aspects of bandwidth, modular propagation, dispersion relations and impedances pertaining to waveguides are given in detail.

10.1 Guided Wave Analysis

This section is important for more than one reason: the method and results of the analysis are going to be useful in the study of transverse electromagnetic (TEM) wave carrying transmission lines and transverse electric (TE) or transverse magnetic (TM) carrying waveguides.

Here we find the field distribution in the region that exists in between two infinitely large parallel conducting plates. The procedure involves in expressing the transverse components in terms of longitudinal components with the help of the Maxwell's curl equations in order to reduce the volume of the problem to one third: now, the number of unknowns is two. These two unknowns are then found by solving their respective wave equations.

10.1.1 Fields in between Conducting Plates

Let us suppose that the two conducting plates, infinite in extant, are parallel, one lying over the *yz*-plane and another at a height of *a* from the bottom plate (Figure 10.1). Let us also suppose that there exists a travelling wave in the positive *z*-direction in between the plates.

The plates are made with perfect conductors, i.e. conductivity σ of the walls is ∞ and that hollow region is a perfect dielectric, i.e. its conductivity σ is zero.

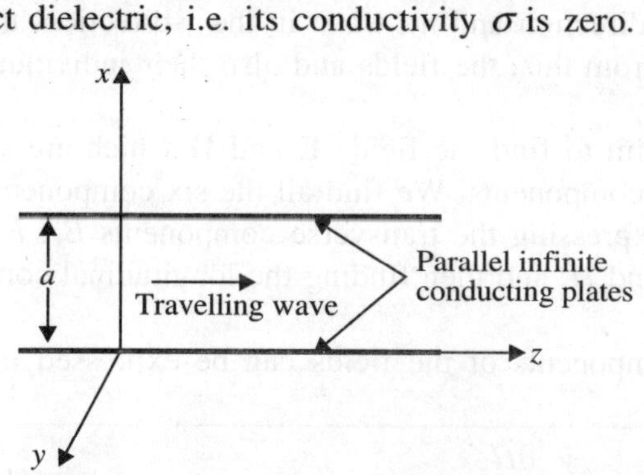

FIGURE 10.1 Parallel infinite conducting plates with travelling wave.

Let us also suppose the time variations of the field quantities are exponential i.e. $e^{j\omega t}$. If the time variations are exponential, then the fields must vary in the same manner, i.e. exponentially along the direction of the propagation of the wave, according to transmission line theory. So the fields must vary with z as $e^{\bar{\gamma} z}$ where $\bar{\gamma}$ is known as the propagation constant which in general is a complex quantity $\bar{\alpha}+j\bar{\beta}$.

The propagation constant $\bar{\gamma}$ is an important parameter describing the behaviour of the medium with respect to the wave. The real part $\bar{\alpha}$ is known as the attenuation constant representing the attenuation offered to the wave by the medium. The imaginary part $\bar{\beta}$ is known as the phase shift constant representing the phase change in the wave motion. If the medium offers no attenuation to the wave motion, then the attenuation constant $\bar{\alpha}=0$ making the propagation constant $\bar{\gamma}$ a pure imaginary quantity, i.e. $\bar{\gamma}+j\bar{\beta}$. If the medium does not allow the wave into it behaving like a pure attenuator, then $\bar{\beta}=0$ making the propagation constant $\bar{\gamma}$ a pure real quantity i.e. $\bar{\gamma}=\bar{\alpha}$. In general, when the wave is travelling through the medium, some amount of attenuation is offered by the medium; for such a case, the propagation constant is complex $\bar{\gamma}=\bar{\alpha}+j\bar{\beta}$.

In the present case, where the plates are made with perfect conductor and in between the hollow region is a perfect dielectric, the power of the wave cannot be absorbed by any one of them. It is because perfect conductor plates are equivalent to a short circuit load and perfect dielectric region is equivalent to an open circuit load of circuit theory. The net result is un-attenuated transmission of the wave with the propagation constant a pure imaginary quantity, i.e. $\bar{\gamma}=j\bar{\beta}$.

If **E** and **H** are electric and magnetic fields at an arbitrary point P in the hollow region, they must be related through the Maxwell's curl equations.

$$\left.\begin{aligned}\nabla\times\mathbf{H}&=\frac{\partial\mathbf{D}}{\partial t}\\ \nabla\times\mathbf{E}&=-\frac{\partial\mathbf{B}}{\partial t}\end{aligned}\right\}\tag{10.1}$$

In general, the fields are functions of x, y, z and t. In the present case, as the plates are infinite in extant in the y-direction and no wave in the y-direction, the fields are independent of that dimension. Apart from this, the fields and also their individual components separately obey the wave equations.

In the analysis we aim to find the fields **E** and **H** which are vectors and so in general each one must have three components. We find all the six components E_x, E_y, E_z, H_x, H_y and H_z of the fields first by expressing the transverse components E_x, E_y, H_x and H_y in terms of the longitudinal ones, E_z and H_z and then finding the longitudinal components by solving their wave equations.

The longitudinal components of the fields can be expressed in terms of the transverse components.

$$H_x = -\frac{\bar{\gamma}}{h^2}\frac{\partial H_z}{\partial x} \tag{10.2a}$$

$$H_y = -\frac{j\omega\varepsilon}{h^2}\frac{\partial E_z}{\partial x} \tag{10.2b}$$

$$E_x = -\frac{\bar{\gamma}}{h^2}\frac{\partial E_z}{\partial x} \tag{10.2c}$$

$$E_y = +\frac{j\omega\mu}{h^2}\frac{\partial H_z}{\partial x} \tag{10.2d}$$

where $h^2 = \bar{\gamma} + \omega^2 \mu\varepsilon$ is called the *characteristic equation*. The constant h is also denoted frequently as k_c and called *cut-off wave number*. In the above relations, all the field quantities are functions of x only.

Proof: The fields **E** and **H** at point P obey the Maxwell's curl equations. In the phasor form, i.e. if the time variations are assumed to be exponential, then they become $\nabla \times \mathbf{H} = j\omega\varepsilon\mathbf{E}$ and $\nabla \times \mathbf{E} = -j\omega\mu\mathbf{H}$. Note in the above equations the fields are independent of time t. Expanding the first equation

$$\nabla \times \mathbf{H} = \begin{vmatrix} \hat{\mathbf{x}} & \hat{\mathbf{y}} & \hat{\mathbf{z}} \\ \dfrac{\partial}{\partial x} & \dfrac{\partial}{\partial y} & \dfrac{\partial}{\partial z} \\ H_x & H_y & H_z \end{vmatrix} = j\omega\varepsilon\mathbf{E} = j\omega\varepsilon\,(\hat{\mathbf{x}}E_x + \hat{\mathbf{y}}E_y + \hat{\mathbf{z}}E_z)$$

If the z-variations of the field quantities are exponential, then $\partial/\partial z = -\bar{\gamma}$. In addition, the space is infinite in extent in y-direction. We can assume a uniform or constant field along the y-direction resulting in derivatives of fields with respect to y zeros. Substituting the above two aspects into the matrix equation above,

$$\nabla \times \mathbf{H} = \begin{vmatrix} \hat{\mathbf{x}} & \hat{\mathbf{y}} & \hat{\mathbf{z}} \\ \dfrac{\partial}{\partial x} & 0 & -\bar{\gamma} \\ H_x & H_y & H_z \end{vmatrix} = j\omega\varepsilon\mathbf{E} = j\omega\varepsilon\,(\hat{\mathbf{x}}E_x + \hat{\mathbf{y}}E_y + \hat{\mathbf{z}}E_z)$$

Equating the field components on both sides,

$$\bar{\gamma} H_y = j\omega\varepsilon E_x \tag{10.3a}$$

$$\frac{\partial H_z}{\partial x} + \bar{\gamma} H_x = -j\omega\varepsilon E_y \tag{10.3b}$$

$$\frac{\partial H_y}{\partial x} = j\omega\varepsilon E_z \tag{10.3c}$$

Similarly from the second equation $\nabla \times \mathbf{E} = -j\omega\mu\mathbf{H}$, another set of three equations can be derived.

$$\bar{\gamma} E_y = -j\omega\mu H_x \tag{10.4a}$$

$$\frac{\partial E_z}{\partial x} + \bar{\gamma} E_x = j\omega\mu H_y \tag{10.4b}$$

$$\frac{\partial E_y}{\partial x} = -j\omega\mu H_z \tag{10.4c}$$

Now to find an expression for H_x, let us consider Eq. (10.3b). It can be recast as

$$-\left(\frac{\partial H_z}{\partial x} + \bar{\gamma} H_x\right)\frac{1}{j\omega\varepsilon} = E_y$$

Substituting this expression for E_y in Eq. (10.4a), we obtain

$$\left(\frac{\partial H_z}{\partial x} + \bar{\gamma} H_x\right)\frac{-\bar{\gamma}}{j\omega\varepsilon} = -j\omega\mu H_x$$

Solving this equation for H_x, we have

$$H_x = -\frac{\bar{\gamma}}{h^2}\frac{\partial H_z}{\partial x}$$

In a similar manner, expressions for other transverse components can be computed by considering one equation from each set.

Critical observations that can be made are: With, $E_z = 0$, $H_z = 0$, simultaneously all the field components become zero indicating a wave with components transverse to the direction of propagation, i.e. transverse electromagnetic wave or TEM wave cannot exist in between the plates. With, $E_z \neq 0$, $H_z = 0$, all the components are not zero indicating the possibility of TM wave in the waveguide. With, $E_z = 0$, $H_z \neq 0$, there exist non-zero field components indicating possibility of TE wave. With $E_z \neq 0$, $H_z \neq 0$, the wave can exist in the guide, as there are non-zero field components. This wave is called *hybrid* or *mixed* wave.

We find E_z and H_z by solving the wave equations which are partial differential equations and hence their particular solution requires boundary conditions.

The boundary conditions to be used in the present context while solving the wave equation are derived from the fact that the tangential components of the electric field at the surface of the perfect conductor are zero. At $x = 0$ and $x = a$ lie the inner faces of the plates; therefore, the tangential field components must be zero, i.e. $E_y = 0 = E_z$.

10.1.2 Transverse Magnetic (TM) Wave

In the case of TM wave,

$$H_z = 0$$

The wave equation for E_z is

$$\nabla^2 E_z = \mu\varepsilon \frac{\partial^2 E_z}{\partial t^2} \tag{10.5}$$

If both the time variations and z-variations are exponential, i.e.

$$E_z(x, y, z, t) = E_z(x, y)e^{-\bar{\gamma}z}e^{j\omega t}$$

then the wave equation becomes

$$\frac{\partial^2 E_z}{\partial x^2} + \frac{\partial^2 E_z}{\partial y^2} + \bar{\gamma}^2 E_z = -\omega^2 \mu\varepsilon E_z \tag{10.6}$$

As the fields do not change along the y-direction, $\partial/\partial y = 0$. Incorporating into the wave equation

$$\frac{\partial^2 E_z}{\partial x^2} + (\bar{\gamma}^2 + \omega^2 \mu\varepsilon) E_z = 0$$

$$\frac{\partial^2 E_z}{\partial x^2} + h^2 E_z = 0 \tag{10.7}$$

The general solution to the above equation is

$$E_z = (A\cosh x + B\sinh x)e^{-\bar{\gamma}z} \tag{10.8}$$

where A and B are arbitrary constants whose values can be fixed with the boundary conditions.

Now applying the boundary conditions $E_z = 0$ at $x = 0$ and at $x = a$, we have

$$A = 0 \text{ and } h = m\pi/a, \qquad m = 0, 1, 2 \ldots$$

And the expression for the field becomes

$$E_z = B\sin\left(\frac{m\pi}{a}\right) x e^{-\bar{\gamma}z} \tag{10.9}$$

Here B is an arbitrary constant and for mathematical convenience we choose its value as $B = \dfrac{jm\pi}{\omega\varepsilon a} C_4$.

Here C_4 is another arbitrary constant. With the expression for E_z, other components can be computed using the relations connecting transverse components to longitudinal ones. After including the exponential time variations and z-variations, the complete set of the field components between the parallel plates becomes

$$H_z = 0 \quad (10.10a)$$

$$E_z = \frac{jm\pi}{\omega\varepsilon a} C_4 \sin\left(\frac{m\pi}{a}x\right) e^{-j\bar{\beta}z} e^{j\omega t} \quad (10.10b)$$

$$E_x = \frac{\bar{\beta}}{\omega\varepsilon} C_4 \cos\left(\frac{m\pi}{a}x\right) e^{-j\bar{\beta}z} e^{j\omega t} \quad (10.10c)$$

$$H_y = C_4 \cos\left(\frac{m\pi}{a}x\right) e^{-j\bar{\beta}z} e^{j\omega t} \quad (10.10d)$$

10.1.3 Transverse Electric (TE) Wave

In the present case, $E_z = 0$ and as the boundary conditions are not available on H_z, we solve the wave equation for E_y

$$\nabla^2 E_y = \mu\varepsilon \frac{\partial^2 E_y}{\partial t^2} \quad (10.11)$$

which leads to

$$\frac{\partial^2 E_y}{\partial x^2} + h^2 E_y = 0 \quad (10.12)$$

The general solution is the same as that for the wave equation of the previous section. Applying boundary conditions to the general solution, we get the particular or specific solution. Here the boundary conditions are $E_y = 0$ at $x = 0$ and $x = a$. The resultant specific solution then is

$$E_y = C_1 \sin\left(\frac{m\pi}{a}\right) x e^{-\bar{\gamma}z}, \quad m = 1, 2, 3, \ldots \quad (10.13)$$

With the expression for E_y, other components can be computed using the relations connecting transverse components to longitudinal ones. After including the exponential time variations and z-variations, the complete set of the field components between the parallel plates becomes

$$E_z = 0 \quad (10.14a)$$

$$H_z = -\frac{\bar{\beta}}{\omega\mu} C_1 \cos\left(\frac{m\pi}{a}x\right) e^{-j\bar{\beta}z} e^{j\omega t} \quad (10.14b)$$

$$E_y = C_1 \sin\left(\frac{m\pi}{a}x\right) e^{-j\bar{\beta}z} e^{j\omega t} \quad (10.14c)$$

$$H_x = \frac{jm\pi}{\omega\mu a} C_1 \sin\left(\frac{m\pi}{a}x\right) e^{-j\bar{\beta}z} e^{j\omega t} \quad (10.14d)$$

10.1.4 Filter Characteristics

The infinite parallel plate arrangement under consideration behaves like a high-pass filter. It admits and allows the wave to propagate through them only if the frequency of the wave is more than certain value known as the cut-off frequency whose value depends upon the dimensions of the guide and the mode of the wave.

Cut-off frequency f_c: It is the frequency above which the frequency of the wave should be in order to get entry into the waveguide for propagating through it. It depends upon the dimensions of the guide as well as on the mode of the wave. Thus,

$$f_c = \frac{m}{2a\sqrt{\mu\varepsilon}} \tag{10.15}$$

Cut-off wavelength λ_c: The wavelength corresponding to the cut-off frequency is called cut-off wavelength. It can be defined as the wavelength below which the wavelength of the wave should be in order to get entry into the waveguide for propagating through it. Its value is related to the dimensions of the guide and mode numbers of the wave through

$$\lambda_c = \frac{2a}{m} \tag{10.16}$$

Proof: We know the propagation constant $\bar{\gamma}$ can be related to the frequency ω of the wave through

$$\bar{\gamma}^2 = h^2 - \omega^2\,\mu\varepsilon \text{ or } \bar{\gamma} = \sqrt{h^2 - \omega^2\,\mu\varepsilon}$$

Depending upon the relation between h^2 and $\omega^2\mu\varepsilon$, the propagation constant $\bar{\gamma}$ can be pure real or imaginary.

When $h^2 > \omega^2\mu\varepsilon$, the propagation constant is a pure real quantity indicating that the waveguide is acting as a pure attenuator without the wave motion refusing entry to the wave.

When $h^2 < \omega^2\mu\varepsilon$, the propagation constant is a pure imaginary quantity indicating that the waveguide is acting as a pure transmission line without any attenuation to the wave. This must be the case to which the waveguide under consideration belongs because the wave is already in and a lossless condition is assumed.

The changeover in the behaviour of the waveguide from pure attenuator to pure transmission line takes place as the frequency is increased from low to high when $h^2 = \omega^2\mu\varepsilon$. The frequency of the wave which satisfies this relation is called cut-off frequency ω_c. Hence $\omega_c^2\mu\varepsilon = h^2$. But

$$h^2 = \left(\frac{m\pi}{a}\right)^2$$

Solving this equation for f_c we get

$$f_c = \frac{1}{2\sqrt{\mu\varepsilon}}\sqrt{\left(\frac{m}{a}\right)^2} = \frac{m}{2a\sqrt{\mu\varepsilon}}$$

The corresponding wavelength, cut-off wavelength, can be found from

$$\lambda_c = \frac{v}{f_c} = \frac{1}{\sqrt{\mu\varepsilon}\,f_c} = \frac{2a}{m}$$

The cut-off frequency f_c and cut-off wavelength λ_c are the same for both the types of the wave, i.e. TE or TM in rectangular waveguides.

10.1.5 Phase Velocity and Group Velocity

The wave phenomenon is described by parameters like wavelength, time period, frequency, etc. Phase and group velocities are also important parameters pertaining to, particularly, waves in non-TEM mode.

The phase velocity is defined as the velocity with which the equiphase surfaces propagate parallel to the incident surface. Let the wave is travelling with a velocity v in the direction of propagation and incident on the surface at an angle θ (Figure 10.2). Also let the wavelength of the wave along the direction of propagation is λ. The wavelength along the incident surface is $\lambda_p = \lambda/\sin\theta$.

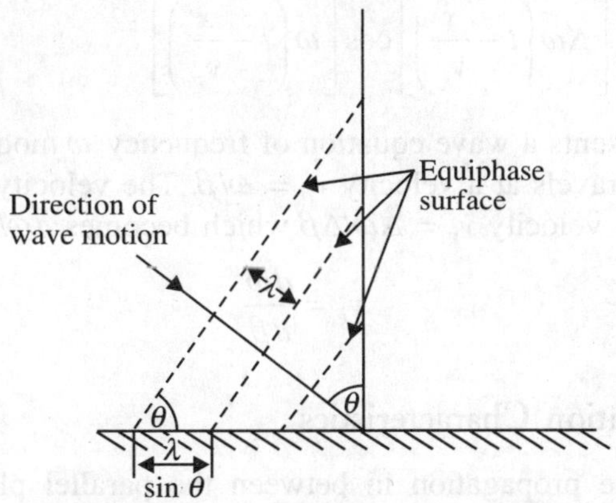

FIGURE 10.2 Illustrating phase velocity.

Then the velocity of the wave along the incident surface becomes

$$v_p = f\lambda_p = \frac{f\lambda}{\sin\theta} = \frac{v}{\sin\theta} \tag{10.17}$$

The phase velocity can also be expressed in another useful form as

$$v_p = \frac{f\lambda}{\sin\theta} = \frac{2\pi f}{\dfrac{2\pi}{\lambda_p}} = \frac{\omega}{\beta} \tag{10.18}$$

Sometimes it is also denoted by $\overline{v}$. Observe from the expression that the phase velocity can be equal to or exceed the velocity of light in the medium. It is a fictitious but convenient quantity capable of measurement without physical reality.

The group velocity is the velocity with which the signal or group of frequencies travels through the system. Energy or intelligence gets transmitted as bands or groups of frequencies. It is given by

$$v_g = \frac{d\omega}{d\beta} \tag{10.19}$$

Consider a signal/intelligence travelling a distance x which contains two frequencies differed by $2\Delta\omega$. Then

$$s(t) = A\{\cos[(\omega - \Delta\omega)t - (\beta - \Delta\beta)x] + \cos[(\omega + \Delta\omega)t - (\beta + \Delta\beta)x]\} \tag{10.20}$$

Let us suppose the medium is dispersive and change in β is $\Delta\beta$ for a change in frequency $\Delta\omega$. Now,

$$s(t) = A\{\cos[(\omega t - \beta x) - (\Delta\omega t - \Delta\beta x)] + \cos[(\omega t - \beta x) + (\Delta\omega t - \Delta\beta x)]\}$$

$$= 2A\cos\left[\Delta\omega\left(t - \frac{\Delta\beta}{\Delta\omega}x\right)\right]\cos\left[\omega\left(t - \frac{\beta}{\omega}x\right)\right]$$

$$= 2A\cos\left[\Delta\omega\left(t - \frac{x}{v_g}\right)\right]\cos\left[\omega\left(t - \frac{x}{v_p}\right)\right] \tag{10.21}$$

The above equation represents a wave equation of frequency ω modulated by a low frequency $\Delta\omega$. The high frequency travels at a velocity $v_p = \omega/\beta$. The velocity of the low frequencies or intelligence travels with a velocity $v_g = \Delta\omega/\Delta\beta$ which becomes $d\omega/d\beta$ in the limit. Therefore,

$$v_g = \frac{d\omega}{d\beta}$$

10.1.6 Modal Propagation Characteristics

The electromagnetic wave propagation in between the parallel plate arrangement is in the form of some definite field patterns known as 'modes' which is an important and special feature of the guided wave propagation. The mode subscripts are denoted by m and 0 in that order as TE_{m0} or TM_{m0}.

- *Subscripts:* For TE wave, m can assume any integer values from 1, 2, 3, etc. For TM wave, m can assume any integer values from 0, 1, 2, 3, etc.
- *Principal mode:* This wave mode is one which has the lowest cut-off frequency. In this mode of propagation, both E_z and H_z are zero resulting in TEM wave. The mode of the wave is TM_{00}.

10.1.7 Characteristics of Guided TM Waves

From the expressions of field components, we can make the following observations:

1. The TE wave is one for which the electric vector is entirely normal to the direction of propagation, i.e. no component along the propagation direction. Therefore, for this type of wave, when the wave propagation is along the z-direction, the fields satisfy $E_z = 0$, $H_z \neq 0$.
2. The TM wave is one for which the magnetic vector is entirely normal to the direction of propagation, i.e. no component along the propagation direction. Therefore, for this type of wave, when the wave propagation is along the z-direction, the fields satisfy $H_z = 0$, $E_z \neq 0$.

3. Hybrid or mixed wave can also exist in between the plates and for this type of wave, both the wave and the field have components along the direction of propagation. Therefore, for this type of wave when the wave propagation is along the z-direction, the fields satisfy $E_z \neq 0$, $H_z \neq 0$.
4. The lowest order transverse electric wave is TE_{10}, whereas in the case of transverse magnetic waves, the lowest order is TM_{00}. For this special case, both the electric and magnetic vectors are normal to the direction of propagation and hence it is called transverse electromagnetic (TEM) wave or principal wave.
5. In the principal wave, the fields are not only entirely transverse, but they are also constant in amplitude across a crosssection normal to the direction of the propagation; of course, their ratio is also a constant.
6. Unlike TE and TM waves, the velocity of the TEM wave is independent of frequency and has a familiar free space value.
7. The cut-off frequency of the TEM wave is zero. Therefore, for TEM waves, all frequencies down to zero can propagate along the guide.
8. The ratio of E to H between the parallel planes for a travelling wave is intrinsic impedance of free space.
9. When the wave is travelling in the z-direction, any x-y plane is an equiphase plane for each of the field components. Table 10.1 lists the guided wave propagation parameters.

Table 10.1 Guided wave propagation

Parameter	$f = f_c$	$f \to \infty$
$\bar{v}$	∞	$\bar{v} = v = \dfrac{1}{\sqrt{\mu\varepsilon}}$
v_g	0	v
$\bar{\lambda}$	∞	$\bar{\lambda} = \lambda = \dfrac{2\pi}{\omega\sqrt{\mu\varepsilon}}$
$\bar{\beta}$	0	$\bar{\beta} = \omega\sqrt{\mu\varepsilon}$

10. For the wave motion to exist, the propagation constant $\bar{\gamma}$ must be pure imaginary equal to $j\bar{\beta}$, where $\bar{\beta}$ is the phase shift constant. It is given by

$$\bar{\beta} = \sqrt{\omega^2\mu\varepsilon - \left(\frac{m\pi}{a}\right)^2} \tag{10.22}$$

11. There will be no wave motion unless the frequency is more than certain value known as the cut-off frequency f_c of the guide. It is given by

$$f_c = \frac{m}{2a\sqrt{\mu\varepsilon}} \tag{10.23}$$

The wavelength corresponding to the cut-off frequency is called the cut-off wavelength. It can be defined as the value below which the wavelength should be in order to propagate in between the plates. An expression for this parameter is derived as

$$\lambda_c = \frac{v}{f_c} = \frac{1}{f_c\sqrt{\mu\varepsilon}} = \frac{2a}{m} \tag{10.24}$$

Here v is the velocity of the wave in an unbounded medium with constants μ and ε and is given by $1/\sqrt{\mu\varepsilon}$.

12. The equiphase surfaces progress along the guide in z-direction with a velocity $\bar{v} = \omega/\bar{\beta}$ where $\bar{\beta}$ is the phase shift constant. It is called phase velocity and is given by

$$\bar{v} = \frac{\omega}{\bar{\beta}} = \frac{\omega}{\sqrt{\omega^2\mu\varepsilon - \left(\frac{m\pi}{a}\right)^2}} \tag{10.25}$$

13. The velocity with which the intelligence travels between the plates along the direction of propagation is called group velocity. It is, by definition, $v_g = \partial\omega/\partial\bar{\beta}$, where $\bar{\beta}$ is the phase shift constant. It is given by

$$v_g = v\left[1 - \frac{1}{\mu_r\varepsilon_r}\left(\frac{\lambda}{\lambda_c}\right)^2\right]^{1/2} = v\left[1 - \left(\frac{f_c}{f}\right)^2\right]^{1/2} \tag{10.26}$$

If the region in between the plates is free space, then

$$v_g = c\left[1 - \left(\frac{\lambda}{\lambda_c}\right)^2\right]^{1/2} = c\left[1 - \left(\frac{f_c}{f}\right)^2\right]^{1/2} \tag{10.27}$$

14. The wavelength $\bar{\lambda}$ is, by definition, the distance between two consecutive points which differ in phase by 2π radians (guide wavelength). Therefore,

$$\bar{\lambda} = \frac{2\pi}{\bar{\beta}} = \frac{2\pi}{\sqrt{\omega^2\mu\varepsilon - \left(\frac{m\pi}{a}\right)^2}} \tag{10.28}$$

10.2 HOLLOW-PIPE WAVEGUIDES

Waveguides are useful to transfer signals of microwave frequencies from one point to another point. Ordinary transmission lines like parallel wirelines or open wirelines are not that much useful at these frequencies because of their heavy radiation losses.

A brief introduction to waveguides and their comparison with coaxial lines and transmission lines is given. A rectangular waveguide is analysed to find the field components and other important parameters like cut-off frequency, guide wavelength, etc. Then circular waveguides are discussed and the relations pertaining to their important parameters and features are derived. While travelling, the wave undergoes attenuation and this particular aspect is then explained. The chapter then concludes with a discussion on impedances.

10.2.1 Introduction

Hollow-pipe waveguides are hollow-conducting metallic pipes used to carry signals of high frequency like microwaves from one point to another point. George C. Southworth and W.L. Barrow invented waveguides independently in the 1930s. Earlier in 1893, Heaviside considered the possibilities of the propagation of the wave inside a closed hollow tube and Lord Raleigh later proved mathematically the possibility of propagation in a waveguide.

Hollow-pipe waveguides are single-conductor systems, usually made with highly conducting metal walls, which can support TE or TM waves. Waveguides can have dimensions that are convenient only when the frequency of operation ranges from 3 to 100 GHz. The waveguides that are used widely have cross sections either of rectangular or of circular. Waveguides with rectangular cross sections are called rectangular waveguides and if the cross sections are circular, they are called circular waveguides. If the cross sections are uniform, then they are called cylindrical waveguides.

The waveguide can be broadly considered as consisting of two regions: one is the highly conducting outside walls and the other one is the highly non-conducting inside dielectric which is usually free space. Both of these two regions have important roles in transferring the signal. The wave travels through the dielectric filling the waveguide after getting reflected from the walls instead of conduction along them. So it is the hollow region through which the wave actually travels and the purpose of the walls is to confine the wave.

Apart from waveguides, coaxial lines are also used at high frequencies. In fact, both can be used in applications at the lower end of microwave range. Waveguides have the advantages of

- Higher power-handling capability and low loss
- Mechanical simplicity, easy to manufacture
- Flash-over less likely
- Higher maximum operating frequency (325 GHz)

But they are bulky and expensive when compared to coaxial lines. The coaxial lines, on the other hand, have very large bandwidth and are convenient for test applications. Low maximum operating frequency (18 GHz) and difficulty in fabricating complex microwave components are two major disadvantages with coaxial lines.

In carrying the signal from one point to another point, waveguides are similar to transmission lines in discharging their fundamental function. But the waveguides differ fundamentally with the transmission lines in several ways which are mentioned here. Waveguides are single-conductor systems, whereas transmission lines are multi-conductor systems. Energy travels in the form of TE or TM or hybrid modes in waveguides, whereas over transmission lines it always travels as TEM wave. The frequency of the wave must be larger than certain value known as cut-off frequency in order to propagate through the waveguide, whereas the wave of any frequency can get transmitted over the transmission line. Modal propagation is an important and special feature of the energy transfer through the waveguides, whereas the field distribution in transmission lines is non-modal in nature.

10.2.2 Fields in RWGs

In this section, we find the field distribution in the hollow-pipe waveguides with rectangular cross sections i.e. rectangular waveguides. As the rectangular waveguides involve rectangular symmetry, Cartesian co-ordination is used in their analysis (Figure 10.3).

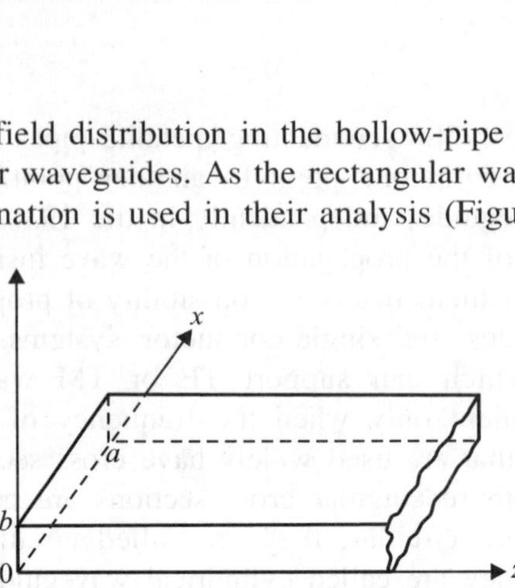

FIGURE 10.3 Structure of an RWG.

We do the analysis in phasor domain. The procedure is a multi-step process. In the first step using the Maxwell's curl equations, the transverse components are expressed in terms of longitudinal components. Here the volume of the problem gets reduced to one third, and the number of unknowns gets reduced from six to two. These two unknowns are then found by solving their respective wave equations. Wave equations are differential equations and they require boundary conditions to get specific solution representing the current problem. From the field values at the interface of dielectric-conductor boundary we derive the necessary boundary conditions.

Let us suppose the waveguide with inner dimensions a and b with $a \geq b$ is lying along the z-axis carrying a travelling wave in the positive z-direction. We assume the walls of the waveguide are made with perfect conductor, i.e. conductivity σ of the walls is ∞ and that hollow region is a perfect dielectric, i.e. its conductivity σ is zero.

Let us also suppose the time variations of the field quantities are exponential, i.e. $e^{j\omega t}$. If the time variations are exponential, then the fields must vary in the same manner, i.e. exponentially along the direction of the propagation of the wave, according to the transmission line theory. So the fields must vary with z as $e^{\bar{\gamma} z}$, where $\bar{\gamma}$ is known as the propagation constant which in general is a complex quantity $\bar{\alpha} + j\bar{\beta}$.

The propagation constant $\bar{\gamma}$ is an important parameter which we need to know about in the analysis of waveguides or transmission lines. It describes the behaviour of the transmission medium with respect to the wave. The real part $\bar{\alpha}$ is known as the attenuation constant representing the attenuation offered to the wave by the medium. The imaginary part $\bar{\beta}$ is known as the phase shift constant representing the phase change in the wave motion.

We will see soon, from the value of the propagation constant, that it is possible to find the status of the wave with respect to the waveguide. If the propagation constant is pure imaginary i.e. $\bar{\gamma} = j\bar{\beta}$, it indicates a travelling wave in the guide without attenuation, i.e. $\bar{\alpha} = 0$. A pure real value of the propagation constant, i.e. $\bar{\gamma} = \bar{\alpha}$, tells that no wave is inside the waveguide and the waveguide is acting as a pure attenuator to the wave, not allowing the wave even to enter into it. An imaginary value of the propagation constant, i.e. $\bar{\gamma} = \bar{\alpha} + j\bar{\beta}$, means the existence of a travelling wave inside the guide with some finite amount of attenuation. The last one represents the most general as well as practically existing situation.

While travelling in the waveguide, the wave comes in contact with both the parts of the waveguide, i.e. conducting walls and hollow loss-free region. In the present case, it is given that the walls of the waveguide are made with perfect conductor and the hollow region is a perfect dielectric. Both the parts of the waveguide cannot absorb the power. The reason is that the perfect conductor walls are equivalent to a short circuit load and the perfect dielectric region is equivalent to an open circuit load. As a result, there is no absorption of energy from the wave and the wave travels un-attenuated along the waveguide. The resultant propagation constant is a pure imaginary quantity, i.e. $\bar{\gamma} = j\bar{\beta}$.

The method we follow is basically applying several restrictions to the Maxwell's equations one after another. First, we select a particular form, i.e. exponential time variations; second, we select the form of propagation, i.e. exponential and lastly we restrict or confine the wave to travel between two sets of two parallel planes each.

If **E** and **H** are electric and magnetic fields at an arbitrary point P in the hollow region, they must be related through the Maxwell's curl equations:

$$\left.\begin{aligned} \nabla \times \mathbf{H} &= \frac{\partial \mathbf{D}}{\partial t} \\ \nabla \times \mathbf{E} &= -\frac{\partial \mathbf{B}}{\partial t} \end{aligned}\right\}$$

Here note that the fields are functions of position and time, i.e. x, y, z and t. We know that these fields and their individual components separately obey wave equations.

In the analysis, we find the fields **E** and **H** which are vectors and so in general each one must have three components. We find all the six components E_x, E_y, E_z, H_x, H_y and H_z of the fields first by expressing the transverse components E_x, E_y, H_x and H_y in terms of the longitudinal ones, E_z and H_z and then finding the longitudinal components by solving their respective wave equations.

Using the phasor form of the Maxwell's curl equations, firstly, we express the transverse run on components of the fields in terms of the longitudinal components as follows:

$$H_x = -\frac{\bar{\gamma}}{h^2}\frac{\partial H_z}{\partial x} + \frac{j\omega\varepsilon}{h^2}\frac{\partial E_z}{\partial y} \tag{10.29a}$$

$$H_y = -\frac{\bar{\gamma}}{h^2}\frac{\partial H_z}{\partial y} - \frac{j\omega\varepsilon}{h^2}\frac{\partial E_z}{\partial x} \tag{10.29b}$$

$$E_x = -\frac{\bar{\gamma}}{h^2}\frac{\partial E_z}{\partial x} - \frac{j\omega\mu}{h^2}\frac{\partial H_z}{\partial y} \tag{10.29c}$$

$$E_y = \frac{\bar{\gamma}}{h^2}\frac{\partial E_z}{\partial y} + \frac{j\omega\mu}{h^2}\frac{\partial H_z}{\partial x} \tag{10.29d}$$

Here $h^2 = \overline{\gamma}^2 + \omega^2 \mu\varepsilon$ which is called the *characteristic equation*. The constant h is also denoted frequently as k_c and called *cut-off wave number*. In the above relations, all the field quantities are phasors and functions of x and y only.

Derivation: The fields **E** and **H** at point P obey the Maxwell's curl equations. In the phasor form, i.e. if the time variations are assumed to be exponential, they become

$$\left.\begin{aligned} \nabla \times \mathbf{H} &= j\omega\varepsilon\mathbf{E} \\ \nabla \times \mathbf{E} &= -j\omega\mu\mathbf{H} \end{aligned}\right\} \tag{10.30}$$

Note that in Eq. (10.30), the fields are independent of time t. Expanding the first part of Eq. (10.30)

$$\nabla \times \mathbf{H} = \begin{vmatrix} \hat{\mathbf{x}} & \hat{\mathbf{y}} & \hat{\mathbf{z}} \\ \partial/\partial x & \partial/\partial y & \partial/\partial z \\ H_x & H_y & H_z \end{vmatrix} = j\omega\varepsilon\mathbf{E} = j\omega\varepsilon\,(\hat{\mathbf{x}}E_x + \hat{\mathbf{y}}E_y + \hat{\mathbf{z}}E_z)$$

As the z-variations of the field quantities are exponential, we can replace $\partial/\partial z = -\overline{\gamma}$ in the expansion. Therefore,

$$\nabla \times \mathbf{H} = \begin{vmatrix} \hat{\mathbf{x}} & \hat{\mathbf{y}} & \hat{\mathbf{z}} \\ \partial/\partial x & \partial/\partial y & -\overline{\gamma} \\ H_x & H_y & H_z \end{vmatrix} = j\omega\varepsilon\mathbf{E} = j\omega\varepsilon(\hat{\mathbf{x}}E_x + \hat{\mathbf{y}}E_y + \hat{\mathbf{z}}E_z)$$

Equating the components of fields on both sides,

$$\frac{\partial H_z}{\partial y} + \overline{\gamma} H_y = j\omega\varepsilon E_x \tag{10.31a}$$

$$\frac{\partial H_z}{\partial x} + \overline{\gamma} H_x = -j\omega\varepsilon E_y \tag{10.31b}$$

$$\frac{\partial H_y}{\partial x} - \frac{\partial H_x}{\partial y} = j\omega\varepsilon E_z \tag{10.31c}$$

Similarly, by expanding the second part of Eq. (10.30), another set of three equations can be derived.

$$\frac{\partial E_z}{\partial y} + \overline{\gamma} E_y = -j\omega\mu H_x \tag{10.32a}$$

$$\frac{\partial E_z}{\partial x} + \overline{\gamma} E_x = j\omega\mu H_y \tag{10.32b}$$

$$\frac{\partial E_y}{\partial x} - \frac{\partial E_x}{\partial y} = -j\omega\mu H_z \tag{10.32c}$$

Using these six equations, we express the transverse components of the fields in terms of longitudinal components. As an example, we start with Eq. (10.31b) to find an expression for H_x.

It is

$$-\left(\frac{\partial H_z}{\partial x}+\overline{\gamma}H_x\right)\frac{1}{j\omega\varepsilon}=E_y$$

Substituting this expression for E_y in Eq. (10.32a), we obtain

$$\frac{\partial E_z}{\partial y}+\left(\frac{\partial H_z}{\partial x}+\overline{\gamma}H_x\right)\frac{-\overline{\gamma}}{j\omega\varepsilon}=-j\omega\mu H_x$$

Now solving the equation for H_x, we get

$$H_x=-\frac{\overline{\gamma}}{h^2}\frac{\partial H_z}{\partial x}+\frac{j\omega\varepsilon}{h^2}\frac{\partial E_z}{\partial y}$$

In a similar manner, the expressions for other transverse components can be computed by considering one equation from each set.

With respect to the expressions in Eqs. (10.29a, b, c, d), certain critical observations can be made. They are:

- With both the longitudinal components zero, i.e. $E_z = 0$, $H_z = 0$, simultaneously, all the field components become zero indicating a wave with components transverse to the direction of propagation, i.e. the transverse electromagnetic wave or TEM wave cannot exist inside a waveguide.
- With $E_z \neq 0$, $H_z = 0$, all the components are not zero indicating the possibility of the existence of a wave in the waveguide. As the magnetic vector is entirely normal to the direction of propagation, this wave is called TM wave.
- With $E_z = 0$, $H_z \neq 0$, there exist non-zero field components indicating the possibility of wave inside the guide. As the electric vector is entirely normal to the direction of propagation, this wave is called TE wave.
- With $E_z \neq 0$, $H_z \neq 0$, also the field components are non-zero indicating the existence of wave in the guide. This wave is often called *hybrid* or *mixed* wave as we can see soon it is actually a combination of TM and TE waves.

Now we move to the next step of the analysis to find E_z and H_z by solving the wave equations.

10.2.3 Boundary Conditions

The wave equations are partial differential equations and their complete solution requires boundary conditions. The boundary conditions that serve the purpose of the present context are derived from the electromagnetic boundary condition at the interface of a perfect dielectric–perfect conductor. They state that the tangential component of the electric field at the surface of the perfect conductor is always zero. At $y = 0$ and $y = b$ lie the inner surfaces of the broader walls; therefore, $E_x = 0 = E_z$. Similarly at $x = 0$ and $x = a$ lie the inner surfaces of the narrow walls resulting in $E_y = 0 = E_z$.

10.2.4 Transverse Magnetic (TM) Wave

In the present case, $H_z = 0$ and the wave equation for E_z is

$$\nabla^2 E_z = \mu\varepsilon \frac{\partial^2 E_z}{\partial t^2} \tag{10.33}$$

If the time variations and z-variations, both, are exponential, i.e.

$$E_z(x, y, z, t) = E_z(x, y)e^{-\bar{\gamma} z} e^{j\omega t}$$

then the wave equation becomes

$$\frac{\partial^2 E_z}{\partial x^2} + \frac{\partial^2 E_z}{\partial y^2} + \bar{\gamma}^2 E_z = -\omega^2 \mu\varepsilon\, E_z$$

$$\frac{\partial^2 E_z}{\partial x^2} + \frac{\partial^2 E_z}{\partial y^2} + (\bar{\gamma}^2 + \omega^2 \mu\varepsilon) E_z = 0$$

$$\frac{\partial^2 E_z}{\partial x^2} + \frac{\partial^2 E_z}{\partial y^2} + h^2 E_z = 0 \tag{10.34}$$

This equation, known as the *Helmholtz equation*, is a second-order, two-dimensional partial differential equation in which E_z is a function of x and y only, i.e. it does not involve z or t.

This equation can be solved using the *method of variable separation.* According to this method, let $E_z = XY$, where X is a function of variable x alone and Y is a function of variable y alone. With this assumption the above wave equation can be decomposed into two second-order ordinary differential equations for which solutions are readily available.

$$\frac{1}{X}\frac{d^2 X}{dx^2} + \frac{1}{Y}\frac{d^2 Y}{dy^2} + h^2 \tag{10.35}$$

Let us suppose

$$\frac{1}{X}\frac{d^2 X}{dx^2} = -A^2 \tag{10.36}$$

where A is an arbitrary constant. Then Eq. (10.35) becomes

$$-A^2 + \frac{1}{Y}\frac{d^2 Y}{dy^2} + h^2 = 0$$

$$\frac{1}{Y}\frac{d^2 Y}{dy^2} = -(h^2 - A^2) = -B^2 \tag{10.37}$$

where B is another arbitrary constant. Here $h^2 = A^2 + B^2$.

The general solutions to the above equations are

$$X = c_1 \cos Ax + c_2 \sin Ax \text{ and}$$
$$Y = c_3 \cos By + c_4 \sin By, \text{ making}$$

$E_z = XY = (c_1 \cos Ax + c_2 \sin Ax)(c_3 \cos By + c_4 \sin By)$, where c_1, c_2, c_3 and c_4 are arbitrary constants whose values can be fixed with the boundary conditions.

Now applying the boundary conditions $E_z = 0$ at $x = 0$, $y = 0$, we get $c_1 = 0$, $c_3 = 0$, whereas the conditions $E_z = 0$ at $x = a$, $y = b$ give $A = m\pi/a$ and $B = n\pi/b$.

Now if the product $c_2c_4 = C$, another arbitrary constant, then

$$E_z = C \sin\left(\frac{m\pi}{a}x\right) \sin\left(\frac{n\pi}{b}y\right) e^{-j\beta z}$$

and

$$h^2 = \left(\frac{m\pi}{a}\right)^2 + \left(\frac{n\pi}{b}\right)^2$$

With these values of H_z and E_z, other components can be computed using the relations connecting transverse components to longitudinal ones. After including the exponential time variations and z-variations, the complete set of the field components in the waveguide are given as follows:

$$H_z = 0; \quad E_z = C \sin\left(\frac{m\pi}{a}x\right) \sin\left(\frac{n\pi}{b}y\right) e^{-j\bar{\beta} z} e^{j\omega t} \tag{10.38a}$$

$$E_x = \left(B\frac{-j\bar{\beta}}{h^2}\right) C \cos\left(\frac{m\pi}{a}x\right) \sin\left(\frac{n\pi}{b}y\right) e^{-j\bar{\beta} z} e^{j\omega t} \tag{10.38b}$$

$$E_y = \left(A\frac{-j\bar{\beta}}{h^2}\right) C \sin\left(\frac{m\pi}{a}x\right) \cos\left(\frac{n\pi}{b}y\right) e^{-j\bar{\beta} z} e^{j\omega t} \tag{10.38c}$$

$$H_x = \left(A\frac{j\omega\varepsilon}{h^2}\right) C \sin\left(\frac{m\pi}{a}x\right) \cos\left(\frac{n\pi}{b}y\right) e^{-j\bar{\beta} z} e^{j\omega t} \tag{10.38d}$$

$$H_y = \left(B\frac{-j\omega\varepsilon}{h^2}\right) C \cos\left(\frac{m\pi}{a}x\right) \sin\left(\frac{n\pi}{b}y\right) e^{-j\bar{\beta} z} e^{j\omega t} \tag{10.38e}$$

10.2.5 Transverse Electric (TE) Wave

In the case of TE wave, the electric vector is entirely normal to the direction of propagation, i.e. $E_z = 0$. The wave equation for H_z is

$$\nabla^2 H_z = \mu\varepsilon \frac{\partial^2 H_z}{\partial t^2} \tag{10.39}$$

When the time variations and z-variations are exponential, i.e.

$$H_z(x, y, z, t) = H_z(x, y)\, e^{-\bar{\gamma} z} e^{j\omega t}$$

the wave equation becomes

$$\frac{\partial^2 H_z}{\partial x^2} + \frac{\partial^2 H_z}{\partial y^2} + \bar{\gamma}^2 H_z = -\omega^2 \mu\varepsilon H_z$$

$$\frac{\partial^2 H_z}{\partial x^2} + \frac{\partial^2 H_z}{\partial y^2} + (\bar{\gamma}^2 + \omega^2 \mu\varepsilon) H_z = 0$$

$$\frac{\partial^2 H_z}{\partial x^2} + \frac{\partial^2 H_z}{\partial y^2} + h^2 H_z = 0 \tag{10.40}$$

In Eq. (10.40), H_z is a function of x and y only. Solving this equation using the *method of variables separation*, we get the general solution as

$$H_z = XY = (c_1 \cos Ax + c_2 \sin Ax)(c_3 \cos By + c_4 \sin By) \tag{10.41}$$

where c_1, c_2, c_3 and c_4 are arbitrary constants whose values can be fixed with the boundary conditions.

As the boundary conditions are not available on H_z but available on E_x and E_y, we can find these transverse components of the electric field using the available expressions for H_z and E_z. Then we can apply the boundary conditions to fix the values of the arbitrary constants.

The conditions $E_y = 0$ at $x = 0$ and $E_x = 0$ at $y = 0$ give $c_2 = 0$, $c_4 = 0$, whereas the conditions $E_y = 0$ at $x = a$ and $E_x = 0$ at $y = b$ give $A = m\pi/a$ and $B = n\pi/b$.

Now if the product $c_1c_3 = C'$, another arbitrary constant, then

$$h^2 = \left(\frac{m\pi}{a}\right)^2 + \left(\frac{n\pi}{b}\right)^2 \quad \text{and}$$

$$H_z = C' \cos\left(\frac{m\pi}{a}x\right)\cos\left(\frac{n\pi}{b}y\right)e^{-j\bar{\beta}z}$$

As the boundary conditions are not available on H_z, with the general solution for H_z, we can obtain E_x, E_y using the relations connecting the transverse components to the lateral components and then fix the values for the constants with the available boundary conditions. With these values of H_z and E_z, other components can be computed using the relations connecting the transverse components to the longitudinal ones as follows:

$$E_z = 0; \quad H_z = C' \cos\left(\frac{m\pi}{a}x\right)\cos\left(\frac{n\pi}{b}y\right)e^{-j\bar{\beta}z}e^{j\omega t} \tag{10.42a}$$

$$E_x = \left(A\frac{j\omega\mu}{h^2}\right)C' \cos\left(\frac{m\pi}{a}x\right)\sin\left(\frac{n\pi}{b}y\right)e^{-j\bar{\beta}z}e^{j\omega t} \tag{10.42b}$$

$$E_y = \left(B\frac{-j\omega\mu}{h^2}\right)C' \sin\left(\frac{m\pi}{a}x\right)\cos\left(\frac{n\pi}{b}y\right)e^{-j\bar{\beta}z}e^{j\omega t} \tag{10.42c}$$

$$H_x = \left(B\frac{j\beta}{h^2}\right)C'\sin\left(\frac{m\pi}{a}x\right)\cos\left(\frac{n\pi}{b}y\right)e^{-j\beta z}e^{j\omega t} \tag{10.42d}$$

$$H_y = \left(A\frac{j\beta}{h^2}\right)C'\sin\left(\frac{m\pi}{a}x\right)\sin\left(\frac{n\pi}{b}y\right)e^{-j\beta z}e^{j\omega t} \tag{10.42e}$$

10.2.6 Electric and Magnetic Flux Lines

The concept and the utility of flux lines or field lines in the representation of the fields are introduced and explained in the second chapter of this book. The electric as well as magnetic field lines never cross each other. The magnetic flux lines are always closed curves. Both the categories of the lines can be sketched from the expressions of the fields in those regions.

Apart from the general usefulness, i.e. graphical representation of fields, the flux line representation of fields in waveguides has another important application. It is used in placing the probe or loop to excite the required mode in the guide. The flux line representation of the fields, both electric and magnetic, is shown in Figure 10.4. The characteristics of standard rectangular waveguides are given in Table 10.2.

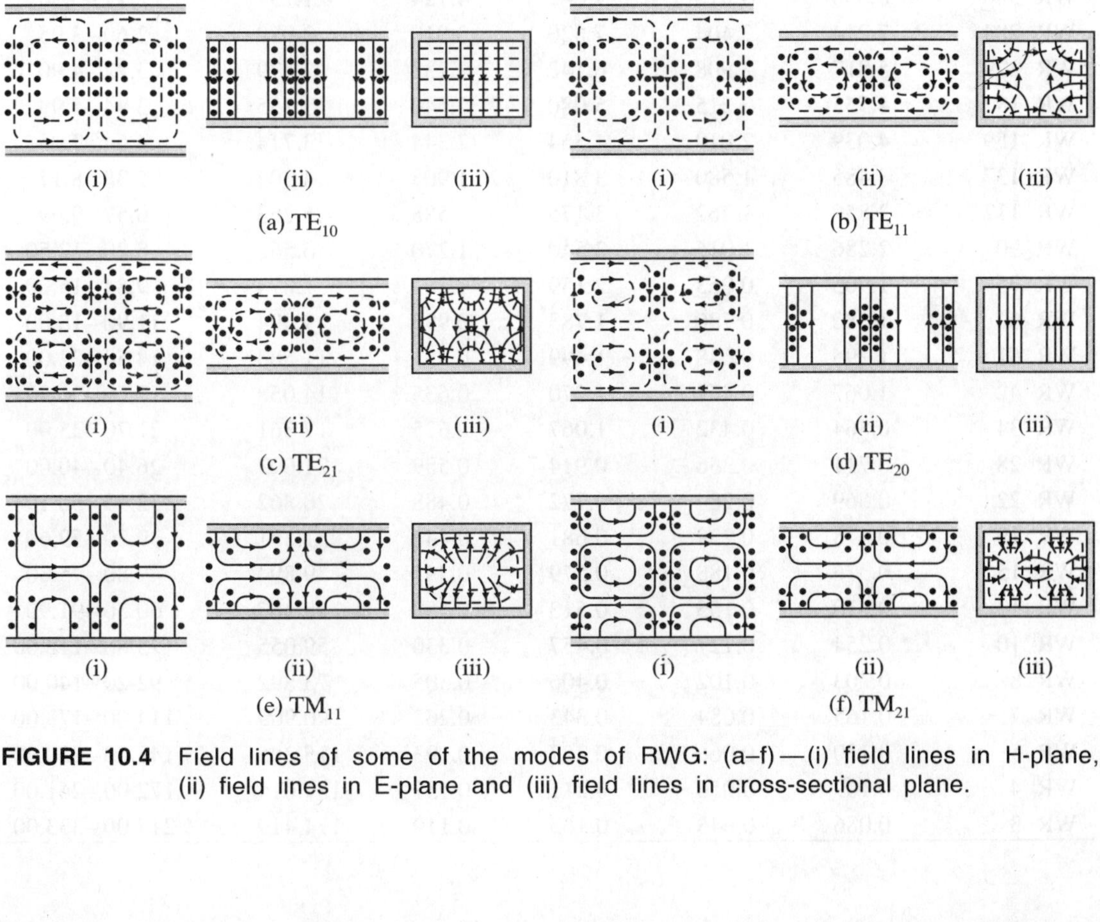

FIGURE 10.4 Field lines of some of the modes of RWG: (a–f)—(i) field lines in H-plane, (ii) field lines in E-plane and (iii) field lines in cross-sectional plane.

Table 10.2 Characteristics of standard rectangular waveguides

EIA designation	Physical dimensions				Cut-off frequency for air-filled waveguide in GHz	Recommended frequency range for TE_{10} mode in GHz
	Inside, in cm		Outside, in cm			
	Width	Height	Width	Height		
WR 2300	58.420	29.210	59.055	29.845	0.257	0.32–0.49
WR 2100	53.340	26.670	53.973	27.305	0.281	0.35–0.53
WR 1800	45.720	22.860	46.350	23.495	0.328	0.41–0.62
WR 1500	38.100	19.050	38.735	19.685	0.394	0.49–0.75
WR 1150	29.210	14.605	29.845	15.240	0.514	0.64–0.98
WR 975	24.765	12.383	25.400	13.018	0.606	0.76–1.15
WR 770	19.550	9.779	20.244	10.414	0.767	0.96–1.46
WR 650	16.510	8.255	16.916	8.661	0.909	1.14–1.73
WR 510	12.954	6.477	13.360	6.883	1.158	1.45–2.20
WR 430	10.922	5.461	11.328	5.867	1.373	1.72–2.61
WR 340	8.636	4.318	9.042	4.724	1.737	2.17–3.30
WR 284	7.214	3.404	7.620	3.810	2.079	2.60–3.95
WR 229	5.817	2.908	6.142	3.233	2.579	3.22–4.90
WR 187	4.755	2.215	5.080	2.540	3.155	3.94–5.99
WR 159	4.039	2.019	4.364	2.344	3.714	4.64–7.05
WR 137	3.485	1.580	3.810	1.905	4.304	5.38–8.17
WR 112	2.850	1.262	3.175	1.588	5.263	6.57–9.99
WR 90	2.286	1.016	2.540	1.270	6.562	8.20–12.50
WR 75	1.905	0.953	2.159	1.207	7.874	9.84–15.00
WR 62	1.580	0.790	1.783	0.993	9.494	11.90–18.00
WR 51	1.295	0.648	1.499	0.851	11.583	14.50–22.00
WR 42	1.067	0.432	1.270	0.635	14.058	17.60–26.70
WR 34	0.864	0.432	1.067	0.635	17.361	21.70–33.00
WR 28	0.711	0.356	0.914	0.559	21.097	26.40–40.00
WR 22	0.569	0.284	0.772	0.488	26.362	32.90–50.10
WR 19	0.478	0.239	0.681	0.442	31.381	39.20–59.60
WR 15	0.376	0.188	0.579	0.391	39.894	49.80–75.80
WR 12	0.310	0.155	0.513	0.358	48.387	60.50–91.90
WR 10	0.254	0.127	0.457	0.330	59.055	73.80–112.00
WR 8	0.203	0.102	0.406	0.305	73.892	92.20–140.00
WR 7	0.165	0.084	0.343	0.262	90.909	114.00–173.00
WR 5	0.130	0.066	0.257	0.193	115.385	145.00–220.00
WR 4	0.109	0.056	0.211	0.157	137.615	172.00–261.00
WR 3	0.086	0.043	0.163	0.119	174.419	217.00–333.00

10.2.7 Filter Characteristics

The hollow-pipe waveguides behave like high-pass filters. They admit and allow the wave to propagate through them only if the frequency of the wave is more than certain value known as the cut-off frequency whose value depends upon the dimensions of the guide and the mode of the wave.

Cut-off frequency f_c: It is the frequency above which the frequency of the wave should be in order to get entry into the wave guide for propagating through it. It depends upon the dimensions of the guide as well as on the mode of the wave. Thus,

$$f_c = \frac{1}{2\sqrt{\mu\varepsilon}}\sqrt{\left(\frac{m}{a}\right)^2 + \left(\frac{n}{b}\right)^2} \tag{10.43}$$

Cut-off wavelength λ_c: The wavelength corresponding to the cut-off frequency is called cut-off wavelength. It can be defined as the wavelength below which the wavelength of the wave should be in order to get entry into the waveguide for propagating through it. Its value is related to the dimensions of the guide and mode numbers of the wave through

$$\lambda_c = \frac{2}{\sqrt{\left(\frac{m}{a}\right)^2 + \left(\frac{n}{b}\right)^2}} \tag{10.44}$$

Proof: We know that the propagation constant $\bar{\gamma}$ can be related to the frequency ω of the wave through

$$\bar{\gamma}^2 = h^2 - \omega^2 \mu\varepsilon \text{ or } \bar{\gamma} = \sqrt{h^2 - \omega^2 \mu\varepsilon}$$

Depending upon the relation between h^2 and $\omega^2\mu\varepsilon$, the propagation constant $\bar{\gamma}$ can be pure real or imaginary.

When $h^2 > \omega^2\mu\varepsilon$, the propagation constant is a pure real quantity indicating that the waveguide is acting as a pure attenuator without wave motion refusing entry to the wave.

When $h^2 < \omega^2\mu\varepsilon$, the propagation constant is a pure imaginary quantity indicating that the waveguide is acting as a pure transmission line without any attenuation to the wave. This must be the case to which the waveguide under consideration belongs because the wave is already in and a lossless condition is assumed.

The changeover in the behaviour of the waveguide from the pure attenuator to the pure transmission line takes place as the frequency is increased from low to high when $h^2 = \omega^2\mu\varepsilon$. The frequency of the wave which satisfies this relation is called the cut-off frequency ω_c. Hence,

$$\omega_c^2 \mu\varepsilon = h^2. \text{ But } h^2 = \left(\frac{m\pi}{a}\right)^2 + \left(\frac{n\pi}{b}\right)^2$$

Solving this equation for f_c we get

$$f_c = \frac{1}{2\sqrt{\mu\varepsilon}}\sqrt{\left(\frac{m}{a}\right)^2 + \left(\frac{n}{b}\right)^2}$$

The corresponding wavelength, cut-off wavelength, can be found from

$$\lambda_c = \frac{v}{f_c} = \frac{1}{\sqrt{\mu\varepsilon}\, f_c} = \frac{2}{\sqrt{\left(\frac{m}{a}\right)^2 + \left(\frac{n}{b}\right)^2}}$$

The relation between the cut-off wavelength, mode and dimensions of the guide can also be put in the following form:

$$\frac{1}{\lambda_c^2} = \left(\frac{m}{2a}\right)^2 + \left(\frac{n}{2b}\right)^2$$

The cut-off frequency f_c and the cut-off wavelength λ_c are the same for both the types of the wave, i.e. TE or TM in rectangular waveguides.

10.2.8 Modal Propagation

The electromagnetic energy propagation along the waveguide is in the form of some definite field patterns, known as *modes.* This is an important and special feature of the energy propagation through the waveguides. These field patterns or modes are described by mentioning the transverse nature of the wave, i.e. TE or TM along with two subscript numbers. The mode subscripts in RWGs are denoted by *m* and *n* in that order such as in TE_{mn} or TM_{mn}. Subscripts *m* and *n* indicate the number of half period variations of the fields along the *x*- and *y*-directions, respectively. For TM wave, *m* and *n* can assume any integer values from 1, 2, 3, etc. For TE wave, *m* and *n* can assume any integer values from 1, 2, 3, etc., but either *m* or *n* can assume zero value at the same time.

There is a mode which is widely used for transmitting the signal through the waveguide, known as *dominant mode* of the waveguide. By definition it is a mode, which has the lowest cut-off frequency. From the expression of Eq. (10.43), we can see that it is TE_{10} for RWGs. The significance of the dominant mode is that a guide of smaller dimensions is sufficient to transmit a given frequency in the dominant mode. Another advantage is, as the frequency can be transmitted in the dominant mode in a guide that is too small to hold the higher-order modes, the energy loss through the generation of spurious modes is prevented.

We can observe from Eq. (10.44) that for this mode, the cut-off wavelength is twice the inner distance between side walls, i.e. $\lambda_c = 2a$. The field components for this mode can be found from Eqs. (10.42a–e) by substituting $m = 1$ and $n = 0$. They are

$$\left.\begin{aligned}
E_z &= 0; & H_z &= C' \cos\left(\frac{\pi}{a}x\right) e^{-j\beta z} e^{j\omega t} \\
E_x &= 0; & H_x &= \left(B\frac{j\beta}{h^2}\right) C' \sin\left(\frac{\pi}{a}x\right) e^{-j\beta z} e^{j\omega t} \\
E_y &= \left(B\frac{-j\omega\mu}{h^2}\right) C' \sin\left(\frac{\pi}{a}x\right) e^{-j\beta z} e^{j\omega t}; & H_y &= 0
\end{aligned}\right\} \quad (10.45)$$

From the expressions for the fields of the dominant mode, we can make certain important observations:

- The electric field is entirely in the *y*-direction, whereas the magnetic field is devoid of a component in the *y*-direction. Regarding the field distribution in the *y*-direction, the electric field is maximum at the middle of the guide and nil at the ends.
- The magnetic field has both the *x*- and *z*-components; so it can only be in the *xz*-plane. Here it is relevant to define two directions or planes which are very much widely used in describing the field directions in waveguides. They are the **E**-plane and the **H**-plane. Any plane containing the **E** vector and parallel to the narrow walls of the waveguide is called the **E**-plane whereas the plane containing the **H** vector and parallel to the broader walls is called the **H**-plane. From the definition, we can observe that any vertical plane parallel to the guide is the **E**-plane and any horizontal plane parallel to the guide is the **H**-plane.

Apart from the dominant mode, higher-order modes, i.e. modes with large *m* and *n* values, can also be used. Several signals can be transmitted simultaneously through the waveguide using different higher-order modes. These modes, as they require the waveguides of larger dimensions, are also capable of transmitting higher frequencies. But they suffer with the difficulty in getting excited and also they require waveguides of larger dimensions to transmit a given frequency.

In addition to the dominant mode and higher-order mode, in the discussion on the modal propagation of waveguides we come across another two important modes. They are degenerate modes and evanescent modes. *Degenerate modes* are different modes of waveguides having the same cut off frequency. Examples of degenerate modes are TE_{mn} and TM_{mn}. *Evanescent modes* are modes whose wavelength is more than the cut-off value and because of this they cannot propagate. There is a possibility for them to get generated but they cannot travel further. These modes just represent the localized field distribution, i.e. induction fields, that can introduce reactive effects but do not carry energy away from the point of origin as does the dominant mode.

10.2.9 Waveguide Dimensions

Waveguide dimensions are almost always in the ratio $a/b \approx 2$ with the whole intention of having single mode propagation over a wide frequency range of operation. It is preferred to operate waveguides so that it is possible only for single pure mode to exist in them. It is to be ensured because the coupling systems and terminations of the guides are designed assuming the presence of single mode in the guide. Waveguides carrying more than one mode are called *overmoded*.

In most of the cases, the single mode that is used is the dominant mode. It is preferred because the guide then has the smallest possible dimensions and the undesired modes can be very easily eliminated by proportioning the guide so that it is large enough to transmit the dominant mode while too small to permit propagation of any other mode. Then any spurious higher modes generated do not travel down the guide, but rather are confined to the region where they are generated.

In rectangular guides, when the guide is so proportioned that $a/b = 2$, we will see soon that there is a 2:1 frequency range over which only the dominant mode propagation exists. In contrast, if the guide were square i.e. $a/b = 1$, the TE_{01} mode would have the same cut-off frequency as the TE_{10} mode, and there would be no frequency range over which only a single mode would propagate. Because of the considerations of this type, rectangular guides are practically always proportioned so that $a/b = 2$, as this ratio gives the best mode separation of all possible proportions. If the ratio $a/b < 2$, then the range frequency over which the single mode propagation is possible gets reduced. If the ratio $a/b > 2$, then the power-handling capability of the waveguide decreases. So the optimum ratio is $a/b = 2$.

10.2.10 Bandwidth of RWGs

The bandwidth of the waveguide generally refers to the range of frequencies over which the single mode propagation is possible. The bandwidth of the waveguide with $a/b = 2$, for all practical purposes, as we are going to see, is less than a 2:1 frequency range.

In rectangular waveguides with $a/b = 2$, the modes with the lowest cut-off frequencies are dominant mode TE_{10} and its degenerate mode TE_{01} (the mode TE_{20} has a cut-off frequency which is equal to that of the TE_{01} mode). In these waveguides, therefore, the single mode propagation is possible only from the cut-off frequency of the mode TE_{10} to that of the mode TE_{01}.

The cut-off wavelength from Eq. (10.44) is

$$\lambda_c = \frac{2}{\sqrt{\left(\frac{m}{a}\right)^2 + \left(\frac{n}{b}\right)^2}}$$

For the mode TE_{10}, the cut-off wavelength becomes $\lambda_c = 2a$ and the corresponding cut-off frequency becomes $f_c = c/\lambda_c = c/2a$. In the case of mode TE_{01}, the cut-off wavelength is $\lambda_c = 2b$ and the corresponding cut-off frequency becomes $f_c = c/\lambda_c = c/2b = c/a$.

Therefore, the single mode propagation is possible only from the frequency of $c/2a$ to the frequency c/a. From these expressions, we can observe that if lower frequencies are to be transmitted, then the ratio $c/2a$ should be low, thus requiring a waveguide of larger dimensions. Similarly to be used for transmitting higher frequencies, c/a should be large, thus requiring a waveguide of smaller dimensions.

The waveguides are components used in communication systems. They require to exhibit larger bandwidths. One method usually adopted to increase the bandwidth of the guide is by lowering the cut-off frequency of the dominant mode. It is done by loading the waveguide with a conducting ridge on the top and/or bottom walls. Apart from increasing the bandwidth, the loading also gives better impedance characteristics, making the guide suitable for impedance matching purposes. The presence of the ridge, however, reduces the power-handling capacity of the waveguide.

The bandwidth of the waveguide is determined by the ratio of transverse dimensions (a/b), ratio whereas the actual frequencies that are allowed to pass through it are decided by the transverse dimensions a and b of the guide.

10.3 EXCITATION OF RWGs

The mode of propagation of the wave is determined by the type and location of the excitation. Although either probes or loops may be used as excitation sources, the probes are normally preferred for their simplicity.

The guide is closed at one end by a conducting wall and an appropriate exciting probe is inserted through the end or side of the guide. The end of the guide serves as a reflector and if the distance between the probe and the wall is properly adjusted, the reflected waves arrive at the probe in phase with the emitted wave, and the two propagate down the guide as one wave.

The probes should coincide with the positions of maximum electric field intensity in the modes they are intended to excite, with attention being given to the proper phasing of the potentials supplied to the probes in accordance with the phasing of the fields to be excited. If the loop is used for excitation, the plane of the loop will be made normal to the magnetic field and the loop will be located at a point of maximum magnetic field intensity.

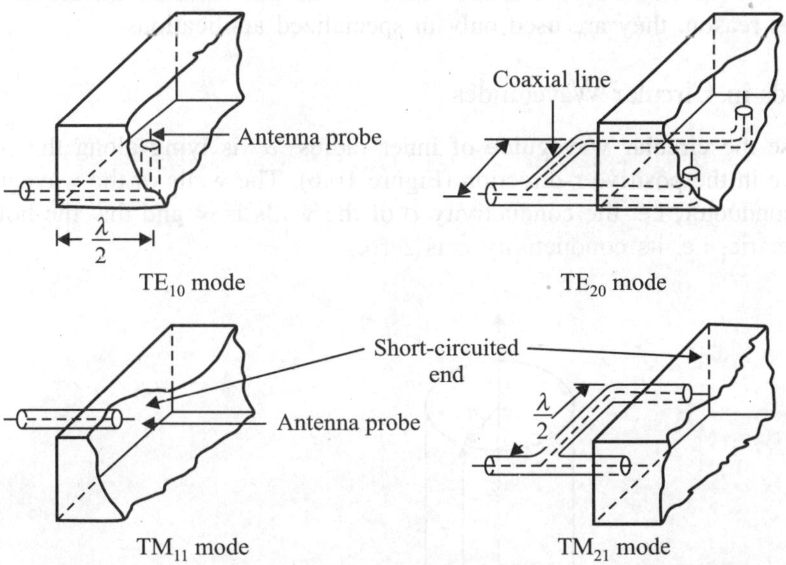

FIGURE 10.5 Methods of excitation of different modes in RWG.

The sources excite not only the desired modes but also the higher-order unwanted modes. But for choosing the guide dimensions appropriately, it is possible to have only the desired wave above the cut-off frequency, the other waves then being attenuated and not propagated. The method of excitation of different modes in RWG are shown in Figure 10.5.

10.4 CIRCULAR WAVEGUIDES

Circular Waveguides (CWGs) are also hollow pipes with uniform circular cross section. It is this cross-sectional shape that finds large application after rectangular. The behaviour of the wave is almost the same in waveguides independent of the cross-sectional shape. As far as the wave behaviour in circular waveguides is concerned, it can be predicted from the analysis results we have for the rectangular waveguides. However, the circular waveguides have geometry, mode designation and applications different from that of the rectangular waveguides. As far as the orientation of the electric field is concerned, the rectangular waveguides have the plane of the polarization of the wave uniquely defined with electric vector **E** across the narrow dimension, but the circular waveguides because of their circular cross section, are difficult to define or maintain the plane of polarization.

The circular waveguides have certain advantages like easy to join and easy to manufacture, but they have also certain disadvantages like requirement of larger cross-sectional area than the rectangular waveguides for a given frequency and rotation of polarization plane. Another disadvantage with them is that there is only a very narrow range between the cut-off wavelength of the dominant mode and the cut-off wavelength of the next higher mode. Thus, the frequency range over which the pure mode operation is assured is relatively limited. In addition, there is a very serious drawback associated with them: arbitrary and unpredictable orientation of electric vector when the wave is travelling through. And because of this reason, they are used only in specialized applications.

10.4.1 Fields in Circular Waveguides

Let us suppose the circular waveguide of inner radius 'a' is lying along the z-axis carrying travelling wave in the positive z-direction (Figure 10.6). The walls of the waveguide are made with perfect conductor, i.e. the conductivity σ of the walls is ∞ and that the hollow region is a perfect dielectric, i.e. its conductivity σ is zero.

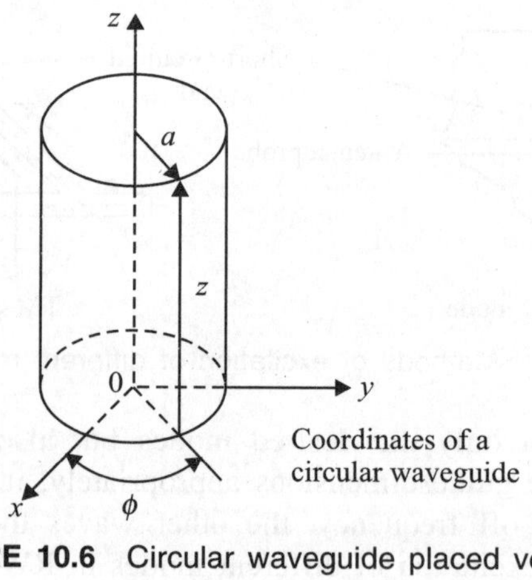

FIGURE 10.6 Circular waveguide placed vertical.

Let us also suppose the time variations of the field quantities are exponential, i.e. $e^{j\omega t}$. If the time variations of the field quantities are exponential, then the fields must vary in the same manner, i.e. exponentially along the direction of the propagation of the wave, according to the transmission line theory. So let us assume the fields vary with z as $e^{\overline{\gamma} z}$. Here $\overline{\gamma}$ is the propagation constant along the direction of propagation, i.e. z-direction, which is in general a complex quantity $\bar{\alpha}+\bar{\beta}$. Its significance in the waveguide theory has been explained in Section 10.2.2.

According to Maxwell, time-varying fields are always accompanied by electromagnetic wave. And the converse is also true. As the wave is inside the waveguide, fields must also be there. If **E** and **H** are fields at an arbitrary point P in the hollow region, they must be related through the Maxwell's curl equations. As we consider this problem in cylindrical coordinate system, the fields are functions of ρ, ϕ, z and t.

The longitudinal components of the fields can be expressed in terms of the transverse components as follows:

$$H_p = -\frac{\bar{\gamma}}{h^2}\frac{\partial H_z}{\partial \rho} + \frac{j\omega\varepsilon}{h^2}\frac{\partial E_z}{\rho\partial\phi} \tag{10.46a}$$

$$H_\phi = -\frac{\bar{\gamma}}{h^2}\frac{\partial H_z}{\rho\partial\phi} - \frac{j\omega\varepsilon}{h^2}\frac{\partial E_z}{\partial \rho} \tag{10.46b}$$

$$E_\rho = -\frac{\bar{\gamma}}{h^2}\frac{\partial E_z}{\partial \rho} - \frac{j\omega\mu}{h^2}\frac{\partial H_z}{\rho\partial\phi} \tag{10.46c}$$

$$E_\phi = -\frac{\bar{\gamma}}{h^2}\frac{\partial E_z}{\rho\partial\phi} + \frac{j\omega\mu}{h^2}\frac{\partial H_z}{\partial \rho} \tag{10.46d}$$

where

$$h^2 = \overline{\gamma^2} + \omega^2\mu\varepsilon \tag{10.47}$$

The constant h is also denoted frequently as k_c and called the cut-off wave number.

The boundary conditions to be used while solving the wave equation are derived based on the fact that the tangential component of the electric field at the surface of the perfect conductor is zero. At $\rho = a$, the inner surface of the CWG exists, so the tangential component of the electric field must be zero, i.e. $E_\phi = 0 = E_z$ at $\rho = a$. The roots of equations $J_n(x) = 0$ and $J_n'(x) = 0$ are shown in Tables 10.3 and 10.4.

Table 10.3 Roots of equation $J_n(x) = 0$

m	$n = 0$	$n = 1$	$n = 2$	$n = 3$	$n = 4$
1	2.40	3.83	5.14	6.38	7.59
2	5.52	7.02	8.42	9.76	11.06
3	8.65	10.17	11.62	13.02	14.37

Table 10.4 Roots of equation $J_n'(x) = 0$

m	$n = 0$	$n = 1$	$n = 2$	$n = 3$	$n = 4$
1	3.83	1.84	3.05	4.20	5.32
2	7.02	5.33	6.71	8.02	9.28
3	10.17	8.54	9.97	11.35	12.68

$J_0'(x) = -J_1(x)$

10.4.2 Transverse Magnetic (TM) Wave

If the magnetic wave is entirely normal to the direction of propagation, the magnetic vector along the direction of propagation is zero, i.e. $H_z = 0$. The wave equation for E_z is

$$\nabla^2 E_z = \mu\varepsilon \frac{\partial^2 E_z}{\partial t^2} \tag{10.48}$$

When time variations and z-variations are exponential, i.e.

$$E_z(\rho, \phi, z, t) = E_z(\rho, \phi)e^{-\bar{\gamma} z} e^{j\omega t}$$

then the wave equation becomes

$$\frac{\partial^2 E_z}{\partial \rho^2} + \frac{\partial^2 E_z}{\rho^2 \partial \phi^2} + \bar{\gamma}^2 E_z + \frac{\partial E_z}{\rho \partial \rho} = -\omega^2 \mu\varepsilon E_z \tag{10.49}$$

In this equation, E_z is a function of ρ and ϕ only. This equation can be solved using the *method of variable separation.* According to this method, let $E_z = PQ$, where P is a function of variable ρ alone and Q is a function of variable ϕ alone. With this substitution, the above wave equation can be decomposed into two second-order differential equations for which solutions are readily available.

$$\frac{1}{Q}\frac{d^2 Q}{dx^2} = -n^2 \tag{10.50a}$$

$$\frac{\partial^2 P}{\partial \rho^2} + \frac{\partial P}{\rho \partial \rho} + \left(h^2 - \frac{n^2}{\rho^2}\right) P = 0 \tag{10.50b}$$

where n is an arbitrary *integer* constant.

The first one is the standard second-order differential equation for which the solution is available.

$$Q = A_n \cos n\phi + B_n \sin n\phi$$

The second equation is known as the Bessel equation. Its solutions which can represent the physical fields are known as Bessel functions $J_n(\rho h)$. These are called Bessel functions of first kind and n order (Figure 10.7). Now

$$E_z = PQ = (A_n \cos n\phi + B_n \sin n\phi)\, J_n(\rho h) \tag{10.51}$$

This is the general solution and to fix the arbitrary constants, we use the boundary conditions. Applying $E_z = 0$ at $\rho = a$, we get

$$h = \frac{(ha)_{nm}}{a} = h_{nm} \tag{10.52}$$

where $(ha)_{nm}$ are the roots of the equation $J_n(ha) = 0$ in the increasing order of their magnitude. Hence

$$E_z = (A_n \cos n\phi + B_n \sin n\phi)\, J_n(\rho h_{nm})e^{-j\bar{\beta}z} \tag{10.53}$$

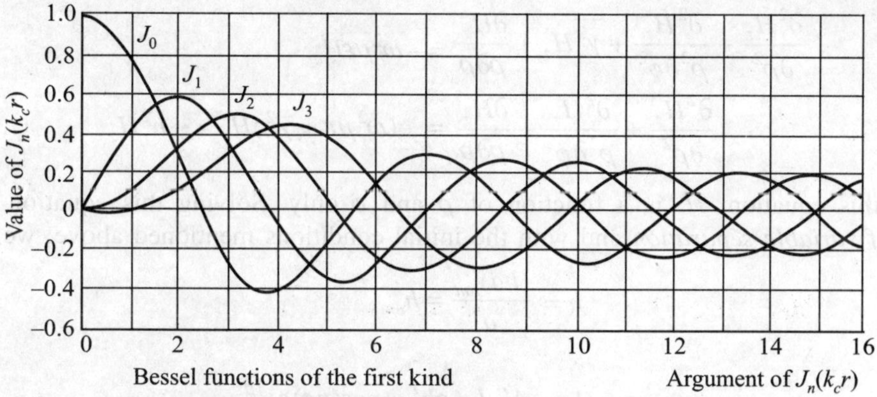

FIGURE 10.7 Illustrating Bessel functions.

It is also possible to orient the reference to the ϕ-axis so that $B_n = 0$. Therefore,

$$E_z = A_n \cos n\phi \, J_n(\rho h_{nm}) e^{-j\bar{\beta}z} \tag{10.54}$$

With these values of H_z and E_z, the other components can be computed using the relations connecting the transverse components to the longitudinal ones. And attaching the time variations, they become

$$E_z = A_n J_n(\rho h_{nm}) \cos(n\phi) e^{-j\bar{\beta}z} e^{j\omega t} \tag{10.55a}$$

$$H_\rho = -\frac{j\omega\varepsilon}{\rho h^2_{nm}} n A_n J_n(\rho h_{nm}) \sin(n\phi) e^{-j\bar{\beta}z} e^{j\omega t} \tag{10.55b}$$

$$E_\phi = -\frac{\bar{\beta}}{\omega\varepsilon} H_\rho \tag{10.55c}$$

$$H_\phi = -\frac{j\omega\varepsilon}{h_{nm}} A_n J'_n(\rho h_{nm}) \cos(n\phi) e^{-j\bar{\beta}z} e^{j\omega t} \tag{10.55d}$$

$$E_\rho = \frac{\bar{\beta}}{\omega\varepsilon} H_\phi \tag{10.55e}$$

10.4.3 Transverse Electric (TE) Wave

In the case of TE wave, $E_z = 0$ and the wave equation for E_z is

$$\nabla^2 H_z = \mu\varepsilon \frac{\partial^2 H_z}{\partial t^2}$$

When both the time and z-variations are exponential, i.e. $H_z(\rho, \phi, z, t) = H_z(\rho, \phi) e^{-\bar{\gamma}z} e^{j\omega t}$, the wave equation becomes

$$\frac{\partial^2 H_z}{\partial \rho^2} + \frac{\partial^2 H_z}{\rho^2 \partial \phi^2} + \overline{\gamma}^2 H_z + \frac{\partial H_z}{\rho \partial \rho} = -\omega^2 \mu \varepsilon H_z$$

$$\frac{\partial^2 H_z}{\partial \rho^2} + \frac{\partial^2 H_z}{\rho^2 \partial \phi^2} + \frac{\partial H_z}{\rho \partial \rho} = -(\omega^2 \mu \varepsilon + \overline{\gamma}^2) H_z = -h^2 H_z \tag{10.56}$$

Note in this equation, H_z is a function of ρ and ϕ only. Solving this equation using the *method of variable separation* and with the initial conditions mentioned above, we obtain

$$h = \frac{(ha)'_{nm}}{a} = h'_{nm} \tag{10.57}$$

and

$$H_z = A'_n \, J_n(\rho h'_{nm}) \cos(n\phi) e^{-j\bar{\beta} z} \tag{10.58}$$

As the boundary conditions are not available on H_z, with the general solution for H_z, we obtain E_x, E_y using the relations connecting the transverse components to the lateral components and then fix the values for the constants with the available boundary conditions.

With these values of H_z and E_z, the other components can be computed using the relations connecting the transverse components to the longitudinal ones as follows:

$$H_z = A'_n \, J_n(\rho h'_{nm}) \cos(n\phi) e^{-j\bar{\beta} z} \tag{10.59a}$$

$$H_\rho = \frac{-j\bar{\beta}}{h'_{nm}} A'_n J_n(\rho h'_{nm}) \cos(n\phi) e^{-j\bar{\beta} z} \tag{10.59b}$$

$$E_\rho = \frac{\omega\mu}{\bar{\beta}} H_\phi \tag{10.59c}$$

$$H_\phi = \frac{j\bar{\beta} n}{\rho {h'_{nm}}^2} A'_n J_n(\rho h'_{nm}) \sin(n\phi) e^{-j\bar{\beta} z} \tag{10.59d}$$

$$E_\phi = -\frac{\omega\mu}{\bar{\beta}} H_\rho \tag{10.59e}$$

10.4.4 Filter Characteristics

Cut-off frequency f_c: The frequency of the wave should be more than a certain value, known as cut-off frequency, in order to get admitted into the waveguide to propagate further. Its value depends upon the dimensions of the guide as well as on the mode of the wave. Thus, it is

$$f_c = \frac{h_{nm}}{2\pi\sqrt{\mu\varepsilon}} \quad \text{for TM wave} \tag{10.60a}$$

and

$$f_c = \frac{h'_{nm}}{2\pi\sqrt{\mu\varepsilon}} \quad \text{for TE wave} \tag{10.60b}$$

***Cut-off wavelength* λ_c:** The wavelength corresponding to the cut-off frequency is the cut-off wavelength and it can be formally defined as the wavelength below which the wavelength of the wave should be in order to get admitted into the guide to propagate further. Its value depends upon the dimensions of the guide as well as on the mode of the wave. Thus, it is

$$\lambda_c = \frac{2\pi a}{(ha)_{nm}} \quad \text{for TM wave} \tag{10.61a}$$

and

$$\lambda_c = \frac{2\pi a}{(ha)'_{nm}} \quad \text{for TE wave} \tag{10.61b}$$

Proof: We know that the propagation $\overline{\gamma}$ constant can be related to the frequency ω of the wave through

$$\overline{\gamma}^2 = h^2 - \omega^2 \mu\varepsilon \text{ or } \gamma = \sqrt{h^2 - \omega^2 \mu\varepsilon}$$

Depending upon the relation between h^2 and $\omega^2\mu\varepsilon$ the propagation constant $\overline{\gamma}$ can be pure real or imaginary.

When $h^2 > \omega^2\mu\varepsilon$, the propagation constant is a pure real quantity indicating that the waveguide is acting as a pure attenuator without the wave motion refusing entry to the wave.

When $h^2 < \omega^2\mu\varepsilon$, the propagation constant is a pure imaginary quantity indicating that the waveguide is acting as a pure transmission line without any attenuation to the wave. This must be the case to which the waveguide under consideration belongs because the wave is already in and the lossless condition is assumed.

The changeover in the behaviour of the waveguide from the pure attenuator to the pure transmission line takes place as the frequency is increased from low to high when $h^2 = \omega^2\mu\varepsilon$. The frequency of the wave which satisfies this relation is called cut-off frequency ω_c. Hence, $\omega_c^2\mu\varepsilon = h^2$. But $h^2 = h_{nm}^2$ in case of TM wave and $h^2 = h_{nm}'^2$ in the case of TE wave.

Solving for f_c, we get

$$f_c = \frac{h_{nm}}{2\pi\sqrt{\mu\varepsilon}} \quad \text{for TM wave}$$

and

$$f_c = \frac{h'_{nm}}{2\pi\sqrt{\mu\varepsilon}} \quad \text{for TE wave}$$

Note here that $h_{nm} = (ha)_{nm}/a$ and $h'_{nm} = (ha)'_{nm}/a$

The corresponding wavelength, the cut-off wavelength, can be found from

$$\lambda_c = \frac{2\pi a}{(ha)_{nm}} \quad \text{for TM wave}$$

and

$$\lambda_c = \frac{2\pi a}{(ha)'_{nm}} \quad \text{for TE wave}$$

In the case of the circular waveguides, both the cut-off frequency and the cut-off wavelength depend upon the type of the wave, i.e. their values are different for TE and TM waves in circular waveguides.

10.4.5 Electric and Magnetic Flux Lines

The flux line representation of the fields, both electric and magnetic, in circular waveguides, is shown in Figure 10.8. Apart from the general usefulness, i.e. graphical representation of fields, the flux line representation of fields in waveguides has another important application. It is useful in indicating the location of the probe or loop over the waveguide to excite the required mode in the guide.

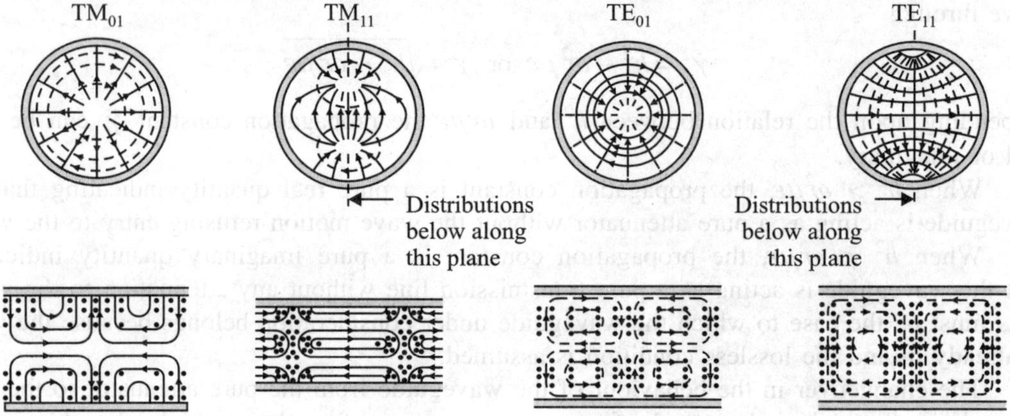

FIGURE 10.8 Various modes of circular waveguides.

10.4.6 Modal Propagation Characteristics

- The mode subscripts in CWG are denoted by n and m in that order as TE_{nm} or TM_{nm}.
- n indicates the order of the Bessel function and m indicates the roots of the equation $J_n(x) = 0$ or $J_n'(x) = 0$ in the order of the magnitude.
- The subscript n can assume any integer values 0, 1, 2, 3, etc., but for the subscript m zero value is forbidden and so it can assume any integer values 1, 2, 3, etc. Note TE_{10} and TM_{00} are not possible. The mode TE_{10} cannot exist because there is no root at the origin for $J_1'(x) = 0$.
- The dominant mode of a waveguide is one, which has the lowest cut-off frequency. It is TE_{11} for CWG.
- The degenerate modes are different modes of waveguides having the same cut-off frequency, for example, TE_{0m} and TM_{1m}.

The characteristics of standard circular waveguides are given in Table 10.5.

Table 10.5 Characteristics of standard circular waveguides

EIA designation	Inside diameter in cm	Cut-off frequency for air-filled waveguide in GHz	Recommended frequency range for TE_{11} mode in GHz
WC 992	25.184	0.698	0.80–1.10
WC 847	21.514	0.817	0.94–1.29
WC 724	18.377	0.957	1.10–1.51
WC 618	15.700	1.120	1.29–1.76
WC 528	13.411	1.311	1.51–2.07
WC 451	11.458	1.534	1.76–2.42
WC 385	9.787	1.796	2.07–2.83
WC 329	8.362	2.102	2.42–3.31
WC 281	7.142	2.461	2.83–3.88
WC 240	6.104	2.880	3.31–4.54
WC 205	5.199	3.381	3.89–5.33
WC 175	4.445	3.955	4.54–6.23
WC 150	3.810	4.614	5.30–7.27
WC 128	3.254	5.402	6.21–8.51
WC 109	2.779	6.326	7.27–9.97
WC 94	2.383	7.377	8.49–11.60
WC 80	2.024	8.685	9.97–13.70
WC 69	1.748	10.057	11.60–15.90
WC 59	1.509	11.649	13.40–18.40
WC 50	1.270	13.842	15.90–21.80
WC 44	1.113	15.794	18.20–24.90
WC 38	0.953	18.446	21.20–29.10
WC 33	0.833	21.103	24.30–33.20
WC 28	0.714	24.620	28.30–38.80
WC 25	0.635	27.683	31.80–43.60
WC 22	0.556	31.617	36.40–49.80
WC 19	0.478	36.776	42.40–58.10
WC 17	0.437	40.227	46.30–63.50
WC 14	0.358	49.103	56.60–77.50
WC 13	0.318	55.280	63.50–87.20
WC 11	0.277	63.462	72.70–99.70
WC 9	0.239	73.552	84.80–116.00

10.5 DISPERSION CHARACTERISTICS

The media in which the velocity of the wave depends upon its frequency are called *dispersive media.* Otherwise they are called non-dispersive media. The velocities of the wave, both phase and group velocities, in the waveguide vary with the frequency and hence the hollow-pipe waveguide is a dispersive medium. Also the TE/TM waves carried by these media are *dispersive* waves.

Dispersion of the signal spoils the original phase relation between different frequency components as it travels down the guide leading to signal distortion.

- *Guide wavelength* λ_g: Guide wavelength is an important parameter of waveguides which can be defined in several different but equivalent ways. According to one definition, it is the axial length corresponding to one cycle of variation of the field configuration in the axial direction. It can also be defined as the distance between two consecutive equiphase planes along the length of the waveguide. Another definition for it is the distance that a wave travels down the guide when undergoing a phase shift of 2π radians.

  It is related to the cut-off wavelength through the relation

$$\lambda_g = \frac{\lambda}{\sqrt{1 - \left(\frac{\lambda}{\lambda_c}\right)^2}} \tag{10.62}$$

  The above standard expression for guide wavelength can be put in a more popular form which connects all the three wavelengths as follows:

$$\frac{1}{\lambda_g^2} + \frac{1}{\lambda_c^2} = \frac{1}{\lambda^2} \tag{10.63}$$

  If the non-conducting region has constants ε_r and μ_r, then

$$\lambda_g = \frac{\lambda}{\sqrt{\varepsilon_r \mu_r - \left(\frac{\lambda}{\lambda_c}\right)^2}} \tag{10.64}$$

- *Phase/wave velocity* v_p: It is the velocity at which the phase of the wave changes along the length of the guide. It is more than or equal to the velocity of the wave in free space. It is, by definition, equal to

$$v_p = \left(\frac{\omega}{\beta}\right) = \frac{v}{\sqrt{1 - \left(\frac{\lambda}{\lambda_c}\right)^2}} \tag{10.65}$$

- *Group velocity* v_g: It is the velocity with which a narrow band signal travels in the guide. It is always less than the velocity of the wave in free space. It is given by

$$v_g = \left(\frac{d\bar{\beta}}{d\omega}\right)^{-1} = v\sqrt{1 - \left(\frac{\lambda}{\lambda_c}\right)^2} \tag{10.66}$$

Proof: We know that for the hollow-pipe waveguide, either rectangular or circular, the propagation constant is

$$\overline{\gamma} = \sqrt{h^2 - \omega^2 \mu\varepsilon}$$

In the present case, it is given that the waveguide is lossless and the wave is inside, so the propagation constant $\overline{\gamma}$ must be pure imaginary, i.e.

$$\overline{\gamma} = \sqrt{h^2 - \omega^2 \mu\varepsilon} = j\overline{\beta}$$

And hence the phase shift constant becomes

$$\overline{\beta} = \sqrt{\omega^2 \mu\varepsilon - h^2}$$

But we have $h^2 = \omega_c^2 \mu\varepsilon$. So the phase shift constant becomes

$$\overline{\beta} = \sqrt{\omega^2 \mu\varepsilon - \omega_c^2 \mu\varepsilon}$$

$$= \omega\sqrt{\mu\varepsilon}\sqrt{1 - \left(\frac{\omega_c}{\omega}\right)^2} = \omega\sqrt{\mu\varepsilon}\sqrt{1 - \left(\frac{f_c}{f}\right)^2} = \omega\sqrt{\mu\varepsilon}\sqrt{1 - \left(\frac{\lambda}{\lambda_c}\right)^2}$$

The above expression for phase shift constant can be put in terms of wavelengths also as follows:

$$f_c = \frac{v}{\lambda_c} = \frac{1}{\sqrt{\mu\varepsilon}}\frac{1}{\lambda_c}$$

and

$$f = \frac{c}{\lambda} = \frac{1}{\sqrt{\mu_0 \varepsilon_0}}\frac{1}{\lambda}$$

$$\frac{f_c}{f} = \frac{\sqrt{\mu_0 \varepsilon_0}}{\sqrt{\mu\varepsilon}}\frac{\lambda}{\lambda_c} = \frac{1}{\sqrt{\mu_r \varepsilon_r}}\frac{\lambda}{\lambda_c}$$

Substituting in the expression for the phase shift constant, we obtain

$$\overline{\beta} = \omega\sqrt{\mu\varepsilon}\sqrt{1 - \frac{1}{\mu_r \varepsilon_r}\left(\frac{\lambda}{\lambda_c}\right)^2}$$

$$= \omega\sqrt{\mu_0 \varepsilon_0}\sqrt{\mu_r \varepsilon_r - \left(\frac{\lambda}{\lambda_c}\right)^2}$$

Phase shift constant (units are rad/m)

In the case of TE/TM wave in a guide filled with dielectric of ε and μ

$$\overline{\beta} = \omega\sqrt{\mu\varepsilon}\sqrt{1 - \left(\frac{\lambda}{\lambda_c}\right)^2} = \frac{2\pi}{\lambda_g}$$

(Contd...)

(Contd...)

$$= \left(\frac{2\pi}{\lambda}\right)\sqrt{\mu_r\varepsilon_r - \left(\frac{\lambda}{\lambda_c}\right)^2} \quad \text{in the case of dielectric-filled guide}$$

$$= \omega\sqrt{\mu_0\varepsilon_0}\sqrt{1-\left(\frac{\lambda}{\lambda_c}\right)^2} = \left(\frac{2\pi}{\lambda}\right)\sqrt{1-\left(\frac{\lambda}{\lambda_c}\right)^2} \quad \text{in the case of air-filled guide}$$

In the case of TEM wave

$$\bar{\beta} = \omega\sqrt{\mu\varepsilon} = \frac{2\pi}{\lambda} = \omega\sqrt{\mu_0\varepsilon_0\mu_r\varepsilon_r}$$

$$= \omega\sqrt{\mu_0\varepsilon_0} \quad \text{in the case of free space or air}$$

To be precise, $\bar{\gamma}$ is the propagation constant along the z-direction, i.e. along the length of the waveguide. So $\bar{\beta}$ must be the phase shift constant along the length of the waveguide. By definition, the phase shift constant is the phase shift per unit length and in this case the length has to be considered along the z-direction.

Hence,

$$\bar{\beta} = \frac{2\pi}{\lambda_g}$$

where λ_g is the wavelength along the z-direction called guide wavelength. Now,

$$\lambda_g = \frac{2\pi}{\bar{\beta}}$$

$$= \frac{2\pi}{\omega\sqrt{\mu_0\varepsilon_0}\sqrt{\mu_r\varepsilon_r - \left(\frac{\lambda}{\lambda_c}\right)^2}} = \frac{2\pi c}{2\pi f\sqrt{\mu_r\varepsilon_r - \left(\frac{\lambda}{\lambda_c}\right)^2}}$$

$$= \frac{f\lambda}{f\sqrt{\mu_r\varepsilon_r - \left(\frac{\lambda}{\lambda_c}\right)^2}} = \frac{\lambda}{\sqrt{\mu_r\varepsilon_r - \left(\frac{\lambda}{\lambda_c}\right)^2}}$$

We can arrive at an expression for guide wavelength in terms of frequencies also. We have

$$\frac{f_c}{f} = \frac{1}{\sqrt{\mu_r\varepsilon_r}}\frac{\lambda}{\lambda_c} \Rightarrow \frac{\lambda}{\lambda_c} = \frac{f_c}{f}\sqrt{\mu_r\varepsilon_r}$$

$$\lambda_g = \frac{\lambda}{\sqrt{\mu_r\varepsilon_r - \mu_r\varepsilon_r\left(\frac{f_c}{f}\right)^2}} = \frac{\dfrac{\lambda}{\sqrt{\mu_r\varepsilon_r}}}{\sqrt{1-\left(\frac{f_c}{f}\right)^2}}$$

We can arrive at an expression for guide wavelength in terms of frequencies also. We have

Guide wavelength λ_g

- Guide filled with a medium

$$\lambda_g = \frac{\lambda}{\sqrt{\mu_r\varepsilon_r - \left(\frac{\lambda}{\lambda_c}\right)^2}} = \frac{\frac{\lambda}{\sqrt{\mu_r\varepsilon_r}}}{\sqrt{1-\left(\frac{f_c}{f}\right)^2}}$$

- Guide with free space

$$\lambda_g = \frac{\lambda}{\sqrt{\mu_r\varepsilon_r - \left(\frac{\lambda}{\lambda_c}\right)^2}} = \frac{\lambda}{\sqrt{1-\left(\frac{f_c}{f}\right)^2}}$$

The phase velocity of the wave, from its basic definition, is

$$v_p = \frac{\omega}{\bar{\beta}}$$

Substituting

$$\bar{\beta} = \omega\sqrt{\mu\varepsilon}\sqrt{1-\left(\frac{\lambda}{\lambda_c}\right)^2}$$

we get

$$v_p = \frac{v}{\sqrt{1-\left(\frac{\lambda}{\lambda_c}\right)^2}}$$

Similarly from its basic definition, the group velocity of the wave can be related to free space wavelength and guide wavelength by

$$v_g = v\left[1-\left(\frac{f_c}{f}\right)^2\right]^{1/2}$$

$$= v\left[1-\frac{1}{\mu_r\varepsilon_r}\left(\frac{\lambda}{\lambda_c}\right)^2\right]^{1/2}$$

Derivation:

$$\bar{\beta} = \omega\sqrt{\mu\varepsilon}\sqrt{1-\left(\frac{f_c}{f}\right)^2} = \sqrt{\mu\varepsilon\omega}\sqrt{1-\left(\frac{\omega_c}{\omega}\right)^2}$$

$$\frac{d\beta}{d\omega} = \sqrt{\mu\varepsilon}\sqrt{1-\left(\frac{\omega_c}{\omega}\right)^2} + \sqrt{\mu\varepsilon}\,\omega\frac{1}{2}\left[1-\left(\frac{\omega_c}{\omega}\right)^2\right]^{-1/2} \frac{-2\omega_c}{\omega}\frac{-\omega_c}{\omega^2}$$

$$= \sqrt{\mu\varepsilon}\sqrt{1-\left(\frac{\omega_c}{\omega}\right)^2} + \sqrt{\mu\varepsilon}\left[1-\left(\frac{\omega_c}{\omega}\right)^2\right]^{-1/2}\left(\frac{\omega_c}{\omega}\right)^2$$

$$= \sqrt{\mu\varepsilon}\left[1-\left(\frac{\omega_c}{\omega}\right)^2+\left(\frac{\omega_c}{\omega}\right)^2\right]\left[1-\left(\frac{\omega_c}{\omega}\right)^2\right]^{-1/2}$$

$$= \sqrt{\mu\varepsilon}\left[1-\left(\frac{\omega_c}{\omega}\right)^2\right]^{-1/2} = \sqrt{\mu\varepsilon}\left[1-\left(\frac{f_c}{f}\right)^2\right]^{-1/2}$$

$$= \frac{1}{v}\left[1-\left(\frac{f_c}{f}\right)^2\right]^{-1/2}$$

Therefore, the group velocity becomes

$$v_g = v\left[1-\left(\frac{f_c}{f}\right)^2\right]^{1/2}$$

Here

$$v = \frac{c}{\sqrt{\mu_r\varepsilon_r}}$$

The group velocity can also be expressed in terms of the wavelengths. We have already calculated that

$$\frac{f_c}{f} = \frac{1}{\sqrt{\mu_r\varepsilon_r}}\frac{\lambda}{\lambda_c}$$

Therefore, the group velocity becomes

$$v_g = v\left[1-\frac{1}{\mu_r\varepsilon_r}\left(\frac{\lambda}{\lambda_c}\right)^2\right]^{1/2}$$

If the hollow region is the free space, then the group velocity becomes

$$v_g = c\left[1-\left(\frac{\lambda}{\lambda_c}\right)^2\right]^{1/2}$$

As the frequency of the wave is increased from the cut-off value to infinity., the guide wavelength and phase velocity vary from infinity to their free space value, whereas the group velocity varies from zero to its free space value.

Also note that the phase velocity and group velocity are the same for TEM wave. Neither of these two depends upon the frequency and so TEM wave is non-dispersive.

10.6 IMPEDANCES OF WAVEGUIDES

Oliver Heavyside first coined the term *impedance* in the nineteenth century to describe the complex ratio **V/I** in AC circuits. S.A. Schelkunoff extended this concept to electromagnetic fields in a systematic way to describe the ratio of electric to magnetic fields and noted that impedance should be considered as a characteristic of the type of field as well as the medium.

While dealing with waveguides, we come across two types of impedances: one is wave impedance and the other is characteristic impedance. The first impedance is borrowed from the wave theory and it is related to the ratio of electric field to magnetic field of the travelling wave, whereas the second impedance is related to the description of the power flow along the length of the guide. Both of these impedances are described hereunder.

10.6.1 Wave Impedance

Wave impedance is usually denoted by Z_z and for a waveguide, it is defined as the ratio of transverse electric field strength to transverse magnetic field strength. In the case of a rectangular waveguide, when it is lying along the z-axis, the x- and y-components of the fields are normal to the propagation direction and hence are the transverse components. Therefore, in the case of rectangular waveguides

$$Z_z = \frac{E_{\text{trans}}}{H_{\text{trans}}} = \frac{\sqrt{E_x^2 + E_y^2}}{\sqrt{H_x^2 + H_y^2}} \tag{10.67}$$

It can be shown that its value depends upon the mode of the wave travelling in the guide and its values are given by

$$Z_z = \frac{\eta}{\sqrt{1-\left(\frac{\lambda}{\lambda_c}\right)^2}} \quad \text{for TE mode wave} \tag{10.68a}$$

$$Z_z = \eta\sqrt{1-\left(\frac{\lambda}{\lambda_c}\right)^2} \quad \text{for TM mode wave} \tag{10.68b}$$

In the case of a circular waveguide, when it is lying along the z-axis, the ρ- and ϕ-components of the fields are normal to the propagation direction and hence are the transverse components. Therefore, in the case of circular waveguides

$$Z_z = \frac{E_{\text{trans}}}{H_{\text{trans}}} = \frac{\sqrt{E_\rho^2 + E_\phi^2}}{\sqrt{H_\rho^2 + H_\phi^2}} \tag{10.69}$$

It can be shown that in the case of circular guides also, its value depends upon the mode of the wave travelling in the guide and its values are the same as given in Eqs. (10.68a) and (10.68b).

Here η is intrinsic impedance and it is related to the constants of the loss-free region of the guide through $\eta = \sqrt{\mu/\varepsilon}$ ohms.

Proof: Let us first consider the rectangular waveguides. When the guide is lying along the *z*-axis, carrying a travelling wave in the TE mode, the transverse field components of the wave are related to the longitudinal components through Eqs. (10.29a–d) given by

$$E_x = -\frac{j\omega\mu}{h^2}\frac{\partial H_z}{\partial y}; \qquad E_y = +\frac{j\omega\mu}{h^2}\frac{\partial H_z}{\partial x}$$

$$H_x = -\frac{\bar{\gamma}}{h^2}\frac{\partial H_z}{\partial x}; \qquad H_y = -\frac{\bar{\gamma}}{h^2}\frac{\partial H_z}{\partial y}$$

Now the magnitude of the transverse electric component is

$$E_{\text{trans}} = \sqrt{E_x^2 + E_y^2}$$

$$= \left(\frac{j\omega\mu}{h^2}\right)\sqrt{\left[\left(\frac{\partial H_z}{\partial y}\right)^2 + \left(\frac{\partial H_z}{\partial x}\right)^2\right]}$$

The transverse magnetic component is

$$H_{\text{trans}} = \sqrt{H_x^2 + H_y^2}$$

$$= \left(\frac{\bar{\gamma}}{h^2}\right)\sqrt{\left[\left(\frac{\partial H_z}{\partial x}\right)^2 + \left(\frac{\partial H_z}{\partial y}\right)^2\right]}$$

$$\frac{E_{\text{trans}}}{H_{\text{trans}}} = \frac{\left(\dfrac{j\omega\mu}{h^2}\right)}{\left(\dfrac{\bar{\gamma}}{h^2}\right)} = \frac{j\omega\mu}{\bar{\gamma}}$$

But as the wave is in the loss-free region, the propagation constant is pure imaginary, i.e. $\bar{\gamma} = j\bar{\beta}$, resulting in

$$\frac{E_{\text{trans}}}{H_{\text{trans}}} = \frac{j\omega\mu}{\bar{\gamma}} = \frac{j\omega\mu}{j\bar{\beta}} = \frac{\omega\mu}{\bar{\beta}}$$

Therefore, with $\bar{\beta} = 2\pi/\lambda_g$, the wave impedance becomes

$$Z_z = \frac{E_{\text{trans}}}{H_{\text{trans}}} = \frac{\omega\mu}{\bar{\beta}}$$

But we know that

$$\bar{\beta} = \omega\sqrt{\mu\varepsilon}\sqrt{1-\left(\frac{\lambda}{\lambda_c}\right)^2}$$

Therefore,

$$Z_z = \frac{\omega\mu}{\omega\sqrt{\mu\varepsilon}\sqrt{1-\left(\frac{\lambda}{\lambda_c}\right)^2}}$$

$$= \sqrt{\frac{\mu}{\varepsilon}}\frac{1}{\sqrt{1-\left(\frac{\lambda}{\lambda_c}\right)^2}} = \eta\frac{1}{\sqrt{1-\left(\frac{\lambda}{\lambda_c}\right)^2}}$$

Following similar steps, it can be shown, for TM mode

$$Z_z = \frac{\bar{\beta}}{\omega\varepsilon}$$

$$Z_z = \frac{\omega\sqrt{\mu\varepsilon}\sqrt{1-\left(\frac{\lambda}{\lambda_c}\right)^2}}{\omega\varepsilon}$$

$$= \sqrt{\frac{\mu}{\varepsilon}}\sqrt{1-\left(\frac{\lambda}{\lambda_c}\right)^2} = \eta\sqrt{1-\left(\frac{\lambda}{\lambda_c}\right)^2}$$

The relations for the circular waveguides can be derived by following the procedure similar to that for the rectangular waveguides.

In general, the wave impedance is a characteristic of the particular type of wave (TEM, TE, and TM) and it usually depends upon the type of line or guide, the material and the operating frequency. This parameter of the waveguide medium can be likened, conceptually, to the intrinsic impedance of the free space medium.

10.6.2 Characteristic Impedance Z_0

The concept of characteristic impedance is brought into waveguides from the theory of transmission lines. It is defined in several different ways for a finite line in transmission line theory using line voltages and currents and power through the line. Some of them are given as under.

Voltage-current formula

$$Z_0(V, I) = \frac{V}{I} \tag{10.70a}$$

Power-current formula

$$Z_0(P, I) = \frac{2P}{II^*} \tag{10.70b}$$

Power-voltage formula

$$Z_0(P, V) = \frac{VV^*}{2P} \tag{10.70c}$$

where V and I are voltage and current, respectively and P is the power flowing over the line when extended to infinity, all representing peak phasors.

All the above definitions give the same value of the characteristic impedance when applied to a low frequency line carrying TEM wave, but different values when applied to a waveguide, which carrys TE and TM waves. The values also differ from mode to mode in waveguides. The main reason for the variation is that voltages and currents cannot be defined uniquely in waveguides.

- In case of rectangular waveguides, for dominant mode
  Voltage-current formula gives

$$Z_0(V, I) = \frac{\pi}{2}\frac{b}{a} Z_{TE} \tag{10.71a}$$

Power-current formula gives

$$Z_0(P, I) = \frac{\pi}{4} Z_0\,(V, I) \tag{10.71b}$$

Power-voltage formula gives

$$Z_0(P, V) = \frac{4}{\pi} Z_0\,(V, I) \tag{10.71c}$$

- In the case of circular waveguides, for dominant mode:
  Voltage-current formula gives

$$Z_0(V, I) = 520\frac{\lambda_g}{\lambda} \tag{10.72a}$$

Power-current formula gives

$$Z_0(P, I) = 354\frac{\lambda_g}{\lambda} \tag{10.72b}$$

Power-voltage formula gives

$$Z_0(P, V) = 764\frac{\lambda_g}{\lambda} \tag{10.72c}$$

The characteristic impedance also depends on the mode of the wave. The reason is, since voltages and impedances are not uniquely defined for TE and TM waves, their characteristic impedance cannot be unique and for such waves it can be defined in several ways.

To match the waveguide to a uniquely defined impedance, the usual practice is to use the impedance value that gives the best agreement between theory and experimental data.

10.7 ATTENUATION IN WAVEGUIDES

Till now, we have assumed loss-free conditions to simplify the analysis. The walls of the guide are assumed to be made of a perfect conductor which is loss free. The hollow region is assumed to be free space which is again loss free. But in practice, the walls of the guide are made with good but finite conductors and the hollow region is a good dielectric but not a perfect one. As a consequence, the wave while travelling through the guides undergoes some amount of attenuation which we consider in this section.

The attenuation of the wave due to a waveguide is divided into two categories: One is reflective attenuation and the other is dissipative attenuation. Quantity-wise the first one, reflective attenuation, is very large and the dissipation attenuation is very small. The reflective attenuation is due to the frequency of the wave being not high enough to get allowed into the waveguide. The dissipative attenuation, which is actually the attenuation the wave undergoes while travelling through the waveguide, has two components: one is called *dielectric loss* and the other is *conductor loss*. The dielectric loss is due to the absorption of power from the wave by the lossy dielectric region of the guide, whereas the conductor loss is due to the absorption of power from the wave by the finite conductivity walls of the guide.

10.7.1 Reflective Attenuation

When the frequency of the wave is lesser than the cut-off frequency ($f < f_c$), it cannot enter into the waveguide. This behaviour of the waveguide is described mathematically by ascribing a large amount of attenuation to the guide known as the '*reflective attenuation*'. The mechanism and important features of this attenuation are as follows:

- When a guide is excited with a wave whose frequency is lesser than its cut-off value, the electric and magnetic fields of the wave decay in the guide exponentially with distance at a very rapid rate than is accounted for by the dissipative losses.
- Moreover the rate of attenuation depends only on the ratio λ/λ_c i.e. the free space wavelength to the cut-off wavelength. This attenuation is independent of the properties of waveguide walls or loss-free region of the guide. This feature is very much unlike the wave of wavelength shorter than the cut-off wavelength, where the attenuation is dependent upon the material of the guide walls and properties of the loss-free region.
- The exact relation for attenuation per unit length in dB is

$$\alpha = \frac{54.6}{\lambda_c}\sqrt{1-\left(\frac{\lambda_c}{\lambda}\right)^2} \tag{10.73}$$

- When the wavelength is much greater than the cut-off, the above formula can be approximated giving

$$\alpha \approx \frac{54.6}{\lambda_c} \tag{10.74}$$

- These relations apply to all modes of propagation in all types of waveguides. And one important observation that can be made from these relations is, when λ/λ_c is large, the attenuation is substantially independent of frequency. In fact, waveguides are often operated at wavelengths greater than cut-off as attenuators in signal generators and in such cases they are called waveguide attenuators.

10.7.2 Dissipative Attenuation

The energy of the wave when it is travelling through the waveguide ($f > f_c$) gets absorbed by the walls due to its finite conductivity and also by the hollow region due to its non-zero conductivity resulting in *dissipative attenuation*.

The attenuation the wave undergoes while travelling through the guide due to energy absorption from the wave by the lossy dielectric region of the guide is called *dielectric loss* and it is given by

$$\alpha_d = \frac{k^2 \tan\delta}{2\overline{\beta}} \text{ Np/m} \tag{10.75}$$

Proof: Consider a perfect dielectric, i.e. with conductivity zero, then its permittivity is $\varepsilon = \varepsilon_0\varepsilon_r$.

For an imperfect dielectric, i.e. with finite conductivity, the permittivity is

$$\varepsilon = \varepsilon_0\dot{\varepsilon} = \varepsilon_0\left(\varepsilon_r - j\frac{\sigma}{\omega\varepsilon_0}\right) = \varepsilon_0\,(\varepsilon_r - j\varepsilon_r\tan\delta)$$

Here $\tan\delta$ is the loss tangent of the dielectric and it is equal to

$$\tan\delta = \frac{\sigma}{\omega\varepsilon_0\varepsilon_r}$$

If the waveguide hollow region is filled with perfect dielectric material, then its propagation constant is

$$\overline{\gamma} = j\overline{\beta} = \sqrt{h^2 - \omega^2\mu\varepsilon} = \sqrt{h^2 - \omega^2\mu_0\varepsilon_0\varepsilon_r} \quad \text{(assuming } \mu_r = 1)$$

If the waveguide hollow region is filled with imperfect dielectric material, then

$$\overline{\gamma} = \sqrt{h^2 - \omega^2\mu_0\varepsilon_0\left(\varepsilon_r - j\frac{\sigma}{\omega\varepsilon_0}\right)} = \sqrt{h^2 - \omega^2\mu_0\varepsilon_0\varepsilon_r + j\omega^2\mu_0\varepsilon_0\varepsilon_r\tan\delta}$$

(once again assuming $\mu_r = 1$)

Now let $h^2 - \omega^2\mu_0\varepsilon_0\varepsilon_r = a^2$ and $j\omega^2\mu_0\varepsilon_0\varepsilon_r\tan\delta = x^2$, then the propagation constant becomes

$$\begin{aligned}\overline{\gamma} &= \sqrt{h^2 - \omega^2\mu_0\varepsilon_0\varepsilon_r + j\omega^2\mu_0\varepsilon_0\varepsilon_r\tan\delta}\\ &= \sqrt{a^2 + x^2}\\ &= a + j\frac{1}{2}\frac{x^2}{a} \quad \text{for } x \ll a.\end{aligned}$$

Therefore, the propagation constant becomes

$$\bar{\gamma} = \sqrt{h^2 - \omega^2 \mu_0 \varepsilon_0 \varepsilon_r} + j \frac{\omega^2 \mu_0 \varepsilon_0 \varepsilon_r \tan\delta}{2\sqrt{h^2 - \omega^2 \mu_0 \varepsilon_0 \varepsilon_r}} = j\bar{\beta} + \frac{\omega^2 \mu_0 \varepsilon_0 \varepsilon_r \tan\delta}{2\bar{\beta}}$$

We know from the definition of propagation constant that its real part is the attenuation constant. Therefore, the attenuation constant due to dielectric loss is

$$\alpha_d = \frac{\omega^2 \mu_0 \varepsilon_0 \varepsilon_r \tan\delta}{2\bar{\beta}} \text{ Np/m for TE or TM wave} \qquad (10.76a)$$

In the case of TEM wave, $\omega/\bar{\beta} = \omega/\beta = v = 1/\sqrt{\mu\varepsilon}$. Using this relation, the attenuation constant can be expressed as

$$\alpha_d = \omega\sqrt{\mu_0 \varepsilon_0 \varepsilon_r}\,\frac{\tan\delta}{2} \text{ Np/m} \qquad (10.76b)$$

It can be observed that the dielectric loss becomes nil when the conductivity or loss tangent of the dielectric is zero. The above relations for the attenuation constant due to the dielectric loss are most general and can be used with both the rectangular and circular waveguides.

The wave also undergoes some amount of attenuation while travelling through the guide due to energy absorption by the conducting walls of the guide. The attenuation due to the conductor loss is given by

$$\alpha_c = \frac{\text{Power dissipated/unit length}}{2 \times \text{Power flow down the waveguide}} \text{ Np/m} \qquad (10.77)$$

Note that if the walls are made with perfect conductors, this type of attenuation would not have been there. It is due to the finite conductivity of the walls of the waveguide.

Proof: To derive the above formula, we need to look into the transmission line theory. Consider a finite length transmission line and let us suppose the voltage and current phasors along the line when it is extended to infinity are

$$V = V_0 e^{-\alpha z} e^{-j\beta z} \quad \text{and} \quad I = I_0 e^{-\alpha z} e^{-j\beta z}$$

Then the average power transmitted is

$$P_{\text{av}} = \frac{1}{2}\text{Re}\,(VI^*) = \frac{1}{2}\text{Re}\,(V_0 I_0^*)\,e^{-2\alpha z}$$

The rate of decrease of the transmitted power along the line will be

$$-\frac{\partial P_{\text{av}}}{\partial z} = +2\alpha P_{\text{av}}$$

The decrease of the transmitted power per unit length of line is

$$-\Delta P_{\text{av}} = 2\alpha P_{\text{av}}$$

And this is the power dissipated per unit length. Solving the above equation for the attenuation constant, we obtain

$$2\alpha = \frac{-\Delta P_{\text{av}}}{P_{\text{av}}} = \frac{\text{Power lost per unit length}}{\text{Power transmitted}}$$

It leads to the expression for attenuation constant

$$\alpha = \frac{\text{Power lost per unit length}}{2 \times \text{Power transmitted}}$$

This relation is the most general one and it can be used with both the waveguides: rectangular and circular.

In the case of rectangular guides, the power lost per unit length is equal to the power dissipated by four walls of the guide. Out of four walls, the top and bottom are identical, so they consume equal amount of power. Similarly the side walls are identical, so each one consumes equal amount of power. The total power dissipated or lost in unit length of the guide, therefore, is twice the sum of the powers absorbed in unit length of side wall and broader wall. It is also equal to twice the sum of the powers absorbed in the side wall of area b and broader wall of area a.

We know, from the basics of electromagnetic theory, if the surface resistance is R_s, the power dissipated per unit area of the slab is $R_s K^2/2$, where K is the surface current density and is equal to $K = H_{\tan}$ over the surface. Then the power dissipated in both the narrow walls of unit length and width b becomes

$$P_{nw} = 2 \times \frac{R_s}{2} \int_{\substack{\text{over narrow} \\ \text{wall of area } b}} H_{\tan}^2 \, da$$

In the case of narrow walls lying parallel to the yz-plane, the y- and z-components of field are tangential, therefore, resulting in

$$H_{\tan}^2 = H_y^2 + H_z^2$$

And the differential area $da = dydz$. With these substitutions,

$$P_{nw} = 2 \times \frac{R_s}{2} \int_{\substack{\text{over narrow} \\ \text{wall of area } b}} [H_y^2 + H_z^2] dydz$$

$$= 2 \times \frac{R_s}{2} \int_{\text{area } b} [H_y^2 + H_z^2]_{\text{at } x=0} \, dydz$$

$$= 2 \times \frac{R_s}{2} \int_0^1 \int_0^b [H_y^2 + H_z^2]_{\text{at } x=0} \, dydz \tag{10.78a}$$

Similarly, the power dissipated in both the broader walls of unit length within a width a becomes

$$P_{nw} = 2 \times \frac{R_s}{2} \int_0^1 \int_0^a [H_z^2 + H_x^2]_{\text{at } y=0}\, dxdz \tag{10.78b}$$

Note that in arriving at the above expressions, the integration is performed over the walls which are located at $x = 0$ and $y = 0$. Now the total power lost per unit length is the sum of Eqs. (10.78a) and (10.78b). Therefore, the power lost per unit length is

$$P_{\text{lost}} = 2 \times \frac{R_s}{2} \left[\int_0^1 \int_0^b [H_y^2 + H_z^2]_{\text{at } x=0}\, dydz + \int_0^1 \int_0^a [H_z^2 + H_x^2]_{\text{at } y=0}\, dxdz \right] \tag{10.79}$$

By following similar steps, for circular waveguides with an inner radius of a, it can be shown that the power lost per unit length

$$P_{\text{lost}} = \frac{R_s}{2} \int |H_{\tan}|^2\, da$$

$$= \frac{R_s}{2} \int_0^1 \int_0^{2\pi} [H_\phi^2 + H_z^2]_{\text{at } \rho=a}\, \rho d\phi dz \tag{10.80}$$

The power flow down the guide can be computed using the Poynting vector. In the case of waveguides lying along the z-axis, it is given by

$$P_{\text{trans}} = \frac{1}{2} \text{Re} \int \mathbf{E} \times \mathbf{H}^* \cdot \hat{\mathbf{z}} da \tag{10.81}$$

In the case of rectangular waveguides, lying along the z-axis, it becomes

$$P_{\text{trans}} = \frac{1}{2} \text{Re} \iint (E_x H_y^* - E_y H_x^*) dxdy = \frac{1}{2} Z_z \int |H_{\text{trans}}|^2\, dxdy \tag{10.82}$$

In the case of circular waveguides, lying along the z-axis, it becomes

$$P_{\text{trans}} = \frac{1}{2} \text{Re} \iint (E_\rho H_\phi^* - E_\phi H_x^*) \rho d\rho d\phi = \frac{1}{2} Z_z \int |H_{\text{trans}}|^2\, \rho d\rho d\phi \tag{10.83}$$

10.8 MICROSTRIP TRANSMISSION LINES

Microstrip transmission lines are one of the most popular types of planar transmission lines. They consist of a conductive ribbon attached to a dielectric sheet with conductive backing. As they can be easily integrated with active as well as passive microwave devices, they are widely used in both the microwave technologies. Because such lines are easily fabricated by photolithographic processes, they have economic and technical merits. The structure of a microstrip line is shown in Figure 10.9.

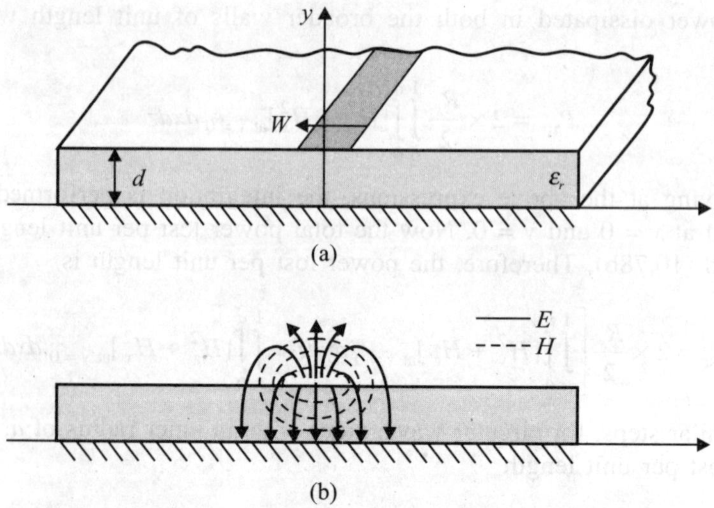

FIGURE 10.9 Microstrip lines: (a) geometry and (b) flux lines.

It can be viewed as a two-wire line consisting of two flat strip conductors of width w separated by a distance $2d$ in the absence of the dielectric. The microstrip line cannot support a pure TEM wave. In most of the practical applications, the dielectric substrate is electrically very thin, i.e. $d \ll \lambda$ and so the fields are quasi-TEM. The microstrip line has the following features:

- The phase velocity $v_p = c/\sqrt{\varepsilon_e}$.
- The propagation constant $\beta = \omega\sqrt{\mu_0 \varepsilon_0}\sqrt{\varepsilon_e}$.
- The microstrip line has most of its field lines in the dielectric region and some fraction are in the air region above the substrate.
- It is an asymmetrical type strip transmission line.

10.8.1 Shielded Strip Lines

The features of the shielded strip line are as follows:

- Planar type transmission line is widely used in microwave integrated circuitry. It is a sort of 'flattened out' coaxial line.
- It has two conductors and a homogeneous dielectric. It supports TEM wave and this is the usual mode of operation.
- The strip line can also support higher-order TM and TE modes but these are suppressed with shorting screws between the ground planes and restricting b to less than $\lambda/4$.
- The analysis of the strip line is a quite complex and difficult process. With the application of the Laplace equation, a reasonable approximation to the exact results can be obtained.

The structure of strip line transmission line is shown in Figure 10.10.

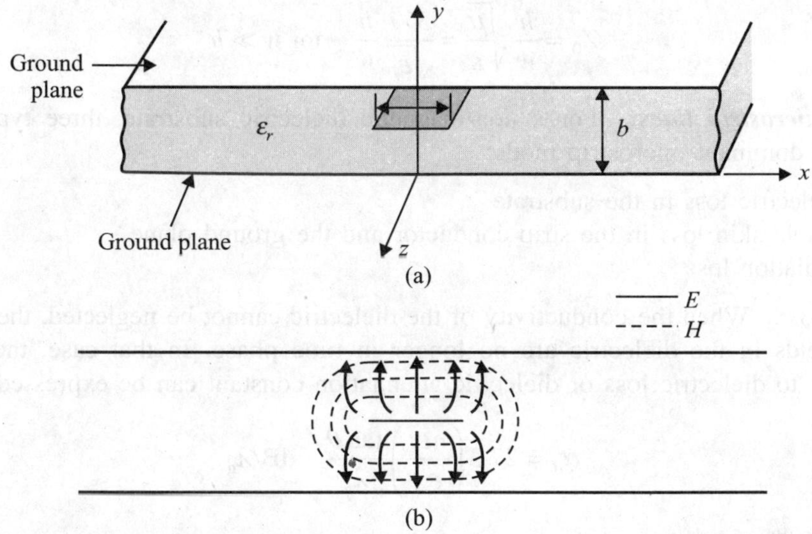

FIGURE 10.10 Strip line transmission line: (a) geometry and (b) flux lines.

10.8.2 Important Parameters

Effective dielectric constant ε_{re}**:** The effective relative dielectric constant for a microstrip line can be related to the relative dielectric constant of the board material. The empirical relation is expressed as

$$\varepsilon_{re} = 0.475\varepsilon_r + 0.67 \tag{10.84}$$

where ε_r is the relative dielectric constant of the board material and ε_{re} is the effective relative dielectric constant for a microstrip line.

Characteristic impedance Z_0**:** The characteristic impedance of a microstrip line is a function of the strip line width, the strip line thickness, the distance between the line and the ground plane and the homogeneous dielectric constant of the board material. Several different methods for determining the characteristic impedance of the microstrip line have been developed. The field equation method and the comparative method are two well-known methods widely in use.

The characteristic impedance of a narrow microstrip line is given by

$$Z_0 = \frac{87}{\sqrt{\varepsilon_r + 1}} \ln\left[\frac{5.98h}{0.8w + t}\right] \quad \text{for } h < 0.8w \tag{10.85}$$

where

ε_r = relative dielectric constant of the board material
h = height from the microstrip line to the ground
w = width of the microstrip line
t = thickness of the microstrip line

For a wide microstrip line, the characteristic impedance is given by

$$Z_0 = \frac{h}{w}\sqrt{\frac{\mu}{\varepsilon}} = \frac{377}{\sqrt{\varepsilon_r}}\frac{h}{w} \quad \text{for } w \gg h \tag{10.86}$$

Losses in microstrip lines: For a non-magnetic dielectric substrate, three types of losses occur in the dominant microstrip mode:

1· Dielectric loss in the substrate
2· Ohmic skin loss in the strip conductor and the ground plane
3· Radiation loss

Dielectric loss: When the conductivity of the dielectric cannot be neglected, the electric and magnetic fields in the dielectric are no longer in time phase. In that case, the attenuation constant due to dielectric loss or dielectric attenuation constant can be expressed as

$$\alpha_d = 27.3\left(\frac{q\varepsilon_r}{\varepsilon_{re}}\right)\frac{\tan\theta}{\lambda_g} \quad \text{dB}/\lambda_g \tag{10.87}$$

where $\lambda_g = \dfrac{\lambda_0}{\sqrt{\varepsilon_{re}}}$ and λ_0 is the wavelength in free space.

The dielectric loss or dielectric attenuation per unit wavelength is independent of frequency. Moreover, if the substrate conductivity is independent of frequency, as for semiconductor, the dielectric attenuation per unit length is also independent of frequency.

Ohmic skin loss: In a microstrip line over a low-loss dielectric substrate, the predominant sources of losses at microwave frequencies are the non-perfect conductors. The current density in the conductors of a microstrip line is concentrated in a sheet that is approximately a skin-depth thick inside the conductor surface and exposed to the electric field. The current density in the strip conductor and the ground conductor is not uniform in the transverse plane. The microstrip conductor contributes the major part of the ohmic loss.

Because of mathematical complexity, exact expressions for the current density of a microstrip line with non-zero thickness have never been derived. For simplicity, if the current distribution is assumed uniform and equal to I/w in both the conductors and confined to the region $|x| < w/2$, then the conducting attenuation constant of a wide microstrip line is given by

$$\alpha_c \simeq \frac{8.686R_s}{Z_0 w} \text{ dB/cm} \quad \text{for } \frac{w}{h} > 1 \tag{10.88}$$

where $R_s = \sqrt{\dfrac{\pi f \mu}{\sigma}}$ is the surface skin resistance in Ω/square.

For a narrow microstrip line with $w/h < 1$, the above equation is not valid because of the non-uniform current distribution.

Radiation loss: The radiation loss depends on the substrate's thickness and dielectric constant as well as its geometry. Lewin's results show that the ratio of radiated power to total dissipated power for an open-circuited microstrip line is given by

$$\frac{P_{rad}}{P_t} = \frac{R_r}{Z_0} \tag{10.89}$$

where Z_0 is the characteristic impedance of the line and R_r is the radiation resistance of an open-circuited microstrip line. It shows that the radiation loss decreases when the characteristic impedance increases. For lower dielectric constant substrates, radiation is significant at higher impedance levels. For higher dielectric constant substrates, radiation becomes significant until very low impedance levels are reached.

Quality factor Q of microstrip line: Many microwave integrated circuits require very high quality resonant circuits. The quality factor of a microstrip line is very high but it is limited by the radiation losses of the substrates and with low dielectric constant. For a wide microstrip line,

- the quality factor Q_c due to ohmic losses has been found to be $Q_c = 0.63h\sqrt{\sigma f}_{\text{GHz}}$, where σ is the conductivity of the dielectric substrate board in ℧/m
- the quality factor Q_d due to dielectric losses has been deducted as $Q_d \simeq 1/\tan\theta$, where $\tan\theta$ is the dielectric loss tangent.

Waveguide formulae

Cut-off frequency (RWG):

$$f_c = \frac{1}{2\sqrt{\mu\varepsilon}}\sqrt{\left(\frac{m}{a}\right)^2 + \left(\frac{n}{b}\right)^2}$$

Cut-off wavelength (RWG):

$$\lambda_c = \frac{2}{\sqrt{\left(\frac{m}{a}\right)^2 + \left(\frac{n}{b}\right)^2}}$$

$$\frac{1}{\lambda_c^2} = \left(\frac{m}{2a}\right)^2 + \left(\frac{n}{2b}\right)^2$$

Cut-off frequency (CWG):

$$f_c = \frac{h_{nm}}{2\pi\sqrt{\mu\varepsilon}} \quad \text{for TM wave}$$

$$f_c = \frac{h'_{nm}}{2\pi\sqrt{\mu\varepsilon}} \quad \text{for TE wave}$$

(Contd...)

(*Contd...*)

Cut-off wavelength (CWG):

$$\lambda_c = \frac{2\pi a}{(ha)_{nm}} \quad \text{for TM wave}$$

$$\lambda_c = \frac{2\pi a}{(ha)'_{nm}} \quad \text{for TE wave}$$

Guide wavelength:

$$\lambda_g = \frac{\lambda}{\sqrt{\varepsilon_r \mu_r - \left(\frac{\lambda}{\lambda_c}\right)^2}}$$

$$\frac{1}{\lambda_g^2} + \frac{1}{\lambda_c^2} = \frac{1}{\lambda^2}$$

Phase velocity:

$$v_p = \left(\frac{\omega}{\beta}\right) = \frac{v}{\sqrt{1 - \left(\frac{\lambda}{\lambda_c}\right)^2}}$$

Group velocity:

$$v_g = \left(\frac{d\beta}{d\omega}\right)^{-1} = v\sqrt{1 - \left(\frac{\lambda}{\lambda_c}\right)^2}$$

Waveguide impedance:

$$Z_z = \frac{\eta}{\sqrt{1 - \left(\frac{\lambda}{\lambda_c}\right)^2}} \quad \text{for TE mode wave}$$

$$= \eta\sqrt{1 - \left(\frac{\lambda}{\lambda_c}\right)^2} \quad \text{for TM mode wave}$$

Waveguide impedance:

$$Z_0(V, I) = \frac{\pi}{2}\frac{b}{a} Z_{TE} = 520\frac{\lambda_g}{\lambda}$$

Attenuation:

$$\alpha = \frac{54.6}{\lambda_c}\sqrt{1 - \left(\frac{\lambda_c}{\lambda}\right)^2}$$

$$\alpha_d = \frac{k^2 \tan\delta}{2\beta} \text{ Np/m}$$

$$\alpha_c = \frac{\text{Power lost per unit length}}{2 \times \text{Power transmitted}}$$

SOLVED PROBLEMS

1. Find the four lowest cut-off frequencies for an air-filled rectangular waveguide when $a/b = 2$ with $a = 4$ cm. Also find the modes that can be used to transmit 8GHz.

Solution: The cut-off frequency for mode with mode numbers m and n is

$$f_{c,mn} = \frac{1}{2\sqrt{\mu\varepsilon}}\sqrt{\left(\frac{m}{a}\right)^2 + \left(\frac{n}{b}\right)^2}$$

$$= \frac{1}{2a\sqrt{\mu\varepsilon}}\sqrt{m^2 + \left(\frac{na}{b}\right)^2}$$

For dominant mode $m = 1$ and $n = 0$, it is

$$f_{c,10} = \frac{1}{2a\sqrt{\mu\varepsilon}}$$

Therefore, the general expression for cut-off frequency is

$$f_{c,mn} = f_{c,10}\sqrt{m^2 + \left(\frac{na}{b}\right)^2}$$

Consider TE mode with $mn = 10, 01, 11, 12, 21$. The cut-off frequencies for various modes when $a/b = 2$ with $a = 4$ cm are

$$f_{c,10} = \frac{1}{2a\sqrt{\mu\varepsilon}} = \frac{3\times10^{10}}{2\times4} = 3.75\text{ GHz}$$

$$f_{c,01} = f_{c,10}\sqrt{m^2 + \left(\frac{na}{b}\right)^2} = 2f_{c,10}$$

$$f_{c,11} = f_{c,10}\sqrt{m^2 + \left(\frac{na}{b}\right)^2} = f_{c,10}\sqrt{1+4} = 2.23f_{c,10}$$

$$f_{c,20} = f_{c,10}\sqrt{m^2 + \left(\frac{na}{b}\right)^2} = 2f_{c,10}$$

$$f_{c,02} = f_{c,10}\sqrt{m^2 + \left(\frac{na}{b}\right)^2} = 4f_{c,10}$$

$$f_{c,12} = f_{c,10}\sqrt{m^2 + \left(\frac{na}{b}\right)^2} = f_{c,10}\sqrt{1+16} = 4.12f_{c,10}$$

$$f_{c,21} = f_{c,10}\sqrt{m^2 + \left(\frac{na}{b}\right)^2} = f_{c,10}\sqrt{4+4} = 2.8 f_{c,10}$$

Consider TM mode with mn = 11, 12, 21 (mn = 00, 01, 10 are not allowed for TM mode). The cut-off frequencies for various modes when $a/b = 2$ with $a = 4$ cm are

$$f_{c,11} = f_{c,10}\sqrt{m^2 + \left(\frac{na}{b}\right)^2} = f_{c,10}\sqrt{1+4} = 2.23 f_{c,10}$$

$$f_{c,12} = f_{c,10}\sqrt{m^2 + \left(\frac{na}{b}\right)^2} = f_{c,10}\sqrt{1+16} = 4.12 f_{c,10}$$

$$f_{c,21} = f_{c,10}\sqrt{m^2 + \left(\frac{na}{b}\right)^2} = f_{c,10}\sqrt{4+4} = 2.8 f_{c,10}$$

By comparing the above values, the modes with four lowest cut-off frequencies are TE_{01}, TE_{20}, TE_{11}, TM_{11}.

Now we have to find the modes that can be used to transmit the frequency 8 GHz. We can use any one of the modes whose cut-off frequencies are less than 8 GHz or $8f_{c,10}/3.75 = 2.13 f_{c,10}$. Therefore, only two modes TE_{01} and TE_{20}, apart from the dominant mode, are possible modes to transmit this frequency.

2. Consider a rectangular waveguide with $a/b = 2$, $a = 2.29$ cm, $\mu_r = 1$ and $\varepsilon_r = 2.5$. Find out the cut-off wavelength and cut-off frequency for TE_{10}, TE_{20} and TM_{11} modes.

Solution: The cut-off wavelength for mode with mode numbers m and n is

$$\lambda_{c,mn} = \frac{2a}{\sqrt{m^2 + \left(\frac{na}{b}\right)^2}}$$

$\lambda_{c,10} = 2a = 2 \times 2.29 = 4.58$ cm in the case of TE_{10} mode

$\lambda_{c,20} = a = 2.29$ cm in the case of TE_{20} mode

$$\lambda_{c,11} = \frac{2ab}{\sqrt{a^2+b^2}} = \frac{2 \times 2.29 \times 1.145}{\sqrt{2.29^2 + 1.145^2}} = 2.04 \text{ cm in the case of } TM_{11} \text{ mode}$$

The cut-off frequency for mode with mode numbers m and n is

$$f_c = \frac{1}{2\sqrt{\mu\varepsilon}}\sqrt{\left(\frac{m}{a}\right)^2 + \left(\frac{n}{b}\right)^2}$$

$$f_{c,mn} = \frac{1}{2a\sqrt{\mu\varepsilon}}\sqrt{m^2 + \left(\frac{na}{b}\right)^2}$$

But

$$\mu\varepsilon = \mu_0\mu_r\varepsilon_0\varepsilon_r = \mu_0\varepsilon_0\mu_r\varepsilon_r$$

$$= \frac{1}{2a\sqrt{\mu\varepsilon}}\sqrt{m^2 + \left(\frac{na}{b}\right)^2} = \frac{1}{2a\sqrt{\mu_0\varepsilon_0}\sqrt{\mu_r\varepsilon_r}}\sqrt{m^2 + \left(\frac{na}{b}\right)^2}$$

$$= \frac{3\times 10^{10}}{2a\sqrt{\mu_r\varepsilon_r}}\sqrt{m^2 + \left(\frac{na}{b}\right)^2}$$

$$f_{c,10} = \frac{3\times 10^{10}\, m}{2a\sqrt{\mu_r\varepsilon_r}} = \frac{3\times 10^{10}}{2\times 2.29\sqrt{2.5}} = 4.14 \text{ GHz in the case of } TE_{10} \text{ mode}$$

$$f_{c,20} = \frac{3\times 10^{10}\, m}{2a\sqrt{\mu_r\varepsilon_r}} = \frac{3\times 10^{10}\times 2}{2\times 2.29\sqrt{2.5}} = 8.28 \text{ GHz in the case of } TE_{20} \text{ mode}$$

$$f_{c,11} = \frac{3\times 10^{10}}{2a\sqrt{\mu_r\varepsilon_r}}\sqrt{1+(2)^2} = \frac{3\times 10^{10}\times\sqrt{5}}{2\times 2.29\sqrt{2.5}} = 9.25 \text{ GHz in the case of } TM_{11} \text{ mode}$$

3. Consider a rectangular waveguide with $a/b = 2$, $a = 5.817$ cm, $\mu_r = 1$ and $\varepsilon_r = 2.5$. Assuming dominant mode propagation, calculate the guide wavelength and characteristic impedance at 3.0 GHz.

Solution: The cut-off wavelength for mode with mode numbers $m = 1$ and $n = 0$ is

$$\lambda_{c,mn} = \frac{2a}{\sqrt{m^2 + \left(\frac{na}{b}\right)^2}}$$

$$\lambda_{c,10} = 2a = 2\times 5.817 = 11.634 \text{ cm in the case of } TE_{10} \text{ mode}$$

The free space wavelength is $\lambda = 3\times 10^{10}/3\times 10^9 = 10$ cm
The guide wavelength then becomes

$$\lambda_g = \frac{\lambda}{\sqrt{\varepsilon_r\mu_r - \left(\frac{\lambda}{\lambda_c}\right)^2}}$$

$$= \frac{10}{\sqrt{2.5 - \left(\frac{10}{11.634}\right)^2}} = 7.535 \text{ cm}$$

The characteristic impedance is

$$Z_0(V, I) = \frac{\pi}{2}\frac{b}{a} Z_{TE} = 377\frac{b}{a}\sqrt{\frac{\mu_r}{\varepsilon_r}}\frac{\lambda_g}{\lambda}$$

$$= 377\frac{1}{2}\sqrt{\frac{1}{2.5}}\frac{7.535}{10} = 89.83\ \Omega$$

4. Consider a rectangular waveguide with $a/b = 2$, $a = 3.0$ cm, $\mu_r = 1$ and $\varepsilon_r = 2.25$. Assuming dominant mode propagation,

(a) Find out the attenuation constant in dB/cm at 2.5 GHz for the dielectric filled guide.

(b) Also find the total attenuation at frequencies much less than the cut-off frequencies.

Solution:

(a) The cut-off frequency in the case of dominant mode is

$$\frac{3\times 10^{10} m}{2a\sqrt{\mu_r \varepsilon_r}} = \frac{3\times 10^{10}}{2\times 3\sqrt{2.25}} = 3.33\text{ GHz}$$

At 2.5 GHz frequency, the wave is below the cut-off and therefore the attenuation is given by

$$\alpha = \frac{54.6}{\lambda_c}\sqrt{1-\left(\frac{\lambda_c}{\lambda}\right)^2}$$

(b) The cut-off wavelength is

$$\lambda_{c,10} = 2a = 2 \times 3 = 6\text{ cm}$$

$$\alpha = \frac{54.6}{6}\sqrt{1-\left(\frac{2.5}{3.33}\right)^2} = 6.01\text{ dB/cm}$$

When the frequency of the wave is well below the cut-off, then

$$\alpha = \frac{54.6}{\lambda_c} = \frac{54.6}{6} = 9.1\text{ dB/cm}$$

5. Find the cut-off frequencies for TM_{12} mode in an air-filled waveguide whose dimensions a and b respectively are

(a) 1.0 × 0.5 cm,

(b) 0.5 × 1.0 cm,

(c) 1.0 × 1.0 cm and

(d) 10 × 10 cm.

Solution: The cut-off frequency for mode with mode numbers *m* and *n* is

$$f_c = \frac{1}{2\sqrt{\mu\varepsilon}}\sqrt{\left(\frac{m}{a}\right)^2 + \left(\frac{n}{b}\right)^2}$$

$$f_{c,mn} = \frac{1}{2a\sqrt{\mu\varepsilon}}\sqrt{m^2 + \left(\frac{na}{b}\right)^2}$$

$$f_{c,12} = \frac{3\times 10^{10}}{2a}\sqrt{1+\left(\frac{2a}{b}\right)^2}$$

$$= \frac{3\times 10^{10}}{2}\sqrt{1+(2\times 2)^2} = 6.18\times 10^{10} \text{ Hz in the case of (a) } 1.0 \times 0.5 \text{ cm}$$

$$= \frac{3\times 10^{10}}{2\times 0.5}\sqrt{1+(2\times 0.5)^2} = 4.24\times 10^{10} \text{ Hz in the case of (b) } 0.5 \times 1.0 \text{ cm}$$

$$= \frac{3\times 10^{10}}{2}\sqrt{1+(2\times 1)^2} = 3.54\times 10^{10} \text{ Hz in the case of (c) } 1.0 \times 1.0 \text{ cm}$$

$$= \frac{3\times 10^{10}}{2\times 10}\sqrt{1+(2\times 1)^2} = 0.354\times 10^{10} \text{ Hz in the case of (d) } 10 \times 10 \text{ cm}$$

6. For an air-filled waveguide with dimensions 2.290 × 1.02 cm, determine the cut-off frequencies for TE_{11}, TE_{12} and TE_{21} modes.

Solution: The cut-off frequency for mode with mode numbers *m* and *n* is

$$f_c = \frac{1}{2\sqrt{\mu\varepsilon}}\sqrt{\left(\frac{m}{a}\right)^2 + \left(\frac{n}{b}\right)^2}$$

$$f_{c,mn} = \frac{1}{2a\sqrt{\mu\varepsilon}}\sqrt{m^2 + \left(\frac{na}{b}\right)^2}$$

Now,

$$f_{c,11} = \frac{3\times 10^{10}}{2\times 2.290}\sqrt{1+\left(\frac{2.290}{1.02}\right)^2} = 16 \text{ GHz}$$

$$f_{c,12} = \frac{3\times 10^{10}}{2\times 2.290}\sqrt{1+\left(\frac{2\times 2.290}{1.02}\right)^2} = 30.01 \text{ GHz}$$

$$f_{c,21} = \frac{3\times 10^{10}}{2\times 2.290}\sqrt{2^2+\left(\frac{2.290}{1.02}\right)^2} = 19.7 \text{ GHz}$$

7. Determine the cut-off wavelength and wave impedance in the case of RWG with dimensions of 5 × 2 cm propagating TM_{11} mode at $f = 9$ GHz.

Solution: The cut-off wavelength for mode with mode numbers m and n is

$$\lambda_{c,mn} = \frac{2a}{\sqrt{m^2 + \left(\frac{na}{b}\right)^2}}$$

$$\lambda_{c,11} = \frac{2 \times 5}{\sqrt{1^2 + \left(\frac{5}{2}\right)^2}} = 3.71 \text{ cm}$$

The wave impedance for TM wave mode

$$Z_z = \eta\sqrt{1 - \left(\frac{\lambda}{\lambda_c}\right)^2}$$

It requires free space wavelength and intrinsic impedance.
The free space wavelength is

$$\lambda = \frac{3 \times 10^{10}}{9 \times 10^9} = 3.33 \text{ cm}$$

The intrinsic impedance is $\eta = 120\pi\ \Omega$
Therefore, the wave impedance is

$$Z_z = \eta\sqrt{1 - \left(\frac{\lambda}{\lambda_c}\right)^2}$$

$$= 120\pi\sqrt{1 - \left(\frac{3.33}{3.71}\right)^2} = 165.5\ \Omega$$

8. A rectangular waveguide has a cross section of 1.5 × 0.8 cm, $\sigma = 0$, $\mu = \mu_0$ and $\varepsilon = 4\varepsilon_0$. The magnetic field component is given by

$$H_x = 2\sin\left(\frac{\pi x}{a}\right)\cos\left(\frac{3\pi y}{b}\right)\sin(\pi \times 10^{11} t - \beta z) \text{ A/m}$$

Determine the mode of operation, cut-off frequency, phase shift constant, propagation constant and the wave impedance.

Solution: Observing the field expression, $m = 1$ and $n = 3$.
The cut-off frequency is

$$f_{c,mn} = \frac{1}{2a\sqrt{\mu\varepsilon}}\sqrt{m^2 + \left(\frac{na}{b}\right)^2}$$

Given that $\mu = \mu_0$ and $\varepsilon = 4\varepsilon_0$. Therefore,

$$f_{c,13} = \frac{3\times10^{10}}{2\times2\times1.5}\sqrt{1+\left(\frac{3\times1.5}{0.8}\right)^2} = 2.85 \text{ GHz}$$

The angular frequency is
$\omega = \pi 10^{11}$ rad/s and f = 50 GHz
Therefore, the phase shift constant is

$$\bar{\beta} = \omega\sqrt{\mu\varepsilon}\sqrt{1-\left(\frac{\lambda}{\lambda_c}\right)^2}$$

$$= \frac{\pi\times10^{11}}{1.5\times10^{10}}\sqrt{1-\left(\frac{2.85}{50}\right)^2} = 20.9 \text{ rad/cm}$$

The propagation constant is

$$\bar{\gamma} = j\bar{\beta} - j20.9$$

The wave impedance is

$$Z_z = \frac{\eta}{\sqrt{1-\left(\frac{\lambda}{\lambda_c}\right)^2}} \quad \text{for TE mode wave}$$

$$= \eta\sqrt{1-\left(\frac{\lambda}{\lambda_c}\right)^2} \quad \text{for TM mode wave}$$

In this case $\eta = \sqrt{\mu/\varepsilon} = 60\pi$

$$Z_z = \frac{60\pi}{\sqrt{1-\left(\frac{2.85}{50}\right)^2}} = 188.87\ \Omega \quad \text{for TE mode wave}$$

$$= 60\pi\sqrt{1-\left(\frac{2.85}{50}\right)^2} = 188.18\ \Omega \quad \text{for TM mode wave}$$

9. An air-filled circular waveguide has radius of 5 cm and acts as being operated at 3 GHz. Find its cut-off frequency, cut-off wavelength, guide wavelength and phase velocity for TE_{11} and TM_{01}.

Solution: For mode TE_{11}

$$h'_{11} = \frac{(ha)'_{11}}{a} = \frac{1.84}{5\times10^{-2}} = 0.368\times10^2$$

$$f_c = \frac{h'_{nm}}{2\pi\sqrt{\mu\varepsilon}}$$

$$= \frac{0.368 \times 10^2 \times 3 \times 10^8}{2\pi} = 1.757 \text{ GHz}$$

$$\lambda_c = \frac{c}{f_c} = \frac{3 \times 10^8}{1.757 \times 10^9} = 17.07 \text{ cm}$$

$$\lambda_g = \frac{\lambda}{\sqrt{1 - \left(\frac{f_c}{f}\right)^2}} = \frac{\frac{3 \times 10^{10}}{3 \times 10^9}}{\sqrt{1 - \left(\frac{1.757 \times 10^9}{3 \times 10^9}\right)^2}} = 12.34 \text{ cm}$$

$$v_p = \frac{c}{\sqrt{1 - \left(\frac{f_c}{f}\right)^2}} = \frac{3 \times 10^8}{\sqrt{1 - \left(\frac{1.757 \times 10^9}{3 \times 10^9}\right)^2}} = 3.7 \times 10^8 \text{ m/s}$$

For mode TM_{01}

$$h'_{11} = \frac{(ha)'_{11}}{a} = \frac{2.40}{5 \times 10^{-2}} = 0.48 \times 10^2$$

$$f_c = \frac{h_{nm}}{2\pi\sqrt{\mu\varepsilon}}$$

$$= \frac{0.48 \times 10^2 \times 3 \times 10^8}{2\pi} = 2.292 \text{ GHz}$$

$$\lambda_c = \frac{c}{f_c} = \frac{3 \times 10^8}{2.292 \times 10^9} = 13.08 \text{ cm}$$

$$\lambda_g = \frac{\lambda}{\sqrt{1 - \left(\frac{f_c}{f}\right)^2}} = \frac{3 \times 10^8}{\sqrt{1 - \left(\frac{2.292 \times 10^9}{3 \times 10^9}\right)^2}} = 4.65 \times 10^8 \text{ m/s}$$

10. An air-filled circular waveguide with an inner radius of 1.2 cm is operating in TM_{01} mode. Determine its cut-off frequency. If it is operating at a frequency of 10 GHz, then find the wavelength in the waveguide.

Solution: The cut-off frequency of the guide when working in TM mode is

$$f_c = \frac{h_{nm}}{2\pi\sqrt{\mu\varepsilon}}$$

with $h_{nm} = (ha)_{nm}/a$. Therefore, with $a = 1.2$ cm, $h_{01} = 2.405/1.15 \times 10^{-2}$, the cut-off frequency becomes

$$f_c = \frac{2.405 \times 3 \times 10^8}{2\pi \times 1.15 \times 10^{-2}} = 9.57 \text{ GHz}$$

The wavelength in guide or guide wavelength is given by

$$\lambda_g = \frac{\lambda}{\sqrt{1 - \left(\frac{f_c}{f}\right)^2}}$$

At 10 GHz, the wavelength is $\lambda = 3 \times 10^8/10 \times 10^9 = 3 \times 10^{-2}$ m. Substituting these values in the above expression, we get

$$\lambda_g = \frac{3 \times 10^{-2}}{\sqrt{1 - \left(\frac{9.57 \times 10^9}{10 \times 10^9}\right)^2}} = 10.34 \text{ cm}$$

11. An X-band waveguide filled with a dielectric with $\varepsilon_r = 2.25$ is operating at $f = 9$ GHz. Calculate the phase and group velocities in waveguide.

Solution: In the present case

$$v = \frac{3 \times 10^8}{\sqrt{2.25}} = 2 \times 10^8 \text{ m/s}$$

$$f_c = \frac{1}{2\sqrt{\mu\varepsilon}} \sqrt{\left(\frac{m}{a}\right)^2 + \left(\frac{n}{b}\right)^2}$$

Given that it is an X-band waveguide. Therefore, its dimensions are 2.286 × 1.016 cm. Assuming dominant mode propagation, the cut-off frequency becomes

$$f_{c,10} = \frac{3 \times 10^{10}}{2 \times 2.286\sqrt{2.25}} = 4.37 \text{ GHz}$$

$$v_p = \left(\frac{\omega}{\beta}\right) = \frac{v}{\sqrt{1 - \left(\frac{\lambda}{\lambda_c}\right)^2}} = \frac{v}{\sqrt{1 - \left(\frac{f_c}{f}\right)^2}}$$

$$= \frac{2 \times 10^8}{\sqrt{1 - \left(\frac{4.37}{9}\right)^2}} = 2.28 \times 10^8 \text{ m/s}$$

$$v_g = \left(\frac{d\beta}{d\omega}\right)^{-1} = v\sqrt{1-\left(\frac{\lambda}{\lambda_c}\right)^2}$$

$$= 2\times10^8 \times \sqrt{1-\left(\frac{4.37}{9}\right)^2} = 1.75\times10^8 \text{ m/s}$$

EXERCISE SECTION

Essay Type Questions

1. Differentiate hollow-pipe waveguides from transmission lines.
2. Describe the structure of commonly used waveguides and compare them with each other.
3. Derive the expressions for the fields in rectangular waveguides in the case of
   (a) Transverse magnetic (TM) wave $H_z = 0$ and
   (b) Transverse electric (TE) wave $E_z = 0$
4. Prove that the rectangular waveguide is a high-pass filter. Define the terms cut-off frequency and cut-off wavelength.
5. What is the meaning of mode of the wave? Give a complete description of modal propagation characteristics of waves in RWGs.
6. How is the bandwidth of waveguides defined? How are the dimensions of the rectangular waveguides selected to give maximum bandwidth?
7. Describe the various methods of excitation of the rectangular waveguides.
8. Derive the expressions for the fields in Circular waveguides in the case of
   (a) Transverse magnetic (TM) wave $H_z = 0$ and
   (b) Transverse electric (TE) wave $E_z = 0$
9. Derive an expression for the bandwidth of the circular waveguide.
10. Describe the modal propagation characteristics in the circular waveguides. How is the mode description of CWGs different from that of RWGs?
11. Prove that the velocity of the wave in waveguides is a function of frequency. What is guide wavelength?
12. Define the different types of impedances of the waveguides. Derive the expressions for the impedances in the case of
    (a) RWGs and
    (b) CWGs.
13. What are the two types of attenuations we come across in waveguides? Derive the general expressions for both the types of attenuations.

14. Explain the following terms with respect to microstrip transmission lines:
    (a) Effective dielectric constant ε_{re}
    (b) Characteristic impedance Z_0
    (c) Losses in microstrip lines
    (d) Quality factor Q of microstrip line
    (e) Shielded strip lines

Numerical Problems

1. Consider a rectangular waveguide with $a/b = 1.5$, $a = 2.3$ cm, $\mu_r = 1$ and $\varepsilon_r = 3.0$.
    (a) Find out the cut-off wavelength, cut-off frequency for TE_{10}, TE_{11}, TE_{02} and TM_{11} modes.
    (b) Calculate the guide wavelength and characteristic impedance at 3.2 GHz.
    (c) Find out the attenuation constant in dB/cm at 3.2 GHz for the dielectric-filled guide.
    (d) Also find the total attenuation at frequencies much less than the cut-off frequencies.
2. Consider a rectangular waveguide with $a/b = 2$, $a = 5.817$ cm, $\mu_r = 1$ and $\varepsilon_r = 2.5$. Assuming dominant mode propagation, calculate the guide wavelength and characteristic impedance at 3.0 GHz.
3. Consider a rectangular waveguide with $a/b = 2$, $a = 3.0$ cm, $\mu_r = 1$ and $\varepsilon_r = 2.25$. Assuming dominant mode propagation,
    (a) find out the attenuation constant in dB/cm at 2.5 GHz for the dielectric-filled guide and
    (b) also find the total attenuation at frequencies much less than the cut-off frequencies.
4. Find the cut-off frequencies for TM_{12} mode in an air-filled waveguide whose dimensions a and b respectively are
    (a) 1.0×0.5 cm,
    (b) 0.5×1.0 cm,
    (c) 1.0×1.0 cm and
    (d) 10×10 cm.
5. For an air-filled waveguide with dimensions 2.290×1.02 cm, determine the cut-off frequencies for TE_{11}, TE_{12} and TE_{21} modes.
6. Determine the cut-off wavelength and wave impedance in the case of RWG with dimensions of 5×2 cm propagating TM_{11} mode at $f = 9$ GHz.
7. An X-band rectangular wave guide of $1' \times 0.5'$ has a signal propagating at frequency 10 GHz. Determine all the propagation characteristics for the dominant mode, if the guide is
    (a) air filled and
    (b) filled with a dielectric of $\varepsilon_r = 2.25$. What are your observations?

8. An air dielectric L-band rectangular waveguide has $a/b = 2$ and a dominant mode cut-off frequency of 0.908 GHz. If the measured guide wavelength is 40 cm, find the operating frequency, the guide dimensions and wave number. For the above problem, also find the lowest frequency at which a TE_{21} mode would propagate.
9. A rectangular waveguide has cross-section dimensions $a = 7$ cm and $b = 4$ cm. Determine all the modes which will propagate at a frequency of
   (a) 3000 MHz and
   (b) 5000 MHz.

   With the above data, also find the phase velocity, cut-off wavelength and guide wavelength for all the possible modes.
10. A lossless air dielectric waveguide for an S-band radar has inside dimensions $a = 7.214$ cm and $b = 3.404$ cm. For the TM_{11} mode propagating at an operating frequency that is 1.1 times the cut-off frequency of the mode, calculate
    (a) critical wave number,
    (b) cut-off frequency,
    (c) operating frequency,
    (d) cut-off wavelength,
    (e) operating wavelength,
    (f) guide wavelength,
    (g) phase velocity,
    (h) wave impedance, and
    (i) group velocity.
11. A rectangular waveguide has a cross section of 1.5 × 0.8 cm, $\sigma = 0$, $\mu = \mu_0$ and $\varepsilon = 4\varepsilon_0$. The magnetic field component is given by

$$H_x = 2\sin\left(\frac{\pi x}{a}\right)\cos\left(\frac{3\pi y}{b}\right)\sin(\pi \times 10^{11} t - \beta z) \text{ A/m}$$

    Determine the mode of operation, cut-off frequency, phase shift constant, propagation constant and wave impedance.
12. An air-filled circular waveguide with an inner radius of 1.2 cm is operating in TM_{01} mode. Determine its cut-off frequency. If it is operating at a frequency of 10 GHz, then find the wavelength in the waveguide.
13. An X-band waveguide filled with a dielectric with $\varepsilon_r = 2.25$ is operating at $f = 9$ GHZ. Calculate the phase and group velocities in the waveguide.

Fill in the Blanks

1. Transmissions lines carry waves in _________ mode, whereas waveguides carry waves in _________ TEM mode.
2. A pure real value of propagation constant indicates _________ attenuation and _________ wave motion.
3. A pure imaginary value of propagation constant indicates _________ attenuation and _________ wave motion.
4. In TE wave, the electric vector is _________ transverse to the direction of propagation of wave.

5. In TE wave, the magnetic vector has ________ component along the direction of propagation of wave.
6. In TM wave, the magnetic vector is entirely ________ to the direction of propagation of wave.
7. In TM wave, the electric vector has a ________ along the direction of propagation of wave.
8. In TEM wave, both the electric and magnetic vectors are entirely ________ the direction of propagation of wave.
9. In mixed or hybrid wave, both the electric and the magnetic vectors have components ________ the direction of propagation of wave.
10. The cut-off frequency of a rectangular waveguide is ________.
11. The cut-off wavelength of a rectangular waveguide is ________ and its guide wavelength is ________.
12. Mode subscripts m and n in rectangular waveguides indicate the number of half wavelengths along ________ directions respectively.
13. Dominant mode in rectangular guides is ________, whereas in circular waveguides it is ________.
14. Cut-off wavelength for dominant mode is equal to $2a$, where a is the ________ distance between the sidewalls of the waveguide.
15. Degenerate modes of waveguides are ________ modes having the same cut-off frequency.
16. Rectangular waveguides are dimensioned with ratio a/b approximately equal to ________.
17. Cut-off frequency in circular waveguides for TM mode is ________.
18. Cut-off frequency in circular waveguides for TE mode is ________.
19. Cut-off wavelength in circular waveguides for TM mode is ________.
20. Cut-off wavelength in circular waveguides for TE mode is ________.
21. The order of mode subscripts m and n in rectangular guides is ________, whereas in circular guides it is ________.
22. The phase velocity in waveguides is ________, whereas the group velocity is ________.
23. In waveguides, the wave impedance is ________ for TE mode and it is ________ for TM mode.
24. Reflective attenuation comes into being when the wave frequency is ________ than the cut-off frequency.
25. Dissipative attenuation comes into being when the wave frequency is ________ than the cut-off frequency.

Answers

1. TEM, non
2. only, no
3. no, only
4. entirely

5. a
6. transverse
7. component
8. transverse to
9. along
10. $\frac{1}{2\sqrt{\mu\varepsilon}}\sqrt{\left(\frac{m}{a}\right)^2+\left(\frac{n}{b}\right)^2}$
11. $\frac{2}{\sqrt{\left(\frac{m}{a}\right)^2+\left(\frac{n}{b}\right)^2}}$, $\frac{\lambda}{\sqrt{1-\left(\frac{\lambda}{\lambda_c}\right)^2}}$
12. *x* and *y*
13. TE_{10}, TE_{11}
14. inner
15. different
16. 2
17. $\frac{h_{nm}}{2\pi\sqrt{\mu\varepsilon}}$
18. $\frac{h_{nm}}{2\pi\sqrt{\mu\varepsilon}}$
19. $\frac{2\pi a}{(ha)_{nm}}$
20. $\frac{2\pi a}{(ha)'_{nm}}$
21. *mn*, *nm*
22. $\frac{c}{\sqrt{1-\left(\frac{\lambda}{\lambda_c}\right)^2}}$, $c\sqrt{1-\left(\frac{\lambda}{\lambda_c}\right)^2}$
23. $\frac{\eta}{\sqrt{1-\left(\frac{\lambda}{\lambda_c}\right)^2}}$, $\eta\sqrt{1-\left(\frac{\lambda}{\lambda_c}\right)^2}$
24. less
25. more

Match the Following

Set I

1. H_x	(a) $-\frac{\bar{\gamma}}{h^2}\frac{\partial H_z}{\partial x}$
2. H_y	(b) $-\frac{j\omega\varepsilon}{h^2}\frac{\partial E_z}{\partial x}$
3. E_x	(c) $-\frac{\bar{\gamma}}{h^2}\frac{\partial E_z}{\partial x}$
4. E_y	(d) $-\frac{j\omega\mu}{h^2}\frac{\partial H_z}{\partial x}$
5. E_z, H_z	(e) zero

Set II

1. H_x	(a) $\overline{\gamma}^2 + \omega^2 \mu\varepsilon$
2. H_y	(b) $-\dfrac{\overline{\gamma}}{h^2}\dfrac{\partial E_z}{\partial y} + \dfrac{j\omega\mu}{h^2}\dfrac{\partial H_z}{\partial x}$
3. E_x	(c) $-\dfrac{\overline{\gamma}}{h^2}\dfrac{\partial E_z}{\partial x} - \dfrac{j\omega\mu}{h^2}\dfrac{\partial H_z}{\partial y}$
4. E_y	(d) $-\dfrac{\overline{\gamma}}{h^2}\dfrac{\partial H_z}{\partial y} - \dfrac{j\omega\varepsilon}{h^2}\dfrac{\partial E_z}{\partial x}$
5. h^2	(e) $-\dfrac{\overline{\gamma}}{h^2}\dfrac{\partial H_z}{\partial x} + \dfrac{j\omega\varepsilon}{h^2}\dfrac{\partial E_z}{\partial y}$

Set III

1. a	(a) inner distance between broader walls
2. b	(b) inner distance between side walls
3. m	(c) number of half waves along b dimension
4. n	(d) number of half waves along a dimension
5. Loss free region	(e) conductivity zero

Set IV

1. TE	(a) E entirely normal to the direction of propagation
2. TM	(b) E and H with components along the direction of propagation
3. TEM	(c) E and H entirely normal to the direction of propagation
4. TE + TM	(d) H entirely normal to the direction of propagation
5. $m = 1, n = 0$	(e) dominant mode of RWG

Set V

1. Phase velocity	(a) $\dfrac{1}{2\sqrt{\mu\varepsilon}}\sqrt{\left(\dfrac{m}{a}\right)^2 + \left(\dfrac{n}{b}\right)^2}$
2. Group velocity	(b) $\dfrac{\lambda}{\sqrt{\varepsilon_r\mu_r - \left(\dfrac{\lambda}{\lambda_c}\right)^2}}$

3. Free space velocity	(c) $\dfrac{1}{\sqrt{\mu_0 \varepsilon_0}}$
4. Guide wavelength	(d) $v\sqrt{1-\left(\dfrac{\lambda}{\lambda_c}\right)^2}$
5. Cut-off frequency	(e) $\dfrac{v}{\sqrt{1-\left(\dfrac{\lambda}{\lambda_c}\right)^2}}$

Set VI

1. Guide wavelength	(a) $\dfrac{2}{\sqrt{\left(\dfrac{m}{a}\right)^2+\left(\dfrac{n}{b}\right)^2}}$
2. Cut-off wavelength	(b) $\lambda_g = \dfrac{\lambda}{\sqrt{\varepsilon_r \mu_r - \left(\dfrac{\lambda}{\lambda_c}\right)^2}}$
3. Free space wavelength	(c) $\dfrac{1}{f\sqrt{\mu_0 \varepsilon_0}}$
4. Z_z for TE	(d) $\eta\sqrt{1-\left(\dfrac{\lambda}{\lambda_c}\right)^2}$
5. Z_z for TM	(e) $\dfrac{\eta}{\sqrt{1-\left(\dfrac{\lambda}{\lambda_c}\right)^2}}$

Set VII

1. Dominant mode of CWG	(a) $\dfrac{\text{Picture lost per unit length}}{2 \times \text{Power transmitted}}$
2. Cut-off wavelength for TE	(b) $\dfrac{k^2 \tan\delta}{2\bar{\beta}}$ Np/m
3. Cut-off wavelength for TM	(c) $\lambda_c = \dfrac{2\pi a}{(ha)_{nm}}$
4. Dielectric attenuation	(d) $\lambda_c = \dfrac{2\pi a}{(ha)'_{nm}}$
5. Conductor attenuation	(e) TE_{11}

Answers

Set I:	1. (a)	2. (b)	3. (c)	4. (d)	5. (e)
Set II:	1. (e)	2. (d)	3. (c)	4. (b)	5. (a)
Set III:	1. (b)	2. (a)	3. (d)	4. (c)	5. (e)
Set IV:	1. (e)	2. (d)	3. (c)	4. (b)	5. (a)
Set V:	1. (b)	2. (a)	3. (c)	4. (e)	5. (d)
Set VI:	1. (a)	2. (d)	3. (c)	4. (b)	5. (e)
Set VII:	1. (e)	2. (d)	3. (c)	4. (b)	5. (a)

Multiple Choice Questions

1. The propagation constant pure real means
   (a) wave without attenuation (b) no wave motion
   (c) wave with attenuation (d) none of these
2. The propagation constant pure imaginary means
   (a) wave without attenuation (b) no wave motion
   (c) wave with attenuation (d) none of these
3. The propagation constant complex means
   (a) wave without attenuation (b) no wave motion
   (c) wave with attenuation (d) none of these
4. The lowest order TE wave in between parallel conducting plates is
   (a) TE_{10} (b) principal wave
   (c) both (a) and (b) (d) none of these
5. The principal wave is
   (a) TM_{00} (b) TEM wave
   (c) both (a) and (b) (d) none of these
6. The nature of the wave normal to plates is
   (a) pure standing (b) pure travelling
   (c) impure travelling (d) none of these
7. The nature of the wave parallel to plates is
   (a) pure standing (b) pure travelling
   (c) impure travelling (d) none of these
8. The phase shift constant $\bar{\beta}$ of the guided wave is
   (a) $\bar{\beta}=\sqrt{\omega^2\mu\varepsilon-\left(\frac{m\pi}{a}\right)^2}$ (b) $\bar{\beta}=\omega\sqrt{\mu\varepsilon}\sqrt{1+\left(\frac{f_c}{f}\right)^2}$
   (c) $\bar{\beta}=\omega\sqrt{\mu\varepsilon}\sqrt{1+\left(\frac{\lambda}{\lambda_c}\right)^2}$ (d) all of these

9. The cut off frequency of the guided wave is
   (a) $f_c = \frac{2a}{m}$ (b) $f_c = \frac{m}{2a\sqrt{\mu\varepsilon}}$
   (c) $f_c = \frac{a}{2m\sqrt{\mu\varepsilon}}$ (d) none of these
10. The cut-off wavelength of the guided wave is
   (a) $\lambda_c = \frac{2a}{m}$ (b) $f_c = \frac{m}{2a\sqrt{\mu\varepsilon}}$
   (c) $f_c = \frac{a}{2m\sqrt{\mu\varepsilon}}$ (d) none of these
11. The guide wavelength $\bar{\lambda}$ is given by
   (a) $\bar{\lambda} = \frac{2\pi}{\bar{\beta}}$ (b) $\bar{\lambda} = \frac{2\pi}{\sqrt{\omega^2\mu\varepsilon - \left(\frac{m\pi}{a}\right)^2}}$
   (c) $\bar{\lambda} = \frac{2\pi}{\omega\sqrt{\mu\varepsilon}\sqrt{1+\left(\frac{\lambda}{\lambda_c}\right)^2}}$ (d) all of these
12. The phase velocity is given by
   (a) $\bar{v} = \frac{\omega}{\bar{\beta}}$ (b) $\bar{v} = \frac{\omega}{\sqrt{\omega^2\mu\varepsilon - \left(\frac{m\pi}{a}\right)^2}}$
   (c) $\bar{v} = \frac{1}{\sqrt{\mu\varepsilon}\sqrt{1+\left(\frac{\lambda}{\lambda_c}\right)^2}}$ (d) all of these
13. The velocity of the wave is independent of frequency in the case of
   (a) TEM wave (b) TE wave
   (c) TM wave (d) none of these
14. The cut-off frequency of TEM wave is
   (a) infinite (b) zero
   (c) $\frac{1}{2\pi\sqrt{\mu\varepsilon}}$ (d) none of these
15. The ratio of **E** to **H** is free space intrinsic impedance only in case of
   (a) TEM wave (b) TE wave
   (c) TM wave (d) none of these

16. The group velocity is given by

(a) $v_g = \dfrac{c^2}{v}$ (b) $v_g = c\sqrt{1-\left(\dfrac{m\lambda}{2a}\right)^2}$

(c) both (a) and (b) (d) none of these

17. The dominant TE mode in a rectangular waveguide is

(a) TE_{01} (b) TE_{11}

(c) TE_{20} (d) TE_{10}

18. Waveguides can carry

(a) TE mode (b) TM mode

(c) mixed mode (d) all of these

19. The cut-off wavelength of circular waveguide in TM mode is

(a) $\dfrac{2}{\sqrt{\left(\dfrac{m}{a}\right)^2+\left(\dfrac{n}{b}\right)^2}}$ (b) $\dfrac{2\pi a}{(ha)_{nm}}$

(c) $\dfrac{2a\pi}{(ha)_{nm}}$ (d) none of these

20. The cut-off wavelength of a rectangular waveguide is

(a) $\dfrac{2}{\sqrt{\left(\dfrac{m}{a}\right)^2+\left(\dfrac{n}{b}\right)^2}}$ (b) $\dfrac{2}{\sqrt{\left(\dfrac{m}{b}\right)^2+\left(\dfrac{n}{a}\right)^2}}$

(c) both (a) and (b) (d) none of these

21. The guide wavelength of cylindrical wave guide is

(a) $\dfrac{\lambda}{\sqrt{1-\left(\dfrac{\lambda}{\lambda_0}\right)^2}}$ (b) $\dfrac{\lambda}{\sqrt{1-\left(\dfrac{f_0}{f}\right)^2}}$

(c) both (a) and (b) (d) none of these

22. The wave impedance of a waveguide in TE mode can be

(a) $\dfrac{\eta}{\sqrt{1-\left(\dfrac{\lambda}{\lambda_0}\right)^2}}$ (b) $\eta\sqrt{1-\left(\dfrac{\lambda}{\lambda_0}\right)^2}$

(c) both (a) and (b) (d) none of these

23. The dominant mode in a cylindrical waveguide is

(a) TE_{11} (b) TE_{10}

(c) either (a) or (b) (d) none of these

24. Dominant modes in regular cavity resonators can be
(a) TE_{111} (b) TE_{110}
(c) either (a) or (b) (d) none of these
25. The non-existent mode in circular waveguides is
(a) TE_{10} (b) TE_{00}
(c) both (a) and (b) (d) none of these
26. Resonant frequency of rectangular cavity resonator is
(a) $\dfrac{1}{2\sqrt{\mu\varepsilon}}\sqrt{\left(\dfrac{m}{a}\right)^2+\left(\dfrac{n}{b}\right)^2+\left(\dfrac{p}{d}\right)^2}$ (b) $\dfrac{1}{2\sqrt{\mu\varepsilon}}\sqrt{\left(\dfrac{m}{a}\right)+\left(\dfrac{n}{b}\right)+\left(\dfrac{p}{d}\right)}$
(c) $\dfrac{1}{2\pi\sqrt{\mu\varepsilon}}\sqrt{\left(\dfrac{m}{a}\right)^2+\left(\dfrac{n}{b}\right)^2+\left(\dfrac{p}{d}\right)^2}$ (d) none of these
27. Attenuation constant due to dielectric loss is (in N/m)
(a) $\dfrac{\bar{\beta}^2 \tan\delta}{2\beta}$ (b) $\dfrac{2\bar{\beta}}{\beta^2}\tan\delta$
(c) $\dfrac{\omega^2}{2\beta}\tan\delta$ (d) $\dfrac{\omega^2\mu_0\varepsilon_0\varepsilon_r}{2\bar{\beta}}\tan\delta$
28. Attenuation constant due to conductor loss is
(a) $\dfrac{\text{power dissipated/unit length}}{2\times\text{power flow}}$ (b) $\dfrac{2\times\text{power flow}}{\text{dissipated power/unit length}}$
(c) $\dfrac{2\times\text{power dissi/unit length}}{\text{power flow}}$ (d) none of these
29. Reflective attenuation comes into being when the frequency of the wave is
(a) less than the cut-off frequency (b) more than the cut-off frequency
(c) both (a) and (b) (d) none of these
30. In RWG, the mode subscripts m and n indicate
(a) number of half wave patterns (b) number of full wave patterns
(c) number of the zeros of the field (d) none of these
31. In RWG, for dominant mode, the cut-off wavelength is
(a) $2a$ (b) $2b$
(c) a (d) none of these
32. The wave whose frequency is 1.5 GHz falls in the band
(a) L (b) S
(c) C (d) none of these
33. The wave whose frequency is 2.5 GHz falls in the band
(a) L (b) S
(c) C (d) none of these

34. The wave whose frequency is 7 GHz falls in the band
(a) L (b) 5
(c) C (d) none of these
35. The wave whose frequency is 9 GHz falls in the band
(a) L (b) X
(c) C (d) none of these
36. Degenerate modes in circular waveguides are
(a) TE_{01} and TM_{11} (b) TE_{22} and TM_{22}
(c) both (a) and (b) (d) none of these
37. At infinite frequency, the guide wavelength is
(a) infinite (b) free space wavelength
(c) cut-off wavelength (d) none of these
38. An air-filled rectangular waveguide has dimensions of 6 × 4 cm. Its cut-off frequency for the TE_{10} mode is
(a) 2.5 GHz (b) 25 GHz
(c) 25 MHz (d) 5 GHz
39. The phase velocity of the guided wave at a frequency of 3.0 GHz in TE_{10} for Problem 38 is
(a) 0.1 m/s (b) 5.42×10^8 m/s
(c) 5.4×10^6 m/s (d) 3.78×10^8 m/s
40. In Problem 38, the group velocity is
(a) 1.659×10^8 m/s (b) 5.42×10^8 m/s
(c) 0.185×10^8 m/s (d) 3.78×10^8 m/s
41. In Problem 38, the wave impedance in the waveguide at 3 GHz is
(a) 120π (b) 681.72
(c) 300 (d) 600
42. In Problem 38, the phase constant in the waveguide is
(a) 69 rad/m (b) 100 rad/m
(c) 34.5 rad/m (d) 50 rad/m
43. The cut-off frequency of a waveguide depends on
(a) dimensions of the waveguide
(b) the dielectric property of the medium in the waveguide
(c) wave mode
(d) all of these
44. In hollow rectangular waveguides,
(a) the phase velocity is greater than the group velocity
(b) the phase velocity is greater than the velocity of light in free space
(c) both (a) and (b)
(d) none of these

45. In a waveguide, the subscripts m and n of the modes TE/TM denote
    (a) half wavelength of **E** field and full wavelength of **H** field
    (b) half wavelengths of **E** and **H** fields
    (c) full wavelength of **E** field and half wavelength of **H** field
    (d) half wavelengths of **H** and **E** fields
46. The waves in a waveguide
    (a) travel along the border walls of the waveguide
    (b) are reflected from side walls but do not travel along them
    (c) travel through the dielectric without touching the walls
    (d) travel along all the four walls
47. Which of the following modes is not supported by RWG
    (a) TE_{10} (b) TE_{11}
    (c) TM_{11} (d) TM_{10}
48. For a wave of finite frequency in an air-filled RWG,
    (a) guide-wavelength is never less than the free space wavelength
    (b) wave impedance is never equal to the free space impedance
    (c) TEM is possible
    (d) all of these
49. The dominant mode in a circular waveguide is
    (a) TE_{10} (b) TE_{11}
    (c) TE_{01} (d) TE_{12}
50. In TE_{20} of RWG, the number of half waves in direction
    (a) 2 (b) 1
    (c) 4 (d) 0
51. In a cylindrical waveguide, Z_{TE} is
    (a) $\dfrac{\beta}{\omega\mu}$ (b) $\dfrac{\omega\mu}{\beta}$
    (c) $\dfrac{\omega\beta}{\mu}$ (d) $\omega\mu\beta$
52. Theoretically the number of modes that can exist in cylindrical waveguides is
    (a) zero (b) one
    (c) two (d) infinite
53. The primary mode in a rectangular resonant cavity is
    (a) TE_{111} (b) TE_{101}
    (c) TE_{100} (d) TE_{001}

54. The real power transmitted in a rectangular waveguide is
   (a) $\mathbf{E} \times \mathbf{H}^*$
   (b) $\mathbf{H} \times \mathbf{E}^*$
   (c) $\frac{1}{2}\text{Re}(\mathbf{H} \times \mathbf{E}^*)$
   (d) $\frac{1}{2}\text{Re}(\mathbf{E} \times \mathbf{H}^*)$
55. A disadvantage of the microstrip transmission line compared with the strip line is
   (a) does not readily lend itself to printed circuit technique
   (b) more likely to radiate
   (c) bulkier
   (d) complex and expensive
56. The transmission system using two ground planes is
   (a) microstrip
   (b) rectangular waveguide
   (c) circular waveguide
   (d) strip line
57. A disadvantage of the strip line waveguide is
   (a) it is smaller bulk
   (b) greater bandwidth
   (c) higher power-handling capability
   (d) greater compatibility with solid-state devices
58. A disadvantage of strip line over microstrip is its
   (a) easier integration with semiconductor devices
   (b) lesser tendency to radiate
   (c) higher isolation between circuits
   (d) higher Q

Answers

1. (b)	2. (a)	3. (c)	4. (a)	5. (c)
6. (a)	7. (b)	8. (d)	9. (b)	10. (a)
11. (d)	12. (d)	13. (a)	14. (b)	15. (a)
16. (c)	17. (d)	18. (d)	19. (b)	20. (a)
21. (c)	22. (a)	23. (c)	24. (a)	25. (b)
26. (a)	27. (d)	28. (a)	29. (a)	30. (a)
31. (a)	32. (a)	33. (b)	34. (c)	35. (b)
36. (a)	37. (b)	38. (a)	39. (b)	40. (a)
41. (b)	42. (c)	43. (d)	44. (c)	45. (b)
46. (b)	47. (d)	48. (d)	49. (b)	50. (a)
51. (b)	52. (d)	53. (b)	54. (d)	55. (b)
56. (d)	57. (b)	58. (b)		

UNIT VI

TRANSMISSION LINE THEORY

11 Line Equations and Impedance

Defeat is not the worst of failures. Not to have tried is the true failure.
—George Edward Woodberry

Transmission lines are physical conductor systems which are useful to transfer electrical energy from one point to another point. They take several different forms, shapes and sizes depending upon the frequency of operation and application. Parallel wire transmission lines and coaxial lines are some of the lines which are seen everywhere.

Transmission lines are also useful, other than energy transmission, at higher frequencies. For example, they are used as lumped circuit elements like inductors, capacitors, etc. and as resonators.

In this chapter, the transmission line concepts are introduced by giving definition, equivalent circuit, primary constants and secondary constants, telegraphers' equations and wave equations governing the electrical phenomenon. In the later sections, important quantities such as line impedance, input impedance, etc. are discussed in detail.

11.1 Definition and Types

By definition, transmission lines are metallic conductor systems with two or more conductors separated by insulators and used primarily to transfer either low-frequency electrical energy or high-frequency telecom signal, in TEM form, from one point to another which are usually separated by large distances.

The hollow-pipe waveguides carrying non-TEM waves and fibre-optic cables which are non-conductors and the free space which is not a conductor, are all excluded from the transmission lines by the definition. The general configuration of a transmission line is shown in Figure 11.1.

Prominent examples of transmission lines are multi-conductor parallel wire systems carrying low-frequency power to run heaters, fans, motors, etc. Another example is the coaxial cable that carries high-frequency video signals to the domestic TV sets.

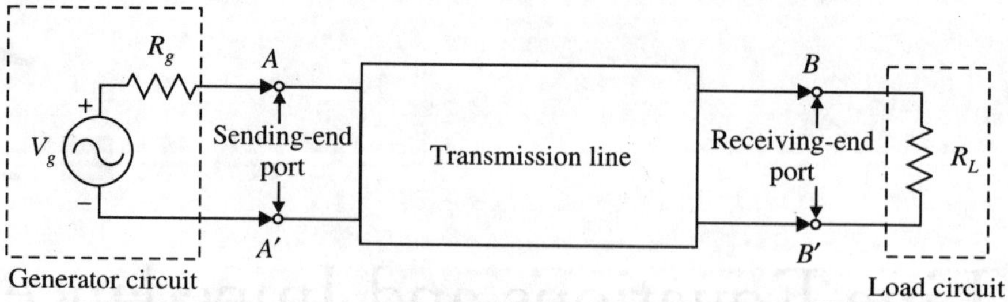

FIGURE 11.1 General configuration involving a transmission line.

The primary application of the transmission lines is to carry electrical energy from one point to another. They have some secondary applications which include tuned circuits in amplifiers and oscillators and also load matching elements.

Depending upon the structure and nature, the lines may be divided into several types. One classification is parallel conductor lines and coaxial lines. Another division is balanced and unbalanced lines. Different types of transmission lines are shown in Figure 11.2.

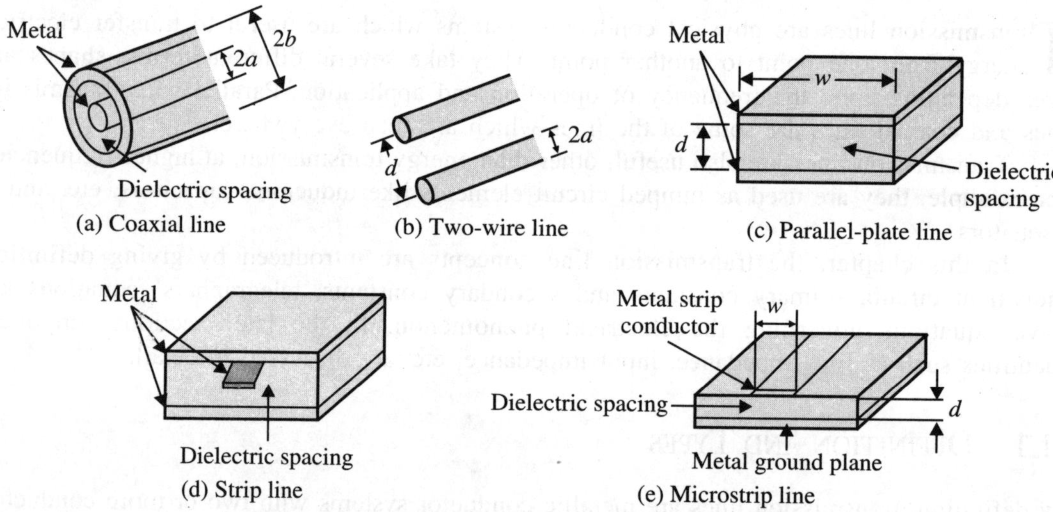

FIGURE 11.2 Different types of transmission lines.

The important features of these lines are mentioned below.

Parallel conductor transmission lines:

- Suitable for low-frequency applications.
- At high frequencies, they become useless as radiation and dielectric losses increase
- Susceptible for noise pick up.
- Example: Open wire TL, twin lead (ribbon) cable, twisted pair cable, etc.

Coaxial or concentric transmission lines:

- Extensively used for high-frequency applications as they give low radiation and dielectric losses.

- Also they give shield against external interference.
- Example: Solid flexible (low losses), rigid air filled (relatively expensive), must be used in unbalanced mode and expensive.

Balanced or differential transmission lines:

- Both the conductors in this line carry signal currents of equal magnitude with respect to the electrical ground but in opposite directions.
- Any pair of wires can be operated in balanced mode provided that neither wire is at ground potential.
- A balanced wire pair has the advantage of noise interference getting cancelled at the load due to high Common Mode Rejection Ratio of 40 to 70 dB.

Unbalanced or single-ended transmission lines:

- One conductor is at the ground potential, whereas the other one is at the signal potential.
- Because one wire being at the ground potential, the probability of noise being induced is more.
- The standard two conductor coaxial cable is an unbalanced line as the second conductor or shield is generally connected to the ground.

Line terminology

- *Baluns:* These are circuit devices used to connect a balanced TL to an unbalanced load like antenna or unbalanced TL such as a coaxial cable. Example: Transformer balun, bazooka balun, etc.
- *Uniform line:* It is a line that exhibits the same properties all along its length, i.e. its primary constants and consequently secondary constants should be the same.
- *Resonant line:* It is lossless line terminated over a short or open circuit. The significant feature of the resonant line is that it carries pure standing waves.
- *Non-resonant line:* A line terminated in its characteristic impedance is called non-resonant or flat or smooth line. This line carries pure travelling waves. The voltage and current over such a line are constant throughout its length if it is lossless and decreases exponentially if the line is lossy.
- *Stub:* It is a piece of the transmission line whose input impedance is pure reactance. Normally short-circuited stubs are used as open-circuited stubs tend to radiate.
- *Infinite line:* It is a line of infinite length. It is characterized by the absence of reflected wave and the current over the line depends only on its characteristic impedance, not on the termination.
- *Velocity factor:* The velocity factor of a dielectric substance or a cable is the velocity reduction ratio. It is given by $v_f = 1/\sqrt{\varepsilon_r}$, where ε_r is the dielectric constant of the medium.

11.2 EQUIVALENT CIRCUIT

The transmission line is a passive circuit. Its equivalent circuit, in general, is made up of several cascaded sections, each one consisting of a series combination of resistance R and inductance L and a shunt combination of capacitance C and conductance G.

The parameters R, L, G and C are called *primary constants* of the line. They represent various electrical quantities of the line that are distributed throughout the length of the line and hence they are called *distributed parameters*. The important aspects that influence the values of the primary constants are: the configuration of the conductors in the line, the material with which the conductors are made, the properties of the dielectric which separates the conductors, etc. All the quantities R, L, G and C are proportional to the length of the line and unless measured and quoted per unit length basis, they are meaningless. The general equivalent circuit of a transmission line is shown in Figure 11.3.

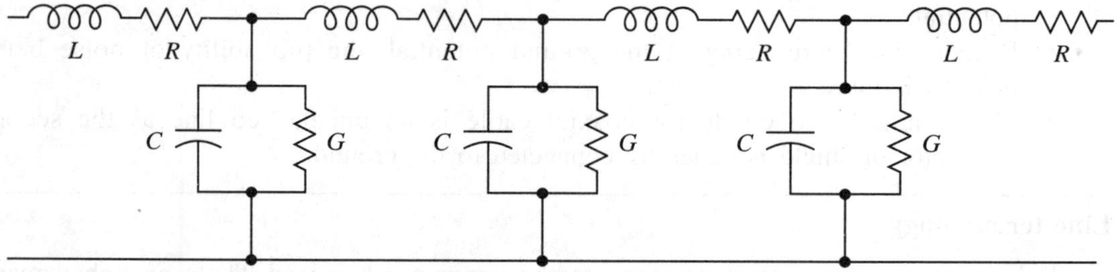

FIGURE 11.3 General equivalent circuit of a transmission line.

The line is made up of two or more conductors and each conductor possesses certain amount of resistance because they have a certain length and diameter. The parameter R represents the sum total of the resistances of the conductors in the line of unit length.

When the current flows through a conductor of the line, it sets up its own magnetic flux. This flux links the very current because of which it exists resulting in certain amount of self-inductance. The self-inductance of the line per unit length is the sum of the self-inductances of all the conductors in the line and is denoted by the parameter L.

In the line, the conductors are separated by dielectric material and in general it is not a perfect one resulting in leakage currents across the line. The leakage currents and the resulting conductance per unit length of the line are represented by the parameter G.

When the line is charged, there exist conductors with charges of opposite polarities separated by a dielectric material resulting in a capacitive effect. The parameter C represents all the capacitances of the line in a unit length.

Lossless line: Lines can be divided into lossy and lossless categories. If the primary constants R and G are negligible in comparison to the reactance ωL and susceptance ωC, respectively, then it is called a lossless line. If R and G are considerable in values, then the line is called lossy line.

When the line is operated at high frequencies like radio or microwave frequencies, the series inductive reactance dominates the series resistance and the shunt capacitive susceptance dominates the shunt conductance. Under these circumstances, the resistance R and conductance G of the line can be neglected. The line then is called lossless line and the equivalent circuit of the line consists of only a series inductance L and a shunt capacitance C as shown in Figure 11.4.

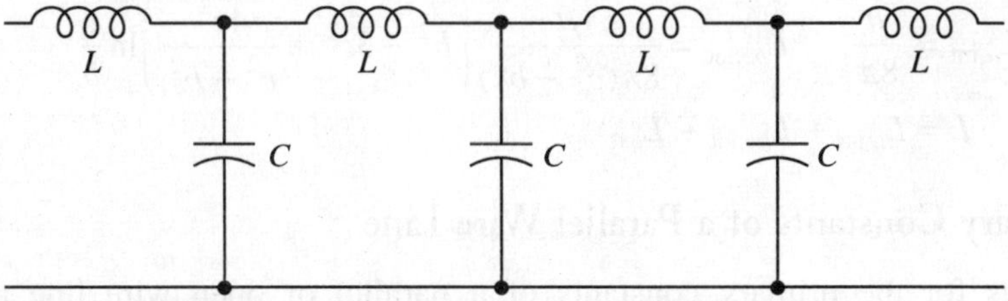

FIGURE 11.4 Equivalent circuit of lossless line.

Unit-length line: The equivalent circuit of a unit-length line consists of only one section which is made up of a series combination of resistance R and inductance L and a shunt combination of capacitance C and conductance G. If the line is lossless, then R and G are absent.

11.2.1 Primary Constants of a Coaxial Line

A coaxial line is used both at high frequencies and at low frequencies. It can carry TEM as well as non-TEM waves. It consists of an inner conductor of radius a and an outer enclosing coaxial conductor with an inner radius of b. Both these conductors are interspersed with dielectric material. The primary constants can be found using their basic definitions.

At high frequencies:

$$C = \frac{2\pi\varepsilon'}{\ln\left(\frac{b}{a}\right)} \tag{11.1}$$

where a and b are inner radius and outer radius of the dielectric, respectively and ε' is its permittivity.

$$G = \frac{2\pi\sigma}{\ln\left(\frac{b}{a}\right)} \tag{11.2}$$

where σ is the conductivity of the dielectric between the conductors.

$$L = \frac{\mu}{2\pi}\ln\left(\frac{b}{a}\right) \tag{11.3}$$

where μ is the permittivity of the dielectric usually μ_0.

$$R = \frac{1}{2\pi\delta\sigma_c}\left(\frac{1}{a}+\frac{1}{b}\right) \tag{11.4}$$

where δ and σ_c are the skin depth and conductivity of the conductors, respectively.

At low frequencies: The values of C and G remain the same as above.

$$R = \frac{1}{\pi\sigma_c}\left(\frac{1}{a^2} + \frac{1}{c^2 - b^2}\right) \tag{11.5}$$

$$L_{a,\text{int}} = \frac{\mu}{8\pi}, \quad L_{bc,\text{int}} = \frac{\mu}{8\pi(c^2 - b^2)}\left(b^2 - 3c^2 + \frac{4c^2}{c^2 - b^2}\right)\ln\frac{c}{b} \tag{11.6}$$

$$L = L_{a,\text{int}} + L_{bc,\text{int}} + L_{\text{ext}} \tag{11.7}$$

11.2.2 Primary Constants of a Parallel Wire Line

The expressions for the primary constants of a parallel or open wire line are given here. They can be derived using the basic electromagnetic theory principles and in fact they are available in different sections of electromagnetic theory part of this book.

At high frequencies:

$$C = \frac{2\pi\varepsilon'}{\ln\left(\frac{d}{a}\right)} \tag{11.8}$$

where a is the radius of the conductors, d is the separation distance between the conductors and ε' is the permittivity of the dielectric between the conductors.

$$G = \frac{\pi\sigma}{\cosh^{-1}\left(\frac{d}{2a}\right)} \tag{11.9}$$

where σ is the conductivity of the dielectric between the conductors.

$$L_{\text{ext}} = \frac{\mu}{\pi}\ln\left(\frac{d}{a}\right) \tag{11.10}$$

where is permittivity of the dielectric usually

$$R = \frac{1}{\pi a\delta\sigma_c} \tag{11.11}$$

where δ and σ_c are the skin depth and conductivity of the conductors.

At low frequencies:

The values of C and G remain the same as above

$$R = \frac{2}{\pi a^2\sigma_c} \tag{11.12}$$

$$L = \frac{\mu}{\pi}\left[\frac{1}{4} + \cosh^{-1}\left(\frac{d}{2a}\right)\right] \tag{11.13}$$

The derivation of the relations given here for primary constants of the lines can be found in the electromagnetics part of this book.

11.3 LOSSES IN LINES

Losses in transmission lines occur in the following three types:

- Radiation losses
- Losses due to conductor heating
- Losses due to dielectric heating

Radiation losses exist in lines which are open to space and they are absent in enclosed lines like coaxial lines. This type of losses arises because the transmission line can act as an antenna in open space radiating the energy which it is supposed to transmit if the separation between the conductors is an appreciable fraction of wavelength.

The conductors of the transmission line, while carrying currents dissipate some amount of power as ohmic losses resulting in conductor heating. The conductor heating or I^2R loss is proportional to the current through the conductor and therefore is inversely proportional to the characteristic impedance. It also increases with frequency because of the skin effect.

The dielectric that exists between the conductors of the line is not usually a perfect one. As a result, it allows the currents to flow from the conductor of one polarity to the conductor of another polarity. The current is proportional to the voltage across the dielectric and it results in heating of the dielectric. Therefore, the dielectric heating is proportional to the voltage across the dielectric and inversely proportional to the characteristic impedance. It increases with frequency.

11.4 SECONDARY CONSTANTS

The transmission line is basically characterized by *R*, *L*, *G* and *C* which are called primary constants. Another set of constants known as secondary constants is also widely used in the analysis of transmission lines. These constants are two in number: one is characteristic impedance $\mathbf{Z}_0$ and the second is propagation constant γ. Both are in general complex quantities. The secondary constants are dependent upon the primary constants of the line. We study about them in a more detailed way.

Characteristic impedance $\mathbf{Z}_0$**:** The characteristic impedance of a transmission line, $\mathbf{Z}_0$, can be defined in several different but equivalent ways. All of them give the same value to this parameter when applied to a line. According to one definition, it is the impedance measured at the input of the line when its length is extended to infinity. Another definition says it as the terminating impedance of a transmission line which can give an input impedance equal to the terminating impedance. It can also be defined as the ratio of the voltage amplitude to the current amplitude for each of the travelling waves individually with an additional minus sign in the case of the reflected wave.

$$\mathbf{Z}_0 = \frac{\mathbf{V}_s}{\mathbf{I}_s} \tag{11.14}$$

when the length of the line is infinite or when the line is terminated over the characteristic impedance $\mathbf{Z}_0$.

It is related to the primary constants of the line through

$$\mathbf{Z}_0 = \sqrt{\frac{(R + j\omega L)}{(G + j\omega C)}} \tag{11.15}$$

We can observe that it is, in general, a complex quantity with unit of ohm. But in the case of lossless line, i.e. for which $R = 0 = G$, it is pure real. As it is a complex quantity, we represent it by a boldface letter in time roman font and its magnitude by a normal italic letter, i.e. $Z_0 = |\mathbf{Z}_0|$.

- For a two-wire parallel transmission line at high frequencies, the characteristic impedance is

$$\mathbf{Z}_0 = \sqrt{\frac{L_{\text{ext}}}{C}} = \frac{1}{\pi}\sqrt{\frac{\mu}{\varepsilon'}}\cosh^{-1}\left(\frac{d}{2a}\right) \tag{11.16}$$

- For a coaxial line at high frequencies, the characteristic impedance is

$$\mathbf{Z}_0 = \sqrt{\frac{L_{\text{ext}}}{C}} = \frac{1}{2\pi}\sqrt{\frac{\mu}{\varepsilon'}}\ln\left(\frac{b}{a}\right) \tag{11.17}$$

Propagation constant: Propagation constant is also a complex quantity and is denoted by γ. As it is a complex quantity, in general it can have both real and imaginary parts, i.e. $\gamma = \alpha + j\beta$. The real part α is known as the attenuation constant with unit of Np/m and β is known as the phase shift constant with unit of rad/m. The attenuation constant of a line is defined as the natural logarithm of the magnitude of the ratio of the sending-end current to the receiving-end current,

$$\alpha = \ln\left|\frac{\mathbf{I}_s}{\mathbf{I}_r}\right| \text{Np/m} \tag{11.18}$$

and the phase shift constant β is defined as the phase difference between the sending-end current and the receiving-end current,

$$\beta = \frac{\angle \mathbf{I}_s}{\mathbf{I}_r} \tag{11.19}$$

where $\mathbf{I}_s$ and $\mathbf{I}_r$ are currents located at a unit distance apart over the line extended to infinity. Therefore, the propagation constant in terms of currents at unit distance apart over the line extended to infinity is

$$\gamma = \ln\left|\frac{\mathbf{I}_s}{\mathbf{I}_r}\right| + \frac{j\angle \mathbf{I}_s}{\mathbf{I}_r} \tag{11.20}$$

It is related to the primary constants through

$$\gamma = \sqrt{ZY} = \sqrt{(R + j\omega L)(G + j\omega C)} \tag{11.21}$$

As it is a complex quantity, we represent it by a boldface letter in time roman font and its magnitude by a normal italic letter, i.e. $\gamma = |\boldsymbol{\gamma}|$.

Sometimes it is useful to have the expressions for attenuation and phase shift constants in terms of the primary constants of the line. Now we try to derive those relations from the basics.

$$\boldsymbol{\gamma} = a + j\beta = \sqrt{(R + j\omega L)(G + j\omega C)}$$

$$\gamma^2 = |\boldsymbol{\gamma}|^2 = \alpha^2 - \beta^2 + 2j\alpha\beta = (R + j\omega L)\ (G + j\omega C)$$

$$= RG - \omega^2 LC + j\omega(RC + LG)$$

This gives

$$\alpha^2 - \beta^2 = RG - \omega^2 LC \tag{11.22}$$

$$2\alpha\beta = \omega(RC + LG)$$

Now the magnitude of the γ^2 is

$$\gamma^2 = \sqrt{(R^2 + \omega^2 L^2)(G^2 + \omega^2 C^2)}$$

and the magnitude square of the γ becomes

$$\gamma^2 = \alpha^2 + \beta^2$$

As the magnitude of the γ^2 is the same as the magnitude square of the γ, we can write

$$\alpha^2 + \beta^2 = (R^2 + \omega^2 L^2)\ (G^2 + \omega^2 C^2) \tag{11.23}$$

By adding Eqs. (11.22) and (11.23), we get

$$2\alpha^2 = RG - \omega^2 LC + \sqrt{(R^2 + \omega^2 L^2)(G^2 + \omega^2 C^2)}$$

It gives the attenuation constant as

$$\alpha = \sqrt{\frac{1}{2}\left[(RG - \omega^2 LC) + \sqrt{(R^2 + \omega^2 L^2)(G^2 + \omega^2 C^2)}\right]} \tag{11.24}$$

By subtracting Eq. (3.22) from Eq. (3.23), we get

$$2\beta^2 = \omega^2 LC - RG + \sqrt{(R^2 + \omega^2 L^2)(G^2 + \omega^2 C^2)}$$

It results in a relation for the phase shift constant as

$$\beta = \sqrt{\frac{1}{2}\left[(\omega^2 LC - RG) + \sqrt{(R^2 + \omega^2 L^2)(G^2 + \omega^2 C^2)}\right]} \tag{11.25}$$

11.4.1 Microwave Frequencies

At microwave frequencies, i.e. when the frequency is high lying in GHz range, it can be observed that $R \ll \omega L$ and $G \ll \omega C$. Here we try to find out expressions for secondary constants at these frequencies.

Propagation constant: Now consider the propagation constant

$$\gamma = \sqrt{(R + j\omega L)(G + j\omega C)}$$

$$= \sqrt{(j\omega)^2 LC}\sqrt{\left(1 + \frac{R}{j\omega L}\right)\left(1 + \frac{G}{j\omega C}\right)}$$

Using the binomial expansion,

$$\gamma \approx (j\omega)\sqrt{LC}\left(1 + \frac{1}{2}\frac{R}{j\omega L}\right)\left(1 + \frac{1}{2}\frac{G}{j\omega C}\right)$$

$$\approx (j\omega)\sqrt{LC} + \frac{1}{2}\left(R\sqrt{\frac{C}{L}} + G\sqrt{\frac{L}{C}}\right)$$

We know that the real part of the propagation constant is the attenuation constant and the imaginary part is the phase shift constant. Therefore, the attenuation constant and phase shift constant are

$$\alpha \approx \frac{1}{2}\left(R\sqrt{\frac{C}{L}} + G\sqrt{\frac{L}{C}}\right) \tag{11.26}$$

and

$$\beta \approx \omega\sqrt{LC} \tag{11.27}$$

Characteristic impedance: Similarly the characteristic impedance can be approximated as

$$\mathbf{Z}_0 = \sqrt{\frac{(R + j\omega L)}{(G + j\omega C)}}$$

$$= \sqrt{\frac{L}{C}}\left(1 + \frac{R}{j\omega L}\right)^{1/2}\left(1 + \frac{G}{j\omega C}\right)^{-1/2}$$

$$\approx \sqrt{\frac{L}{C}}\left(1 + \frac{1}{2}\frac{R}{j\omega L}\right)\left(1 - \frac{1}{2}\frac{G}{j\omega C}\right)$$

$$\approx \sqrt{\frac{L}{C}}\left[1 + \frac{1}{2}\left(\frac{R}{j\omega L} - \frac{G}{j\omega C}\right)\right]$$

$$\approx \sqrt{\frac{L}{C}} \tag{11.28}$$

Phase velocity: The phase velocity, also known as the wave velocity, of wave over the line, by definition, is the velocity with which the phase of the wave changes along the line. From the basic definition of wave, the phase of the wave travels a distance of one wavelength λ in a duration of one time period T. Therefore, the wave velocity is

$$v \triangleq \frac{\lambda}{T} \tag{11.29}$$

It can be manipulated to give a more useful relation.

$$v = \frac{\lambda}{T} = \frac{2\pi\lambda}{2\pi T} = \frac{2\pi f}{\frac{2\pi}{\lambda}}$$

$$= \frac{\omega}{\beta} \tag{11.30}$$

But from the basic electromagnetic wave theory, the phase velocity of a wave travelling in a medium of constants ε and μ is

$$v = \frac{1}{\sqrt{\mu\varepsilon}} \tag{11.31}$$

We can use either one of the above two relations to find the wave velocity over the line. Substituting the expressions for frequency and phase shift constant in the defining relation, we get

$$v = \frac{\omega}{\beta} \approx \frac{\omega}{\omega\sqrt{LC}}$$

$$\approx \frac{1}{\sqrt{LC}} \tag{11.32}$$

Here note that the product LC is independent of the size and separation of the conductors and it depends only on the permeability and permittivity of the insulating medium. If the separating insulation is air, free space parameters are to be assumed.

In such a case, it results in

$$v \approx \frac{1}{\sqrt{LC}} = \frac{1}{\sqrt{\mu_0\varepsilon_0}} = 3 \times 10^{-8} \text{ m/s}$$

If the dielectric separating the conductors is not air, then the phase velocity becomes smaller. It is

$$v \approx \frac{1}{\sqrt{LC}} = \frac{1}{\sqrt{\mu\varepsilon}} = \frac{3 \times 10^{-8}}{\sqrt{\mu_r\varepsilon_r}}$$

Velocity factor: The velocity factor of a cable is the velocity reduction ratio. It is defined as

$$\text{Velocity factor } v_r = \frac{\text{actual phase velocity}}{\text{free space velocity of light}} \tag{11.33}$$

Therefore,

$$v_r = \frac{v}{3 \times 10^{-8}} = \frac{1}{\sqrt{\mu_r\varepsilon_r}} \tag{11.34}$$

- Also note that for a total lossless line, $R = 0 = G$, leading to

$$\text{Attenuation constant } \alpha = 0 \tag{11.35a}$$

and

$$\text{Phase shift constant } \beta = \omega\sqrt{LC} \tag{11.35b}$$

- In the case of telephone cable, $Z = R$ and $Y = j\omega C$, resulting in

$$\alpha = \beta = \sqrt{\frac{\omega CR}{2}} \tag{11.36a}$$

$$v = \sqrt{\frac{2\omega}{CR}} \tag{11.36b}$$

As the attenuation constant α and the wave velocity v are functions of frequency, distortions of both types take place while the signaling is travelling over the line.

11.5 ANALYSIS OF UNIFORM LINE

The line can be analysed both in time and in phasor domains. But it is the phasor domain analysis that has practical as well as theoretical importance. All the critical parameters of the line such as secondary constants, reflection coefficient, etc. are defined in phasor or frequency domain. And because of its significance, a larger part of this chapter as well as the next two chapters is dedicated to the phasor analysis.

In the present section, we start the line analysis in time domain and derive the well-known transmission line or telegrapher's equations. And then we move to the phasor domain analysis to derive the fundamental line equations known as wave equations.

Time domain: The behaviour of a line when a voltage source is connected is described by a set of partial differential equations known as transmission line equations. These are the first-order differential equations which describe the voltage profile of the line as well as the current carried by the line as a function of time t and spatial position x.

Let us consider a differential length Δx of the line and using the lumped-element model of the line, we can write down its equivalent circuit as shown in Figure 11.5.

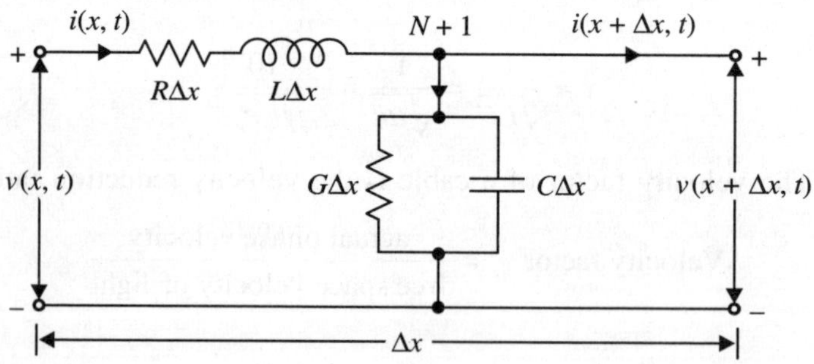

FIGURE 11.5 Equivalent circuit of a differential length line.

The quantities $v(x, t)$ and $i(x, t)$ denote the instantaneous voltage and current at the left end of the differential section (node N) and similarly $v(x + \Delta x, t)$ and $i(x + \Delta x, t)$ denote the same quantities at the right end (node $N + 1$).

The application of the Kirchhoff's voltage law gives

$$v(x, t) - R\Delta x i(x, t) - L\Delta x \frac{\partial i(x, t)}{\partial t} - v(x + \Delta x, t) = 0 \tag{11.37}$$

Upon dividing all the terms by Δx and rearranging the terms, we obtain

$$-\left[\frac{v(x + \Delta x, t) - v(x, t)}{\Delta x}\right] = Ri(x, t) + L\frac{\partial i(x, t)}{\partial t} \tag{11.38}$$

In the limit as $\Delta x \to 0$, it becomes a differential equation as

$$-\frac{\partial v(x, t)}{\partial x} = Ri(x, t) + L\frac{\partial i(x, t)}{\partial t} \tag{11.39}$$

Similarly the application of the Kirchhoff's current law at node $N + 1$ gives

$$i(x, t) - G\Delta x v(x + \Delta x, t) - C\Delta x \frac{\partial v(x + \Delta x, t)}{\partial t} - i(x + \Delta x, t) = 0 \tag{11.40}$$

Upon dividing all the terms by Δx and taking the limit as $\Delta x \to 0$, we get a second differential equation

$$-\frac{\partial i(x, t)}{\partial x} = Gv(x, t) + C\frac{\partial v(x, t)}{\partial t} \tag{11.41}$$

These two equations are known as 'transmission line equations' or 'telegrapher's equations'. These are in time domain as the time is visible explicitly in them.

Transmission line equations

$$-\frac{\partial v}{\partial x} = Ri(x, t) + L\frac{\partial i}{\partial t}$$

$$-\frac{\partial i}{\partial x} = Gv(x, t) + C\frac{\partial v}{\partial t}$$

Frequency domain: Most of the times, we are interested in the performance of transmission lines under sinusoidal steady-state conditions, i.e. when they are connected to sinusoidal sources. To this end, we undertake the analysis in the phasor domain. Let us consider a line of finite length l lying along the x-direction. Over this line, let an arbitrary point P be at a distance of x from the sending end side and at a distance of d from the receiving end side. If we consider a line of differential length dx at this point, then the series impedance of this section will be $(R + j\omega L)dx$ and the shunt admittance will be $(G + j\omega C)dx$. Let the voltage and current at the sending end of the differential length be $\mathbf{V}$ and $\mathbf{I}$, respectively and those at the receiving end be $\mathbf{V} + d\mathbf{V}$ and $\mathbf{I} + d\mathbf{I}$, respectively. Then,

$$\mathbf{V} - (\mathbf{V} + d\mathbf{V}) = \mathbf{I}(R + j\omega L)dx; \quad \mathbf{I} - (\mathbf{I} + d\mathbf{I}) = \mathbf{V}(G + j\omega C)dx$$

$$\Rightarrow \quad -\frac{d\mathbf{V}}{dx} = (R + j\omega L)\mathbf{I} \quad \text{and} \quad -\frac{d\mathbf{I}}{dx} = (G + j\omega C)\mathbf{V}$$

$$-\frac{d\mathbf{V}}{dx} = \mathbf{ZI} \tag{11.42}$$

$$-\frac{d\mathbf{I}}{dx} = \mathbf{YV} \tag{11.43}$$

Equations (11.42) and (11.43) are called '*telegrapher's equations*' in the phasor form.

Now,

$$\frac{d^2\mathbf{V}}{dx^2} = -\mathbf{Z}\frac{d\mathbf{I}}{dx}$$

and

$$\frac{d^2\mathbf{I}}{dx^2} = -\mathbf{Y}\frac{d\mathbf{V}}{dx}$$

here $\mathbf{Z} = R + j\omega L$ and $\mathbf{Y} = G + j\omega C$

$$\frac{d^2\mathbf{V}}{dx^2} = \mathbf{ZYV} \quad \text{and} \quad \frac{d^2\mathbf{I}}{dx^2} = \mathbf{ZYI}$$

The constant

$$\sqrt{\mathbf{ZY}} = \sqrt{(R + j\omega L)(G + j\omega C)}$$

is equal to the propagation constant γ. Therefore,

$$\frac{d^2\mathbf{V}}{dx^2} = \gamma^2\mathbf{V} \tag{11.44}$$

and

$$\frac{d^2\mathbf{I}}{dx^2} = \gamma^2\mathbf{I} \tag{11.45}$$

where γ = complex propagation constant = $\sqrt{\mathbf{ZY}} = \sqrt{(R + j\omega L)(G + j\omega C)}$.

The above two second-order differential equations for voltage and current at an arbitrary point over a uniform line are called *wave equations* and they are useful in describing the voltage and current profiles over the line.

Wave equations

$$\frac{d^2\mathbf{V}}{dx^2} = \gamma^2\mathbf{V}$$

$$\frac{d^2\mathbf{I}}{dx^2} = \gamma^2\mathbf{I}$$

The complex propagation constant γ describes the wave propagation characteristics of the medium. In general, it is a complex quantity with both real and imaginary parts. The real part, called the attenuation constant α, describes the attenuation characteristic and the imaginary part, called the phase shift constant β, gives the phase shift per unit length. We choose the square root values that give positive values for α and β as the transmission lines we consider are passive type.

The sending end need not necessarily be the input point of the line where the source is connected and the receiving end need not necessarily be the output point where the load is connected.

The general solution to the above two equations can be expressed in either one of the two forms: a form involving exponential functions or another form involving hyperbolic functions. We discuss each one of them.

11.5.1 Solutions of Wave Equations

The wave equations of the lines are the second-order ordinary differential equations and their solutions are given by

$$\mathbf{V} = ae^{-\gamma x} + be^{\gamma x} \tag{11.46}$$

$$\mathbf{I} = ce^{-\gamma x} + de^{\gamma x} \tag{11.47}$$

Here *a*, *b*, *c* and *d* are arbitrary constants, which can be complex and out of the four only two are independent.

Using

$$\frac{d\mathbf{V}}{dx} = -\mathbf{ZI}$$

the current can be expressed as

$$\mathbf{I} = -\sqrt{\frac{(G + j\omega C)}{R + j\omega L}}(-ae^{-\gamma x} + be^{\gamma x}) \tag{11.48}$$

The fraction as we will see soon is equal to the inverse of the characteristic impedance $\mathbf{Z}_0$ of the line. Hence,

$$\mathbf{I} = \frac{-1}{\mathbf{Z}_0}(-ae^{-\gamma x} + be^{\gamma x}) \tag{11.49}$$

Each of the above two expressions involve two exponentials: the term involving the exponential $e^{-\gamma x}$ represents $+x$ travelling wave or forward travelling wave, whereas the exponential involving $e^{-\gamma x}$ represents $-x$ travelling wave or reverse travelling wave. So the first terms in the above expressions represent the voltage and current in the forward travelling wave, whereas the second terms represent those pertaining to the reverse travelling wave.

We can make an important observation from Eqs. (11.47) and (11.49) regarding the ratio of voltage to current. In the case of incident wave, the ratio is $\mathbf{Z}_0$ and it is $-\mathbf{Z}_0$ for the reflected wave.

Therefore, we can conclude that there exist, in general, two waves over the line travelling in opposite directions and it results in a standing wave.

11.5.2 Evaluation of the Arbitrary Constants

The evaluation of the arbitrary constants that arise in the general solution of a differential equation requires boundary conditions or initial conditions. The number of required initial conditions must always be equal to the number of arbitrary constants for which values are to be fixed. In the case of transmission lines, the initial conditions are given or available in the form of voltages, currents and impedances, sometimes those at the sending end and sometimes those at the receiving end. Figure 11.6 shows the analysis of a uniform transmission line.

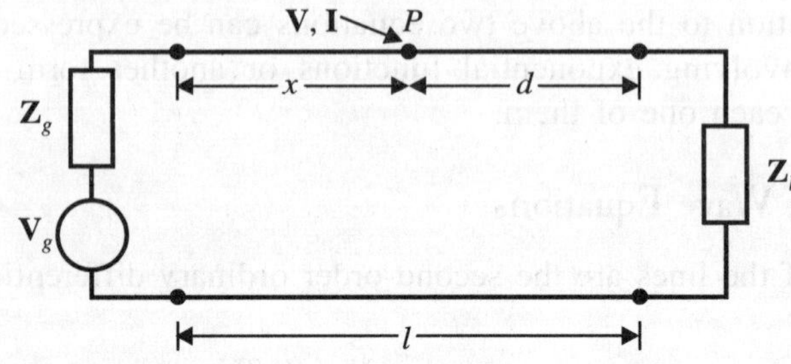

FIGURE 11.6 Analysis of uniform transmission line.

To begin with, we proceed assuming the initial conditions as sending end voltage and current and try to evaluate the arbitrary constants a and b in terms of the given initial conditions. By definition, the sending end voltage and current are

$$\mathbf{V}_s = \mathbf{V}_{x=0} \quad \text{and} \quad \mathbf{I}_s = \mathbf{I}|_{x=0}$$

Applying these definitions to the wave equations, we obtain

$$\mathbf{V}_s = a + b \quad \text{and} \quad \mathbf{I}_s = \frac{a}{\mathbf{Z}_0} - \frac{b}{\mathbf{Z}_0} \quad \text{or} \quad \mathbf{I}_s\mathbf{Z}_0 = a - b$$

Therefore, the arbitrary constants become

$$a = \frac{\mathbf{V}_s + \mathbf{I}_s\mathbf{Z}_0}{2} \quad \text{and} \quad b = \frac{\mathbf{V}_s - \mathbf{I}_s\mathbf{Z}_0}{2}$$

Substituting the above values in the wave equations, we get the expression for voltage as

$$\mathbf{V} = \frac{\mathbf{V}_s + \mathbf{I}_s\mathbf{Z}_0}{2} e^{-\gamma x} + \frac{\mathbf{V}_s - \mathbf{I}_s\mathbf{Z}_0}{2} e^{\gamma x}$$

$$= \frac{\mathbf{V}_s(e^{-\gamma x} + e^{\gamma x})}{2} + \frac{\mathbf{I}_s\mathbf{Z}_0(e^{-\gamma x} - e^{\gamma x})}{2}$$

$$= \mathbf{V}_s \cosh \gamma x - \mathbf{I}_s\mathbf{Z}_0 \sinh \gamma x$$

Similarly for current, the expression becomes

$$\mathbf{I} = \mathbf{I}_s \cosh \gamma x - \left(\frac{\mathbf{V}_s}{\mathbf{Z}_0}\right) \sinh \gamma x$$

These are the expressions for voltage and currents in terms of sending end voltage and currents. They can also be expressed in terms of sending end current and impedance in the following manner.

$$\mathbf{V} = \mathbf{I}_s \frac{\mathbf{Z}_s + \mathbf{Z}_0}{2} e^{-\gamma x} + \mathbf{I}_s \frac{\mathbf{Z}_s - \mathbf{Z}_0}{2} e^{\gamma x}$$

$$= \frac{\mathbf{I}_s}{2}\left[(\mathbf{Z}_s + \mathbf{Z}_0)e^{-\gamma x} + (\mathbf{Z}_s - \mathbf{Z}_0)e^{\gamma x}\right]$$

Similarly for current,

$$\mathbf{I} = \frac{\mathbf{I}_s}{2\mathbf{Z}_0}\left[(\mathbf{Z}_s + \mathbf{Z}_0)e^{-\gamma x} + (\mathbf{Z}_s - \mathbf{Z}_0)e^{\gamma x}\right]$$

When the receiving end parameters are given, then the arbitrary constants can be expressed as

$$\mathbf{V}_r = \mathbf{V}|_{x=l} \quad \text{and} \quad \mathbf{I}_r = \mathbf{I}|_{x=l}$$

$$\mathbf{V}_r = ae^{-\gamma l} + be^{\gamma l} \quad \text{and} \quad \mathbf{I}_r\mathbf{Z}_\gamma = ae^{-\gamma l} - be^{\gamma l}$$

Therefore,

$$a = \frac{\mathbf{V}_r + \mathbf{I}_r\mathbf{Z}_0}{2} e^{\gamma l} \quad \text{and} \quad b = \frac{\mathbf{V}_r - \mathbf{I}_r\mathbf{Z}_0}{2} e^{-\gamma l}$$

Substituting the above values of the arbitrary constants, now we obtain the expression for voltage as

$$\mathbf{V} = \frac{\mathbf{V}_r + \mathbf{I}_r\mathbf{Z}_0}{2} e^{\gamma l} e^{-\gamma x} + \frac{\mathbf{V}_r - \mathbf{I}_r\mathbf{Z}_0}{2} e^{-\gamma l} e^{\gamma x}$$

$$= \frac{\mathbf{V}_r\left[e^{\gamma(l-x)} + e^{-\gamma(l-x)}\right]}{2} + \frac{\mathbf{I}_r\mathbf{Z}_0\left[e^{\gamma(l-x)} - e^{-\gamma(l-x)}\right]}{2}$$

$$= \mathbf{V}_r \cosh \gamma(l - x) + \mathbf{I}_r\mathbf{Z}_0 \sinh \gamma(l - x)$$

$$= \mathbf{V}_r \cosh \gamma d + \mathbf{I}_r\mathbf{Z}_0 \sinh \gamma d, \text{ where } (l - x) = d$$

Similarly for current

$$\mathbf{I} = \mathbf{I}_r \cosh \gamma(l - x) + \left(\frac{\mathbf{V}_r}{\mathbf{Z}_0}\right)\mathbf{I}_r \sinh \gamma(l - x)$$

$$= \mathbf{I}_r \cosh \gamma d + \left(\frac{\mathbf{V}_r}{\mathbf{Z}_0}\right)\mathbf{I}_r \sinh \gamma d$$

They can also be expressed in terms of receiving end current and impedance in the following manner:

$$\mathbf{V} = \mathbf{I}_r \frac{\mathbf{Z}_r + \mathbf{Z}_0}{2} e^{\gamma l} e^{-\gamma x} + \mathbf{I}_r \frac{\mathbf{Z}_r - \mathbf{Z}_0}{2} e^{-\gamma l} e^{\gamma x}$$

$$= \frac{\mathbf{I}_r}{2}\left[(\mathbf{Z}_r + \mathbf{Z}_0)e^{\gamma(l-x)} + (\mathbf{Z}_r - \mathbf{Z}_0)e^{-\gamma(l-x)}\right]$$

$$= \frac{\mathbf{I}_r}{2}\left[(\mathbf{Z}_r + \mathbf{Z}_0)e^{\gamma d} + (\mathbf{Z}_r - \mathbf{Z}_0)e^{-\gamma d}\right]$$

Similarly for current

$$\mathbf{I} = \frac{\mathbf{I}_r}{2\mathbf{Z}_0}\left[(\mathbf{Z}_r + \mathbf{Z}_0)e^{\gamma d} - (\mathbf{Z}_r - \mathbf{Z}_0)e^{-\gamma d}\right]$$

Voltage-current profiles over a line

- In terms of sending end current and impedance, I_s, Z_s

$$\mathbf{V} = \frac{\mathbf{I}_s}{2}\left[(\mathbf{Z}_s + \mathbf{Z}_0)e^{-\gamma x} + (\mathbf{Z}_s - \mathbf{Z}_0)e^{\gamma x}\right] \tag{11.50a}$$

$$\mathbf{I} = \frac{\mathbf{I}_s}{2\mathbf{Z}_0}\left[(\mathbf{Z}_s + \mathbf{Z}_0)e^{-\gamma x} + (\mathbf{Z}_s - \mathbf{Z}_0)e^{\gamma x}\right] \tag{11.50b}$$

- In terms of receiving end current and impedance

$$\mathbf{V} = \frac{\mathbf{I}_r}{2}\left[(\mathbf{Z}_r + \mathbf{Z}_0)e^{\gamma d} + (\mathbf{Z}_r - \mathbf{Z}_0)e^{-\gamma d}\right] \tag{11.51a}$$

$$\mathbf{I} = \frac{\mathbf{I}_r}{2\mathbf{Z}_0}\left[(\mathbf{Z}_r + \mathbf{Z}_0)e^{\gamma d} - (\mathbf{Z}_r - \mathbf{Z}_0)e^{-\gamma d}\right] \tag{11.51b}$$

- In terms of sending end voltage and currents

$$\mathbf{V} = \mathbf{V}_s \cosh \gamma x - \mathbf{I}_s\mathbf{Z}_0 \sinh \gamma x \tag{11.52a}$$

$$\mathbf{I} = \mathbf{I}_s \cosh \gamma x - \left(\frac{\mathbf{V}_s}{\mathbf{Z}_0}\right) \sinh \gamma x \tag{11.52b}$$

- In terms of receiving end voltage and currents

$$\mathbf{V} = \mathbf{V}_r \cosh \gamma d + \mathbf{I}_r\mathbf{Z}_0 \sinh \gamma d \tag{11.53a}$$

$$\mathbf{I} = \mathbf{I}_r \cosh \gamma d + \left(\frac{\mathbf{V}_r}{\mathbf{Z}_0}\right) \sinh \gamma d \tag{11.53b}$$

In the above expressions, the first terms represent the forward travelling waves or incident waves, whereas the second terms represent reverse travelling waves or reflected waves. The interference of these two waves results in a standing wave pattern.

Another important observation here is that the voltage and current are varying in exponential manner along the line when the time variations of the source are exponential. So we can conclude that the voltage and current vary along the line in a manner which is the same as that of the time variations of the source, which, of course, is a result of linearity.

11.6 STANDING WAVE PATTERN

The reflected wave is generated at the load as a result of reflection of the incident wave by the load impedance. As a result of reflection, now we have two waves over the line: one travelling towards load, incident wave and the other travelling towards the source, reflected wave. As both are travelling over the same path in opposite directions, interference between the waves takes place resulting in a standing wave pattern with maxima and minima. The voltage waveforms over a lossless line are shown in Figure 11.7.

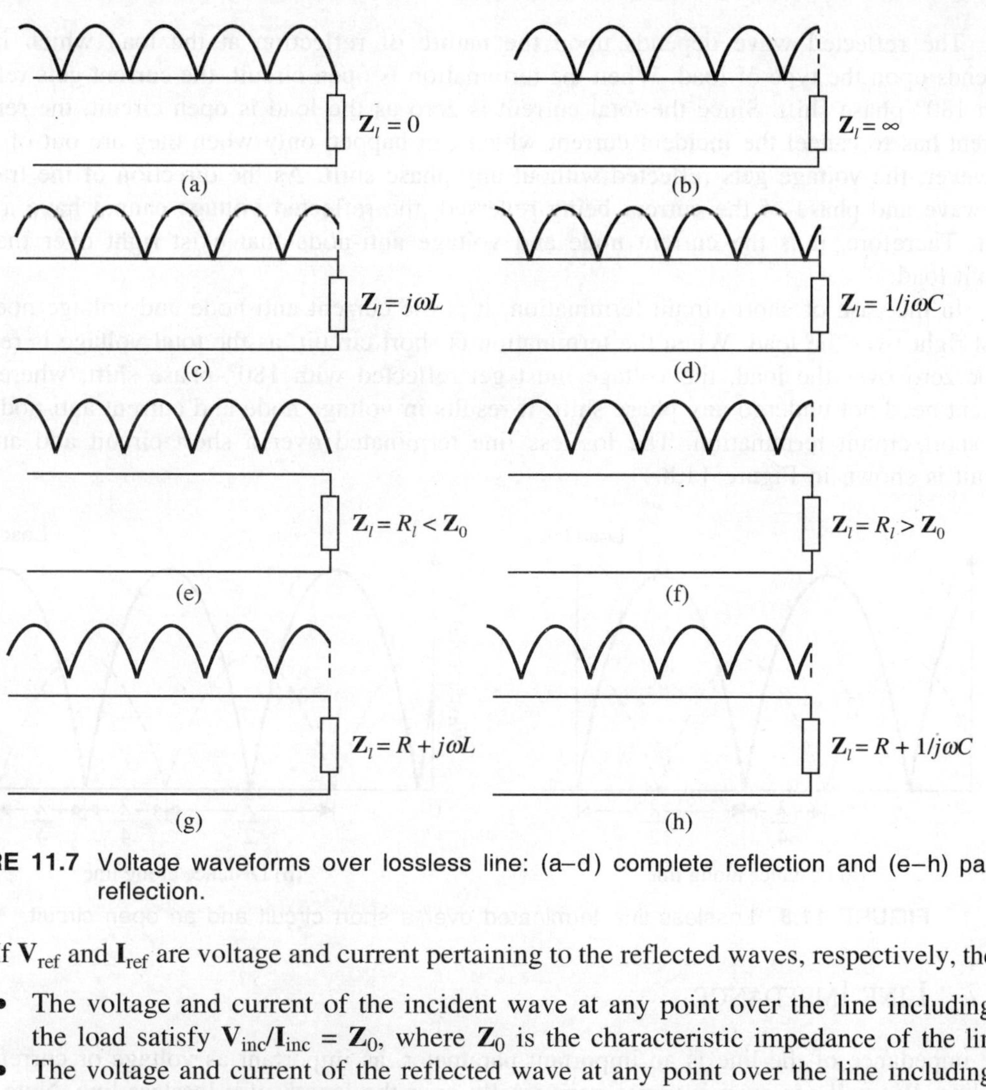

FIGURE 11.7 Voltage waveforms over lossless line: (a–d) complete reflection and (e–h) partial reflection.

If $\mathbf{V}_{\text{ref}}$ and $\mathbf{I}_{\text{ref}}$ are voltage and current pertaining to the reflected waves, respectively, then

- The voltage and current of the incident wave at any point over the line including at the load satisfy $\mathbf{V}_{\text{inc}}/\mathbf{I}_{\text{inc}} = \mathbf{Z}_0$, where $\mathbf{Z}_0$ is the characteristic impedance of the line.
- The voltage and current of the reflected wave at any point over the line including at the load satisfy $\mathbf{V}_{\text{ref}}/\mathbf{I}_{\text{ref}} = -\mathbf{Z}_0$.
- The total voltage and current at any point over the line are the sum of the voltages and currents of incident and reflected waves, i.e. $\mathbf{V} = \mathbf{V}_{\text{inc}} + \mathbf{V}_{\text{ref}}$ and $\mathbf{I} = \mathbf{I}_{\text{inc}} + \mathbf{I}_{\text{ref}}$, respectively.
- The load voltage V_l is the sum of the voltages of the incident and reflected waves at the load, that is, $\mathbf{V}_l = \mathbf{V}_{\text{inc}}\big|_{\text{at load}} + \mathbf{V}_{\text{ref}}\big|_{\text{at load}}$.
- The load current I_l is the sum of the currents of the incident and reflected waves at the load, that is, $\mathbf{I}_l = \mathbf{I}_{\text{inc}}\big|_{\text{at load}} + \mathbf{I}_{\text{ref}}\big|_{\text{at load}}$.
- The vector ratio of **V/I** at any point over the load gives the line impedance at that point. If the point under consideration is right over the load, the ratio becomes equal to the load impedance, i.e. $\mathbf{V}_l/\mathbf{I}_l = \mathbf{Z}_l$, where $\mathbf{Z}_l$, where $\mathbf{Z}_l$ is the load impedance.

The reflected wave depends upon the nature of reflection at the load which in turn depends upon the type of load. When the termination is open circuit, the current gets reflected with 180° phase shift. Since the total current is zero as the load is open circuit, the reflected current has to cancel the incident current which can happen only when they are out of phase. However, the voltage gets reflected without any phase shift. As the direction of the travel of the wave and phase of the current being reversed, the reflected voltage cannot have a phase shift. Therefore, it is the current node and voltage anti-node that exist right over the open circuit load.

In the case of short circuit termination, it is the current anti-node and voltage node that exist right over the load. When the termination is short circuit, as the total voltage is required to be zero over the load, the voltage must get reflected with 180° phase shift, whereas the current need not undergo any phase shift. It results in voltage node and current anti-node over the short circuit termination. The lossless line terminated over a short circuit and an open circuit is shown in Figure 11.8.

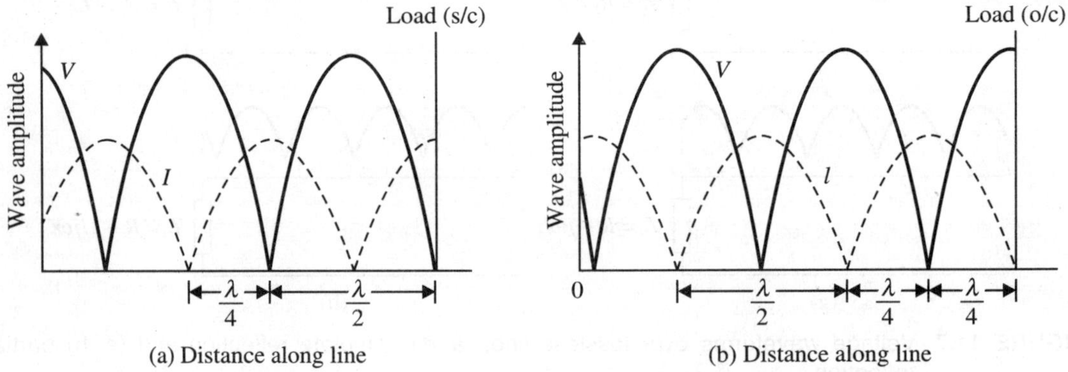

FIGURE 11.8 Lossless line terminated over a short circuit and an open circuit.

11.7 LINE IMPEDANCE

The impedance of the line is an important parameter, as important as voltage or current over the line. We will see soon it varies periodically over the length of a lossless line. Note that it is equal to the load impedance at the terminating end of the line and equal to the input impedance at the sending end of the line. Now we define the line impedance formally and try to find an expression for it in terms of characteristic and terminating impedances.

The line impedance at a point over the line is P defined as the ratio of the complex phasor voltage **V** to the complex phasor current **I** at that point. Thus,

$$\mathbf{Z}(P) = \frac{\mathbf{V}(P)}{\mathbf{I}(P)} \tag{11.54}$$

It is a complex quantity and we represent it by a boldface letter in time roman font and its magnitude by a normal italic letter, i.e. $Z = |\mathbf{Z}|$.

The line impedance at the load becomes the load impedance and at the sending end it is equal to the sending end impedance or input impedance. By substituting the appropriate expressions for voltage and current in the defining relation, we get the following expressions for the line impedance:

The line impedance at a distance of x from the sending end, in terms of sending end impedances $\mathbf{Z}_s$ and $\mathbf{Z}_0$, is

$$\mathbf{Z} = \mathbf{Z}_0 \frac{\mathbf{Z}_s - \mathbf{Z}_0 \tanh \gamma x}{\mathbf{Z}_0 - \mathbf{Z}_s \tanh \gamma x} \tag{11.55a}$$

For lossless line, $\gamma = j\beta$ and $\tanh \gamma x = \tanh j\beta x = j \tan \beta x$, resulting in

$$\mathbf{Z} = \mathbf{Z}_0 \frac{\mathbf{Z}_s - \mathbf{Z}_0\, j \tan \beta x}{\mathbf{Z}_0 - \mathbf{Z}_s\, j \tan \beta x} \tag{11.55b}$$

The line impedance at a distance of d from the receiving end, in terms of receiving end impedances $\mathbf{Z}_l$ and $\mathbf{Z}_0$, is

$$\mathbf{Z} = \mathbf{Z}_0 \frac{\mathbf{Z}_l + \mathbf{Z}_0 \tanh \gamma d}{\mathbf{Z}_0 + \mathbf{Z}_l \tanh \gamma d} \tag{11.56a}$$

For lossless line,

$$\mathbf{Z} = \mathbf{Z}_0 \frac{\mathbf{Z}_l + \mathbf{Z}_0\, j \tan \beta d}{\mathbf{Z}_0 + \mathbf{Z}_l\, j \tan \beta d} \tag{11.56b}$$

We observe from the above expressions for line impedance that

$$\mathbf{Z}\left(x + \frac{\lambda}{2}\right) = \mathbf{Z}(x) \tag{11.57}$$

$$\mathbf{Z}\left(d + \frac{\lambda}{2}\right) = \mathbf{Z}(d) \tag{11.58}$$

It implies that the line impedance is periodic along the length of the line with a periodicity of half wavelength.

Derivation

To derive the expressions in terms of sending end parameters, consider the following expressions for voltage and current from Eqs. (11.52) and (11.53).

$$\mathbf{V} = \mathbf{V}_s \cosh \gamma x - \mathbf{I}_s \mathbf{Z}_0 \sinh \gamma x$$

$$\mathbf{I} = \mathbf{I}_s \cosh \gamma x - \left(\frac{\mathbf{V}_s}{\mathbf{Z}_0}\right) \sinh \gamma x$$

$$\mathbf{V} = \mathbf{V}_r \cosh \gamma d + \mathbf{I}_r \mathbf{Z}_0 \sinh \gamma d$$

$$\mathbf{I} = \mathbf{I}_r \cosh \gamma d + \left(\frac{\mathbf{V}_r}{\mathbf{Z}_0}\right) \sinh \gamma d$$

Now consider the ratio of voltage to current resulting in

$$\mathbf{Z} = \frac{\mathbf{V}}{\mathbf{I}} = \frac{\mathbf{V}_s \cosh \gamma x - \mathbf{I}_s \mathbf{Z}_0 \sinh \gamma x}{\mathbf{I}_s \cosh \gamma x - \left(\frac{\mathbf{V}_s}{\mathbf{Z}_0}\right) \sinh \gamma x}$$

$$= \mathbf{Z}_0 \frac{\left(\frac{\mathbf{V}_s}{\mathbf{I}_s}\right) \cosh \gamma x - \mathbf{Z}_0 \sinh \gamma x}{\mathbf{Z}_0 \cosh \gamma x - \left(\frac{\mathbf{V}_s}{\mathbf{I}_s}\right) \sinh \gamma x} = \mathbf{Z}_0 \frac{\mathbf{Z}_s - \mathbf{Z}_0 \tanh \gamma x}{\mathbf{Z}_0 - \mathbf{Z}_s \tanh \gamma x}$$

In the case of lossless lines, $\gamma = j\beta$ and $\tanh \gamma x = j \tan \beta x$, resulting in an expression for line impedance of the lossless line as

$$\mathbf{Z} = \mathbf{Z}_0 \frac{\mathbf{Z}_s - \mathbf{Z}_0\, j \tan \beta x}{\mathbf{Z}_0 - \mathbf{Z}_s\, j \tan \beta x}$$

To derive the expressions for line impedance in terms of receiving end parameters, consider the following expressions for voltage and current from Eqs. (11.51a and b):

$$\mathbf{V} = \mathbf{V}_r \cosh \gamma d + \mathbf{I}_r \mathbf{Z}_0 \sinh \gamma d$$

$$\mathbf{I} = \mathbf{I}_r \cosh \gamma d + \left(\frac{\mathbf{V}_r}{\mathbf{Z}_0}\right) \sinh \gamma d$$

Now consider the ratio of voltage to current resulting in

$$\mathbf{Z} = \frac{\mathbf{V}}{\mathbf{I}} = \frac{\mathbf{V}_r \cosh \gamma d + \mathbf{I}_r \mathbf{Z}_0 \sinh \gamma d}{\mathbf{I}_r \cosh \gamma d + \left(\frac{\mathbf{V}_r}{\mathbf{Z}_0}\right) \sinh \gamma d}$$

$$\mathbf{Z} = \mathbf{Z}_0 \frac{\left(\frac{\mathbf{V}_r}{\mathbf{I}_r}\right) \cosh \gamma d + \mathbf{Z}_0 \sinh \gamma d}{\mathbf{Z}_0 \cosh \gamma d + \left(\frac{\mathbf{V}_r}{\mathbf{I}_r}\right) \sinh \gamma d} = \mathbf{Z}_0 \frac{\mathbf{Z}_l + \mathbf{Z}_0 \tanh \gamma d}{\mathbf{Z}_0 + \mathbf{Z}_l \tanh \gamma d}$$

In the case of lossless lines, $\gamma = j\beta$ and $\tanh \gamma d = j \tan \beta d$, resulting in an expression for line impedance of the lossless line as

$$\mathbf{Z} = \mathbf{Z}_0 \frac{\mathbf{Z}_l + \mathbf{Z}_0\, j \tan \beta d}{\mathbf{Z}_0 + \mathbf{Z}_l\, j \tan \beta d}$$

The line impedance for a uniform line, in general, we can observe, is a complex quantity varying periodically along the length of the line. Its periodicity is $\lambda_g/2$.

11.8 INPUT IMPEDANCE

The input impedance of a line is the impedance offered by the line at its input terminals. As the source is connected to the line at its input terminals, this quantity is of some importance in selecting a source of proper internal impedance so that the maximum power is pushed over to the line. It is useful to compute the power flowing down the line when a generator of voltage with internal impedance is connected to the line.

Formally we can define the input impedance of the line as the ratio of the complex phasor voltage to the complex phasor current at the input terminals of the line. Thus,

$$\mathbf{Z}_{\text{in}} = \frac{\mathbf{V}_{\text{in}}}{\mathbf{I}_{\text{in}}} = \frac{\text{Input voltage}}{\text{Input current}} \tag{11.59}$$

It is a complex quantity and its value depends upon the configuration, length and termination of the line. For a line of length terminated over an impedance $\mathbf{Z}_l$, it is given by

$$\mathbf{Z}_{\text{in}} = \mathbf{Z}_0 \frac{\mathbf{Z}_l + \mathbf{Z}_0 \tanh \gamma l}{\mathbf{Z}_0 + \mathbf{Z}_l \tanh \gamma l} \tag{11.60}$$

Derivation

$$\mathbf{Z}_{\text{in}} = \mathbf{Z}\big|_{d=l} = \frac{\mathbf{V}_s}{\mathbf{I}_s} = \frac{\mathbf{V}}{\mathbf{I}}\bigg|_{d=l}$$

$$= \frac{\mathbf{V}_r \cosh \gamma d + \mathbf{I}_r \mathbf{Z}_0 \sinh \gamma d}{\mathbf{I}_r \cosh \gamma d + \left(\dfrac{\mathbf{V}_r}{\mathbf{Z}_0}\right) \sinh \gamma d}\Bigg|_{d=l} = \frac{\mathbf{V}_r \cosh \gamma l + \mathbf{I}_r \mathbf{Z}_0 \sinh \gamma l}{\mathbf{I}_r \cosh \gamma l + \left(\dfrac{\mathbf{V}_r}{\mathbf{Z}_0}\right) \sinh \gamma l}$$

$$= \frac{\left(\dfrac{\mathbf{V}_r}{\mathbf{I}_r}\right) + \mathbf{Z}_0 \tanh \gamma l}{1 + \left(\dfrac{1}{\mathbf{Z}_0}\right) \tanh \gamma l}$$

but $\mathbf{V}_r/\mathbf{I}_r = \mathbf{Z}_l$ as the line is terminated with $\mathbf{Z}_l$. Therefore,

$$\mathbf{Z}_{\text{in}} = \frac{\mathbf{Z}_l + \mathbf{Z} \tanh \gamma l}{1 + \left(\dfrac{1}{\mathbf{Z}_0}\right) \tanh \gamma l} \quad \text{or} \quad = \mathbf{Z}_0 \frac{\mathbf{Z}_l + \mathbf{Z}_0 \tanh \gamma l}{\mathbf{Z}_0 + \mathbf{Z}_l \tanh \gamma l}$$

In case the line is lossless, $\gamma = j\beta$ and $\tanh \gamma l = \tanh j\beta l = j\beta l = j \tan \beta l$ resulting in

$$\mathbf{Z}_{\text{in}} = \mathbf{Z}_0 \frac{\mathbf{Z}_l + j\mathbf{Z}_0 \tan \beta l}{\mathbf{Z}_0 + j\mathbf{Z}_l \tan \beta l} \tag{11.61}$$

The normalized input impedance, defined as the ratio of input impedance to characteristic impedance, can be expressed as

$$\mathbf{z}_{\text{in}} = \frac{\mathbf{Z}_{\text{in}}}{\mathbf{Z}_0} = \frac{\mathbf{z}_l + j\tan\beta l}{1 + j\mathbf{z}_l \tan\beta l} \tag{11.62a}$$

The normalized input admittance, which is the inverse of the normalized impedance, can be expressed as

$$\mathbf{y}_{\text{in}} = \frac{1}{\mathbf{z}_{\text{in}}} = \frac{\mathbf{y}_l + j\tan\beta l}{1 + j\mathbf{y}_l \tan\beta l} \tag{11.62b}$$

11.9 MEASUREMENT OF SECONDARY CONSTANTS

The characteristic impedance and propagation constant of a line are secondary constants of the line and they are also measurable quantities. The procedures and necessary formulae are discussed hereunder.

The characteristic impedance is an important measurable parameter pertaining to lines. Using the network analyser, it is possible to measure the input impedances $\mathbf{Z}_{\text{sc}}$ and $\mathbf{Z}_{\text{oc}}$ of the line when the terminations are short circuit and open circuit, respectively. Then the characteristic impedance of the line and its phase shift constant can be computed by using the following formula:

$$\mathbf{Z}_0 = \sqrt{\mathbf{Z}_{\text{sc}}\mathbf{Z}_{\text{oc}}} \tag{11.63}$$

Derivation

Consider a finite length l line whose characteristic impedance is to be measured. When the termination is short circuit, its input impedance will be

$$\mathbf{Z}_{\text{in}}\Big|_{\mathbf{Z}_l=0} = \mathbf{Z}_{\text{sc}} = \mathbf{Z}_0 \left.\frac{\mathbf{Z}_l + \mathbf{Z}_0 \tanh \gamma l}{\mathbf{Z}_0 + \mathbf{Z}_l \tanh \gamma l}\right|_{\mathbf{Z}_l=0} = \mathbf{Z}_0 \tanh \gamma l$$

With open circuit termination, its input impedance will be

$$\mathbf{Z}_{\text{in}}\Big|_{\mathbf{Z}_l\to\infty} = \mathbf{Z}_{\text{oc}} = \mathbf{Z}_0 \left.\frac{\mathbf{Z}_l + \mathbf{Z}_0 \tanh \gamma l}{\mathbf{Z}_0 + \mathbf{Z}_l \tanh \gamma l}\right|_{\mathbf{Z}_l\to\infty} = \mathbf{Z}_0 \coth \gamma l$$

Multiplying the above two equations, we get $\mathbf{Z}_{\text{sc}}\mathbf{Z}_{\text{oc}} = \mathbf{Z}_0^2$ resulting in

$$\mathbf{Z}_0 = \sqrt{\mathbf{Z}_{\text{sc}}\mathbf{Z}_{\text{oc}}}$$

Using the network analyser it is possible to measure the input impedances of the line both on open circuit and on short circuit. Then the characteristic impedance of the line can be computed by using the following formula:

$$\mathbf{Z}_0 = \sqrt{\mathbf{Z}_{\text{sc}}\mathbf{Z}_{\text{oc}}}$$

The propagation constant can also be computed from the measured values of $\mathbf{Z}_{sc}$ and $\mathbf{Z}_{oc}$ using

$$\gamma = \frac{1}{l}\tanh^{-1}\sqrt{\frac{\mathbf{Z}_{sc}}{\mathbf{Z}_{oc}}} \tag{11.64}$$

Proof: For lossless line

$$\mathbf{Z}_{sc} = \mathbf{Z}_0 \tanh \gamma l$$
$$\mathbf{Z}_{oc} = \mathbf{Z}_0 \coth \gamma l$$

$$\frac{\mathbf{Z}_{sc}}{\mathbf{Z}_{oc}} = \tanh^2 \gamma l$$

$$\therefore \qquad \gamma = \frac{1}{l}\tanh^{-1}\sqrt{\frac{\mathbf{Z}_{sc}}{\mathbf{Z}_{oc}}}$$

The phase shift constant can also be computed from the measured values of $\mathbf{Z}_{sc}$ and $\mathbf{Z}_{oc}$ when the line is lossless using

$$\beta = \frac{1}{l}\tan^{-1}\sqrt{\frac{-\mathbf{Z}_{sc}}{\mathbf{Z}_{oc}}}$$

Because of the phase ambiguity associated with the tangent function, the length should be less than or equal to $\lambda/2$ to provide an unambiguous result.

11.10 LINE SECTIONS

Small sections of the lines are used for some important purposes other than power transmission. Their applications include load measurement, impedance transformation, load matching, etc. Now we analyse and study important properties of some of the widely used line sections.

11.10.1 Eighth-wave Transmission Line

- It is a lossless uniform line of length $\lambda/8$.
- The magnitude of its input impedance is equal to its characteristic impedance $\mathbf{Z}_0 = R_0$ when termination is over a pure resistance.

  Proof: The input impedance of a lossless line of length $\lambda/8$ is

$$\mathbf{Z}_{in}\big|_{l=\lambda/8} = \mathbf{Z}_0 \left.\frac{\mathbf{Z}_l + j\mathbf{Z}_0 \tan \beta l}{\mathbf{Z}_0 + j\mathbf{Z}_l \tan \beta l}\right|_{l=\lambda/8}$$

$$= \mathbf{Z}_0 \frac{\mathbf{Z}_l + j\mathbf{Z}_0}{\mathbf{Z}_0 + j\mathbf{Z}_l} \tag{11.65a}$$

  If the line is terminated in a pure resistance $\mathbf{Z}_l = R_l$, then,

$$\mathbf{Z}_{in}\big|_{l=\lambda/8} = R_0 \frac{R_l + jR_0}{R_0 + jR_l} \tag{11.65b}$$

The numerator and denominator of the above expression have identical magnitudes, so that $|\mathbf{Z}_{in}|_{l=\lambda/8} = R_0$, a pure real quantity. Therefore, the magnitude of its input impedance is equal to its characteristic impedance $\mathbf{Z}_0 = R_0$ when termination is over a pure resistance.

Thus, an eighth-wave line can be used to transform any resistance into an impedance whose magnitude is equal to the characteristic impedance $Z_0 = R_0$ of the line and also to obtain a magnitude match between a resistance of any value and a source of internal resistance R_0.

11.10.2 Quarter-wave Transmission Line

- It is a lossless uniform line of length $\lambda/4 + n\lambda/2$, $n = 0, 1, 2, \ldots$.
- Its input impedance is inversely proportional to its terminating impedance.

Proof: The input impedance of a lossless line of length $\lambda/4$ is

$$\mathbf{Z}_{in}\Big|_{l=\lambda/4} = \mathbf{Z}_0 \left.\frac{\mathbf{Z}_l + j\mathbf{Z}_0 \tan \beta l}{\mathbf{Z}_0 + j\mathbf{Z}_l \tan \beta l}\right|_{l=\lambda/4}$$

As $\beta = \dfrac{2\pi}{\lambda}$, $\beta l = \left(\dfrac{2\pi}{\lambda}\right)\left(\dfrac{\lambda}{4}\right) = \dfrac{\pi}{2}$

$$\mathbf{Z}_{in}\Big|_{l=\lambda/4} = \mathbf{Z}_0 \frac{\mathbf{Z}_l + j\mathbf{Z}_0 \tan\left(\dfrac{\pi}{2}\right)}{\mathbf{Z}_0 + j\mathbf{Z}_l \tan\left(\dfrac{\pi}{2}\right)}$$

$$= \mathbf{Z}_0 \frac{\dfrac{\mathbf{Z}_l}{\tan\left(\dfrac{\pi}{2}\right)} + j\mathbf{Z}_0}{\dfrac{\mathbf{Z}_0}{\tan\left(\dfrac{\pi}{2}\right)} + j\mathbf{Z}_l}$$

$$= \frac{\mathbf{Z}_0^2}{\mathbf{Z}_l} \tag{11.66a}$$

Also

$$\frac{\mathbf{Z}_{in}\big|_{l=\lambda/4}}{\mathbf{Z}_0} = \frac{\mathbf{Z}_0}{\mathbf{Z}_l}$$

or

$$\mathbf{Z}_{in}\Big|_{l=\lambda/4} = \mathbf{y}_l \tag{11.66b}$$

- Thus, the quarter-wave section acts as an impedance transformer or impedance inverter. The input impedance is large when the terminal impedance is small and vice versa. A pure resistance termination gets transformed into a pure resistance. If the output impedance consists of a resistance in series with an inductive reactance, the input impedance becomes a resistance in parallel with a capacity reactance. Its input impedance is inductive if the termination is capacitive and vice versa provided the characteristic impedance is resistive.

- It can also be used to step up the voltage. As long as it is lossless, the ratio between output and input voltages is just the square root of the ratio of the output and input impedances being matched.

  Derivation: The voltage equation for the lossless line is

$$\mathbf{V} = \mathbf{V}_r \left[\cos \beta d + j\left(\frac{\mathbf{Z}_0}{\mathbf{Z}_r}\right) \sin \beta d \right]$$

For a quarter-wave line, i.e. $d = \lambda/4$, the input voltage $\mathbf{V}_s$ is

$$\mathbf{V}_s = |\mathbf{V}|_{d=\lambda/4} = j\mathbf{I}_l \mathbf{Z}_0$$

and hence

$$\frac{\mathbf{V}_s}{\mathbf{V}_l} = j\frac{\mathbf{I}_l}{\mathbf{V}_l}\mathbf{Z}_0 = j\frac{1}{\mathbf{Z}_l}\sqrt{\mathbf{Z}_l \mathbf{Z}_s} = j\sqrt{\frac{\mathbf{Z}_s}{\mathbf{Z}_l}}$$

Therefore, the voltage step-up, i.e. the ratio of output voltage to input voltage, becomes

$$\left|\frac{\mathbf{V}_l}{\mathbf{V}_s}\right| = \sqrt{\frac{Z_l}{Z_s}} \tag{11.67a}$$

In the case of an open-circuited quarter wave line, according to the above equation, the step-up is infinite but it is for an ideal case of an absolute lossless line. To find the exact step-up for the practical case of a low loss line, we need to consider the equation which takes the losses and the consequent attenuation into account. Consider the voltage equation

$$\mathbf{V} = \mathbf{V}_r \cosh \gamma l + \mathbf{I}_r \mathbf{Z}_0 \sinh \gamma l$$

As the line is open-circuited, $\mathbf{I}_l = 0$ and the second term becomes zero. As the line is quarter wave in length, $l = \lambda/4$ and $\gamma l = \alpha l + j\beta l = [\alpha l + \beta(\lambda/4)] = \alpha l + j\pi/2$. Incorporating these aspects into the voltage equation, we get

$$\mathbf{V}_s = j\mathbf{V}_l \sinh \alpha l$$

$$\frac{\mathbf{V}_l}{\mathbf{V}_s} = \frac{1}{\sinh \alpha l} \approx \frac{1}{\alpha l} = \frac{2\mathbf{Z}_0}{Rl}$$

Here we have used an expression for the attenuation constant of a low loss line given by

$$\alpha = \frac{R}{2\mathbf{Z}_0}$$

Its derivation is given in one of the chapter-end problems.

Therefore, for a quarter-wave line, i.e. $\lambda = l/4 = v/4f$, the voltage step-up is given by

$$\left|\frac{\mathbf{V}_l}{\mathbf{V}_s}\right| = \frac{2Z_0}{Rl} = \frac{8Z_0}{R\lambda} = \frac{8Z_0 f}{Rv} \tag{11.67b}$$

Incidentally, for a three quarter-wave section, the voltage step-up is

$$\left|\frac{\mathbf{V}_l}{\mathbf{V}_s}\right| = \frac{8Z_0 f}{3Rv}$$

- The sensitivity to frequency change is its main drawback.
- Applications:
  - It acts as an *impedance transformer* or *inverter* as it can step up or step down the impedance.
  - It can be used as a voltage step-up transformer.
  - It is used for load-matching purposes.
  - Another application of the short circuit quarter-wave line is as an insulator to support an open-wire line or the centre conductor of a coaxial line. This application makes use of the fact that the input impedance of the quarter-wave-shorted line is very high.

11.10.3 Half-wave Transmission Line

- It is a lossless uniform line of length $n\lambda/2$, $n = 1, 2, \ldots$.
- Its input impedance is equal to the terminating impedance; this property is independent of the characteristic impedance $\mathbf{Z}_0$ but it is frequency dependent.

  Proof: The input impedance of a lossless line of length $\lambda/2$ is

$$\mathbf{Z}_{\text{in}}\big|_{l=\lambda/2} = \mathbf{Z}_0 \frac{\mathbf{Z}_l + j\mathbf{Z}_0 \tan \beta l}{\mathbf{Z}_0 + j\mathbf{Z}_l \tan \beta l}\bigg|_{l=\lambda/2}$$

$$= \mathbf{Z}_0 \frac{\mathbf{Z}_l + j\mathbf{Z}_0 \tan \pi}{\mathbf{Z}_0 + j\mathbf{Z}_l \tan \pi} = \mathbf{Z}_0 \frac{\mathbf{Z}_l}{\mathbf{Z}_0} = \mathbf{Z}_l \tag{11.68}$$

- Applications:
  - It has its greatest utility in connecting a load to a source in cases where the load and source cannot be made adjacent.
  - The short-circuited $\lambda/2$ line can act as a band stop filter; it can be used to measure velocity factor and dielectric constant of medium.
  - Half-wave line is also used to measure the impedance that is not accessible physically.

11.11 ULTRA HIGH FREQUENCY (UHF) LINES AS CIRCUIT ELEMENTS

Apart from the transfer of energy, the transmission lines have other applications also. At UHF, sections of lines are used as circuit elements. In the frequency range of 0.15 GHz to 3 GHz, the physical size of the sections of the transmission lines is convenient enough to use them as circuit elements. However, above 3 GHz, the physical size of the lines becomes too small to be of any use as circuit elements and instead of line sections, in their place, pieces of waveguides are used.

Some of the widely used line sections are shorted and opened lines with lengths less than, equal to and more than a quarter wave length. The magnitude of the input reactance of these sections can be computed from the expression we have for input impedance of the lossless line by the substitution of appropriate values for load impedance, i.e. $\mathbf{Z}_l = 0$ for shorted sections and $\mathbf{Z}_l = \infty$ for opened out sections.

$$\mathbf{Z}_{\text{in}}\Big|_{\mathbf{Z}_l=0} = \mathbf{Z}_{\text{sc}} = \mathbf{Z}_0 \left.\frac{\mathbf{Z}_l + \mathbf{Z}_0 \tanh \gamma l}{\mathbf{Z}_0 + \mathbf{Z}_l \tanh \gamma l}\right|_{\mathbf{Z}_l=0} = \mathbf{Z}_0 \tanh \gamma l = j\mathbf{Z}_0 \tanh \beta l \qquad (11.69a)$$

$$\mathbf{Z}_{\text{in}}\Big|_{\mathbf{Z}_l\to\infty} = \mathbf{Z}_{\text{oc}} = \mathbf{Z}_0 \left.\frac{\mathbf{Z}_l + \mathbf{Z}_0 \tanh \gamma l}{\mathbf{Z}_0 + \mathbf{Z}_l \tanh \gamma l}\right|_{\mathbf{Z}_l\to\infty} = \mathbf{Z}_0 \coth \gamma l = -\,j\mathbf{Z}_0 \coth \beta l \qquad (11.69b)$$

Thus, it can be observed that for line lengths less than a quarter of a wavelength, the shorted section is capacitive and open section is inductive. An important aspect that should be noted about these inductances and capacitances is their frequency dependency. Various resonant sections and their equivalents are shown in Figure 11.9.

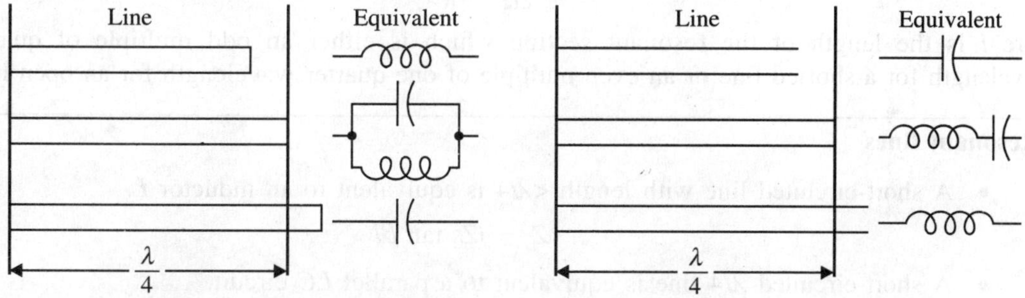

FIGURE 11.9 Various resonant sections and their equivalents.

In general, the resistive component of the input impedance is almost negligible for low loss line sections when they are used as circuit elements at higher frequency ranges like UHF. However, in the case of shorted quarter-wave line and open half-wave line, the input reactance goes to infinity and it corresponds to a parallel resonant circuit whose impedance is infinite if resistance is neglected. For a practical line, the resistive component of the input impedance exists and it is required to be taken into account. For both these line sections, the actual input impedance when the series resistance is neglected is a pure resistance of a very high value. In the case of a line of length l, its input resistance R_{ar} at a resonant length is given by

$$R_{ar} = \frac{2\mathbf{Z}_0^2}{Rl} \tag{11.70}$$

Here R is the resistance, primary constant of the line and l is the length of the resonant section which is either an odd multiple of a quarter wavelength for a shorted line or an even multiple of one-quarter wavelength for an open line.

The shorted quarter-wave section is equivalent to the parallel resonant circuit and it produces the resonant rise of voltage or current which exists in such circuits. A voltage step-up up of several hundred times is possible depending upon the Q of the line. When the quarter wave section is tapped at some point along its length, it can be shown that the impedance seen at the tap point is a pure resistance that varies as the square of the sine of the angular distance from the shorted end. It is given by

$$R_x = \frac{2\mathbf{Z}_0^2}{Rl} \sin^2 \beta x \tag{11.71}$$

The proof for the above relation is given in one of the chapter-end problems.

One of the important properties of resonant transmission line sections is their frequency selectivity or ability to pass freely some frequencies but to discriminate against others. The selectivity of a resonant circuit is usually stated in terms of quality factor Q. It can be shown for resonant line sections of length l, the Q is given by

$$Q = \frac{\beta}{2\alpha} = \frac{\omega l}{R} \tag{11.72}$$

Here l is the length of the resonant section which is either an odd multiple of quarter wavelength for a shorted line or an even multiple of one-quarter wavelength for an open line.

Resonant lines

- A short-circuited line with length $<\lambda/4$ is equivalent to an inductor L.

$$Z_s = jZ_0 \tan \beta l$$

- A short-circuited $\lambda/4$ line is equivalent to a parallel LC circuit.

$$Z_s = \frac{Z_0^2}{\tanh \alpha l} \approx \frac{2Z_0^2}{Rl}$$

- A short-circuited line with length $>\lambda/4$ is equivalent to a capacitor C.

$$Z_s = -jZ_0 \cot \beta l$$

- An open-circuited line with length $<\lambda/4$ is equivalent to a capacitor C.

$$Z_s = -jZ_0 \cot \beta l$$

(Contd...)

(Contd...)

- An open-circuited $\lambda/4$ line is equivalent to a series LC circuit.
- An open-circuited line with length $>\lambda/4$ is equivalent to an inductor L.

$$Z_s = jZ_0 \tan \beta l$$

- A quarter-wave lossless shorted transmission line is equivalent to a parallel LC resonant circuit; on the low frequency side it is inductive and on the high frequency side it is capacitive.
- A quarter-wave lossless open-circuited transmission line is equivalent to a series LC resonant circuit; on the low frequency side it is capacitive and on the high frequency side it is inductive.

SOLVED PROBLEMS

1. At 8 MHz the characteristic impedance of a transmission line is $(40 - j2)\ \Omega$ and the propagation constant is $(0.01 + j0.18)$ per metre. Find the primary constants.

Solution: In finding the primary constants, we use the following relations:

$$\boldsymbol{\gamma}\mathbf{Z}_0 = R + j\omega L$$

$$\frac{\boldsymbol{\gamma}}{\mathbf{Z}_0} = G + j\omega C$$

From the given values,

$$\boldsymbol{\gamma}\mathbf{Z}_0 = (0.01 + j0.18)\ (40 - j2) - 0.76 + j7.18$$

$$\frac{\boldsymbol{\gamma}}{\mathbf{Z}_0} = \frac{(0.01 + j0.18)}{(40 - j2)} = 2.49 \times 10^{-5} + j4.5 \times 10^{-3}$$

Therefore,

$$R = 0.76\ \Omega$$

$$\omega L = 7.18 \quad \Rightarrow \quad L = \frac{7.18}{2\pi \times 8 \times 10^6} = 0.142\ \mu\text{H}$$

$$G = 2.49 \times 10^{-5}\ ℧$$

$$\omega C = 4.5 \times 10^{-3} \quad \Rightarrow \quad C = \frac{4.5 \times 10^{-3}}{2\pi \times 8 \times 10^6} = 0.089 \times 10^{-9}\ \text{F}$$

2. A telephone line has $R = 10\ \Omega$/km, $L = 0.0037$ H/km, $C = 0.0083\ \mu$F/km and $G = 0.4 \times 10^{-6}$ ℧/km. Determine Z_0, α and β at 1 KHz.

Solution:

$$\mathbf{Z}_0 = \sqrt{\frac{R + j\omega L}{G + j\omega C}} = \sqrt{\frac{10 + j2\pi \times 10^3 \times 0.0037}{0.4 \times 10^{-6} + j2\pi \times 10^3 \times 0.0083 \times 10^{-6}}}$$

$$= 10^3 \sqrt{\frac{10 + j2\pi \times 3.7}{0.4 + j2\pi \times 8.3}} = 10^3 \sqrt{0.485\angle -22.83} = 696.4\angle -11.41°$$

$$\gamma = \sqrt{(R + j\omega L)(G + j\omega C)}$$

$$= \sqrt{(10 + j2\pi \times 10^3 \times 0.0037)(0.4 \times 10^{-6} + j2\pi \times 10^3 \times 0.0083 \times 10^{-6})}$$

$$= 10^{-3}\sqrt{(10 + j2\pi \times 3.7)(0.4 + j2\pi \times 8.3)} = 10^{-3}\sqrt{1319.825\angle 156.285^\circ}$$

$$= 36.33 \times 10^{-3}\angle 78.1425^\circ = 7.465 \times 10^{-3} + j0.0355$$

The real part of the propagation constant represents the attenuation constant and the imaginary part represents the phase shift constant. Therefore,

$$\alpha = 7.465 \times 10^{-3} \quad \text{and} \quad \beta = 0.0355$$

3. Determine the propagation velocity v and the characteristic impedance of a lossless cable with L = 0.50 μH/m and C = 150 pF/m. Also find these parameters when ε_r = 2.5 and μ_r = 1.

Solution: The propagation velocity is related to the primary constants through

$$v = \frac{1}{\sqrt{LC}} = \frac{1}{\sqrt{0.5 \times 10^{-6} \times 150 \times 10^{-12}}} = 1.15 \times 10^8 \text{ m/s}$$

The characteristic impedance is related to the primary constants through

$$\mathbf{Z}_0 = \sqrt{\frac{L}{C}} = \sqrt{\frac{0.5 \times 10^{-6}}{150 \times 10^{-12}}} = 57.7\ \Omega$$

In the second case, when ε_r = 2.5 and mr = 1, the propagation velocity is

$$v = \frac{c}{\sqrt{\varepsilon_r \mu_r}} = \frac{3 \times 10^8}{\sqrt{2.5}} = 1.897 \times 10^8 \text{ m/s}$$

and the characteristic impedance is

$$\mathbf{Z}_0 = \frac{1}{vC} = \frac{1}{1.897 \times 10^8 \times 150 \times 10^{-12}} = 35.13\ \Omega$$

4. A lossy coaxial cable, operated at a frequency of 500 MHz, has the following primary constants: R = 2.25 Ω/m, L = 1 μH/m, C = 100 pF/m and G = 0 ℧/m Determine

(a) the propagation constant and the attenuation constant and

(b) the attenuation in dB of a 100 ft length cable.

Solution:

(a) The propagation constant in terms of the primary constants of the line is

$$\gamma = \sqrt{ZY} = \sqrt{(R + j\omega L)(G + j\omega C)}$$

Substituting the given values of primary constants, we obtain

$$\gamma = \sqrt{(2.25 + j2\pi \times 500 \times 10^6 \times 10^{-6})(j2\pi \times 500 \times 10^6 \times 100 \times 10^{-12})}$$

$$= \sqrt{(2.25 + j\pi \times 10^3)(j\pi \times 10^{-1})}$$

$$= \sqrt{986.96\angle 179.95^\circ}$$

$$\gamma = 31.42\angle 89.975^\circ \text{ m}^{-1} = 1.125 \times 10^{-2} + j31.42 \text{ m}^{-1}$$

(b) The real part of the propagation constant is the attenuation constant. Therefore, the attenuation constant of the given line is

$$\alpha = 1.125 \times 10^{-2} \text{ Np/m} = 3.43 \times 10^{-3} \text{ Np/ft}$$

Therefore, the attenuation in dB of l = 100 ft length cable is given by

$$3.43 \times 10^{-3} \times 100 = 0.343 \text{ Np}$$

$$= 0.343 \times 8.686 = 2.98 \text{ dB}$$

5. Consider a lossless transmission line terminated over a pure resistance. Show that the maximum or minimum of the standing pattern exists right over the termination.

Solution: The voltage and current over the lossless line, in terms of receiving end voltage and currents, are given by

$$\mathbf{V} = \mathbf{V}_r \cosh \gamma d + \mathbf{I}_r\mathbf{Z}_0 \sinh \gamma d$$

$$= \mathbf{V}_r \cosh j\beta d + \mathbf{I}_r\mathbf{Z}_0 \sinh j\beta d$$

$$= \mathbf{V}_r \cos \beta d + \mathbf{I}_r\mathbf{Z}_0\, j \sin \beta d$$

$$= \mathbf{V}_r \left[\cos \beta d + j\left(\frac{\mathbf{Z}_0}{\mathbf{Z}_r}\right)\sin \beta d\right]$$

$$\mathbf{I} = \mathbf{I}_r \cosh \gamma d + \left(\frac{\mathbf{V}_r}{\mathbf{Z}_0}\right)\sinh \gamma d$$

$$= \mathbf{I}_r \cosh j\beta d + \left(\frac{\mathbf{V}_r}{\mathbf{Z}_0}\right)\sinh j\beta d$$

$$= \mathbf{I}_r \left[\cos \beta d + \left(\frac{\mathbf{Z}_r}{\mathbf{Z}_0}\right) j \sin \beta d\right]$$

As the line is lossless, $\mathbf{Z}_0$ is real and as the termination is resistance, the $\mathbf{Z}_r$ is also real. The magnitudes of the voltage and current are

$$|\mathbf{V}| = |\mathbf{V}_r|\sqrt{\cos^2 \beta d + \left(\frac{\mathbf{Z}_0}{\mathbf{Z}_r}\right)^2 \sin^2 \beta d}$$

$$|\mathbf{I}| = |\mathbf{I}_r|\sqrt{\cos^2 \beta d + \left(\frac{\mathbf{Z}_r}{\mathbf{Z}_0}\right)^2 \sin^2 \beta d}$$

Now consider the expression for voltage when $\mathbf{Z}_0 > \mathbf{Z}_r$. From the expression, we can observe that the voltage assumes maximum for the value of d that makes the sine squared term maximum, i.e. at $d = \lambda/4$, $3\lambda/4$, etc. The current assumes maximum for the value of that makes the cosine squared term maximum, i.e. at $d = 0$, $\lambda/2$, $3\lambda/4$, etc. Note in this case, the current maximum and voltage minimum are over the load.

For the case $\mathbf{Z}_0 < \mathbf{Z}_r$, from the expression, we can observe that the voltage assumes maximum for the value of d that makes the cosine squared term maximum, i.e. at $d = 0$, $\lambda/2$, $3\lambda/4$, etc. The current assumes maximum for the value of d that makes the sine squared term maximum, i.e. at $d = \lambda/4$, $3\lambda/4$,, etc. In this case, the voltage maximum and current minimum are over the load.

Therefore, in both the cases, i.e. $\mathbf{Z}_0 > \mathbf{Z}_r$ and $\mathbf{Z}_0 < \mathbf{Z}_r$, load heads maxima or minima.

6. Consider a lossless transmission line terminated over a pure resistance R. Show that the terminating resistance $R = \rho R_0$ or R_0/ρ, or where ρ is the ratio of maximum to minimum voltage over the line, called SWR (standing wave ratio).

Solution: The magnitudes of the voltage over the lossless line is

$$|\mathbf{V}| = |\mathbf{V}_r|\sqrt{\cos^2 \beta d + \left(\frac{\mathbf{Z}_0}{\mathbf{Z}_r}\right)^2 \sin^2 \beta d}$$

We need to consider two cases, one for $\mathbf{Z}_0 > \mathbf{Z}_r$ and another for $\mathbf{Z}_0 < \mathbf{Z}_r$ separately. For $\mathbf{Z}_0 > \mathbf{Z}_r$, the voltage maximum is

$$|\mathbf{V}|_{\max} = |\mathbf{V}|_{\sin \beta d=1} = |\mathbf{V}_r|\frac{\mathbf{Z}_0}{\mathbf{Z}_r}$$

And the voltage minimum is

$$|\mathbf{V}|_{\min} = |\mathbf{V}|_{\cos \beta d=1} = |\mathbf{V}_r|$$

The ratio of maximum to minimum voltage gives

$$\frac{|\mathbf{V}|_{\max}}{|\mathbf{V}|_{\min}} = \frac{\mathbf{Z}_0}{\mathbf{Z}_r} \rightarrow \mathbf{Z}_r = \frac{\mathbf{Z}_0}{\rho}$$

For $\mathbf{Z}_0 < \mathbf{Z}_r$, the voltage minimum is

$$|\mathbf{V}|_{\min} = |\mathbf{V}|_{\sin \beta d=1} = |\mathbf{V}_r|\frac{\mathbf{Z}_0}{\mathbf{Z}_r}$$

And the voltage maximum is

$$|\mathbf{V}|_{\max} = |\mathbf{V}|_{\cos \beta d=1} = |\mathbf{V}_r|$$

The ratio of maximum to minimum voltage gives

$$\frac{|\mathbf{V}|_{\max}}{|\mathbf{V}|_{\min}} = \frac{\mathbf{Z}_r}{\mathbf{Z}_0} \rightarrow \mathbf{Z}_r = \rho\mathbf{Z}_0$$

It can also be shown, in a similar manner, that the ratio of maximum to minimum current is also ρ for both types of terminations.

7. Find the input impedance for a lossless line with characteristic impedance of 75 Ω and terminating impedance of $(45 + j60)\Omega$ when frequency and lengths are

(a) $f = 15$ MHz, $l = 5$ m
(b) $f = 50$ MHz, $l = 3$ m and
(c) $f = 37.5$ MHz, $l = 5$ m.

Solution: We use the following formula to find out the input impedance:

$$\mathbf{Z}_{\text{in}} = \mathbf{Z}_0 \frac{\mathbf{Z}_l + j\mathbf{Z}_0 \tan \beta l}{\mathbf{Z}_0 + j\mathbf{Z}_l \tan \beta l}$$

It can also be done via reflection coefficient route by using the following formulae:

$$\mathbf{Z}_{\text{in}} = \frac{1 - \Gamma_l}{1 + \Gamma_l}$$

$$\Gamma_l = \frac{\mathbf{Z}_l - \mathbf{Z}_0}{\mathbf{Z}_l + \mathbf{Z}_0}$$

(a) $f = 15$ MHz, $l = 5$ m

$$\beta = \frac{2\pi}{\lambda} = \frac{2\pi \times f}{c} = \frac{2\pi \times 15 \times 10^6}{3 \times 10^8} \text{ rad/m}$$

$$\beta l = \frac{2\pi \times 15 \times 10^6 \times 5}{3 \times 10^8} = 1.57 \text{ rad}$$

$$\tan \beta l = \tan 1.57 = 1255.76$$

$$\begin{aligned}
\mathbf{Z}_{\text{in}} &= \mathbf{Z}_0 \frac{\mathbf{Z}_l + j\mathbf{Z}_0 \tan \beta l}{\mathbf{Z}_0 + j\mathbf{Z}_l \tan \beta l} \\
&= 75 \frac{45 + j60 + j75 \times 1255.76}{75 + j(45 + j60) \times 1255.76} \\
&= 75 \frac{45 + j94242}{-75270.6 + j56509.2} \\
&= 75\angle{-53.13^\circ} = 45 - j59.99\ \Omega
\end{aligned}$$

(b) $f = 50$ MHz, $l = 3$ m

$$\beta = \frac{2\pi}{\lambda} = \frac{2\pi \times f}{c} = \frac{2\pi \times 50 \times 10^6}{3 \times 10^8} \text{ rad/m}$$

$$\beta l = \frac{2\pi \times 50 \times 10^6 \times 3}{3 \times 10^8} = 3.14 \text{ rad}$$

$$\tan \beta l = \tan 3.14 = -1.592 \times 10^{-3}$$

$$\mathbf{Z}_{\text{in}} = \mathbf{Z}_0 \frac{\mathbf{Z}_l + j\mathbf{Z}_0 \tan \beta l}{\mathbf{Z}_0 + j\mathbf{Z}_l \tan \beta l}$$

$$= 75 \frac{45 + j60 + j75 \times (-1.592 \times 10^{-3})}{75 + j(45 + j60) \times (-1.592 \times 10^{-3})}$$

$$= 75 \frac{45 + j59.88}{75.0955 - j0.07164}$$

$$= 44.88 + j59.84\ \Omega$$

(c) $f = 37.5$ MHz, $l = 5$ m.

$$\beta = \frac{2\pi}{\lambda} = \frac{2\pi \times f}{c} = \frac{2\pi \times 37.5 \times 10^6}{3 \times 10^8} \text{rad/m}$$

$$\beta l = \frac{2\pi \times 37.5 \times 10^6 \times 5}{3 \times 10^8} = 0.785 \text{ rad}$$

$$\tan \beta l = \tan 0.785 = 0.999$$

$$\mathbf{Z}_{\text{in}} = \mathbf{Z}_0 \frac{\mathbf{Z}_l + j\mathbf{Z}_0 \tan \beta l}{\mathbf{Z}_0 + j\mathbf{Z}_l \tan \beta l}$$

$$= 75 \frac{45 + j60 + j75 \times 0.999}{75 + j(45 + j60) \times 0.999}$$

$$= 75 \frac{45 + j134.9}{15.06 + j44.955}$$

$$= 224.96 + j0.287\ \Omega$$

8. Find the input impedance of a lossless line of 50 Ω with the following parameters: source voltage $\mathbf{V}_g = 100\angle 0$, source internal impedance $\mathbf{Z}_g = 30 + j40\ \Omega$, terminating impedance $\mathbf{Z}_l = 100\ \Omega$ and length of the line $l = 5\lambda/8$. Also find the time average power delivered to the load.

Solution: We use the following formula to find out the input impedance:

$$\mathbf{Z}_{\text{in}} = \mathbf{Z}_0 \frac{\mathbf{Z}_l + j\mathbf{Z}_0 \tan \beta l}{\mathbf{Z}_0 + j\mathbf{Z}_l \tan \beta l}$$

$$\beta l = \frac{2\pi}{\lambda} \times \frac{5\lambda}{8} = 1.25\,\pi$$

$$\tan \beta l = \tan 1.25\,\pi = 1$$

$$\mathbf{Z}_{\text{in}} = \mathbf{Z}_0 \frac{\mathbf{Z}_l + j\mathbf{Z}_0 \tan \beta l}{\mathbf{Z}_0 + j\mathbf{Z}_l \tan \beta l}$$

$$= 50 \frac{100 + j50}{50 + j100} = 40 - j30\ \Omega$$

The current into the line is

$$\mathbf{I}_s = \frac{\mathbf{V}_g}{\mathbf{Z}_g + \mathbf{Z}_{\text{in}}} = \frac{100\angle 0}{30 + j40 + 40 - j30}$$

$$= \frac{100\angle 0}{70 + j10} = 1.414\angle -0.142^\circ = 1.4 - j0.2$$

Now, we find out the sending end voltage of the line.

$$\mathbf{V}_s = \mathbf{I}_s\mathbf{Z}_{\text{in}} = (1.4 - j0.2)\ (40 - j30)$$

$$= 50 - j50 = 70.71\angle -0.785^\circ$$

Next, we calculate the power into the line. As the line is lossless, it is the power into the load from the line.

$$P_{\text{ave}} = \frac{1}{2} \times 1.414 \times 70.71 \times \cos(0.6432)$$

$$= \frac{1}{2} \times 1.414 \times 70.71 \times 0.8 = 40\ \text{W}$$

Note that the angle 0.6432 rad is the angle between the voltage and the current.

9. Find the load impedance of a lossless line of 100 Ω for maximum power transfer when the source voltage $\mathbf{V}_g = 100\angle 0$, source internal impedance $\mathbf{Z}_g = 25 + j50\ \Omega$, and length of the line $l = 0.2\lambda$. Also find the time average power delivered to the load in such circumstances.

Solution: For maximum power transfer, the source internal impedance and input impedance of the line must be related through a complex conjugate relation. When the relation is satisfied, maximum power enters into the line and as the line is lossless, maximum power then moves into the load.
Therefore, for maximum power transfer,

$$\mathbf{Z}_{\text{in}} = \mathbf{Z}_g^* = 25 - j50\ \Omega$$

$$\beta l = \frac{2\pi}{\lambda} \times 0.2\lambda = 0.4\pi$$

$$\tan \beta l = \tan 0.4\pi = 0.37$$

$$\mathbf{Z}_{\text{in}} = \mathbf{Z}_0 \frac{\mathbf{Z}_l + j\mathbf{Z}_0 \tan \beta l}{\mathbf{Z}_0 + j\mathbf{Z}_l \tan \beta l}$$

Substituting the available or given values

$$25 - j50 = 100\frac{\mathbf{Z}_l + j100 \times 3.07}{100 + j\mathbf{Z}_l \times 3.07}$$

$$(25 - j50)\ (100 + j\mathbf{Z}_l \times 3.07) = 100(\mathbf{Z}_l + j100 \times 3.07)$$

$$\mathbf{Z}_l(-1 + 1.535 + j0.7675) = -25 + j357$$

$$\mathbf{Z}_l = \frac{-25 + j357}{0.535 + j0.7675} = 297.76 + j240.13\ \Omega$$

Next, we calculate the power into the line. As the line is lossless, it is also the power into the load from the line.

The current into the line is

$$\mathbf{I}_s = \frac{\mathbf{V}_g}{\mathbf{Z}_g + \mathbf{Z}_{\text{in}}} = \frac{100\angle 0}{25 + j50 + 25 - j50} = 2\angle 0$$

Now, we find out the sending end voltage of the line.

$$\mathbf{V}_s = \mathbf{I}_s\mathbf{Z}_{\text{in}} = 2(25 - j50) = 118\angle{-63.43^\circ}$$

$$P_{\text{ave}} = \frac{1}{2} \times 2 \times 111.8 \times \cos(63.43) = 50\ \text{W}$$

Note that the angle 63.43°is the angle between the voltage and the current.

10. A transmission line of length 70 m is terminated in an impedance of $Z_r = 125 + j48$. If the frequency is 3 MHz and the characteristic impedance is 230 Ω, the find the sending impedance.

Solution: The relation we use to find the solution is

$$\mathbf{Z}_{\text{in}} = \mathbf{Z}_0 \frac{\mathbf{Z}_l + j\mathbf{Z}_0 \tan \beta l}{\mathbf{Z}_0 + j\mathbf{Z}_l \tan \beta l}$$

Given that

$$l = 70\ \text{m},\ \mathbf{Z}_r = \mathbf{Z}_l = 125 + j48,\ \beta = \frac{2\pi}{\lambda} = \frac{2\pi \times 3 \times 10^6}{3 \times 10^8} = 0.02\pi\ \text{rad/m and}\ \mathbf{Z}_0 = 230\ \Omega$$

$$\mathbf{Z}_{\text{in}}\big|_{l=70} = 230 \frac{125 + j48 + j230 \tan(0.02\pi \times 70)}{230 + j(125 + j48) \tan(0.02\pi \times 70)}$$

$$= 230 \frac{125 + j48 + j230 \times 3.077}{230 + j(125 + j48) \times 3.077}$$

$$= 230 \frac{125 + j755.71}{82 + j384.625} = 447.97\angle 2.64^\circ$$

This is the required sending end impedance.

11. A 100 Ω line of 1 km is terminated by a 200 Ω load and is fed by a generator of 10 V and internal impedance of 50 Ω. Find the load voltage and load power at 3×10^5 rad/s.

Solution: The relation we use to find the solution is

$$\mathbf{Z}_{\text{in}} = \mathbf{Z}_0 \frac{\mathbf{Z}_l + j\mathbf{Z}_0 \tan \beta l}{\mathbf{Z}_0 + j\mathbf{Z}_l \tan \beta l}$$

Given that

$$l = 1000 \text{ m}, \; \mathbf{Z}_r = \mathbf{Z}_l = 200 \text{ W}, \; \beta = \frac{2\pi}{\lambda} = \frac{3\times10^5}{3\times10^8} = 10^{-3} \text{ rad/m and } \mathbf{Z}_0 = 100\ \Omega$$

$$\mathbf{Z}_{\text{in}}\big|_{l=1000} = 100\frac{200 + j100\tan(10^{-3}\times10^3)}{100 + j200\tan(10^{-3}\times10^3)}$$

$$= 100\frac{200 + j100\times1.557}{100 + j200\times1.557} = 100\frac{200 + j155.7}{100 + j311.481}$$

$$= 100\frac{200 + j155.7}{100 + j311.481} = 77.47\angle-34.30$$

This impedance is in series with the source resistance across the voltage source. The current through this impedance and the voltage across it are sending end voltage and currents of the line. They can be calculated as

$$\mathbf{I}_s = \frac{10\angle0°}{50\angle0° + 77.47\angle-34.30°} = 0.0819\angle20.95°$$

$$\mathbf{V}_s = \mathbf{Z}_{\text{in}}\mathbf{I}_s = 77.47\angle-34.30 \times 0.0819\angle20.95° = 6.344\angle-13.35°$$

With the availability of the sending end current and voltages, the receiving end current and voltages can be computed using the following formulae:

$$\mathbf{V} = \frac{\mathbf{I}_s}{2}\left[(\mathbf{Z}_s + \mathbf{Z}_0)e^{-\gamma x} + (\mathbf{Z}_s - \mathbf{Z}_0)e^{\gamma x}\right]$$

$$\mathbf{I} = \frac{\mathbf{I}_s}{2\mathbf{Z}_0}\left[(\mathbf{Z}_s + \mathbf{Z}_0)e^{-\gamma x} + (\mathbf{Z}_s - \mathbf{Z}_0)e^{\gamma x}\right]$$

In the above expressions, $\gamma x = j\beta x = j \times 10^{-3} \times 10^3 = j$

$$\mathbf{V}_l = \frac{0.0819\angle20.95°}{2}\left[(77.47\angle-34.30 + 100)e^{-j} + (77.47\angle-34.30 - 100)e^{j}\right]$$

$$= \frac{0.0819\angle20.95°}{2}(169.7\angle-72.195° + 56.58\angle-72.21°)$$

$$= 19.26\angle-51.24°$$

This the value of the voltage across the load.

$$\mathbf{I}_l = \frac{0.0819\angle20.95°}{2\times100}\left[(77.47\angle-34.30 + 100)e^{-j} - (77.47\angle-34.30 - 100)e^{j}\right]$$

$$= \frac{0.0819\angle20.95°}{2\times100}(169.7\angle-72.195° - 56.58\angle-72.21°)$$

$$= 0.0463\angle-51.24°$$

The power consumed in the load becomes

$$P = \frac{1}{2}\text{Re}\,\mathbf{V}_l\mathbf{I}_l^* = \frac{1}{2}|\mathbf{V}_l||\mathbf{I}_l|\cos 0 = \frac{1}{2}\times 19.26\times 0.0463 = 0.4458\text{ W}$$

12. Derive, from its definition, the relation between characteristic impedance and primary constants.

Solution: It can be shown from its basic definition that the characteristic impedance of the line is related to its primary constants through

$$\mathbf{Z}_0 = \sqrt{\frac{(R + j\omega L)}{(G + j\omega C)}}$$

Consider the relation for the input impedance

$$\mathbf{Z}_{\text{in}} = \mathbf{Z}_0 \frac{\mathbf{Z}_l + \mathbf{Z}_0 \tanh \gamma l}{\mathbf{Z}_0 + \mathbf{Z}_l \tanh \gamma l}$$

When the termination is $\mathbf{Z}_{\text{in}}$, its input impedance is, by definition, its characteristic impedance.

$$\mathbf{Z}_{\text{in}}\Big|_{\mathbf{Z}_l=\mathbf{Z}_{\text{in}}} \triangleq \text{Characteristic impedance (CI)}$$

$$= \mathbf{Z}_0 \frac{\mathbf{Z}_l + \mathbf{Z}_0 \tanh \gamma l}{\mathbf{Z}_0 + \mathbf{Z}_l \tanh \gamma l}\Bigg|_{\mathbf{Z}_l=\mathbf{Z}_{\text{in}}}$$

$$\text{(CI)} = \mathbf{Z}_0 \frac{\text{(CI)} + \mathbf{Z}_0 \tanh \gamma l}{\mathbf{Z}_0 + \text{(CI)} \tanh \gamma l}$$

$$\text{(CI)}\,[\mathbf{Z}_0 + \text{(CI)}\tanh \gamma l] = \mathbf{Z}_0[\text{(CI)} + \mathbf{Z}_0 \tanh \gamma l]$$

$$\text{(CI)}^2 = \mathbf{Z}_0^2 \quad \Rightarrow \quad \text{(CI)} = \pm\mathbf{Z}_0$$

The negative values cannot be realized physically. Hence, the characteristic impedance must be positive only.

$$\text{(CI)} = \mathbf{Z}_0 = \sqrt{\frac{(R + j\omega L)}{(G + j\omega C)}}$$

13. Show that in the case of a shorted quarter-wave line and open half-wave line, the input impedance is given by

$$R_{\text{ar}} = \frac{2\mathbf{Z}_0^2}{Rl}$$

Solution:

$$\mathbf{V} = \mathbf{V}_r \cosh \gamma d + \mathbf{I}_r\mathbf{Z}_0 \sinh \gamma d$$

$$\mathbf{I} = \mathbf{I}_r \cosh \gamma d + \left(\frac{\mathbf{V}_r}{\mathbf{Z}_0}\right)\sinh \gamma d$$

For a shorted line of length l, the input voltage and current are

$$\mathbf{V}_s = \mathbf{I}_r\mathbf{Z}_0 \sinh \gamma l$$
$$\mathbf{I}_s = \mathbf{I}_r \cosh \gamma l$$

Dividing the voltage equation by the current equation, we can get the input impedance of the short-circuited line.

$$\mathbf{Z}_s = \mathbf{Z}_0 \tanh \gamma l = \mathbf{Z}_0 \frac{\sinh \gamma l}{\cosh \gamma l}$$
$$= \mathbf{Z}_0 \frac{\sinh \alpha l \cos \beta l + j \cosh \alpha l \sin \beta l}{\cosh \alpha l \cos \beta l + j \sinh \alpha l \sin \beta l}$$

For line of lengths equal to an odd multiple of a quarter wavelength, $\sin \beta l = \pm 1$ and $\cos \beta l = 0$. Thus,

$$\mathbf{Z}_s = \mathbf{Z}_0 \frac{\cosh \alpha l}{\sinh \alpha l}$$

As the line attenuation is small when working at high frequencies, αl is small, and in general $\cosh \alpha l \approx 1$ and $\sin \alpha l = \alpha l$ so that

$$Z_s \approx \frac{Z_0}{\alpha l}$$

We have already seen that for a low loss line, the attenuation constant is

$$\alpha = \frac{1}{2}\left(R\sqrt{\frac{C}{L}} + G\sqrt{\frac{L}{C}}\right)$$

In the case of air, dielectric lines, the losses due to conductance can be neglected and the attenuation constant becomes

$$\alpha = \frac{R}{2}\left(\sqrt{\frac{C}{L}}\right) = \frac{R}{2Z_0}$$

Substituting this in the above expression, we get the required relation for the input impedance of the line. An identical expression can be derived for an open ended section that is a multiple of a half-wave long.

14. Show that the input resistance variation is as the square of the distance from the shorted end over the length of a shorted quarter-wave line.

Solution: Let us suppose, V_s and R_s are the input voltage and line resistance to a shorted quarter-wave line when the source is connected at the open end. Then the input power is V_s^2/R_s and this power goes to meet the copper losses. When the source is connected across the line at a distance of x from the shorted end V_x and R_x and are the input voltage and line resistance, then the power into the line becomes V_x^2/R_x and this power also goes to meet the copper losses. As the copper losses remain the same independent of the point of source connection point, both the powers must be equal.

$$\frac{V_x^2}{R_x} = \frac{V_s^2 \sin^2 \beta x}{R_x} = \frac{V_s^2}{R_s} = \frac{V_s^2 Rl}{2Z_0^2}$$

which gives

$$R_x = \frac{2Z_0^2 \sin^2 \beta x}{Rl}$$

Therefore, we can say that the input resistance varies as the square of the sine of the distance from the shorted end.

15. Show that the Q of a resonant line section is given by

$$Q = \frac{\beta}{2\alpha} = \frac{\omega L}{R}$$

Solution: The input impedance of any shorted line section is

$$\mathbf{Z}_{\text{in}}\Big|_{\mathbf{Z}_l=0} = \mathbf{Z}_{\text{sc}} = \mathbf{Z}_0 \left.\frac{\mathbf{Z}_l + \mathbf{Z}_0 \tanh \gamma l}{\mathbf{Z}_0 + \mathbf{Z}_l \tanh \gamma l}\right|_{\mathbf{Z}_l=0} = \mathbf{Z}_0 \tanh \gamma l$$

$$= \mathbf{Z}_0 \frac{\sinh \gamma l}{\cosh \gamma l} = \mathbf{Z}_0 \frac{\sinh \alpha l \cos \beta l + j \cosh \alpha l \sin \beta l}{\cosh \alpha l \cos \beta l + j \sinh \alpha l \sin \beta l}$$

In the above expansion, we used the relation $\gamma = \alpha + j\beta$.
At resonant frequency f_0, the angle $\beta l = n\pi/2$ (n is odd), $\cos \beta l = 0$ and $\sin \beta l = \pm 1$. Substituting these relations,

$$\mathbf{Z}_{\text{in}} = \mathbf{Z}_0 \frac{\cosh \alpha l}{\sinh \alpha l} = \mathbf{Z}_0 \frac{1}{\tanh \alpha l} = \frac{\mathbf{Z}_0}{\alpha l}$$

At a frequency $f = f_0 + \delta f$, which is slightly off resonance, then

$$\beta l = \frac{2\pi f}{v} l = \frac{2\pi(f + \delta f)}{v} l = \frac{n\pi}{2} + \frac{2\pi l \delta f}{v}$$

and

$$\cos \beta l = -\sin \frac{2\pi l \delta f}{v}$$

$$\sin \beta l = \cos \frac{2\pi l \delta f}{v}$$

$$\mathbf{Z}_{\text{in}} = \mathbf{Z}_0 \frac{-\sinh \alpha l \sin\left(\frac{2\pi l \delta f}{v}\right) + j \cosh \alpha l \cos\left(\frac{2\pi l \delta f}{v}\right)}{-\cosh \alpha l \sin\left(\frac{2\pi l \delta f}{v}\right) + j \sinh \alpha l \cos\left(\frac{2\pi l \delta f}{v}\right)}$$

For moderately high Q circuits, $\cosh \alpha l \approx 1$ and $\sinh \alpha l \approx \alpha l$, $\cos(2\pi l\delta f/v) \approx 1$ and $\sin(2\pi l\delta f/v) \approx 2\pi l\delta f/v$. Substituting these values, we obtain

$$\mathbf{Z}_{in} = \frac{\mathbf{Z}_0}{\alpha l + j\left(\dfrac{2\pi l\delta f}{v}\right)}$$

When the imaginary term in the denominator is equal to the real term, the input impedance $\mathbf{Z}_{in}$ becomes 70.7% of its value for a resonant length, and the frequency shift required to give is, let us say, $\Delta f/2$. Therefore,

$$\frac{2\pi l}{2v}\Delta f = \alpha l$$

$$\Delta f = \frac{\alpha v}{\pi} = \frac{2\alpha f_0}{\beta}$$

The Q of the resonant section is therefore

$$Q = \frac{f_0}{\Delta f} = \frac{\beta}{2\alpha} = \frac{\pi f_0}{\alpha v} = \frac{2\pi f_0 \mathbf{Z}_0}{Rv} = \frac{\omega L}{R}$$

16. A lossless line with L = 0.50 μH/m and C = 150 pF/m is operated at a frequency of 10 MHz. Determine

(a) the shortest length of the line at which it acts as a 150 pF capacitor on an open circuit and a short circuit and

(b) the shortest length of the line at which it acts as a 2 μH inductor on an open circuit and a short circuit.

Solution: In further calculations we need the propagation velocity and the characteristic impedance of the line. We can find them from the given data as follows:

$$v = \frac{1}{\sqrt{LC}} = \frac{1}{\sqrt{0.5\times10^{-6}\times150\times10^{-12}}} = 1.15\times10^{8} \text{ m/s}$$

$$\mathbf{Z}_0 = \sqrt{\frac{L}{C}} = \sqrt{\frac{0.5\times10^{-6}}{150\times10^{-12}}} = 57.7\ \Omega$$

(a) The shortest length of the line at which it acts as a 150 pF capacitor:
On a short circuit the input impedance of lossless line ($\gamma = j\beta$) is

$$\mathbf{Z}_{in}\big|_{\mathbf{Z}_l=0} = \mathbf{Z}_{sc} = \mathbf{Z}_0 \tanh \gamma l$$

$$= j\mathbf{Z}_0 \tan \beta l = j\mathbf{Z}_0 \tan\left(\frac{2\pi}{\lambda}\right)l$$

Now we have to find out the smallest value of l for which

$$\frac{1}{j\omega C} = j\mathbf{Z}_0 \tan\left(\frac{2\pi}{\lambda}\right)l = j\mathbf{Z}_0 \tan\left(\frac{2\pi f}{v}\right)l \text{ as } \lambda = \frac{v}{f}$$

$$\tan\left(\frac{2\pi f}{v}\right)l = -\frac{1}{\omega C\mathbf{Z}_0}$$

$$= -\frac{1}{2\pi(10\times10^6)\times150\times10^{-12}\times57.7} = -1.839$$

Therefore,

$$\left(\frac{2\pi f}{v}\right)l = \tan^{-1}(-1.839) = n\pi - 1.07$$

Note that in the above expression, we used the fact that tan is periodic with π. As we need the shortest length we have to consider $n = 1$. It results in

$$\left(\frac{2\pi f}{v}\right)l = \pi - 1.07 = 2.07$$

Substituting the given values and solving for length, we get the minimum value for the length as

$$l = \frac{2.07\times1.15\times10^8}{2\pi(10\times10^6)} = 3.79 \text{ m}$$

With an open circuit termination, its input impedance will be

$$\mathbf{Z}_{\text{in}}\big|_{\mathbf{Z}_l\to\infty} = \mathbf{Z}_{\text{oc}} = \mathbf{Z}_0 \coth \gamma l$$

$$= -j\mathbf{Z}_0 \cot \beta l = -j\mathbf{Z}_0 \cot\left(\frac{2\pi}{\lambda}\right)l$$

If the same line is acting as the capacitor with the open circuit, then

$$\frac{1}{j\omega C} = -j\mathbf{Z}_0 \cot\left(\frac{2\pi}{\lambda}\right)l \rightarrow l = \frac{v}{2\pi f}\cot^{-1}\frac{1}{\omega C\mathbf{Z}_0} = \frac{v}{2\pi f}\tan^{-1}\omega C\mathbf{Z}_0$$

Note that $\cot^{-1}(x) = \tan^{-1}(1/x)$. After substituting the given values in the expressions, the minimum value of the length in this case can be found as

$$l = \frac{v}{2\pi f}\tan^{-1}\omega C\mathbf{Z}_0$$

$$l = \frac{1.15\times10^8}{2\pi(10\times10^6)}\tan^{-1}\left[2\pi(10\times10^6)\times150\times10^{-12}\times57.7\right]$$

$$= 1.83\times\tan^{-1}\frac{2\pi\times15\times57.7}{10^4} = 1.83\times0.498 = 0.91$$

(b) The shortest length of the line at which it acts as a 2 μH inductor:
If the same short-circuited line is to act as an inductor, then

$$j\omega L = j\mathbf{Z}_0 \tan\left(\frac{2\pi f}{v}\right)l \quad \Rightarrow \quad l = \frac{v}{2\pi f}\tan^{-1}\frac{\omega L}{\mathbf{Z}_0}$$

After substituting the given values in the expressions, the minimum value of the length in this case can be found as

$$l = \frac{1.15\times10^8}{2\pi(10\times10^6)}\tan^{-1}\frac{2\pi(10\times10^6)\times2\times10^{-6}}{57.7}$$

$$= 1.83 \times 1.14 = 2.08 \text{ m}$$

If the open-circuited line is acting as the inductor, then

$$j\omega L = -j\mathbf{Z}_0 \cot\left(\frac{2\pi f}{v}\right)l \quad \Rightarrow \quad l = \frac{v}{2\pi f}\cot^{-1}\frac{-\omega L}{\mathbf{Z}_0} = \frac{v}{2\pi f}\tan^{-1}\frac{-\mathbf{Z}_0}{\omega L}$$

After substituting the given values in the expressions, the minimum value of the length in this case can be found as

$$l = \frac{1.15\times10^8}{2\pi(10\times10^6)}\tan^{-1}\frac{-57.7}{2\pi(10\times10^6)\times2\times10^{-6}}$$

$$= 1.83 \times (\pi - 0.43) = 4.96 \text{ m}$$

EXERCISE SECTION

Essay Type Questions

1. Define a transmission line and list out its applications.
2. What are primary constants of a transmission line? Sketch equivalent circuits of different types of transmission lines using them.
3. What are secondary constants of a transmission line? Define and express them in terms of primary constants.
4. Define attenuation and phase shift constants of the line and express them in terms of primary constants.
5. Derive the transmission line equations in time domain and phasor domain.
6. Derive the wave equations of a transmission line. Write down their general solutions.
7. Express the voltage and current profiles of a line in hyperbolic form as well as in exponential form.
8. Sketch the equivalent circuits of nearly quarter and half-wave lines.
9. Derive an expression for the input impedance of a lossless line which is terminated by

(a) a load Z_l,
(b) an open circuit, and
(c) a short circuit
(Draw the suitable sketches.)

10. What are the properties and applications of an eighth-wave line, quarter-wave line and half-wave line? Give a list of their applications.

Numerical Problems

1. Determine the propagation velocity and the characteristic impedance of a lossless cable with $L = 0.75$ μH/m and $C = 150$ pF/m. Also find these parameters when $\varepsilon_r = 3.0$ and $\mu_r = 1$.
2. A telephone line has $R = 10$ Ω/km, $L = 0.004$ H/km $C = 0.0084$ μF/km and $G = 0.6 \times 10^{-6}$ ℧/km. Determine Z_0, α and β at 1 kHz.
3. A lossless line with $L = 0.65$ mH/m and $C = 135$ pF/m is operated at a frequency of 18 MHz. Determine
   (a) the shortest length of the line at which it acts as a 200 pF capacitor on an open circuit and a short circuit and
   (b) the shortest length of the line at which it acts as a 3.5 μH inductor on an open circuit and a short circuit.
4. A lossy coaxial cable operated at a frequency of 600 MHz, has the following primary constants: $R = 2.52$ Ω/m, $L = 1.55$ μH/m, $C = 155$ pF/m and $G = 0$ ℧/m. Determine the input impedance of a 20 m length cable on short circuit/open circuit and terminated with a load of 125 Ω
5. Find the input impedance for a lossless line with characteristic impedance of 75 Ω and terminating impedance of $(40 + j25)$Ω when frequencies and lengths are
   (i) $f = 21$ MHz, $l = 5$ m,
   (ii) $f = 57$ MHz, $l = 3$ m and
   (iii) $f = 37.5$ MHz, $l = 6$ m.
6. Find the load impedance of a lossless line of 100 Ω for maximum power transfer when source voltage $\mathbf{V}_g = 180\angle 0$ source internal impedance $\mathbf{Z}_g = 35 + j70$ Ω, and length of the line $l = 0.3\ \lambda$. Also find the time average power delivered to the load in such circumstances.
7. A 150 Ω line of 1 km is terminated by a 250 Ω load and is fed by a generator of 20 V and internal impedance of 50 Ω. Find the load voltage and load power at 4×10^5 rad/s.
8. A transmission line of length 80 m is terminated in an impedance of $\mathbf{Z}_r = 150 + j50$. If the frequency is 4 MHz and the characteristic impedance is 250 Ω, then find the sending impedance.

Fill in the Blanks

1. The primary constants of the transmission line are __________ constants and must always be specified __________ basis.
2. The secondary constants of the line are __________ and __________.
3. Time domain transmission line equations are __________.
4. Wave equations are __________.
5. For an HF line, the characteristic impedance is __________ and propagation constant is __________
6. The real part of a propagation constant is __________ and its imaginary part is __________.
7. For an HF line, the attenuation constant is __________ and the phase shift constant is __________.
8. Infinite line carries only __________ travelling wave.
9. The input impedance of the line when it is extended to infinity is __________.
10. The input impedance of the line is given by __________.
11. The input impedance of the shorted line is __________, whereas that of an opened line is __________.
12. The characteristic impedance of the line and its phase shift constant can be computed by using the formula __________.
13. Positions at which the input impedance is real are at __________.
14. An eighth-wave line transforms any resistance to impedance with a magnitude equal to __________ of the line.
15. __________ wave line acts as an impedance transformer or inverter.
16. An open circuited line with length $<\lambda/4$ is equivalent to __________.
17. An open circuited line $\lambda/4$ is equivalent to __________ circuit.
18. An open circuited line with length $>\lambda/4$ is equivalent to an __________.
19. The impedance of a quarter wave line along its length is pure __________.

Answers

1. distributed, per unit length
2. characteristic impedance, propagation constant
3. $-\dfrac{\partial v}{\partial x} = Ri(x,t) + L\dfrac{\partial i}{\partial t}, \quad -\dfrac{\partial i}{\partial x} = Gv(x,t) + C\dfrac{\partial v}{\partial t}$
4. $\dfrac{d^2\mathbf{V}}{dx^2} = \gamma^2\mathbf{V}, \quad \dfrac{d^2\mathbf{I}}{dx^2} = \gamma^2\mathbf{I}$
5. $\sqrt{\dfrac{L}{C}}, \quad \omega\sqrt{LC}$

6. attenuation constant, phase shift constant
7. zero, $j\omega\sqrt{LC}$
8. forward
9. charactertistic impedance
10. $\mathbf{Z}_{\text{in}} = \mathbf{Z}_0 \dfrac{\mathbf{Z}_l + \mathbf{Z}_0 \tanh \gamma l}{\mathbf{Z}_0 + \mathbf{Z}_l \tanh \gamma l}$
11. $\mathbf{Z}_{\text{in}}^{\text{sc}} = j\mathbf{Z}_0 \tan \beta l,\ \ \mathbf{Z}_{\text{in}}^{\text{oc}} = -j\mathbf{Z}_0 \cot \beta l$
12. $\mathbf{Z}_0 = \sqrt{\mathbf{Z}_{\text{sc}}\mathbf{Z}_{\text{oc}}}\ \ \beta = \dfrac{1}{l}\tan^{-1}\sqrt{\dfrac{-\mathbf{Z}_{\text{sc}}}{\mathbf{Z}_{\text{oc}}}}$
13. maxima and minima
14. characteristic impedance
15. quarter
16. capacitor C
17. series LC
18. inductor L
19. resistance

Match the Following

Set I

1. Primary constants — (a) real quantity
2. Secondary constants — (b) complex quantity
3. Characteristic impedance — (c) ohms/metre
4. Propagation constant — (d) $\mathbf{Z}_0$ and γ
5. Attenuation constant — (e) distributed parameters

Set II

1. Wave equation — (a) $-\dfrac{d\mathbf{V}}{dx} = \mathbf{ZI}$
2. Transmission line equation — (b) $\dfrac{d^2\mathbf{V}}{dx^2} = \gamma^2\mathbf{V}$
3. Characteristic impedance — (c) $\sqrt{(R + j\omega L)(G + j\omega C)}$
4. Propagation constant — (d) $\sqrt{\dfrac{(R + j\omega L)}{(G + j\omega C)}}$
5. Phase velocity — (e) $\dfrac{1}{\sqrt{LC}}$

Set III

1. Wave equation — (a) $G = 0$ and $R = 0$
2. Transmission line equation — (b) $j\omega\sqrt{LC}$
3. Characteristic impedance — (c) $\sqrt{\dfrac{L}{C}}$

4. Propagation constant	(d) $-\dfrac{\partial v}{\partial x} = Ri(x,t) + L\dfrac{\partial i}{\partial t}$
5. Lossless line	(e) $\dfrac{d^2\mathbf{I}}{dx^2} = \gamma^2\mathbf{I}$

Set IV

1. Velocity factor	(a) $\mathbf{Z}_0 \dfrac{\mathbf{Z}_l + j\mathbf{Z}_0 \tan \beta l}{\mathbf{Z}_0 + j\mathbf{Z}_l \tan \beta l}$
2. Input impedance	(b) $\dfrac{1}{\sqrt{\varepsilon_r}}$
3. Stub	(c) no reflected wave
4. Infinite line	(d) oc or sc lossless line
5. Lossless line	(e) HF line

Set V

1. Eighth-wave line	(a) $\mathbf{Z}_{\text{in}}^{\text{oc}} = -j\mathbf{Z}_0 \cot \beta l$
2. Quarter-wave line	(b) $\mathbf{Z}_{\text{in}}^{\text{sc}} = j\mathbf{Z}_0 \tan \beta l$
3. Half-wave line	(c) band stop filter
4. Short-circuited line	(d) impedance inverter
5. Open-circuited line	(e) resistance to impedance transformation

Answers

Set I:	1. (e)	2. (d)	3. (c)	4. (b)	5. (a)
Set II:	1. (b)	2. (a)	3. (d)	4. (c)	5. (e)
Set III:	1. (e)	2. (d)	3. (c)	4. (b)	5. (a)
Set IV:	1. (b)	2. (a)	3. (d)	4. (c)	5. (e)
Set V:	1. (e)	2. (d)	3. (c)	4. (b)	5. (a)

Multiple Choice Questions

1. The transmission lines carry waves of the form
   (a) TEM (b) TE
   (c) TM (d) hybrid
2. Which one of the following is not a transmission line?
   (a) coaxial line (b) parallel wire line
   (c) free space (d) none of these

3. The parallel wire transmission line is an example of
   (a) a balanced line (b) an unbalanced line
   (c) both (a) and (b) (d) none of these
4. The coaxial line is an example of
   (a) a balanced line (b) an unbalanced line
   (c) both (a) and (b) (d) none of these
5. The radiation losses do not occur in
   (a) parallel wire lines (b) coaxial lines
   (c) both (a) and (b) (d) none of these
6. The primary constants of the line are
   (a) lumped constants (b) distributed constants
   (c) both (a) and (b) (d) none of these
7. The primary constants of the line are
   (a) R, L, G and C (b) R, L, α and β
   (c) R, L, G and α (d) R, L, β and C
8. The secondary constants of the line are
   (a) lumped constants (b) distributed constants
   (c) both (a) and (b) (d) none of these
9. The secondary constants of the line are
   (a) α and β (b) α and γ
   (c) γ and β (d) Z_0 and γ
10. For a lossless line,
    (a) $R = 0, G = 0$ (b) $R = 0, G \neq 0$
    (c) $R = 0, L = 0$ (d) $R = 0, C = 0$
11. For a lossless line,
    (a) $\alpha = 0, \beta = 0$ (b) $\alpha = 0, \beta \neq 0$
    (c) $\alpha \neq 0, \beta = 0$ (d) $\alpha \neq 0, \beta \neq 0$
12. The input impedance of a line terminated over its CI is
    (a) CI (b) terminating impedance
    (c) both (a) and (b) (d) none of these
13. The units of normalized load impedance are
    (a) ohms (b) ohms/metre
    (c) no units (d) none of these
14. The units of attenuation constant are
    (a) nepers (b) nepers/metre
    (c) no units (d) none of these
15. For lossless lines, the propagation constant is
    (a) real (b) complex
    (c) imaginary (d) zero

16. For lossless lines, the characteristic impedance is
    (a) real (b) complex
    (c) imaginary (d) zero
17. For lossless lines, β depends upon
    (a) frequency (b) L, C
    (c) both (a) and (b) (d) none of these
18. For lossless lines, the velocity of wave over the line depends upon
    (a) frequency (b) L, C
    (c) both (a) and (b) (d) none of these
19. The input impedance of a $\lambda/4$ transformer is
    (a) terminal impedance (b) terminal admittance
    (c) characteristic impedance (d) none of these
20. The input impedance of a $\lambda/2$ transformer is
    (a) terminal impedance (b) terminal admittance
    (c) characteristic impedance (d) none of these

Answers

1. (a)	2. (c)	3. (a)	4. (b)	5. (b)
6. (b)	7. (a)	8. (a)	9. (d)	10. (a)
11. (b)	12. (c)	13. (c)	14. (b)	15. (c)
16. (a)	17. (c)	18. (b)	19. (b)	20. (a)

12

Reflection Coefficient and Standing Wave Ratio

Diligence is the mother of good fortune.
—Benjamin Disraeli

Reflection coefficient is an important parameter useful to describe the nature and amount of discontinuity over a transmission line. It is the most general parameter in describing the reflections and discontinuity applicable for both lossy and lossless lines. However, it has a major disadvantage of being difficult to measure. Standing wave ratio is another parameter which also represents the discontinuity of a line but it is defined and relevant only for lossless lines. Its major advantage over the reflection coefficient is its easy measurability. We discuss both these quantities in this chapter.

12.1 Reflection Coefficient

When a line is terminated over impedance and connected to a sinusoidal source, then in steady state, there exist two waves over the line: one towards the load called *incident wave* and the other towards the source called *reflected wave*. In general, the magnitude of the incident wave depends upon the source-line matching condition at the input end and that of the reflected wave depends upon the load-line matching condition at the termination end.

The vector ratio of the voltage of the reflected wave to that of the incident wave at a point over the line is known as *reflection coefficient* (RC) at that point and is usually denoted by $\mathbf{\Gamma}$, i.e.

$$\mathbf{\Gamma} \triangleq \frac{\mathbf{V}_{\text{ref}}}{\mathbf{V}_{\text{inc}}} \tag{12.1}$$

$$\Gamma = |\mathbf{\Gamma}| = \frac{|\mathbf{V}_{\text{ref}}|}{|\mathbf{V}_{\text{inc}}|} \tag{12.2a}$$

$$\angle\mathbf{\Gamma} = \angle\mathbf{V}_{\text{ref}} - \angle\mathbf{V}_{\text{inc}} \tag{12.2b}$$

As the voltages $\mathbf{V}_{ref}$ and $\mathbf{V}_{inc}$ are phasors and in general complex, their ratio is complex. Therefore, reflection coefficient is a complex quantity having both magnitude and phase. In general, it varies from point to point over the line and its range is from –1 to 1 through 0.

From the definition, we can also make two important observations. The magnitudes of the voltages $\mathbf{V}_{ref}$ and $\mathbf{V}_{inc}$ remain the same at all points over a lossless line, resulting in constant magnitude of reflection coefficient since it is the ratio of magnitudes of reflected and incident voltages. Thus, for the lossless line, the magnitude of reflection coefficient remains the same at all points over the line. The phase of reflection coefficient over a lossless line is shown in Figure 12.1.

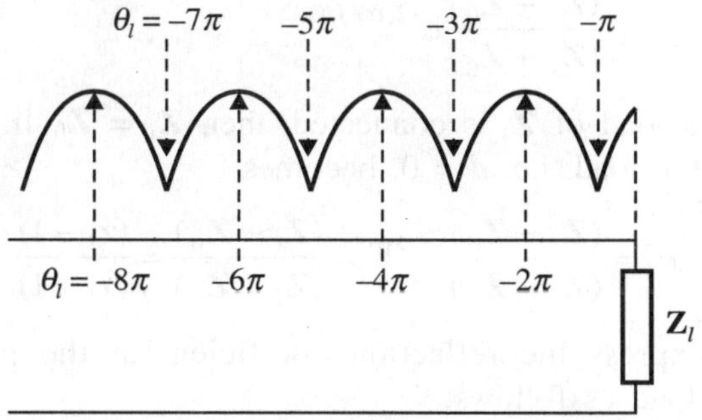

FIGURE 12.1 Phase of reflection coefficient over a lossless line.

The reflection coefficient is pure real at voltage maxima and minima points of the standing wave pattern. Maxima occur due to constructive interference between reflected and incident waves at points where their phase difference is $-2(k-1)\pi$ and minima occur due to destructive interference between waves at points where their phase difference is $-(2k-1)\pi$. In both the cases, from the definition, we can see that the reflection coefficient is pure real.

The vector ratio of currents of reflected wave and incident wave can also be considered which is always negative of the voltage ratio, i.e.

$$\mathbf{\Gamma} \triangleq \frac{\mathbf{V}_{ref}}{\mathbf{V}_{inc}} = -\frac{\mathbf{I}_{ref}}{\mathbf{I}_{inc}} \tag{12.3}$$

As it is a complex quantity, we represent it by a boldface symbol in time roman font and its magnitude by a lightface symbol, i.e. $\Gamma = |\mathbf{\Gamma}|$.

The reflection coefficient at an arbitrary point over the line can be expressed in terms of receiving end impedance $\mathbf{Z}_r$ as well as sending end impedance $\mathbf{Z}_s$. The procedures are listed below.

12.1.1 Reflection Coefficient in Terms of Receiving End Impedance Z_r

If the line is terminated over an impedance $\mathbf{Z}_l$ carrying current $\mathbf{I}_l$ with a voltage $\mathbf{V}_l$ across, then, $\mathbf{V}_r = \mathbf{V}_l$, $\mathbf{I}_r = \mathbf{I}_l$ and $\mathbf{Z}_r = \mathbf{Z}_l$. The voltage and current at a point P located at a distance of d from the receiving end over the line, then, are

$$\mathbf{V} = \frac{\mathbf{I}_r}{2}\left[(\mathbf{Z}_r + \mathbf{Z}_0)e^{\gamma d} + (\mathbf{Z}_r - \mathbf{Z}_0)e^{-\gamma d}\right] = \mathbf{V}_{\text{inc}} + \mathbf{V}_{\text{ref}} \tag{12.4}$$

$$\mathbf{I} = \frac{\mathbf{I}_r}{2\mathbf{Z}_0}\left[(\mathbf{Z}_r + \mathbf{Z}_0)e^{\gamma d} - (\mathbf{Z}_r - \mathbf{Z}_0)e^{-\gamma d}\right] = \mathbf{I}_{\text{inc}} + \mathbf{I}_{\text{ref}} \tag{12.5}$$

Then the reflection coefficient at the point P becomes

$$\mathbf{\Gamma} = \frac{\mathbf{V}_{\text{ref}}}{\mathbf{V}_{\text{inc}}} = \frac{(\mathbf{Z}_r - \mathbf{Z}_0)e^{-\gamma d}}{(\mathbf{Z}_r + \mathbf{Z}_0)e^{\gamma d}} = \frac{(\mathbf{Z}_r - \mathbf{Z}_0)}{(\mathbf{Z}_r + \mathbf{Z}_0)}e^{-2\gamma d}$$

$$= \frac{(\mathbf{Z}_r - \mathbf{Z}_0)}{(\mathbf{Z}_r + \mathbf{Z}_0)}e^{-2(\alpha + j\beta)d} \tag{12.6}$$

At the receiving end if a load of $\mathbf{Z}_l$ is connected, then $\mathbf{Z}_r = \mathbf{Z}_l$. In that case, the reflection coefficient exactly over the load, i.e. $d = 0$, becomes

$$\mathbf{\Gamma}_l = \frac{(\mathbf{Z}_r - \mathbf{Z}_0)}{(\mathbf{Z}_r + \mathbf{Z}_0)}e^{-2\gamma 0} = \frac{(\mathbf{Z}_l - \mathbf{Z}_0)}{(\mathbf{Z}_l + \mathbf{Z}_0)} = \frac{(\mathbf{z}_l - 1)}{(\mathbf{z}_l + 1)} \tag{12.7}$$

It is also possible to express the reflection coefficient at the point P in terms of the reflection coefficient at load as follows:

$$\mathbf{\Gamma} = \frac{(\mathbf{Z}_r - \mathbf{Z}_0)}{(\mathbf{Z}_r + \mathbf{Z}_0)}e^{-2\gamma d} = \mathbf{\Gamma}_l e^{-2\gamma d}$$

$$= |\mathbf{\Gamma}_l| e^{j\theta_l} e^{-2\gamma d} = |\mathbf{\Gamma}_l| e^{-2\alpha d + j(\theta_l - 2\beta d)} \tag{12.8}$$

Here the phase of $\mathbf{\Gamma}_l$ is assumed to be θ_l. Therefore, the reflection coefficient $\mathbf{\Gamma}$ at a distance of d from the load end is $\mathbf{\Gamma} = \mathbf{\Gamma}_l e^{-2\gamma d}$. its magnitude will be $\Gamma = |\mathbf{\Gamma}_l| e^{-2\alpha d}$ with a phase $\phi = (\theta_l - 2\beta d)$.

In the case of a lossless line, the reflection coefficient has the same magnitude everywhere which is equal to that at the load. In the case of a lossy line, the reflected wave becomes smaller and the incident wave larger with increasing distance from the load causing $\mathbf{\Gamma}$ to decrease correspondingly.

The relation between the voltage and current at the load end and the voltages of the incident and reflected waves at the load end can be deduced as

$$\mathbf{V}_{\text{inc}}\big|_{\text{load}} = \frac{\mathbf{V}_l}{1 + \mathbf{\Gamma}_l} = \frac{\mathbf{V}_l + \mathbf{I}_l\mathbf{Z}_0}{2} \tag{12.9}$$

$$\mathbf{V}_{\text{ref}}\big|_{\text{load}} = \mathbf{\Gamma}_l \mathbf{V}_{\text{inc}}\big|_{\text{load}} = \frac{\mathbf{\Gamma}_l\mathbf{V}_l}{1 + \mathbf{\Gamma}_l} = \frac{\mathbf{V}_l - \mathbf{I}_l\mathbf{Z}_0}{2} \tag{12.10}$$

Derivation:

$$\mathbf{V}_l = \mathbf{V}_{\text{inc}}\big|_{\text{load}} + \mathbf{V}_{\text{ref}}\big|_{\text{load}} = \mathbf{V}_{\text{inc}}\big|_{\text{load}} + \mathbf{\Gamma}_l \mathbf{V}_{\text{inc}}\big|_{\text{load}}$$

$$= \mathbf{V}_{\text{inc}}\big|_{\text{load}}(1 + \mathbf{\Gamma}_l)$$

Substituting the expression for the reflection coefficient over the load

$$\mathbf{\Gamma}_l = \frac{(\mathbf{Z}_l - \mathbf{Z}_0)}{(\mathbf{Z}_l + \mathbf{Z}_0)}$$

in the above expression, we get

$$\mathbf{V}_l = \mathbf{V}_{\text{inc}}\big|_{\text{load}}(1 + \mathbf{\Gamma}_l)$$

or

$$\mathbf{V}_{\text{inc}}\big|_{\text{load}} = \frac{\mathbf{V}_l}{(1 + \mathbf{\Gamma}_l)}$$

The first part of the derivation is over. Now let us move to the second part.

$$\begin{aligned}\mathbf{V}_l &= \mathbf{V}_{\text{inc}}\big|_{\text{load}}(1 + \mathbf{\Gamma}_l)\\ &= \mathbf{V}_{\text{inc}}\big|_{\text{load}}\left(1 + \frac{\mathbf{Z}_l - \mathbf{Z}_0}{\mathbf{Z}_l + \mathbf{Z}_0}\right)\\ &= \mathbf{V}_{\text{inc}}\big|_{\text{load}}\left(\frac{2\mathbf{Z}_l}{\mathbf{Z}_l + \mathbf{Z}_0}\right)\end{aligned}$$

It gives

$$\begin{aligned}\mathbf{V}_{\text{inc}}\big|_{\text{load}} &= \mathbf{V}_l\left(\frac{\mathbf{Z}_l + \mathbf{Z}_0}{2\mathbf{Z}_l}\right) = \frac{\mathbf{V}_l\mathbf{Z}_l + \mathbf{V}_l\mathbf{Z}_0}{2\mathbf{Z}_l}\\ &= \frac{\mathbf{V}_l + \mathbf{I}_l\mathbf{Z}_0}{2}\end{aligned}$$

In a similar manner, we can prove for the current also.

12.1.2 Reflection Coefficient in Terms of Sending End Impedance $\mathbf{Z}_s$

The voltage and current at a point P located at a distance of x from the sending end over the line are

$$\mathbf{V} = \frac{\mathbf{I}_s}{2}\left[(\mathbf{Z}_s + \mathbf{Z}_0)e^{-\gamma x} + (\mathbf{Z}_s - \mathbf{Z}_0)e^{\gamma x}\right] = \mathbf{V}_{\text{inc}} + \mathbf{V}_{\text{ref}} \quad (12.11)$$

$$\mathbf{I} = \frac{\mathbf{I}_s}{2\mathbf{Z}_0}\left[(\mathbf{Z}_s + \mathbf{Z}_0)e^{-\gamma x} + (\mathbf{Z}_s - \mathbf{Z}_0)e^{\gamma x}\right] = \mathbf{I}_{\text{inc}} + \mathbf{I}_{\text{ref}} \quad (12.12)$$

Then the reflection coefficient at the point P is

$$\begin{aligned}\mathbf{\Gamma} &= \frac{\mathbf{V}_{\text{ref}}}{\mathbf{V}_{\text{inc}}} = \frac{(\mathbf{Z}_s - \mathbf{Z}_0)e^{\gamma x}}{(\mathbf{Z}_s + \mathbf{Z}_0)e^{-\gamma x}} = \frac{(\mathbf{Z}_s - \mathbf{Z}_0)}{(\mathbf{Z}_s + \mathbf{Z}_0)}e^{2\gamma x}\\ &= \frac{(\mathbf{Z}_s - \mathbf{Z}_0)}{(\mathbf{Z}_s + \mathbf{Z}_0)}e^{2(\alpha + j\beta)x} \quad (12.13)\end{aligned}$$

At the sending end, i.e. exactly over the source $x = 0$, the reflection coefficient becomes

$$\mathbf{\Gamma}_s = \mathbf{\Gamma}_{\text{at } x=0} = \frac{(\mathbf{Z}_s - \mathbf{Z}_0)}{(\mathbf{Z}_s + \mathbf{Z}_0)} \tag{12.14}$$

Then the reflection coefficient at the point P in terms of $\mathbf{\Gamma}_s$ becomes

$$\mathbf{\Gamma} = \frac{(\mathbf{Z}_s - \mathbf{Z}_0)}{(\mathbf{Z}_s + \mathbf{Z}_0)} e^{2\gamma x} = \mathbf{\Gamma}_s e^{2\gamma x} \tag{12.15}$$

$$\mathbf{\Gamma} \triangleq \frac{\mathbf{V}_{\text{ref}}}{\mathbf{V}_{\text{inc}}}$$

$$\mathbf{\Gamma}_l = \frac{(\mathbf{Z}_l - \mathbf{Z}_0)}{(\mathbf{Z}_l + \mathbf{Z}_0)}$$

$$\mathbf{\Gamma}_s = \frac{(\mathbf{Z}_s - \mathbf{Z}_0)}{(\mathbf{Z}_s + \mathbf{Z}_0)}$$

$$\mathbf{\Gamma} = \mathbf{\Gamma}_l e^{-2\gamma d} = \mathbf{\Gamma}_s e^{2\gamma x}$$

12.1.3 Line Impedance–Reflection Coefficient

We know the line impedance at a point over the line is the complex ratio of the voltage **V** to the current **I** at that point, i.e. **Z** = **V**/**I**.

The line impedance at the load becomes the load impedance and at the sending end it is equal to the sending end impedance or input impedance. Consider voltage and current over the line at a distance of d from the receiving end in terms of receiving end current $\mathbf{I}_r$ and impedance $\mathbf{Z}_r$. Thus,

$$\mathbf{V} = \frac{\mathbf{I}_r}{2}\left[(\mathbf{Z}_r + \mathbf{Z}_0)e^{\gamma d} + (\mathbf{Z}_r - \mathbf{Z}_0)e^{-\gamma d}\right] \tag{12.16}$$

$$\mathbf{I} = \frac{\mathbf{I}_r}{2\mathbf{Z}_0}\left[(\mathbf{Z}_r + \mathbf{Z}_0)e^{\gamma d} - (\mathbf{Z}_r - \mathbf{Z}_0)e^{-\gamma d}\right] \tag{12.17}$$

Their ratio, i.e. voltage to current, is the impedance of the line at that point. Thus,

$$\mathbf{Z} = \mathbf{Z}_0 \frac{(\mathbf{Z}_r + \mathbf{Z}_0)e^{\gamma d} + (\mathbf{Z}_r - \mathbf{Z}_0)e^{-\gamma d}}{(\mathbf{Z}_r + \mathbf{Z}_0)e^{\gamma d} - (\mathbf{Z}_r - \mathbf{Z}_0)e^{-\gamma d}}$$

$$= \mathbf{Z}_0 \frac{1 + \dfrac{(\mathbf{Z}_r - \mathbf{Z}_0)}{(\mathbf{Z}_r + \mathbf{Z}_0)} e^{-2\gamma d}}{1 - \dfrac{(\mathbf{Z}_r - \mathbf{Z}_0)}{(\mathbf{Z}_r + \mathbf{Z}_0)} e^{-2\gamma d}} \tag{12.18}$$

If the line is terminated with the impedance $\mathbf{Z}_l$, then $\mathbf{Z}_r = \mathbf{Z}_l$ and

$$\frac{\mathbf{Z}_r - \mathbf{Z}_0}{\mathbf{Z}_r + \mathbf{Z}_0} = \frac{\mathbf{Z}_l - \mathbf{Z}_0}{\mathbf{Z}_l + \mathbf{Z}_0} = \mathbf{\Gamma}_l \tag{12.19}$$

Impedance and reflection coefficient

$$\mathbf{z} = \frac{1+\mathbf{\Gamma}}{1-\mathbf{\Gamma}}$$

$$\mathbf{\Gamma} = \frac{\mathbf{z}-1}{\mathbf{z}+1}$$

$$\mathbf{z}_{\max} = \frac{1+|\mathbf{\Gamma}|}{1-|\mathbf{\Gamma}|} = \rho$$

$$\mathbf{z}_{\min} = \frac{1-|\mathbf{\Gamma}|}{1+|\mathbf{\Gamma}|} = \frac{1}{\rho}$$

With the above equalities incorporating into the expression, the line impedance at a distance of d from the load end becomes

$$\mathbf{Z}(d) = \mathbf{Z}_0 \frac{1+\mathbf{\Gamma}_l e^{-2\gamma d}}{1-\mathbf{\Gamma}_l e^{-2\gamma d}} \tag{12.20}$$

The normalized line impedance then becomes

$$\mathbf{z}(d) = \frac{\mathbf{Z}(d)}{\mathbf{Z}_0} = \mathbf{Z}_0 \frac{1+\mathbf{\Gamma}(d)}{1-\mathbf{\Gamma}(d)} \tag{12.21}$$

For a lossless line,

$$\mathbf{z}(d) = \frac{1+\mathbf{\Gamma}_l e^{-2\gamma d}}{1-\mathbf{\Gamma}_l e^{-2\gamma d}} = \frac{1+\mathbf{\Gamma}_l e^{-2j\beta d}}{1-\mathbf{\Gamma}_l e^{-2j\beta d}} \tag{12.22}$$

The line impedance over a lossless line is shown in Figure 12.2. From the expression of Eq. (12.22) for the line impedance of the lossless line in terms of reflection coefficient, we can make the following observations:

(a) At voltage maxima of the standing wave pattern, the reflection coefficient is pure real and hence $1 + \mathbf{\Gamma}(d)$ and $1 - \mathbf{\Gamma}(d)$ are purely real assuming their maximum and minimum values of $(1+|\mathbf{\Gamma}_l|)$ and $(1-|\mathbf{\Gamma}_l|)$, respectively. Hence, the line impedance is purely real and maximum and if we let it be $\mathbf{z}_{\min}$, then

$$\mathbf{z}_{\max} = \frac{1+|\mathbf{\Gamma}_l|}{1-|\mathbf{\Gamma}_l|} = \rho \tag{12.23}$$

(b) At voltage minima of the standing wave pattern also, the reflection coefficient is pure real and hence $1 + \mathbf{\Gamma}(d)$ and $1 - \mathbf{\Gamma}(d)$ are purely real assuming their minimum

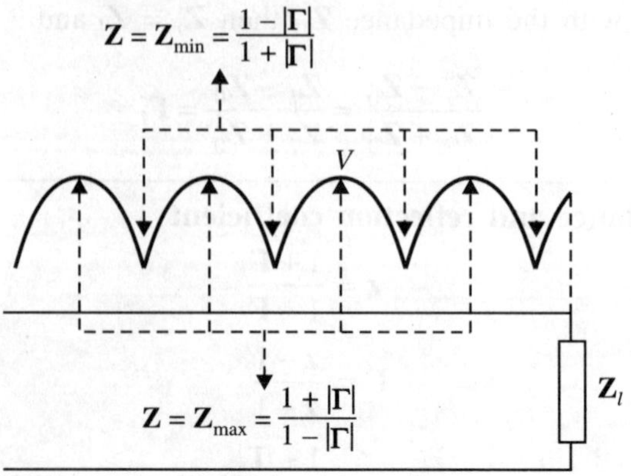

FIGURE 12.2 Line impedance over a lossless line.

and maximum values of $\left(1-\left|\mathbf{\Gamma}_l\right|\right)$ and $\left(1+\left|\mathbf{\Gamma}_l\right|\right)$, respectively. Hence, the line impedance is purely real and minimum and if we let it be $\mathbf{z}_{\text{min}}$, then

$$\mathbf{z}_{\text{min}} = \frac{1-\left|\mathbf{\Gamma}_l\right|}{1+\left|\mathbf{\Gamma}_l\right|} = \frac{1}{\rho} \tag{12.24}$$

(c) In between the voltage maxima and minima of the standing wave pattern, the reflection coefficient is complex and hence both 1 + $\mathbf{\Gamma}(d)$ and 1 − $\mathbf{\Gamma}(d)$ are also complex. Hence, the line impedance $\mathbf{z}(d)$ is complex, with magnitude lying between ρ and $1/\rho$. Thus,

$$\mathbf{z}(d) = \frac{1+\mathbf{\Gamma}(d)}{1-\mathbf{\Gamma}(d)} \tag{12.25}$$

In the above expressions, the parameter ρ represents standing wave ratio, which is the ratio of maximum voltage to minimum voltage over the line. The nature of standing wave pattern over a lossless line is shown in Figure 12.3.

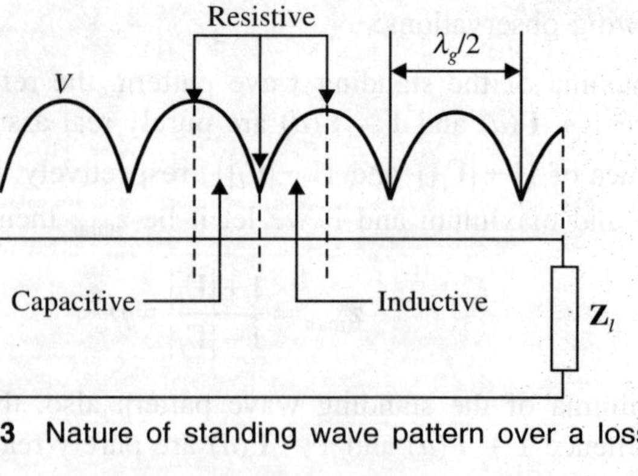

FIGURE 12.3 Nature of standing wave pattern over a lossless line.

(d) Another very important aspect pertaining to the nature of line impedance at or near to a node is that it is capacitive in a distance of $\lambda/4$ to the right side of the voltage node and it is inductive in a distance of $\lambda/4$ to its left side.

(e) It can be shown easily that the impedances at two points A and B are equal if they are separated by a distance equal to the integer multiples of $\lambda/2$. Since $\mathbf{\Gamma}(d \pm n\lambda/2) = \mathbf{\Gamma}(d)e^{\mp j2\beta n\lambda/2} = \mathbf{\Gamma}(d)$, $n = 1, 2, 3, ...,$ the $\mathbf{\Gamma}(d)$ repeats at intervals of $\lambda/2$ and hence $\mathbf{Z}(d)$ repeats at intervals of $\lambda/2$. Therefore, the line impedance, which is a complex quantity varying periodically along the length of the line with a periodicity, is equal to one half wavelength, i.e. $\lambda/2$.

(f) It can also be shown that the impedances at two points A and B have inverse proportionality if the distance is odd multiples of $\lambda/4$. The product of the line impedances at two values of d separated by $\lambda/4$ is given by

$$\mathbf{Z}(d)\,\mathbf{Z}\left(d \pm \frac{\lambda}{4}\right) = \mathbf{Z}_0 \frac{1+\mathbf{\Gamma}(d)}{1-\mathbf{\Gamma}(d)} \mathbf{Z}_0 \frac{1+\mathbf{\Gamma}\left(d \pm \frac{\lambda}{4}\right)}{1-\mathbf{\Gamma}\left(d \pm \frac{\lambda}{4}\right)}$$

$$= \mathbf{Z}_0^2 \frac{1+\mathbf{\Gamma}(d)}{1-\mathbf{\Gamma}(d)} \frac{1+\mathbf{\Gamma}(d)e^{\mp j2\beta\lambda/4}}{1-\mathbf{\Gamma}(d)e^{\mp j2\beta\lambda/4}}$$

$$= \mathbf{Z}_0^2 \frac{1+\mathbf{\Gamma}(d)}{1-\mathbf{\Gamma}(d)} \frac{1-\mathbf{\Gamma}(d)}{1+\mathbf{\Gamma}(d)} = \mathbf{Z}_0^2 \tag{12.26}$$

(g) From considering all of the above points, we can say that the impedance profile is periodic over the line, with maxima and minima repeating with a periodicity of $\lambda/2$. We can also observe that the maxima and minima points are interspersed by a distance of $\lambda/4$. The value of minimum line impedance and maximum value of line admittance are related through $\mathbf{z}_{min} = \mathbf{y}_{max}$.

12.2 STANDING WAVES

Consider a uniform line placed along the x-axis, carrying both the incident and reflected waves. The voltage over an arbitrary point, at a distance of x from the sending end, can be expressed as

$$\begin{aligned} \mathbf{V} &= \mathbf{V}_{inc} + \mathbf{V}_{ref}\ \mathbf{V} = ae^{-\gamma x} + be^{\gamma x} \\ &= ae^{-\alpha x - j\beta x} + be^{\alpha x + j\beta x} \\ &= (ae^{-\alpha x} + be^{\alpha x}) \cos \beta x - j(ae^{-\alpha x} - be^{\alpha x}) \sin \beta x \\ &= Ve^{-j\phi} \end{aligned} \tag{12.27}$$

The total voltage over the line is a complex quantity representing standing wave phenomenon. The amplitude pattern of the wave is

$$V = \left[(ae^{-\alpha x} + be^{\alpha x})^2 \cos^2 \beta x + (ae^{-\alpha x} - be^{\alpha x})^2 \sin^2 \beta x\right]^{\frac{1}{2}} \tag{12.28a}$$

whereas the phase pattern of the wave is

$$\phi = \tan^{-1}\left[\frac{(ae^{-\alpha x} - be^{\alpha x})}{(ae^{-\alpha x} + be^{\alpha x})}\tan \beta x\right] \tag{12.28b}$$

In the case of a lossless line, i.e. $\alpha = 0$, the amplitude and phase patterns for voltage become

$$V = \left[(a+b)^2 \cos^2 \beta x + (a-b)^2 \sin^2 \beta x\right]^{\frac{1}{2}}$$

$$= \left[(a+b)^2 \cos^2 \beta x + (a-b)^2 \sin^2 \beta x\right]^{\frac{1}{2}}$$

$$= \left[a^2 + b^2\ 2ab \cos 2\beta x\right]^{\frac{1}{2}} \tag{12.29a}$$

$$\phi = \tan^{-1}\left[\frac{(a-b)}{(a+b)}\tan \beta x\right] \tag{12.29b}$$

12.2.1 Observations

Voltage maxima: The amplitude of voltage is not constant along the length of the line. It varies assuming maximum at certain points given by

$$\mathbf{V}_{\max} = (ae^{-\alpha x} + be^{\alpha x})$$

$$= ae^{-\alpha x}\left[1 + \left(\frac{be^{\alpha x}}{ae^{-\alpha x}}\right)\right]$$

$$= ae^{-\alpha x}(1 + |\Gamma|) = ae^{-\alpha x}(1 + \Gamma) \tag{12.30}$$

where $|\Gamma| = (b/a)e^{2\alpha x}$ is the magnitude of the reflection coefficient. In the case of a lossless line, i.e. $\alpha = 0$,

$$\mathbf{V}_{\max} = a(1 + |\Gamma|) = a(1 + \Gamma) \tag{12.31}$$

The location of the maxima is given by $\beta x = n\pi$ and the distance between two consecutive maxima is $\lambda/2$.

Voltage minima: At certain points there exist minima over the line and they are given by

$$\mathbf{V}_{\min} = (ae^{-\alpha x} - be^{\alpha x})$$

$$= ae^{-\alpha x}\left[1 - \left(\frac{be^{\alpha x}}{ae^{-\alpha x}}\right)\right]$$

$$= ae^{-\alpha x}(1 - |\Gamma|) = ae^{-\alpha x}(1 - \Gamma) \tag{12.32}$$

In the case of a lossless line,

$$\mathbf{V}_{\min} = a(1 - |\Gamma|) = a(1 - \Gamma) \tag{12.33}$$

The location of the voltage minima is given by $\beta x = (2n - 1)\pi/2$ and the distance between two consecutive minima is $\lambda/2$.

Current maxima and minima: The current also varies over the line and current maxima and minima can be found as

$$\mathbf{I}_{max} = \mathbf{Y}_0 a e^{-\alpha x}(1+|\Gamma|) = \mathbf{Y}_0 a e^{-\alpha x}(1+\Gamma) \tag{12.34a}$$

and

$$\mathbf{I}_{min} = \mathbf{Y}_0 a e^{-\alpha x}(1-|\Gamma|) = \mathbf{Y}_0 a e^{-\alpha x}(1-\Gamma) \tag{12.34b}$$

In the above expressions, $\mathbf{Y}_0$ represents the characteristic admittance, inverse of the characteristic impedance of the line. The current maxima occur at $\beta x = (2n - 1)\pi/2$ where the voltage minima exist. The current minima occur at $\beta x = n\pi$ where the minima exist for voltage. The distance between two consecutive current maxima or minima can be found to be $\lambda/2$, the same as for voltage maxima or minima. Therefore, the distance between two successive nodes or anti-nodes of voltage (or current) is always $\lambda_g/2$. And it is $\lambda_g/4$ between voltage node to current anti-node or voltage anti-node to current node.

Line impedance

- Input impedance: $\mathbf{Z}_{in} = \mathbf{Z}_0\left(\dfrac{\mathbf{Z}_l + j\mathbf{Z}_0 \tan \beta l}{\mathbf{Z}_0 + j\mathbf{Z}_l \tan \beta l}\right)$
- Positions at which $\mathbf{Z}_{in}$ is real: at voltage maxima and minima
- $\mathbf{Z}_{in}$ at voltage maxima: $\mathbf{Z}_{in} = \mathbf{Z}_0\left(\dfrac{1+\Gamma}{1-\Gamma}\right)$
- $\mathbf{Z}_{in}$ at voltage minima: $\mathbf{Z}_{in} = \mathbf{Z}_0\left(\dfrac{1-\Gamma}{1+\Gamma}\right)$
- $\mathbf{Z}_{in}$ of a short-circuited line: $\mathbf{Z}_{in}^{sc} = j\mathbf{Z}_0 \tan \beta l$
- $\mathbf{Z}_{in}$ of an open-circuited line: $\mathbf{Z}_{in}^{oc} = -j\mathbf{Z}_0 \cot \beta l$
- $\mathbf{Z}_{in}$ of a line of length: $l = \dfrac{n\lambda}{2} : \mathbf{Z}_{in} = \mathbf{Z}_l, n = 0, 1, 2, \ldots$
- $\mathbf{Z}_{in}$ of a line of length: $l = \dfrac{\lambda}{4} + \dfrac{n\lambda}{2} : \mathbf{Z}_{in} = \dfrac{\mathbf{Z}_0^2}{\mathbf{Z}_l}, n = 0, 1, 2, \ldots$

Standing wave pattern

- Voltage maximum $\mathbf{V}_{max} = a(1 + \Gamma)$
- Voltage minimum $\mathbf{V}_{min} = a(1 - \Gamma)$
- Current maximum $\mathbf{I}_{max} = a\mathbf{Y}_0 e^{-\alpha x}(1 + \Gamma)$
- Current minimum $\mathbf{I}_{min} = a\mathbf{Y}_0 e^{-\alpha x}(1 - \Gamma)$
- Positions of voltage maxima

$$d_{max} = \frac{\theta_l \lambda}{4\pi} + \frac{(k-1)\lambda}{2}, \quad k = 1, 2, \ldots$$

(Contd...)

(Contd...)

- Position of the first maximum

$$d_{\max} = \begin{cases} \dfrac{\theta_l \lambda}{4\pi} & \text{for } 0 \le \theta_l \le \pi \\ \dfrac{\theta_l \lambda}{4\pi} + \dfrac{\lambda}{2} & \text{for } -\pi \le \theta_l \le 0 \end{cases}$$

- Positions of voltage minima

$$d_{\min} = \frac{\theta_l \lambda}{4\pi} + \frac{(2k-1)\lambda}{4}, \quad k = 1, 2, \ldots$$

- Position of the first minimum

$$d_{\min} = \frac{\lambda}{4}\left(1 + \frac{\theta_l}{\pi}\right)$$

12.2.2 Nature of Waves over Lines

Different types of waves exist over the line depending upon the type of termination. Pure standing wave comes into being when the forward travelling wave is totally reflected, whereas pure travelling wave exists when the forward wave is completely absorbed by the load. Impure travelling wave or standing wave comes into being when the wave is partially reflected by the load. The properties of these three types of waves are given hereunder.

Pure standing wave: It comes into being when the amplitudes of the incident and reflected waves are the same, i.e. $|ae^{-\alpha x}| = |be^{\alpha x}|$.

The amplitude of the total voltage over the line becomes

$$\mathbf{V}_0 = 2a\mathbf{Y}_0 e^{-\alpha x} \cos \beta x \tag{12.35}$$

and that of the total current becomes

$$\mathbf{I}_0 = -2ja\mathbf{Y}_0 e^{-\alpha x} \sin \beta x \tag{12.36}$$

The presence of j in one of the above expressions indicates the existence of a 90° phase difference between the voltage and current over the line. It can also be observed that the magnitude of the reflection coefficient is the same everywhere over the line. Important properties of the pure standing wave are mentioned as follows:

- Standing wave ratio, which is the ratio of maximum voltage to minimum voltage over the line, is infinity indicating the total reflection of the incident wave by the load.
- It occurs when the line is terminated with an open circuit or a short circuit or over a pure reactance.

- The ratio of **V** to **I** is a function of the position along the length of the line.
- In this wave, the total voltage **V** and the total current **I** always differ in phase by 90° at any point over the line.
- At all points between a pair of successive voltage nulls, i.e. in one half cycle, the voltage is in phase. All the points in the next half cycle exhibit a 180° phase difference with the points of previous half cycle. Similar is the case with the current wave form.

Pure travelling wave: It is worth considering the properties of pure travelling wave so that they can be well understood in comparison with the standing waves. Salient features of the pure travelling wave are given as follows:

- Standing wave ratio is unity indicating no reflections from the load and consequently no standing wave over the line.
- It can exist only when the line is terminated with its matched impedance.
- The ratio of **V** to **I** is constant. $\mathbf{Z}_0$ is the characteristic impedance of the line.
- In this wave, the total voltage **V** and the total current **I** are always in phase at any point over the line, i.e. the current and voltage reach maximum or minimum simultaneously.
- In the pure travelling wave, there exists continuous phase variation along the length of the line for voltage and current. But the amplitude of the wave remains constant.

Impure standing wave: Waves on the line need not always be standing wave or travelling wave in pure form. Sometimes it can be a combination of travelling and standing waves which is called impure standing wave. Its salient features are given as follows:

- It is a combination of pure travelling wave and pure standing wave.
- In this case, the total voltage **V** and the total current **I** are neither in phase nor 90° out of phase. In general, it varies along the length of the line.
- It occurs when the termination is different from open circuit, short circuit or pure reactance and matched termination, i.e. the termination is different from the characteristic impedance with a resistive component.

12.3 REFLECTION LOSS

Reflection loss is also known as *return loss*. By definition, it is

$$RL = -10\log\frac{P_i}{P_r} \tag{12.37}$$

Here P_i is the incident power and P_r is the reflected power over an impedance discontinuity. The discontinuity may be an impedance mismatch or some other discontinuity. Now, we try to estimate its value for a line terminated over a mismatched load in terms of the reflection coefficient as well as the SWR. When R_0 is the characteristic impedance of the line, the net power transmitted over the line and delivered to the load is

$$P = \frac{|\mathbf{V}_i|^2 - |\mathbf{V}_r|^2}{R_0} \tag{12.38}$$

In this expression, the first term gives the incident power, whereas the second one represents the power in the reflected wave. Therefore, it can be easily recognized as the difference of two power flows, one being the power transmitted in the incident wave and the other being the power travelling back in the reflected wave. The ratio of the power delivered to the load to the power transmitted by the incident wave is

$$\frac{P}{P_i} = \frac{P_i - P_r}{P_i} = \frac{|\mathbf{V}_i|^2 - |\mathbf{V}_r|^2}{|\mathbf{V}_i|^2} = 1 - \frac{|\mathbf{V}_r|^2}{|\mathbf{V}_i|^2} = 1 - |\Gamma|^2 \tag{12.39a}$$

The above expression can also be given in terms of SWR which is related to the reflection coefficient through

$$\Gamma = |\Gamma| = \frac{\rho - 1}{\rho + 1}$$

Use of the this relation results in

$$\frac{P}{P_i} = 1 - \left(\frac{\rho - 1}{\rho + 1}\right)^2 = \frac{4\rho}{(\rho + 1)^2} \tag{12.39b}$$

Now,

$$1 - \frac{P_r}{P_i} = 1 - |\Gamma|^2$$

or

$$\frac{P_r}{P_i} = |\Gamma|^2 \tag{12.40}$$

The return loss becomes

$$RL = -10 \log \frac{P_i}{P_r} = 10 \log |\Gamma|^2$$

$$= 20 \log \Gamma = 20 \log \left(\frac{\rho - 1}{\rho + 1}\right) \tag{12.41}$$

12.4 STANDING WAVE RATIO

Standing wave ratio (SWR) is one of the most important parameters used to describe the amount of mismatch or reflected wave over a lossless line.

The ratio of the maximum amplitude to the minimum amplitude possessed by the voltage distribution is equal to the current distribution which is known as standing wave ratio, denoted by ρ. It is always a pure real quantity and its value ranges from one to infinity.

Thus,

$$\rho \triangleq \frac{|\mathbf{V}_{max}|}{|\mathbf{V}_{min}|} \tag{12.42}$$

Standing wave ratio is a measure of amplitude ratio of reflected to incident waves. Thus, an SWR of unity denotes the absence of a reflected wave, while a very high SWR indicates that the reflected wave is as large as the incident wave.

Theoretically, for the case of a lossless line, the SWR will be infinite when the termination is open circuit or short circuit or non-dissipative pure reactance.

The SWR is one means expressing the magnitude of the reflection coefficient Γ when the line is lossless. The exact relation between the two is

$$\rho = \frac{1+|\Gamma|}{1-|\Gamma|} = \frac{1+\Gamma}{1-\Gamma} \tag{12.43a}$$

Proof: Standing wave ratio

$$\rho = \frac{|\mathbf{V}_{max}|}{|\mathbf{V}_{min}|} = \frac{|\mathbf{V}_{inc}|+|\mathbf{V}_{ref}|}{|\mathbf{V}_{inc}|-|\mathbf{V}_{ref}|} = \frac{1+\dfrac{|\mathbf{V}_{ref}|}{|\mathbf{V}_{inc}|}}{1-\dfrac{|\mathbf{V}_{ref}|}{|\mathbf{V}_{inc}|}}$$

$$= \frac{1+|\Gamma|}{1-|\Gamma|}$$

From this, it can also be expressed that

$$\Gamma = |\Gamma| = \frac{\rho-1}{\rho+1} \tag{12.43b}$$

From the value of the reflection coefficient, it is possible to find the standing wave ratio. From the standing wave ratio, it is possible to find only the magnitude of reflection coefficient but not its phase.

12.4.1 SWR for Different Types of Loads

The SWRs for different types of loads are given hereunder.

- When the termination is matched, i.e. $\mathbf{Z}_l = \mathbf{Z}_0$, the SWR is one.

  Proof: When the termination is matched, the total power of the incident wave goes into the load and the reflected wave becomes nil. As a result, only the forward travelling wave exists over the line and as it is lossless, the amplitudes of the oscillations at all points over the line remain the same. Therefore, one can say that there are no maxima points or minima points over the line with $|\mathbf{V}_{max}| = |\mathbf{V}_{min}|$, resulting in SWR of unity.

- When the termination is pure reactance, open circuit or short circuit, the SWR is infinity.

  Proof: When the termination is one of pure reactance, open circuit or short circuit, then it cannot absorb any power from the incident wave. Therefore, the reflected wave is as strong as the incident wave. As the amplitudes of the incident and reflected waves are the same, perfect cancellation takes place at the points of minima of the standing wave pattern, making the minimum voltage $\mathbf{V}_{min} = 0$, resulting in SWR of infinity.

- When the termination is pure resistance, i.e. $\mathbf{Z}_0 = R_l$, then the SWR is $\mathbf{Z}_0/R_l$ or $R_l/\mathbf{Z}_0$, depending on whichever is more than one.

Proof: The SWR is related to the magnitude of the reflection coefficient through

$$\rho = \frac{1+|\Gamma|}{1-|\Gamma|}$$

Even though the reflection coefficient varies from point to point over the line, its magnitude remains the same over the line provided the line is lossless. For convenience, consider the magnitude of reflection coefficient over the load.

$$|\Gamma| = |\Gamma_l| = \left|\frac{\mathbf{Z}_l - \mathbf{Z}_0}{\mathbf{Z}_l + \mathbf{Z}_0}\right| = \frac{|\mathbf{Z}_l - \mathbf{Z}_0|}{|\mathbf{Z}_l + \mathbf{Z}_0|}$$

In the present case, the line is lossless resulting in a real value of the characteristic impedance $\mathbf{Z}_0$ and the load is resistance making $\mathbf{Z}_l$ also real. In addition, these two are also positive quantities. Under such circumstances,

$$|\mathbf{Z}_l - \mathbf{Z}_0| = \mathbf{Z}_l - \mathbf{Z}_0 \text{ if } \mathbf{Z}_l > \mathbf{Z}_0$$
$$= \mathbf{Z}_0 - \mathbf{Z}_l \text{ if } \mathbf{Z}_0 > \mathbf{Z}_l$$
$$|\mathbf{Z}_l + \mathbf{Z}_0| = \mathbf{Z}_l + \mathbf{Z}_0$$

Substituting these values into the expression for SWR, we obtain

$$\rho = \frac{1+|\Gamma|}{1-|\Gamma|} = \frac{1+|\Gamma_l|}{1-|\Gamma_l|}$$
$$= \frac{|\mathbf{Z}_l + \mathbf{Z}_0| + |\mathbf{Z}_l - \mathbf{Z}_0|}{|\mathbf{Z}_l + \mathbf{Z}_0| - |\mathbf{Z}_l - \mathbf{Z}_0|}$$

Firstly, we consider the case in which $\mathbf{Z}_l > \mathbf{Z}_0$.

$$\rho = \frac{|\mathbf{Z}_l + \mathbf{Z}_0| + |\mathbf{Z}_l - \mathbf{Z}_0|}{|\mathbf{Z}_l + \mathbf{Z}_0| - |\mathbf{Z}_l - \mathbf{Z}_0|}$$
$$= \frac{\mathbf{Z}_l + \mathbf{Z}_0 + \mathbf{Z}_l - \mathbf{Z}_0}{\mathbf{Z}_l + \mathbf{Z}_0 - \mathbf{Z}_l + \mathbf{Z}_0} = \frac{\mathbf{Z}_l}{\mathbf{Z}_0}$$

Next, we consider the situation where $\mathbf{Z}_0 > \mathbf{Z}_l$.

$$\rho = \frac{|\mathbf{Z}_l + \mathbf{Z}_0| + |\mathbf{Z}_l - \mathbf{Z}_0|}{|\mathbf{Z}_l + \mathbf{Z}_0| - |\mathbf{Z}_l - \mathbf{Z}_0|}$$

$$= \frac{\mathbf{Z}_l + \mathbf{Z}_0 + \mathbf{Z}_0 - \mathbf{Z}_l}{\mathbf{Z}_l + \mathbf{Z}_0 - \mathbf{Z}_0 + \mathbf{Z}_l} = \frac{\mathbf{Z}_0}{\mathbf{Z}_l}$$

Therefore, the SWR is $\mathbf{Z}_0/R_l$ (for $\mathbf{Z}_0 > \mathbf{Z}_l$) or $R_l/\mathbf{Z}_0$ (for $\mathbf{Z}_0 > \mathbf{Z}_l$). In other words, we can say that it is one of these two ratios, depending on whichever is more than one.

- When the termination is a complex impedance, then SWR can be found, first finding the magnitude of the reflection coefficient over the line by

$$\Gamma = \frac{|\mathbf{Z}_l - \mathbf{Z}_0|}{|\mathbf{Z}_l - \mathbf{Z}_0|}$$

And then by using Eq. (12.42a), the SWR can be calculated. Or we can use the Smith chart.

12.4.2 Significance of SWR

The SWR is one of the important paremeters used because of the following reasons:

- The importance of the standing wave ratio is due to the fact that it can be very easily measured experimentally.
- The SWR indicates directly the extent to which reflected waves exist on a system.
- In addition, standing wave measurements provide an important means of measuring impedance.

Standing wave ratio

$$\rho = \frac{1 + \Gamma}{1 - \Gamma}$$

Reflection coefficient

$$\Gamma = \frac{\rho - 1}{\rho + 1}$$

SWR = 1 for matched termination
$\rightarrow \infty$ for open circuit, short circuit or pure reactance
= $\mathbf{Z}_0/R_l$ or $R_l/\mathbf{Z}_0$ for resistance termination

12.5 SWR–REFLECTION COEFFICIENT

Even though both these parameters are used to denote the mismatch and reflections over the line, there exist some fundamental and basic differences between them given as follows:

- The reflection coefficient can be defined for both lossy and lossless lines, whereas the standing wave ratio is meaningful only for lossless lines.
- The reflection coefficient varies from –1 to +1 through 0, whereas the standing wave ratio varies from 1 to ∞.

- The reflection coefficient varies from point to point over the line, whereas the standing wave ratio is designed for the entire length of the line.
- The reflection coefficient of voltages is negative of the reflection coefficient of currents, whereas the standing wave ratio remains the same whether it is voltage ratio or current ratio.
- The chief advantage of SWR over the reflection coefficient is its measurability. The reflection coefficient is difficult to measure, whereas the standing wave ratio is an easily measurable quantity.

12.6 IMPEDANCE MEASUREMENT

Many a number of times we come across a need to know the impedance of the load. At low frequencies, with the help of easily available measuring instruments, it can be found. But at high frequencies, measuring instruments are not that much readily available and hence we look towards alternative methods. One method by which it can be found is by terminating the load over a lossless line and then measuring the SWR over the line.

It is possible to measure the impedance of a device by measuring the distance of the first minimum from the load. In the case of difficulty in locating the first minimum position exactly, the impedance can still be computed by measuring the shift in the position of any one minimum when the load is replaced by a short. Now we investigate the theory behind these two methods.

In the standing wave pattern, minima occur at locations wherever the reflected wave is in phase opposition to the incident wave resulting in mutual cancellation. Therefore, at minima, the phases of reflection coefficient are $-(2n - 1)\pi$, where $n = 1, 2, \ldots$. In the case of the first minimum, i.e. the minimum nearest to the load, it is $-\pi$.

It is also possible to express the reflection coefficient at a distance of d from the load, in terms of the reflection coefficient at load as

$$\mathbf{\Gamma} = \frac{(\mathbf{Z}_l - \mathbf{Z}_0)}{(\mathbf{Z}_l + \mathbf{Z}_0)} e^{-2j\beta d} = \mathbf{\Gamma}_l e^{-2j\beta d} = |\mathbf{\Gamma}_l| e^{j\theta_l} e^{-2j\beta d}$$

$$= |\mathbf{\Gamma}_l| e^{j(\theta_l - 2\beta d)} \qquad (12.44)$$

Here the phase of $\mathbf{\Gamma}_l$ is assumed to be θ_l. It is given by

$$\theta_l = \angle \mathbf{\Gamma}_l = \angle \frac{(\mathbf{Z}_l - \mathbf{Z}_0)}{(\mathbf{Z}_l + \mathbf{Z}_0)} \qquad (12.45)$$

It can assume any value in between $-\pi$ and π. Its value is zero when the load is pure real and greater than $\mathbf{Z}_0$. In this case, a voltage maximum occurs right at the load and the distance of the first minimum is $d_{\min} = \lambda/4$. Its value is $-\pi$ when the load is pure real and less than $\mathbf{Z}_0$. In this case, the first voltage minimum occurs right at the load resulting in $d_{\min} = 0$.

From the above expressions, we can observe that the phase of reflection coefficient $\mathbf{\Gamma}$ over the first minimum which is at a distance of $d_{\min}$ from the load end is $\phi = \theta_l - 2\beta d_{\min}$.

But we have seen that it is equal to $-\pi$. Therefore,

$$\theta_l - 2\beta d_{min} = -\pi \tag{12.46}$$

By measuring d_{min} and the wavelength, it is possible to find θ_l. From the measured value of SWR the magnitude of the reflection coefficient can be obtained.

$$\Gamma_l = \frac{(\rho - 1)}{(\rho + 1)} \tag{12.47}$$

Once the reflection coefficient over the load is available, then the load can be computed using

$$\mathbf{Z}_l = \mathbf{Z}_0 \frac{(1 + \Gamma_l)}{(1 - \Gamma_l)} \tag{12.48}$$

Sometimes it is difficult to locate the first minimum. In that case, the above method is modified slightly so that the impedance measurement is possible without requiring the location of the first minimum.

Consider a minimum, let us say the kth one, at a distance of d_l, in the pattern when the line is terminated with the load. Then the phase of the reflection coefficient over this minimum is

$$\theta_l - 2\beta d_l = -(2k - 1)\pi \tag{12.49}$$

If we replace the load with a short, let us suppose, the minimum identified earlier is shifted to d_s. In this case, as the load is short ($\theta_l = -\pi$), the phase of reflection coefficient over the minimum is

$$\theta_l - 2\beta d_s = -\pi - 2\beta d_s = -(2k - 1)\pi \tag{12.50}$$

Subtracting Eq. (12.49) from (12.50), we get

$$2\beta(d_l - d_s) + \pi = \theta_l \tag{12.51}$$

By measuring the shift in the reference minimum and wave length (to calculate β), using the above equation, it is possible to find θ_l. From the measured value of SWR, the magnitude of the reflection coefficient using Eq. (12.47) can be obtained. Once the reflection coefficient over the load is available, then its impedance can be computed using Eq. (12.48).

Load measurement

- $\mathbf{Z}_l = \mathbf{Z}_0 \dfrac{(1 + \Gamma_l)}{(1 - \Gamma_l)} = \mathbf{Z}_0 \dfrac{(1 + \Gamma_l \angle \theta_l)}{(1 - \Gamma_l \angle \theta_l)}$
- $\Gamma_l = \dfrac{(\rho - 1)}{(\rho + 1)}$
- $\theta_l = -\pi - 2\beta d_{min} \rightarrow$ first minimum method
- $\theta_l = \pi + 2\beta d_{shift} \rightarrow$ minimum shift method
- $d_{shift} = d_l - d_s \rightarrow$ shift in reference minimum when shorted
- Minimum shift towards load $\rightarrow d_{shift}$ becomes +ve $\rightarrow$ load inductive
- Minimum shift towards generator $\rightarrow d_{shift}$ becomes –ve $\rightarrow$ load capacitive

12.7 SPECIAL TYPES OF LOSSLESS LINES

We often encounter situations involving lossless transmission lines with particular terminations like short circuit and open circuit. They find wide applications in making the stubs which are used for load-matching purposes. Here we consider such types of lines and try to find out their properties.

12.7.1 Short-circuited Line

- The voltage reflection coefficient over the load is

$$\Gamma_l\big|_{Z_l=0} = \frac{Z_l - Z_0}{Z_l + Z_0}\bigg|_{Z_l=0} = -1 \tag{12.52}$$

As the line is lossless, the magnitude of the reflection coefficient remains the same at all points over the line, i.e.

$$\Gamma = |\Gamma_l| = |-1| = 1$$

- The standing wave ratio for the line is

$$\rho = \frac{1+\Gamma}{1-\Gamma} = \infty \tag{12.53}$$

- The voltage on the line is

$$\mathbf{V} = a'e^{j\beta d} + b'e^{-j\beta d} \tag{12.54}$$

Here the reflection coefficient is –1, therefore, $a' = -b'$ and hence

$$\mathbf{V}_{sc}(d) = 2a'j \sin \beta d \tag{12.55}$$

- The current on the line is

$$\mathbf{I} = \frac{-1}{\mathbf{Z}_0}(a'e^{j\beta d} - b'e^{-j\beta d}) \tag{12.56}$$

with $a' = -b'$ it becomes

$$\mathbf{I}_{sc}(d) = \frac{-2a'}{\mathbf{Z}_0} \cos \beta d \tag{12.57}$$

- The voltage is zero at the load, i.e. at $d = 0$ as it should be for a short circuit, and its amplitude varies as $\sin \beta d$, whereas the current is maximum at the load and it varies as $\cos \beta d$.
- The input impedance of a lossless line is

$$\mathbf{Z}_{in} = \mathbf{Z}_0 \frac{\mathbf{Z}_l + j\mathbf{Z}_0 \tan \beta l}{\mathbf{Z}_0 + j\mathbf{Z}_l \tan \beta l} \tag{12.58}$$

For the short-circuited load

$$\mathbf{Z}_{in}\big|_{Z_l=0} = \mathbf{Z}_{in}^{sc} = \mathbf{Z}_0\ j \tan \beta l \tag{12.59}$$

In general, the input impedance of a lossless line is complex but from the above expression we can observe that

- The input impedance of a short-circuited line is purely reactive.

$$\mathbf{Z}_{\text{in}}^{\text{sc}} = \mathbf{Z}_0 \; j \tan \beta l = jX_{\text{in}} \tag{12.60}$$

- If $\tan \beta l \geq 0$, the line appears inductive with an inductance of $L = (Z_0 \tan \beta l)/\omega$. The shortest length of the line which can give an inductance of L is

$$l = \frac{1}{\beta} \tan^{-1}\left(\frac{\omega L}{Z_0}\right) \tag{12.61}$$

- If $\tan \beta l \leq 0$, the line appears capacitive with a capacitance of $C = -1/\omega Z_0 \tan \beta l$. The shortest length of the line which can give a capacitance of C is

$$l = \frac{1}{\beta}\left[\pi - \tan^{-1}\left(\frac{1}{\omega C Z_0}\right)\right] \tag{12.62}$$

From these results, we can observe that through the proper choice of the length of a short-circuited line, we can make substitutes for capacitors and inductors with any desired reactance. Such a practice is indeed common in the design of microwave circuits and high-speed integrated circuits because making an actual capacitor or inductor is often more difficult than making a shorted transmission line. The voltage, current and impedance of resonant lines are shown in Figure 12.4.

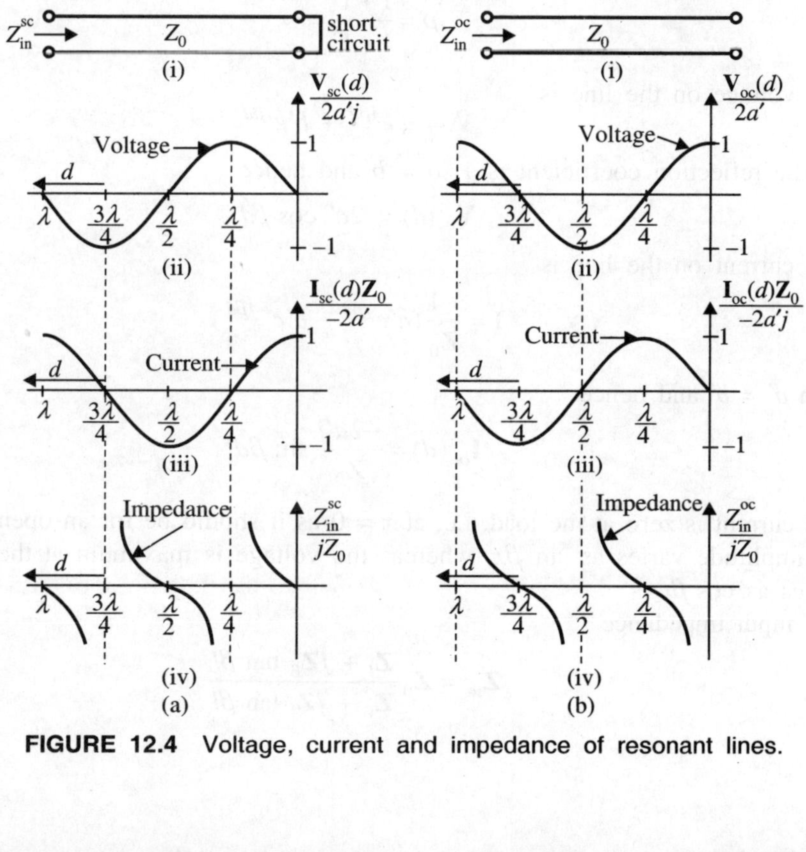

FIGURE 12.4 Voltage, current and impedance of resonant lines.

- At frequencies in the vicinity of $\beta l = (2n - 1)\pi/2$ (an odd number of quarter-wavelengths), the shorted line behaves like a parallel resonant circuit.
- At frequencies in the vicinity of $\beta l = n\pi$ (an integral number of half-wavelengths), the shorted line behaves like a series resonant circuit.

These properties are utilized in the design of band pass and band stop filters at microwave frequencies. It should be noted that unlike an ideal *L–C* circuit, the shorted line has an infinite number of resonances.

12.7.2 Open-circuited Line

- The voltage reflection coefficient over the load

$$\boldsymbol{\Gamma}_l\Big|_{\mathbf{Z}_l\to\infty} = \left.\frac{\mathbf{Z}_l - \mathbf{Z}_0}{\mathbf{Z}_l + \mathbf{Z}_0}\right|_{\mathbf{Z}_l\to\infty} = 1 \tag{12.63}$$

As the line is lossless, the magnitude of the reflection coefficient remains the same at all points over the line, i.e.

$$\Gamma = |\boldsymbol{\Gamma}_l| = 1$$

- The standing wave ratio for the line is

$$\rho = \frac{1+\Gamma}{1-\Gamma} = \infty \tag{12.64}$$

- The voltage on the line is

$$\mathbf{V} = a' e^{j\beta d} + b' e^{j\beta d} \tag{12.65}$$

As the reflection coefficient is 1, $a = b$ and hence

$$\mathbf{V}_{oc}(d) = 2a' \cos \beta d \tag{12.66}$$

- The current on the line is

$$\mathbf{I} = \frac{-1}{\mathbf{Z}_0}(a' e^{j\beta d} - b' e^{-j\beta d}) \tag{12.67}$$

with $a' = b'$ and hence

$$\mathbf{I}_{oc}(d) = \frac{-2a'j}{\mathbf{Z}_0} \sin \beta d \tag{12.68}$$

The current is zero at the load, i.e. at $x = 0$ as it should be for an open circuit, and its amplitude varies as $\sin \beta x$, whereas the voltage is maximum at the load and it varies as $\cos \beta x$.
- The input impedance.

$$\mathbf{Z}_{in} = \mathbf{Z}_0 \frac{\mathbf{Z}_l + j\mathbf{Z}_0 \tan \beta l}{\mathbf{Z}_0 + j\mathbf{Z}_l \tan \beta l} \tag{12.69}$$

For the open-circuited load

$$\mathbf{Z}_{\text{in}}\big|_{\mathbf{Z}_l=\infty} = \mathbf{Z}_{\text{in}}^{\text{oc}} = -j\mathbf{Z}_0 \cot \beta l \tag{12.70}$$

- At frequencies in the vicinity of $\beta l = (2n - 1)\,\pi/2$ (an odd number of quarter-wavelengths), the open-circuited line behaves like a series resonant circuit.
- At frequencies in the vicinity of $\beta l = n\pi$ (an integral number of half-wavelengths), the open-circuited line behaves like a parallel resonant circuit.

  These properties are also used in the design of filters at microwave frequencies. Some of such applications are mentioned here.
- An opened out parallel wire $\lambda/4$ transmission line is used as a wire radiator called 'half wave dipole'.
- An opened out parallel wire transmission line of length less than $\lambda/4$ is used as a wire parasitic radiator called 'director' in Yagi-Uda array. So the director carries capacitive currents. In other words, an opened out line excited at a frequency less than resonant is capacitive.
- An opened out parallel wire transmission line of length more than $\lambda/4$ is used as a wire parasitic radiator called 'reflector' in Yagi-Uda array. So the reflectors carry inductive currents. In other words, an opened out line excited at a frequency more than resonant is inductive.

12.8 TRANSIENTS ON LINES

Until now, we considered the steady-state response of the transmission line system for a sinusoidal input. Now we consider the transient response of the system which also has some interesting applications. Firstly we consider the response of the line system for a step voltage input and then one interesting application where a charged line section is used for the formation of a pulse.

Step input: Consider a lossless line terminated with a mismatched load, i.e. $R_l \neq Z_0$. Let us suppose it is connected to a dc source V_0 with an internal impedance of R_g. In the steady state, the voltage at an arbitrary point over the line or over the load is the sum of the first forward and reflected waves, second forward and reflected waves, etc. Therefore,

$$\begin{aligned}
V_l &= V_1^+ + V_1^- + V_2^+ + V_2^- + \cdots \\
&= V_1^+(1+\Gamma_l+\Gamma_g\Gamma_l+\Gamma_g\Gamma_l^2+\Gamma_g^2\Gamma_l^2+\Gamma_g^2\Gamma_l^3+\cdots) \\
&= V_1^+\left[(1+\Gamma_g\Gamma_l+\Gamma_g^2\Gamma_l^2+\cdots)+\Gamma_l(1+\Gamma_g\Gamma_l+\Gamma_g^2\Gamma_l^2+\cdots)\right] \\
&= V_1^+\left[\frac{1}{1+\Gamma_g\Gamma_l}+\frac{\Gamma_l}{1+\Gamma_g\Gamma_l}\right] \\
&= V_1^+\left[\frac{1+\Gamma_l}{1+\Gamma_g\Gamma_l}\right]
\end{aligned} \tag{12.71}$$

Then in the steady state, it can be shown that the load voltage is

$$V_l = V_1^+\left(\frac{1+\boldsymbol{\Gamma}_l}{1+\boldsymbol{\Gamma}_g\boldsymbol{\Gamma}_l}\right) \tag{12.72}$$

where

$$V_1^+ = \left(\frac{V_0 Z_0}{R_g + Z_0}\right) \tag{12.73}$$

$$\boldsymbol{\Gamma}_l = \left(\frac{R_l - \mathbf{Z}_0}{R_l + \mathbf{Z}_0}\right) \tag{12.74}$$

and

$$\boldsymbol{\Gamma}_g = \left(\frac{\mathbf{Z}_g - \mathbf{Z}_0}{\mathbf{Z}_g + \mathbf{Z}_0}\right) \tag{12.75}$$

In the case of $Z_g = 0$, the reflection coefficient at the source end $\boldsymbol{\Gamma}_g = -1$ and $V_1^+ = V_0$.

Pulse forming line: When an open-circuited transmission line charged to V_0 is connected to a resistor R_g, then a voltage wave V_1^+ is set over the line. It is given by

$$V_1^+ = \left(\frac{-V_0 Z_0}{R_g + Z_0}\right) \tag{12.76}$$

If $R_g = Z_0$, then $V_1^+ = -V_0/2$. The voltage across the resistor becomes $V_R = V_0 + V_1^+ = V_0/2$. And this voltage persists for a time of $T = 2l/v$ in the form of a pulse.

SOLVED PROBLEMS

1. A lossless line of length l = 2 m under sinusoidal steady state is operated at a frequency f = 175 MHz, has a wave velocity v = 200 m/μs, characteristic impedance $\mathbf{Z}_0$ = 100 Ω, source resistance $\mathbf{Z}_s$ = 50 Ω and is terminated over a load $\mathbf{Z}_l$ = (200 – j30) Ω. Find the electrical length of the line and reflection coefficient at load and at input to the line.

Solution: Electrical length, a dimensionless quantity, is a physical length in terms of the wavelength, i.e. $l_e = l/\lambda$. But the wavelength is the ratio of wave velocity over the line to frequency, i.e. $\lambda = v/f$. Therefore, the electrical length $l_e = lf/v$. From the given values,

$$l_e = \frac{fl}{v} = \frac{2\times 175\times 10^6}{2\times 10^8} = 1.75$$

The reflection coefficient at load is

$$\boldsymbol{\Gamma}_l = \frac{\mathbf{Z}_l - \mathbf{Z}_0}{\mathbf{Z}_l + \mathbf{Z}_0} = \frac{(200 - j30) - 100}{(200 - j30) + 100} = 0.346\angle 11°$$

The reflection coefficient at source is

$$\Gamma_s = \Gamma_l e^{-j2\beta l} = 0.346\angle 11^\circ e^{-j2(2\pi/\lambda)l}$$
$$= 0.346\angle 11^\circ e^{-j4\pi le} = 0.346\angle 11^\circ \times 1\angle 180^\circ = 0.34\angle 169^\circ$$

2. A lossy coaxial cable, operated at a frequency of 500 MHz, has the following primary constants: R = 2.25 Ω/m, L = 1 μH/m, C = 100 pF/m and G = 0 ℧/m. Determine the input impedance of a 10 m length cable on short circuit/open circuit terminated with a load of 10 Ω.

Solution: The propagation constant in terms of the primary constants of the line is

$$\gamma = \sqrt{ZY} = \sqrt{(R + j\omega L)(G + j\omega C)}$$

Substituting the given values of primary constants, we get

$$\gamma = \sqrt{(2.25 + j2\pi \times 500 \times 10^6 \times 10^{-6})(j2\pi \times 500 \times 10^6 \times 100 \times 10^{-12})}$$
$$= \sqrt{(2.25 + j\pi \times 10^3)(j\pi \times 10^{-1})}$$
$$= \sqrt{986.96\angle 179.95^\circ}$$

$$\gamma = 31.42\angle 89.975^\circ \text{m}^{-1} = 1.125 \times 10^{-2} + j31.42 \text{ m}^{-1}$$

The real part of the propagation constant is the attenuation constant. Therefore, the attenuation constant of the given line is

$$\alpha = 1.125 \times 10^{-2} \text{ Np/m} = 3.43 \times 10^{-3} \text{ Np/ft}$$

Therefore, attenuation in dB of l = 100 ft length cable is

$$= 3.43 \times 10^{-3} \times 100 = 0.343 \text{ Np}$$
$$= 0.343 \times 8.686 = 2.98 \text{ dB}$$

The characteristic impedance of the line is

$$Z_0 = \sqrt{\frac{(2.25 + j\pi \times 10^3)}{(j\pi \times 10^{-1})}} = \sqrt{\frac{3141.59\angle 89.959}{0.314\angle 90}}$$
$$= 100.02\angle -0.02^\circ \ \Omega$$

The reflection coefficient at the source is related to its value at load through

$$\Gamma_s = \Gamma_l e^{-2\gamma l} = |\Gamma_l| e^{j\theta_l} e^{-2\alpha l} e^{-2j\beta l}$$

Under the open-circuited condition, the reflection coefficient at the load is Γ_l = 1. Therefore, the reflection coefficient at the source becomes

$$\Gamma_s = |\Gamma_l| e^{j\theta_l} e^{-2\alpha l} e^{-2j\beta l}$$
$$= 1e^{j0} e^{-2\times 1.125\times 10^{-2}\times 10} e^{-j2\times 31.42\times 10}$$
$$= e^{-2\times 1.125\times 10^{-2}\times 10}\angle -2 \times 31.42 \times 10 = 0.798\angle 0^\circ$$
$$= 0.798\angle -2 \times 31.42 \times 10 \times 57.3^\circ$$
$$= 0.798\angle -7.32^\circ$$

The input impedance of the line then can be calculated using

$$\mathbf{Z}_{\text{in}} = \mathbf{Z}_0 \frac{1+\mathbf{\Gamma}_s}{1-\mathbf{\Gamma}_s} = 100\frac{1+0.798\angle-7.32°}{1-0.798\angle-7.32°} = 892.5\angle 0°\ \Omega$$

$$= 100\frac{1+0.798\angle-7.32°}{1-0.798\angle-7.32°} = 773\angle-29.24°\ \Omega$$

Under the short-circuited condition, the reflection coefficient at the load is $\mathbf{\Gamma}_l = -1$. The reflection coefficient at the source then becomes

$$\mathbf{\Gamma}_s = \mathbf{\Gamma}_1 e^{-2\gamma l} = -0.798\angle-7.32°$$

The input impedance of the line then can be calculated using

$$\mathbf{Z}_{\text{in}} = \mathbf{Z}_0 \frac{1+\mathbf{\Gamma}_s}{1-\mathbf{\Gamma}_s}$$

$$= 100\frac{1+0.798\angle-7.32°}{1-0.798\angle-7.32°} = 12.93\angle 29.24°\ \Omega$$

Under the terminated condition with $\mathbf{Z}_l = 10\ \Omega$, the reflection coefficient at the load is

$$\mathbf{\Gamma}_l = \frac{\mathbf{Z}_l - \mathbf{Z}_0}{\mathbf{Z}_l + \mathbf{Z}_0}$$

$$= \frac{10-100.02\angle-0.02°\Omega}{10+100.02\angle-0.02°\Omega} = 0.818\angle 179.99°$$

Then the reflection coefficient at source becomes

$$\mathbf{\Gamma}_s = \mathbf{\Gamma}_l e^{-2\gamma l}$$

$$= 0.798\angle-7.32° \times 0.818\angle 179.99°$$

$$= 0.653\angle 172.67°$$

The input impedance of the line then can be calculated using

$$\mathbf{Z}_{\text{in}} = \mathbf{Z}_0 \frac{1+\mathbf{\Gamma}_s}{1-\mathbf{\Gamma}_s}$$

$$= 100\frac{1+0.653\angle 172.67°}{1-0.653\angle 172.67°}$$

$$= 21.945\angle 16.198°\ \Omega$$

3. A 75 Ω line is terminated by a load of (120 + j80) Ω. Find the maximum and minimum impedances over the line.

Solution: We use the following formulae to find the maximum and minimum impedances over the line:

$$z_{max} = \frac{1+|\Gamma|}{1-|\Gamma|}$$

$$z_{min} = \frac{1-|\Gamma|}{1+|\Gamma|}$$

Let us calculate the magnitude of the reflection coefficient.

$$|\Gamma| = \left|\frac{\mathbf{Z}_l - \mathbf{Z}_0}{\mathbf{Z}_l + \mathbf{Z}_0}\right| = \left|\frac{120 + j80 - 75}{120 + j80 + 75}\right| = \left|\frac{45 + j80}{195 + j80}\right| = 0.435$$

Now we compute the maximum and minimum impedances

$$\mathbf{Z}_{max} = \mathbf{Z}_0 \frac{1+|\Gamma|}{1-|\Gamma|} = 75\frac{1+0.435}{1-0.435} = 190.48\ \Omega$$

$$\mathbf{Z}_{min} = \mathbf{Z}_0 \frac{1+|\Gamma|}{1-|\Gamma|} = 75\frac{1-0.435}{1+0.435} = 29.52\ \Omega$$

4. The voltage reflection coefficient due to load connected to a lossless transmission line of characteristic impedance 50 Ω and working at 3 GHz is 0.5∠45°. Assuming the load voltage to be 10 V, calculate the rms voltage and current at intervals of one-eighth wavelength from the load up to a distance of 5 cm.

Solution: The total voltage or current at any point over the line is the sum of that due to incident and reflected waves. It is so even over the load also.

The total voltage over the load is therefore the sum of the incident wave voltage and the reflected wave voltage over the load, i.e.

$$\begin{aligned} \mathbf{V}_l &= \mathbf{V}_{l,\text{inc}} + \mathbf{V}_{l,\text{ref}} \\ &= \mathbf{V}_{l,\text{inc}} + \Gamma_l \mathbf{V}_{l,\text{inc}} = \mathbf{V}_{l,\text{inc}}\,(1 + \Gamma_l) \end{aligned}$$

From the given values

$$\mathbf{V}_{l,\text{inc}} = \frac{\mathbf{V}_l}{(1+\Gamma_l)} = \frac{10\angle 0}{1+0.5\angle 45^\circ} = 7.148\angle -14.63^\circ$$

Similarly the total voltage over the line at a point which is at a distance of d from the load is therefore the sum of the incident wave voltage and reflected wave voltage at that point, i.e.

$$\begin{aligned} \mathbf{V}_d &= \mathbf{V}_{d,\text{inc}} + \mathbf{V}_{d,\text{ref}} \\ &= \mathbf{V}_{d,\text{inc}} + \Gamma_d \mathbf{V}_{d,\text{inc}} = \mathbf{V}_{d,\text{inc}}\,(1 + \Gamma_d) \end{aligned}$$

But $\Gamma_d = \Gamma_l e^{-j2\beta d}$ and $\mathbf{V}_{d,\text{inc}} = \mathbf{V}_{l,\text{inc}} e^{j\beta d}$. With these substitutions

$$\begin{aligned} \mathbf{V}_d &= \mathbf{V}_{d,\text{inc}}\,(1 + \Gamma_d) \\ &= \mathbf{V}_{l,\text{inc}} e^{j\beta d}(1 + \Gamma_l e^{-j2\beta d}) \end{aligned}$$

By following the similar procedure, we can also find out the expression for current as

$$\mathbf{I}_d = \mathbf{I}_{d,\text{inc}}(1 - \mathbf{\Gamma}_d)$$
$$= \mathbf{I}_{l,\text{inc}} e^{j\beta d}(1 - \mathbf{\Gamma}_l e^{-j2\beta d})$$
$$= \left(\frac{\mathbf{V}_{l,\text{inc}}}{\mathbf{Z}_0}\right) e^{j\beta d}(1 - \mathbf{\Gamma}_l e^{-j2\beta d})$$

From the given values

$$\lambda = \frac{3 \times 10^8}{3 \times 10^9} = 10 \text{ cm}$$

We need values up to a distance of 5 cm or $\lambda/2$. At a distance of $\lambda/8$,

$$\beta d = \frac{2\pi}{\lambda}\frac{\lambda}{8} = \frac{\pi}{4} = 0.125\pi = 22.5°$$

$$\mathbf{V}_d = \mathbf{V}_{l,\text{inc}} e^{j\beta d}(1 + \mathbf{\Gamma}_l e^{-j2\beta d})$$
$$= 7.148\angle(-14.63° + 22.5°)\ [1 + 0.5\angle(45° - 45°)]$$
$$= 7.148\angle(-14.63° + 22.5°) \times 1.5$$
$$= 10.722\angle 7.87°$$

$$\mathbf{I}_d = \left(\frac{\mathbf{V}_{l,\text{inc}}}{\mathbf{Z}_0}\right) e^{j\beta d}(1 - \mathbf{\Gamma}_l e^{-j2\beta d})$$
$$= \left(\frac{7.148}{50}\right)\angle(-14.63° + 22.5°)[1 - 0.5\angle(45° - 45°)]$$
$$= \left(\frac{7.148}{50}\right)\angle(-14.63° + 22.5°) \times 0.5$$
$$= 0.0714\angle 7.87°$$

The rms values, then, are

$$V_{\text{rms}} = \frac{10.722}{\sqrt{2}} = 7.58 \text{ V}$$

$$I_{\text{rms}} = \frac{0.0714}{\sqrt{2}} = 0.050 \text{ A}$$

The voltages and currents at other distances can also be computed in similar manner.

5. A lossless line of length l = 10 m with a characteristic impedance of 50 Ω and propagation velocity of 200 m/μs is driven by a 26 MHz source of voltage 100 V with an internal impedance of 50 Ω. The line is terminated with an impedance of 100 + j50 Ω. Determine
   (a) input impedance to the line,
   (b) input voltage for the line and
   (c) load voltage.

Solution:

(a) We evaluate first the reflection coefficient over the load $\mathbf{\Gamma}_l$ using

$$\mathbf{\Gamma}_l = \frac{\mathbf{Z}_l - \mathbf{Z}_0}{\mathbf{Z}_l + \mathbf{Z}_0}$$

And then find the reflection coefficient over the source $\mathbf{\Gamma}_s$ using

$$\mathbf{\Gamma}_s = \mathbf{\Gamma}_l e^{-j2\beta l}$$

Finally we compute the input impedance using

$$\mathbf{Z}_{\text{in}} = \mathbf{Z}_0 \frac{1 + \mathbf{\Gamma}_s}{1 - \mathbf{\Gamma}_s}$$

Now let us do the actual calculations.

$$\mathbf{\Gamma}_l = \frac{\mathbf{Z}_l - \mathbf{Z}_0}{\mathbf{Z}_l + \mathbf{Z}_0} = \frac{50 + j50}{150 + j50} = 0.44\angle 26.6°$$

$$\mathbf{\Gamma}_s = \mathbf{\Gamma}_l e^{-j2\beta l} = 0.44\angle 26.6° \times 1\angle - 216° = 0.44\angle 189.4°$$

In the above calculation, we find the value of the phase shift constant from $\beta = \omega/v = 0.82$ rad/m.

Finally

$$\mathbf{Z}_{\text{in}} = \mathbf{Z}_0 \frac{1 + \mathbf{\Gamma}_s}{1 - \mathbf{\Gamma}_s} = 50 \frac{1 + 0.44\angle 189.4°}{1 - 0.44\angle 189.4°} = 19.53\angle 10.4°\ \Omega$$

(b) The input voltage to the line can be evaluated by using the principle of voltage division in series circuit.

$$\mathbf{V}_{\text{in}} = \frac{\mathbf{Z}_{\text{in}}}{\mathbf{Z}_{\text{in}} + \mathbf{Z}_s}\mathbf{V}_s = \frac{19.53\angle 10.4°}{19.53\angle 10.4° + 50\angle 0°} 100\angle 0° = 28.18\angle 7.5°\ \text{V}$$

(c) The input voltage to the load is

$$\mathbf{V}_l = \mathbf{V}_s e^{-j\beta l} \frac{1 + \mathbf{\Gamma}_l}{1 - \mathbf{\Gamma}_s} = 70.71\angle -100°\ \text{V}$$

6. A low loss transmission line of 100 Ω characteristic impedance is connected to a load of 400 Ω. Calculate

(a) the current reflection coefficient and
(b) standing wave ratio.

Solution:

(a) Reflection coefficient: The voltage reflection coefficient varies from point to point over the line and over the load it is

$$\mathbf{\Gamma}_l = \frac{\mathbf{Z}_l - \mathbf{Z}_0}{\mathbf{Z}_l + \mathbf{Z}_0} = \frac{400 - 100}{400 + 100} = \frac{3}{5} = 0.6$$

The current reflection coefficient is negative of the voltage reflection coefficient. Therefore, in the present case, the current reflection coefficient is –0.6.

(b) VSWR: When the termination is pure resistance, i.e. $Z_l = R_l$, then the SWR is Z_0/R_l or R_l/Z_0, depending on whichever is more than one. Therefore,

$$\text{VSWR} = \frac{140}{100} = 1.4$$

It can be also be calculated using

$$\rho = \frac{1+\Gamma}{1-\Gamma} = \frac{1+0.6}{1-0.6} = 4$$

7. A 100 Ω lossless line connects a signal of 100 kHz to a load of 140 Ω. The load power is 100 mW. Calculate

(a) voltage reflection coefficient,
(b) VSWR and
(c) positions of V_{max}, I_{max}, V_{min} and I_{min}.

Solution:

(a) Voltage reflection coefficient: The voltage reflection coefficient varies from point to point over the line and over the load it is

$$\mathbf{\Gamma} = \frac{\mathbf{Z}_l - \mathbf{Z}_0}{\mathbf{Z}_l + \mathbf{Z}_0} = \frac{140 - 100}{140 + 100} = \frac{40}{240} = 0.167$$

(b) VSWR: When the termination is pure resistance, i.e. $Z_l = R_l$, then the SWR is Z_0/R_l or R_l/Z_0, depending on whichever is more than one. Therefore,

$$\text{VSWR} = \frac{140}{100} = 1.4$$

(c) Positions of V_{max}, I_{max}, V_{min} and I_{min}: As the load $\mathbf{Z}_l$ is purely resistive and $\mathbf{Z}_l > \mathbf{Z}_0$, voltage maximum and current minimum both occur right over the load. And the voltage minimum and current maximum occur at a distance of half wavelength from the load.

8. A lossless line of characteristic impedance equal to 60 Ω is connected to a load consisting of a series combination of R = 30 Ω, L = 1 μH and C = 100 pF. Find SWR and the distance of the first minimum d_{min} from the load for the following frequencies:

(a) $\omega = 10^8$
(b) $\omega = 2 \times 10^8$
(c) $\omega = 0.8 \times 10^8$.

Solution:

(a) $\omega = 10^8$

Firstly we evaluate the terminating impedance.

$$Z_l = R + j\left(\omega L - \frac{1}{\omega C}\right)$$

$$= 30 + j\left(10^8 \times 10^6 - \frac{1}{10^8 \times 100 \times 10^{-12}}\right) = 30\ \Omega$$

Next we find out the reflection coefficient over the load

$$\mathbf{\Gamma}_l = \frac{\mathbf{Z}_l - \mathbf{Z}_0}{\mathbf{Z}_l + \mathbf{Z}_0} = \frac{30 - 60}{30 + 60} = -\frac{1}{3} = 0.33\angle\pi$$

Using the magnitude of the reflection coefficient we evaluate the SWR and with its angle the minimum distance.

$$\rho = \frac{1 + \Gamma_l}{1 - \Gamma_l} = \frac{1 + 0.333}{1 - 0.333} = 1.99$$

Next we calculate the distance of the first minimum using

$$d_{\min} = \frac{\lambda}{4\pi}(\theta_l + \pi)$$

$$= \frac{\lambda}{4\pi}(\pi + \pi) = \frac{\lambda}{2} = 0$$

(b) $\omega = 2 \times 10^8$

$$\mathbf{Z}_l = R + j\left(\omega L - \frac{1}{\omega C}\right)$$

$$= 30 + j\left(2 \times 10^8 \times 10^6 - \frac{1}{2 \times 10^8 \times 100 \times 10^{-12}}\right) = 30 + j150\ \Omega$$

Next we find out the reflection coefficient over the load and from its magnitude the value of SWR.

$$\mathbf{\Gamma}_l = \frac{\mathbf{Z}_l - \mathbf{Z}_0}{\mathbf{Z}_l + \mathbf{Z}_0} = \frac{30 + j150 - 60}{30 + j150 + 60} = 0.875\angle 0.737$$

$$\rho = \frac{1 + \Gamma_l}{1 - \Gamma_l} = \frac{1 + 0.875}{1 - 0.875} = 14.87$$

Next we calculate the distance of the first minimum using

$$d_{\min} = \frac{\lambda}{4\pi}(\theta_l + \pi)$$

$$= \frac{\lambda}{4\pi}(0.737 + \pi) = 0.308\lambda$$

(c) $\omega = 0.8 \times 10^8$

$$\mathbf{Z}_l = R + j\left(\omega L - \frac{1}{\omega C}\right)$$

$$= 30 + j\left(0.8\times10^8 \times10^6 - \frac{1}{0.8\times10^8 \times100\times10^{-12}}\right) = 30 - j45\ \Omega$$

Next we find out the reflection coefficient over the load and from its magnitude the value of SWR.

$$\Gamma_l = \frac{\mathbf{Z}_l - \mathbf{Z}_0}{\mathbf{Z}_l + \mathbf{Z}_0} = \frac{30 - j45 - 60}{30 - j45 + 60} = 0.537\angle -1.69$$

$$\rho = \frac{1+\Gamma_l}{1-\Gamma_l} = \frac{1+0.537}{1-0.537} = 3.319$$

Next we calculate the distance of the first minimum using

$$d_{\min} = \frac{\lambda}{4\pi}(\theta_l + \pi)$$

$$= \frac{\lambda}{4\pi}(-1.69 + \pi) = 0.115\lambda$$

9. A lossless line of characteristic impedance equal to 60 Ω is connected to an unknown impedance. Find its value for the following sets of observations:

(a) SWR = 1.5, the voltage minimum right over the load,
(b) SWR = 3, the first voltage minimum at 3 cm and the next one at 9 cm and
(c) SWR = 2, the first voltage minimum at 3 cm and the next one at 7 cm from the load.

Solution: We use the relations to find the unknown impedance

$$\mathbf{Z}_l = \mathbf{Z}_0 \frac{1+\Gamma_l}{1-\Gamma_l}$$

$$\Gamma_l = \Gamma_l \angle \theta_l$$

$$\Gamma_l = \frac{\rho - 1}{\rho + 1}$$

$$\theta_l = \frac{4\pi}{\lambda} d_{\min} - \pi$$

(a) SWR = 1.5, the voltage minimum right over the load

$$\Gamma_l = \frac{\rho - 1}{\rho + 1}$$

$$= \frac{1.5 - 1}{1.5 + 1} = 0.2$$

In the present case, the first minimum is right over the load: it means $d_{\min} = 0$. Therefore,

$$\theta_l = \frac{4\pi}{\lambda} d_{\min} - \pi = -\pi$$

From the available value of the reflection coefficient, now we compute the impedance value of the load.

$$\mathbf{Z}_l = \mathbf{Z}_0 \frac{1 + \Gamma_l}{1 - \Gamma_l}$$

$$= 60 \frac{1 + 0.2\angle -\pi}{1 - 0.2\angle -\pi} = 40\ \Omega$$

(b) SWR = 3, the first voltage minimum at 3 cm and the next one at 9 cm

$$\Gamma_l = \frac{\rho - 1}{\rho + 1}$$

$$= \frac{3 - 1}{3 + 1} = 0.5$$

Wavelength is twice the distance between two consecutive minima. Therefore, in this case, it is 12 cm. Therefore,

$$\theta_l = \frac{4\pi}{\lambda} d_{\min} - \pi$$

$$= \frac{4\pi}{12} \times 3 - \pi = 0$$

Now we compute the impedance value of the load, from the available value of the reflection coefficient.

$$\mathbf{Z}_l = \mathbf{Z}_0 \frac{1 + \Gamma_l}{1 - \Gamma_l}$$

$$= 60 \frac{1 + 0.5}{1 - 0.5} = 180\ \Omega$$

(c) SWR = 2, the first voltage minimum at 3 cm and the next one at 7 cm from the load.

$$\Gamma_l = \frac{\rho - 1}{\rho + 1}$$

$$= \frac{2 - 1}{2 + 1} = 0.333$$

In the present case, the wavelength is 8 cm. Therefore,

$$\theta_l = \frac{4\pi}{\lambda} d_{\min} - \pi$$

$$= \frac{4\pi}{8} \times 3 - \pi = 0.5\pi$$

Now we compute the impedance value of the load, from the available value of the reflection coefficient.

$$\mathbf{Z}_l = \mathbf{Z}_0 \frac{1 + \Gamma_l}{1 - \Gamma_l}$$

$$= 60 \frac{1 + 0.333\angle 0.5\pi}{1 - 0.333\angle 0.5\pi} = 48.3 + j35.728\ \Omega$$

10. Show that the lossless line terminated complex impedance is given by

$$R = \frac{\rho \mathbf{Z}_0}{\rho^2 \cos^2 \beta l + \sin^2 \beta l}$$

$$X = \frac{-\mathbf{Z}_0 (\rho^2 - 1) \sin \beta l \cos \beta l}{\rho^2 \cos^2 \beta l + \sin^2 \beta l}$$

where ρ is the standing wave ratio and l the distance of the first minimum from the load.

Solution: When the termination is complex impedance, the minimum or maximum cannot occur over the load: they can only happen away from the load. Now let us suppose the first minimum occurred at a distance of l from the load towards the generator.

Now we can observe that the standing wave pattern does not get disturbed and the line sees the same terminating impedance when it is extended, after disconnecting the original impedance, a length of $\lambda/2 - l$ and terminated by a resistance R, the line resistance at voltage minimum. Therefore, we can say that the terminating impedance is equal to the input impedance of the line of length $\lambda/2 - l$ terminated by a resistance R which is

$$\mathbf{Z}_{\text{in}} = \mathbf{Z}_0 \frac{\mathbf{Z}_l + j\mathbf{Z}_0 \tan \beta l}{\mathbf{Z}_0 + j\mathbf{Z}_l \tan \beta l}$$

$$= \mathbf{Z}_0 \frac{R + j\mathbf{Z}_0 \tan \beta\left(\frac{\lambda}{2} - l\right)}{\mathbf{Z}_0 + jR \tan \beta\left(\frac{\lambda}{2} - l\right)} = \mathbf{Z}_0 \frac{R + j\mathbf{Z}_0 \tan(\pi - \beta l)}{\mathbf{Z}_0 + jR \tan(\pi - \beta l)}$$

$$= \mathbf{Z}_0 \frac{R \cos(\pi - \beta l) + j\mathbf{Z}_0 \sin(\pi - \beta l)}{\mathbf{Z}_0 \cos(\pi - \beta l) + jR \sin(\pi - \beta l)} = \mathbf{Z}_0 \frac{-R \cos \beta l + j\mathbf{Z}_0 \sin \beta l}{-\mathbf{Z}_0 \cos \beta l + jR \sin \beta l}$$

Rationalizing, we get

$$\mathbf{Z}_{\text{in}} = \mathbf{Z}_0 \frac{(-R\cos\beta l + j\mathbf{Z}_0 \sin\beta l)(-\mathbf{Z}_0 \cos\beta l - jR\sin\beta l)}{\mathbf{Z}_0^2 \cos^2\beta l + R^2 \sin^2\beta l}$$

$$= \mathbf{Z}_0 \frac{R\mathbf{Z}_0(\cos^2\beta l + \cos^2\beta l) + j(R^2 - \mathbf{Z}_0^2)\sin\beta l \cos\beta l}{\mathbf{Z}_0^2 \cos^2\beta l + R^2 \sin^2\beta l}$$

$$= \mathbf{Z}_0 \frac{R\mathbf{Z}_0 + j(R^2 - \mathbf{Z}_0^2)\sin\beta l \cos\beta l}{\mathbf{Z}_0^2 \cos^2\beta l + R^2 \sin^2\beta l}$$

$$= \mathbf{Z}_0 \frac{\left(\dfrac{\mathbf{Z}_0}{R}\right) + j\left[1 - \left(\dfrac{\mathbf{Z}_0^2}{R^2}\right)\right]\sin\beta l \cos\beta l}{\left(\dfrac{\mathbf{Z}_0^2}{R^2}\right)\cos^2\beta l + \sin^2\beta l}$$

Replacing by $\mathbf{Z}_0/R$ by ρ, we get the standing wave ratio as

$$\mathbf{Z}_{\text{in}} = \mathbf{Z}_0 \frac{\rho + j[1 - \rho^2]\sin\beta l \cos\beta l}{\rho^2 \cos^2\beta l + \sin^2\beta l}$$

Separating real and imaginary parts, we obtain

$$R = \frac{\rho \mathbf{Z}_0}{\rho^2 \cos^2\beta l + \sin^2\beta l}, \quad X = \frac{-\mathbf{Z}_0(\rho^2 - 1)\sin\beta l \cos\beta l}{\rho^2 \cos^2\beta l + \sin^2\beta l}$$

11. A transmission line with a load of Z_l = 90 Ω and a characteristic impedance of Z_0 = 30 Ω is connected to a dc voltage source of 100 V with an internal resistance of 60 Ω at the time instant of t = 0. Find the steady-state load voltage and load current.

Solution: We know that the steady-state voltage is

$$V_1^+ = \left(\frac{V_0 Z_0}{R_g + Z_0}\right) = \left(\frac{30 \times 100}{60 + 30}\right) = 33.33 \text{ V}$$

Reflection coefficient at source is

$$\Gamma_g = \left(\frac{Z_g - Z_0}{Z_g + Z_0}\right) = \left(\frac{60 - 30}{60 + 30}\right) = 0.33$$

Reflection coefficient at load is

$$\Gamma_l = \left(\frac{R_l - Z_0}{R_l + Z_0}\right) = \left(\frac{90 - 30}{90 + 30}\right) = 0.5$$

Steady-state load voltage then becomes

$$V_l = V_1^+\left(\frac{1+\Gamma_l}{1+\Gamma_g\Gamma_l}\right) = 33.33\left(\frac{1+0.5}{1-0.33\times 0.5}\right) = 60 \text{ V}$$

Steady-state load current then is

$$I_l = \frac{V_l}{Z_l} = \frac{60}{90} = 0.667 \text{ A}$$

EXERCISE SECTION

Essay Type Questions

1. Define reflection coefficient and express it in terms of reflection coefficient over the load.
2. Derive a relation for the reflection coefficient over the load.
3. Derive the interrelations between reflection coefficient and line impedance.
4. Analyse and derive an expression for voltage and current of standing waves over the line.
5. What is standing wave ratio? Relate it to the reflection coefficient.
6. Derive the values of standing wave ratio for different types of terminations.
7. Differentiate pure standing wave from pure travelling wave.
8. Describe and discuss important properties of open- and short-circuited lossless lines.

Numerical Problems

1. A low loss transmission line of 100 Ω characteristic impedance is connected to a load of 400 Ω. Calculate the reflection coefficient and standing wave ratio.
2. A loss line is terminated over a 150 Ω resistance and it is energized by a source of 100 Ω and 50 V. The length of the line is 2λ and its characteristic impedance is 75 Ω. Find its input impedance, the magnitude of the instantaneous load voltage and instantaneous power delivered to the load.
3. Calculate the reflection coefficient and VSWR for a 75 Ω line, terminated with
   (a) matched load,
   (b) short circuit,
   (c) $+j50$ Ω load and
   (d) $-j50$ Ω load.
4. A lossless line of characteristic impedance equal to 50 Ω is connected to a load consisting of a series combination of R = 15 Ω, L = 1.0 μH and C = 100 pF. Find SWR and the distance of the first minimum d_{min} from the load for the following frequencies:

(a) $\omega = 1.2 \times 10^9$,
(b) $\omega = 2.5 \times 10^8$ and
(c) $\omega = 1.8 \times 10^9$.

5. A transmission line with a load of $\mathbf{Z}_l = 100\ \Omega$ and a characteristic impedance of $\mathbf{Z}_0 = 50\ \Omega$ is connected to a dc voltage source of 100 V with an internal resistance of 70 Ω at the time instant of $t = 0$. Find the steady-state load voltage and load current.

Fill in the Blanks

1. The reflection coefficient is __________ and its range is __________.
2. The voltage reflection coefficient is __________ of the current reflection coefficient.
3. The reflection coefficient over the load is __________.
4. The magnitude of reflection coefficient remains the same over a __________ line.
5. The line impedance in terms of reflection coefficient is __________.
6. The reflection coefficient in terms of line impedance is __________.
7. The line impedance right of voltage node is __________ and on left side it is __________.
8. The line impedance over the current node is __________ and over the voltage node it is __________.
9. Locations of voltage maxima are __________ and locations of voltage minima are __________.
10. The range of SWR is __________.
11. The SWR is related to the __________ of the reflection coefficient and not its phase.
12. The SWR is related to the reflection coefficient through __________.
13. The SWR is defined as __________.
14. The SWR is infinity when the termination is __________.
15. For matched termination, the SWR value is __________.
16. For pure resistive termination, the SWR value is __________.

Answers

1. complex, –1 to +1
2. negative
3. $\dfrac{(\mathbf{Z}_l - \mathbf{Z}_0)}{(\mathbf{Z}_l + \mathbf{Z}_0)}$
4. lossless
5. $\mathbf{z} = \dfrac{1+\Gamma}{1-\Gamma}$
6. $\Gamma = \dfrac{\mathbf{z}-1}{\mathbf{z}+1}$
7. capacitive, inductive
8. $\mathbf{z}_{\max} = \dfrac{1+|\Gamma|}{1-|\Gamma|}$; $\mathbf{z}_{\min} = \dfrac{1-|\Gamma|}{1+|\Gamma|}$

9. $\beta x = n\pi, \quad \beta x = \dfrac{(2n-1)\pi}{2}$

10. one to infinity

11. magnitude

12. $\rho = \dfrac{1+|\Gamma|}{1-|\Gamma|}$

13. $\rho = \dfrac{|\mathbf{V}_{max}|}{|\mathbf{V}_{min}|} = \dfrac{|\mathbf{V}_{inc}| + |\mathbf{V}_{ref}|}{|\mathbf{V}_{inc}| - |\mathbf{V}_{ref}|}$

14 sc, oc or pure reactance

15. one

16. $\dfrac{\mathbf{Z}_l}{\mathbf{Z}_0}$ or $\dfrac{\mathbf{Z}_0}{\mathbf{Z}_l}$

Match the Following

Set I

1. Reflection coefficient	(a) 0 to ∞
2. Standing wave ratio	(b) –1 to 1
3. Short circuit load	(c) ∞
4. Matched termination	(d) $\dfrac{\mathbf{Z}_0}{R_l}$ or $\dfrac{R_l}{\mathbf{Z}_0}$
5. Resistive termination	(e) 1

Set II

1. Reflection coefficient	(a) $\mathbf{\Gamma} = \mathbf{\Gamma}_l e^{-2\gamma d}$
2. Standing wave ratio	(b) $\dfrac{\mathbf{Z}_l - \mathbf{Z}_0}{\mathbf{Z}_l + \mathbf{Z}_0}$
3. Open circuit load	(c) infinity
4. RC over load $\mathbf{\Gamma}_l$	(d) only for a lossless line
5. RC at a point $\mathbf{\Gamma}$	(e) general line

Set III

1. Reflection coefficient	(a) $\dfrac{1+\Gamma}{1-\Gamma}$
2. Standing wave ratio	(b) $\dfrac{\rho-1}{\rho+1}$
3. Lossless line	(c) Γ constant
4. Pure reactance termination	(d) voltage/current max/min over load
5. Pure resistive termination	(e) infinity

Set IV

1. Open circuit line	(a) variable over the line.
2. Short circuit line	(b) $\beta x = n\pi$
3. Voltage minima	(c) $\beta x = \frac{(2n-1)\pi}{2}$
4. Voltage maxima	(d) –1
5. Phase of reflection coefficient	(e) +1

Set V

1. Voltage	(a) $\mathbf{I}_0 = -2ja\mathbf{Y}_0 e^{-\alpha x} \sin \beta x$
2. Current	(b) $\mathbf{V}_0 = 2a\mathbf{Y}_0 e^{-\alpha x} \cos \beta x$
3. Voltage minima	(c) $\beta x = \frac{(2n-1)\pi}{2}$
4. Voltage maxima	(d) $1 - \lvert\Gamma\rvert^2$
5. Reflection loss	(e) $\beta x = n\pi$

Answers

Set I:	1. (b)	2. (a)	3. (c)	4. (e)	5. (d)
Set II:	1. (e)	2. (d)	3. (c)	4. (b)	5. (a)
Set III:	1. (b)	2. (a)	3. (c)	4. (e)	5. (d)
Set IV:	1. (e)	2. (d)	3. (c)	4. (b)	5. (a)
Set V:	1. (b)	2. (a)	3. (c)	4. (e)	5. (d)

Multiple Choice Questions

1. Multiple reflections occur when mismatch exists at
   (a) load end (b) source end
   (c) both (a) and (b) (d) none of these
2. The line impedance at voltage minimum is
   (a) resistive (b) capacitive
   (c) inductive (d) complex
3. The line impedance just by the source side of voltage minimum is
   (a) resistive (b) capacitive
   (c) inductive (d) complex
4. The line impedance just by the load side of voltage minimum is
   (a) resistive (b) capacitive
   (c) inductive (d) complex
5. The reflection coefficient is the ratio of
   (a) voltages (b) currents
   (c) both (a) and (b) (d) none of these

6. The reflection coefficient, in general, is a
   (a) complex (b) pure real
   (c) pure imaginary (d) none of these
7. The reflection coefficient is useful to describe reflections of
   (a) lossless lines (b) lossy lines
   (c) both (a) and (b) (d) none of these
8. The magnitude of reflection coefficient over a lossless line
   (a) remains constant (b) varies from point to point
   (c) is periodic (d) none of these
9. The phase of reflection coefficient over a lossless line
   (a) remains constant (b) varies from point to point
   (c) is periodic (d) none of these
10. The range of reflection coefficient is
    (a) –1 to 1 through 0 (b) +1 to ∞
    (c) 0 to infinity (d) none of these
11. The voltage and current reflection coefficients are related through
    (a) –1 (b) +1
    (c) 2 (d) none of these
12. The reflection coefficient for matched termination is
    (a) zero (b) one
    (c) infinity (d) none of these
13. The magnitude of RC for sc or oc termination of a lossless line is
    (a) zero (b) one
    (c) infinity (d) none of these
14. The reflection coefficient over the normalized load z_l is
    (a) $\frac{(z_l - 1)}{(z_l + 1)}$ (b) $\frac{(z_l + 1)}{(z_l - 1)}$
    (c) $\frac{(z_s - 1)}{(z_s + 1)}$ (d) none of these
15. The SWR is meaningful for
    (a) lossy lines (b) lossless lines
    (c) both (a) and (b) (d) none of these
16. The SWR can be found from
    (a) magnitude of RC (b) phase of RC
    (c) both (a) and (b) (d) none of these
17. The SWR
    (a) is constant of line and load (b) varies over the line
    (c) both (a) and (b) (d) none of these

18. The range of SWR is
    (a) –1 to 1 through 0 (b) +1 to infinity
    (c) 0 to ∞ (d) none of these
19. The SWR for an sc- or oc-terminated line is
    (a) zero (b) one
    (c) infinity (d) none of these
20. The SWR for a match-terminated line is
    (a) zero (b) one
    (c) infinity (d) none of these
21. The SWR for pure resistance-terminated line is
    (a) $\frac{R_l}{Z_0}$ (b) $\frac{Z_0}{R_l}$
    (c) either (a) or (b) (d) none of these
22. The SWR for a complex impedance-terminated Z_l line is
    (a) $\frac{Z_l}{Z_0}$ (b) $\frac{Z_0}{Z_l}$
    (c) either (a) or (b) (d) none of these
23. The SWR represents the amount of
    (a) mismatch (b) reflection
    (c) both (a) and (b) (d) none of these
24. The SWR is the ratio of
    (a) voltages (b) currents
    (c) both (a) and (b) (d) none of these
25. The advantage of SWR over RC is
    (a) easily measurable (b) more accurate
    (c) both (a) and (b) (d) none of these
26. Unique determination of termination load of a lossless line requires
    (a) RC (b) SW
    (c) both (a) and (b) (d) none of these

Answers

1. (c)	2. (b)	3. (c)	4. (a)	5. (c)
6. (a)	7. (c)	8. (a)	9. (a)	10. (a)
11. (a)	12. (a)	13. (b)	14. (a)	15. (a)
16. (b)	17. (c)	18. (b)	19. (c)	20. (d)
21. (c)	22. (c)	23. (c)	24. (a)	25. (d)
26. (c)				

13

Line Distortion and Line Matching

Don't aim for success if you want it; just do what you love and believe in, and it will come naturally.
—David Frost

When a telecom signal is placed over a line, it is expected to be carried and delivered by the line to the receiver at destination without causing any damage to the waveform. It is also intended to deliver the entire signal to the load/receiver without causing any reflections.

Causing damage to the waveform of the signal by the line while transmitting leads to occurrence of an unwanted phenomenon known as distortion. If the damage is too much, then it may lead to total uselessness of the transmitted signal requiring retransmission. It can be avoided by taking certain measures like equalization, loading, etc. Another type of problem comes into being due to a situation called load mismatch which leads to partial, instead of full, delivery of the signal to load. When the line fails to deliver the total signal to the load, reflections occur, leading to inefficient operation of the system. As remedial measures, to avoid reflections and the consequent inefficient operation, load matching is resorted to by undertaking certain measures known as matching techniques like quarter-wave matching technique, single stub technique, double stub technique and so forth.

In the present chapter, both the aspects of distortion and matching, one pertaining to transmission and the other to delivery are discussed. Smith chart is considered to be the most useful of all the available tools to analyse transmission lines. The basics of this chart are introduced along with several examples illustrating its utility in transmission line theory.

13.1 Line Distortion

The deviation of the waveform at the output of the line from that at its input is called *line distortion*. It is due to the fact that all frequencies in the waveform do not have the same attenuation and same delay during the propagation. The characteristic impedance, attenuation and velocity of propagation on the line being functions of frequency are the causes of distortion.

13.1.1 Types of Distortion

Frequency distortion and delay distortion

Frequency distortion is due to various frequency components of the signal undergoing different amounts of attenuation when the attenuation constant α is a function of frequency. To eliminate this distortion, the attenuation constant α must be made independent of frequency.

Phase or delay distortion is due to different frequency components of the signal undergoing different amounts of phase delays while reaching the destination, thus spoiling the original phase relation between them. To eliminate this, the phase shift constant β must be made proportional to angular frequency ω.

Equalizer

Frequency distortion can be reduced by cascading the lines with networks known as *equalizers*. An equalizer is a network whose attenuation versus frequency characteristic is just opposite to that of the line. Delay distortion can also be reduced with equalizers, but it must be designed in such a way that β for total circuit is proportional to ω. For audio transmission, frequency distortion is a serious problem, whereas for video transmission both are serious.

Distortion-less line

It is a line which transmits the input signal without any distortion. This occurs when the primary constants of the line are related through

$$\frac{R}{L} = \frac{G}{C} \tag{13.1}$$

With this interrelation among the primary constants of the line, we see below that the line is able to transfer the signal from the transmitting end to the receiving end without any distortion.

Proof: For the line to have neither frequency nor delay distortion, its attenuation constant and velocity of propagation cannot be functions of frequency. In view of the fact $v = \omega/\beta$, the phase shift constant must be a direct function of frequency.

If we observe the expression for the phase shift constant, it becomes a direct function of frequency when the term under the second radical is equal to $(RG + \omega^2 LC)^2$. Enforcing this condition on the radical, we obtain

$$(LG - CR)^2 = 0 \quad \Rightarrow \quad LG = CR$$

Then the phase shift constant becomes

$$\begin{aligned}\beta &= \sqrt{\frac{1}{2}\left[(\omega^2 LC - RG) + \sqrt{(R^2 + \omega^2 L^2)(G^2 + \omega^2 C^2)}\right]} \\ &= \omega\sqrt{LC}\end{aligned} \tag{13.2}$$

which is proportional to the angular frequency making the velocity of propagation independent of the frequency, thus eliminating delay distortion.

$$v = \frac{\omega}{\beta} = \frac{\omega}{\omega\sqrt{LC}} = \frac{1}{\sqrt{LC}} \tag{13.3}$$

We can also observe that the expression for attenuation constant becomes independent of frequency when the term under the second radical is equal to $(RG + \omega^2 LC)^2$. Enforcing this condition on the radical, we get

$$(LG - CR)^2 = 0 \quad \Rightarrow \quad LG = CR$$

With this incorporated, the attenuation constant becomes independent of the frequency,

$$\alpha = \sqrt{\frac{1}{2}\left[(RG - \omega^2 LC) + \sqrt{(R^2 + \omega^2 L^2)(G^2 + \omega^2 C^2)}\right]}$$

$$= \sqrt{RG} \tag{13.4}$$

It is also instructive to consider the expression of characteristic impedance for distortion-less line.

$$\mathbf{Z}_0 = \sqrt{\frac{(R + j\omega L)}{(G + j\omega C)}}$$

$$= \sqrt{\frac{L\left(\dfrac{R}{L} + j\omega\right)}{C\left(\dfrac{G}{C} + j\omega\right)}}$$

$$= \sqrt{\frac{L}{C}} \tag{13.5}$$

Table 13.1 summarizes distortion-less line vs. lossless line.

Loading

In the actual lines, the primary constants are such that $R/L \gg G/C$ because of low values of G. To make the line distortionless, usually R/L is decreased instead of going for an increase of G/C for various reasons. The decrease in R/L is achieved by increasing the inductance L instead of going for the decrease of R. This is affected either by changing the line configuration or by connecting highly inductive coils to the line. The method of reducing the distortion by increasing the inductance of the line is called *loading*. It is of two types: continuous loading and Lumped loading.

Continuous loading: The tape of steel or some other magnetic materials such as 'permalloy' or 'mumetal' is wound around the conductor to be loaded. It increases the permeability of the surrounding medium, thereby increasing the inductance. It is costly and used in sub-marine cables only.

Lumped loading: In this method, inductance coils are introduced at definite and uniform intervals along the length of the line to increase its inductance.

Table 13.1 Distortion-less line versus lossless line

	Distortion-less line	Lossless line
Primary constants	$R = G = 0$	$\left(\frac{R}{L}\right) = \left(\frac{G}{C}\right)$
Characteristic impedance	$\mathbf{Z}_0 = \sqrt{\frac{L}{C}}$	$\mathbf{Z}_0 = \sqrt{\frac{L}{C}}$
Attenuation constant	$\alpha = 0$	$\alpha = \sqrt{RG}$
Phase shift constant	$\beta = \omega\sqrt{LC}$	$\beta = \omega\sqrt{LC}$
Wave velocity over line	$v = \frac{1}{\sqrt{LC}}$	$v = \frac{1}{\sqrt{LC}}$

13.2 LOAD MATCHING

A line is said to be matched to the load when the load accepts all the power that has been placed over the line by the generator without any reflections back. We will see soon that it can happen only when the load impedance is equal to the characteristic impedance of the line. Under matched conditions, therefore, the line will be carrying only the forward travelling wave with no reflected wave.

Matched termination:

$$\mathbf{Z}_l = \mathbf{Z}_0 \tag{13.6}$$

Proof: The generation point of the reflected wave is the load and the load exists at a distance of infinity for an infinite line. Therefore, in this case, the forward wave can never reach the load and hence there can never be a reflected wave over the infinite line.

The finite length line terminated over its characteristic impedance can be viewed as a line terminated over the infinite line because the input impedance of the infinite line is equal to its characteristic impedance. Now we can observe that the resultant of termination of the finite line over the infinite line is another infinite line and as such no reflected wave can exist over any portion of it. Hence, the finite length line terminated over its characteristic impedance does not carry the reflected wave.

When the line is designed for the transfer of power to the load, then the matched termination is advantageous because of maximum power transfer, maximum efficiency, lesser peak voltages and consequent lower flashover likelihood, elimination of modulation, distortion, etc.

When the line is terminated over impedance which is different from its characteristic impedance, reflections occur resulting in an inefficient transmission system. The reflected wave over a major part of the line can be eliminated by taking recourse to load-matching techniques.

13.2.1 Quarter-wave Transformer

Load matching with quarter-wave transformer (QWT) technique is based on the fact that the input impedance of a quarter-wave line is inversely proportional to the termination impedance and directly proportional to the square of the characteristic impedance of the transformer.

$$\mathbf{Z}_{\text{in}}\Big|_{l=\lambda/4} = \frac{\mathbf{Z}_{0,\text{trans}}^2}{\mathbf{Z}_l} \tag{13.7}$$

where $\mathbf{Z}_{0,\text{trans}}$ is the characteristic impedance of the transformer line and $\mathbf{Z}_l$ is its termination impedance.

To match a lossless line of characteristic impedance $\mathbf{Z}_0$ to a load of $\mathbf{Z}_l \neq \mathbf{Z}_0$, connect a quarter-wave transformer in between the load and the line. The transformer should have a characteristic impedance equal to

$$\mathbf{Z}_{0,\text{trans}} = \sqrt{\mathbf{Z}_0 \mathbf{Z}_l} \tag{13.8}$$

Proof: To avoid reflections, the termination impedance must be equal to the characteristic impedance of the line. The termination impedance of the line becomes the input impedance of the quarter-wave transformer when the line is terminated over the transformer. The input impedance of the transformer terminated over $\mathbf{Z}_l$ can be varied and made equal to $\mathbf{Z}_0$ by appropriately selecting its characteristic impedance $\mathbf{Z}_{0,\text{trans}}$

$$\mathbf{Z}_{\text{in}}\Big|_{l=\lambda/4} = \frac{\mathbf{Z}_{0,\text{trans}}^2}{\mathbf{Z}_l} = \mathbf{Z}_0$$

$$\mathbf{Z}_{0,\text{trans}}^2 = \mathbf{Z}_0 \mathbf{Z}_l \quad \text{or} \quad \mathbf{Z}_{0,\text{trans}} = \sqrt{\mathbf{Z}_0 \mathbf{Z}_l}$$

This technique works well only if the terminating impedance is pure resistive. Otherwise $\mathbf{Z}_{0,\text{trans}}$ becomes complex and it cannot be achieved as the transformer is lossless and hence its characteristic impedance is pure real.

This transformer is a single frequency or narrow-band device. The bandwidth may be increased by using two or more quarter-wave sections in series, each accomplishing part of the total transformation. The structure of load matching with quarter-wave transformer is shown in Figure 13.1.

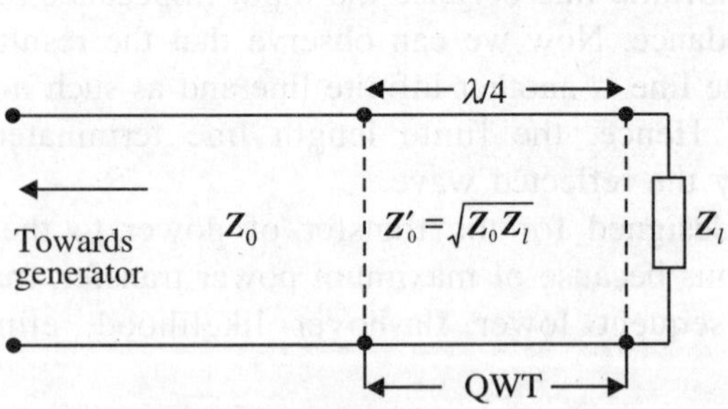

FIGURE 13.1 Load matching with quarter-wave transformer.

In the case of complex load, this technique still can be used but with some minor modifications which are mentioned as follows:

- Convert the load impedance to admittance and find out the susceptance part of it.

- Tune out the susceptance part of the load admittance by connecting a shorted stub across the load. Now the effective load is pure real.
- Find out the characteristic impedance of the transformer using the effective load impedance and the characteristic impedance of the line.
- Connect the quarter wave transformer in between the load and the line.

Another method which is useful to match complex loads involves insertion of a QWT at the location of a voltage maximum or minimum. We know that the line impedance is pure real at voltage maxima and minima. The first minima from the load occurs at a distance of $d_{\min} = (\lambda/4\pi)\,(\theta_l + \pi)$ from the load and the first maxima exists at a distance of $\pm\lambda/4$ from the location of the minima.

	min	**max**
d_q	$\dfrac{\lambda}{4\pi}(\theta_l + \pi)$	$\dfrac{\lambda}{4\pi}(\theta_l + \pi) \pm \dfrac{\lambda}{4}$
$\mathbf{Z}_{0,q}$	$\mathbf{Z}_0\sqrt{\dfrac{(1-\Gamma_l)}{(1+\Gamma_l)}}$	$\mathbf{Z}_0\sqrt{\dfrac{(1+\Gamma_l)}{(1-\Gamma_l)}}$

The line impedance at minima is $\mathbf{Z}_0(1-\Gamma_l)/(1+\Gamma_l)$ and the characteristic impedance of the QWT to be inserted, therefore, should be $\mathbf{Z}_0\sqrt{(1-\Gamma_l)/(1+\Gamma_l)}$ and at maxima which occurs at a distance of $\pm\lambda/4$ from the minima, the line impedance is $\mathbf{Z}_0\,(1+\Gamma_l)/(1-\Gamma_l)$ and hence the characteristic impedance of the QWT should be $\mathbf{Z}_0\sqrt{(1+\Gamma_l)/(1-\Gamma_l)}$.

13.2.2 Single-stub Matching

Let us consider a piece of lossless transmission line with characteristic impedance $\mathbf{Z}_0$ terminated over a load $\mathbf{Z}_l \neq \mathbf{Z}_0$. As the line is terminated over impedance not equal to its characteristic impedance, the reflected wave, most of the times unwanted, exists over the line. It is possible to eliminate the reflected wave from a major portion of the line by connecting a short-circuited stub l_s to the line at a distance of d_s from the load side end (Figure 13.2).

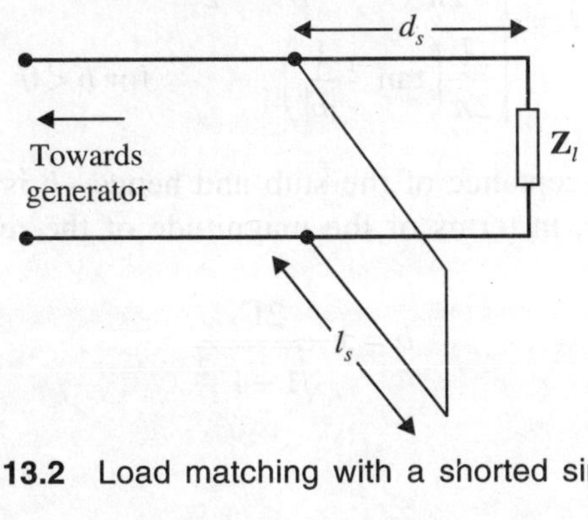

FIGURE 13.2 Load matching with a shorted single stub.

The distance of d_s from the load side is given by

$$d_s = \frac{\lambda}{2\pi}\tan^{-1}\sqrt{\frac{\mathbf{Z}_l}{\mathbf{Z}_0}} \tag{13.9}$$

The stub length l_s is given by

$$l_s = \frac{\lambda}{2\pi}\tan^{-1}\frac{\sqrt{\mathbf{Z}_l\mathbf{Z}_0}}{\mathbf{Z}_l - \mathbf{Z}_0} \tag{13.10a}$$

If the stub is made from a line which is different from the main line, then

$$l_s = \frac{\lambda}{2\pi}\tan^{-1}\left(\frac{\mathbf{Z}_0}{\mathbf{Z}_{0s}}\frac{\sqrt{\mathbf{Z}_l\mathbf{Z}_0}}{\mathbf{Z}_l - \mathbf{Z}_0}\right) \tag{13.10b}$$

The above relations are derived assuming purely resistive termination. The actual process of derivation of the above relations can be found in one of the solved problems given at the end of this chapter. But most of the times the line is terminated over a complex impedance. The relations useful even for such a general situation are available and given.

In the case of a complex impedance termination, the distance d_s from the load and the length of the stub l_s, respectively, are

$$d_s = \begin{cases} \dfrac{\lambda}{4\pi}\left(\theta_l \mp \dfrac{\pi}{2} \mp \tan^{-1}\dfrac{|b|}{2} - 2n\pi\right) \\ \dfrac{\lambda}{4\pi}(\theta_l \mp \pi \pm \cos^{-1}\Gamma_l - 2n\pi) \end{cases} \tag{13.11}$$

In the above expression, the upper sign is to be considered for positive values of b and the lower sign for its negative values. The angle θ_l is the angle of the reflection coefficient over the load. The integer n can assume any value, positive or negative, including zero such that the distance of the stub is within a half wavelength.

$$l_s = \begin{cases} -\dfrac{\lambda}{2\pi}\left(\tan^{-1}\dfrac{1}{b}\right) + \dfrac{\lambda}{2} & \text{for } b > 0 \\ \dfrac{\lambda}{2\pi}\left(\tan^{-1}\dfrac{1}{|b|}\right) & \text{for } b < 0 \end{cases} \tag{13.12}$$

Here b is the normalized susceptance of the stub and hence $-b$ is that of the line at the point of stub connection. Its value, in terms of the magnitude of the reflection coefficient Γ_l, is

$$b = \pm\frac{2\Gamma_l}{\sqrt{1 - \Gamma_l^2}} \tag{13.13}$$

Derivation

Now we consider the derivation of the relations pertaining to the complex impedance termination. After the attachment of the stub, the line admittance $\mathbf{y}_p$ at the point of connection must be one for matching. If b is the susceptance of the stub, then before the connection of the stub, the admittance of the line must be

$$\mathbf{y}'_p = 1 - jb$$

And the corresponding reflection coefficient then is

$$\Gamma_p = \frac{1 - \mathbf{y}_p}{1 + \mathbf{y}_p} = \frac{jb}{2 - jb}$$

We know, for a lossless line, that the reflection coefficient over the load can be written with the knowledge of reflection coefficient over the line at the stub location.

$$\Gamma_l = \Gamma_p e^{j\beta d_s} = \frac{jb}{2 - jb} e^{j\beta d_s}$$

$$= \frac{|b|}{\sqrt{4^2 + b^2}} e^{j\left(\pm\pi/2 \pm \tan^{-1}|b|/2 + 2\beta d_s + 2n\pi\right)}$$

The phase of the reflection coefficient Γ_p is the phase of its numerator ($\pm\pi/2$) subtracted from the phase of the denominator ($\pm\tan^{-1}|b|/2 + 2n\pi$). In the above expression, the upper sign is to be considered for positive values of b and the lower sign for its negative values and n can assume positive or negative integer values (Figure 13.3). Thus, the magnitude of the reflection coefficient over the load is

$$\Gamma_l = |\Gamma_l| = \frac{|b|}{\sqrt{4^2 + |b|^2}}$$

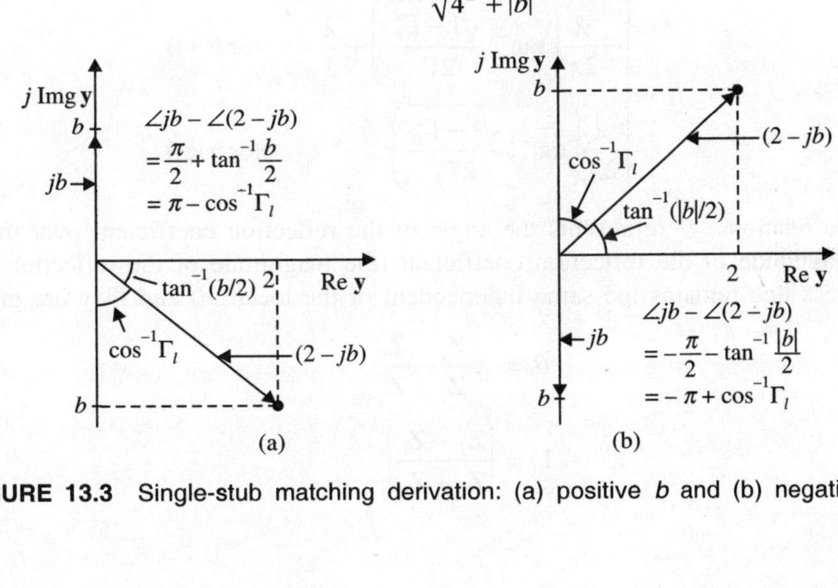

FIGURE 13.3 Single-stub matching derivation: (a) positive b and (b) negative b.

and its phase angle is

$$\theta_l = \left(\pm\frac{\pi}{2} \pm \tan^{-1}\frac{|b|}{2} + 2\beta d_s + 2n\pi\right)$$

Solving the above two equations for b and d_s, we obtain

$$b = \pm\frac{2\Gamma_l}{\sqrt{1-\Gamma_l^2}}$$

$$d_s = \frac{\lambda}{4\pi}\left(\theta_l \mp \frac{\pi}{2} \mp \tan^{-1}\frac{|b|}{2} - 2n\pi\right)$$

Using the information provided in Figure 13.3, the location of the stub can also be expressed in terms of reflection coefficient as

$$d_s = \frac{\lambda}{4\pi}(\theta_l \mp \pi \pm \cos^{-1}\Gamma_l - 2n\pi)$$

Once d_s is available, then the length of the stub can be computed from the condition of nullification of line susceptance by the stub for matching. It results in

$$\cot \beta l_s = -b$$

Solving this equation, we get

$$l_s = \begin{cases} -\dfrac{\lambda}{2\pi}\left(\tan^{-1}\dfrac{1}{b}\right) + \dfrac{\lambda}{2} & \text{for } b > 0 \\ \dfrac{\lambda}{2\pi}\left(\tan^{-1}\dfrac{1}{|b|}\right) & \text{for } b < 0 \end{cases}$$

It can also be expressed in terms of reflection coefficient as

$$l_s = \begin{cases} -\dfrac{\lambda}{2\pi}\left(\tan^{-1}\dfrac{\sqrt{1-\Gamma_l^2}}{2\Gamma_l}\right) + \dfrac{\lambda}{2} & \text{for } b > 0 \\ \dfrac{\lambda}{2\pi}\left(\tan^{-1}\dfrac{\sqrt{1-\Gamma_l^2}}{2\Gamma_l}\right) & \text{for } b < 0 \end{cases}$$

In the above relations, θ_l represents the angle of the reflection coefficient over the load and Γ_l is the magnitude of the reflection coefficient (the magnitude of the reflection coefficient over a lossless line remains the same independent of the location) and they are given by

$$\theta_l = \angle\frac{\mathbf{Z}_l - \mathbf{Z}_0}{\mathbf{Z}_l + \mathbf{Z}_0} \tag{13.14a}$$

$$\Gamma_l = \left|\frac{\mathbf{Z}_l - \mathbf{Z}_0}{\mathbf{Z}_l + \mathbf{Z}_0}\right| \tag{13.14b}$$

An important aspect of the stub-matching technique is the existence of reflected wave over the portion of the line in between the load and the stub. The reflected wave does not exist only in between the sending end of the line and the point where the stub is connected. Therefore, it is always preferred to connect the stub as nearer to the load as possible.

There can be several points over the line satisfying the relation $d_s = (\lambda/2\pi)\tan^{-1}\sqrt{Z_l/Z_0}$. But it is advisable to connect the stub at a point which is nearest to the load so that only a smaller length line is under the reflected wave.

A major disadvantage of the single-stub technique is its narrow bandwidth. With change in frequency, the length and location of the stub have to be changed but it is the change of the location of the stub that is more troublesome. Another disadvantage is that in the final adjustment, the stub requires the movement over the line very minutely that is not possible for coaxial lines, resulting in inaccurate matching.

To overcome these disadvantages, instead of one, two short-circuited stubs whose lengths are adjustable independently but whose positions are fixed are used. The distance of the nearest stub from the load and the inter stub is normally either $\lambda/4$ or $3\lambda/8$. The distance between the farthest stub and the load should be as small as possible because the line within this length undergoes reflection loss, and hence the lesser the length, the lower will be the loss.

13.2.3 Double-stub Matching

The single-stub matching is difficult to implement, particularly in the case of coaxial lines where it is just not possible to place the stub physically over the line in an ideal location. In double-stub matching, two stubs are used whose lengths can be adjusted at will giving more degrees of freedom to the matching designer (Figure 13.4). The stubs can be connected as near to the load as possible and the lengths of the stubs are adjusted to get proper matching.

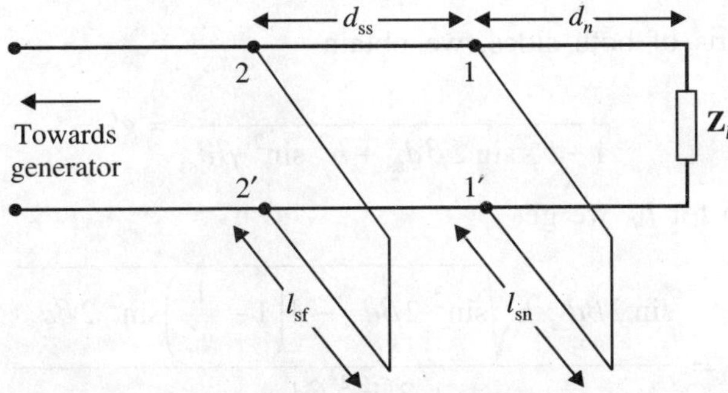

FIGURE 13.4 Load matching with shorted-double stubs.

In the following analysis, we find out analytical expressions for the susceptances, and hence the lengths of the stubs to be connected to the line for matching purpose. Here, we designate inter stub spacing by d_{ss}, distance of the near-stub from the load by d_n, the length of the near-stub by l_{sn} and the length of the far-stub by l_{sf}. After connection of the far-stub, the normalized admittance at that point must be unity to obtain matching. If the far-stub susceptance is b_f, then before its connection, the admittance and reflection coefficient of the line at the point of its connection are

$$\mathbf{y}'_f = 1 - jb_f \tag{13.15}$$

$$\mathbf{\Gamma}'_f = \frac{1-\mathbf{y}'_f}{1+\mathbf{y}'_f} = \frac{jb_f}{2-jb_f} \tag{13.16}$$

Then the reflection coefficient of the line at the location of the near-stub becomes

$$\mathbf{\Gamma}_n = \mathbf{\Gamma}'_f\, e^{j\beta d_{ss}} = \frac{jb_f}{2-jb_f} e^{j\beta d_{ss}} \tag{13.17}$$

The normalized admittance of the line at the point of the near-stub location is

$$\mathbf{y}_n = \frac{1-\mathbf{\Gamma}_n}{1+\mathbf{\Gamma}_n}$$

$$= \frac{4 - j(4b_f \cos 2\beta d_{ss} - 2b_f^2 \sin 2\beta d_{ss})}{4 - 4b_f \sin 2\beta d_{ss} + 4b_f^2 \sin^2 \beta d_{ss}} \tag{13.18}$$

If the near-stub susceptance is b_n, then before its connection, the admittance of the line at the point of its connection is

$$\mathbf{y}'_n = \mathbf{y}_n - jb_n$$

$$= \frac{1}{1 - b_f \sin 2\beta d_{ss} + b_f^2 \sin^2 \beta d_{ss}} + j\left(\frac{b_f^2 \sin 2\beta d_{ss} - 2b_f \cos 2\beta d_{ss}}{2 - 2b_f \sin 2\beta d_{ss} + 2b_f^2 \sin^2 \beta d_{ss}} - b_n\right)$$

$$= g'_n + jb'_n \tag{13.19}$$

Equating the real parts of both sides, we obtain

$$\frac{1}{1 - b_f \sin 2\beta d_{ss} + b_f^2 \sin^2 \beta d_{ss}} = g'_n \tag{13.20}$$

Solving this equation for b_f, we get

$$b_f = \frac{\sin 2\beta d_{ss} \pm \sqrt{\sin^2 2\beta d_{ss} - 4\left(1 - \dfrac{1}{g'_n}\right)\sin^2 2\beta d_{ss}}}{2\sin^2 \beta d_{ss}}$$

$$= \frac{\cos \beta d_{ss} \pm \sqrt{\dfrac{1}{g'_n} - \sin^2 \beta d_{ss}}}{\sin \beta d_{ss}} \tag{13.21}$$

We can observe that the above equation does not have a solution for b_f when $1/\sin^2 \beta d_{ss} < g'_n$, hence it is not possible to arrange matching for loads which results in such kind of inequality. (one simple method by which this problem can be overcome is by increasing the inter stub distance by a quarter wavelength). Otherwise, the b_f has two possible values and the corresponding values of b_n are then given by

$$b_n = \left(\frac{b_f^2 \sin 2\beta d_{ss} - 2b_f \cos 2\beta d_{ss}}{2 - 2b_f \sin 2\beta d_{ss} + 2b_f^2 \sin^2 \beta d_{ss}} - b_n' \right) \tag{13.22}$$

Once b_n and b_f are known, the lengths of the two stubs can be computed as in the case of the single-stub matching technique.

Double-stub matching
• For double-stub matching to be possible, the condition to be satisfied is $g_n' \le 1/\sin^2 \beta d_{ss}$.   • Otherwise, increase d_{ss} by $\lambda/4$ so that the above condition is satisfied.

13.3 SMITH CHART

Smith chart, developed by P. Smith in 1939, is the best known and widely used graphical aid in solving transmission line problems. The real utility of the Smith chart lies in the fact that it can be used to convert the reflection coefficients to normalized impedances or admittances. It is a polar plot of the complex reflection coefficient with the normalized impedance or admittance in a unity circle. The formation of the Smith chart is shown in Figure 13.5.

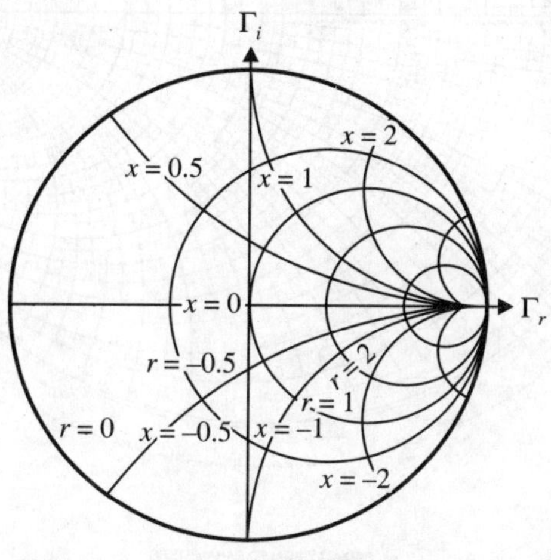

FIGURE 13.5 Formation of Smith chart.

The chart is circular in shape and it consists of two sets of circles or arcs of circles:

- The constant resistance 'r' circles whose centres lie on the straight line of the chart. They represent the normalized resistance along the transmission line.
- The constant reactance 'x' loci, arcs of the circles lie on both sides of the horizontal line. They represent various values of the normalized reactance of the transmission line.

- The constant resistance '*r*' and constant reactance '*x*' loci form two families of circles and all of them pass through the point $\Gamma_r = 1$, $\Gamma_i = 0$. The circles and arcs are orthogonal.
- Upper half of the chart represents inductive reactance/susceptance where as the lower half represents the capacitive reactance/susceptance.
- The Smith chart describes the line of half wavelength only.
- It can be used only with normalized impedances or admittances.
- The movement towards the generator corresponds to clockwise motion on the chart and towards the load corresponds to the anticlockwise motion. A widely used Smith chart is shown in Figure 13.6.

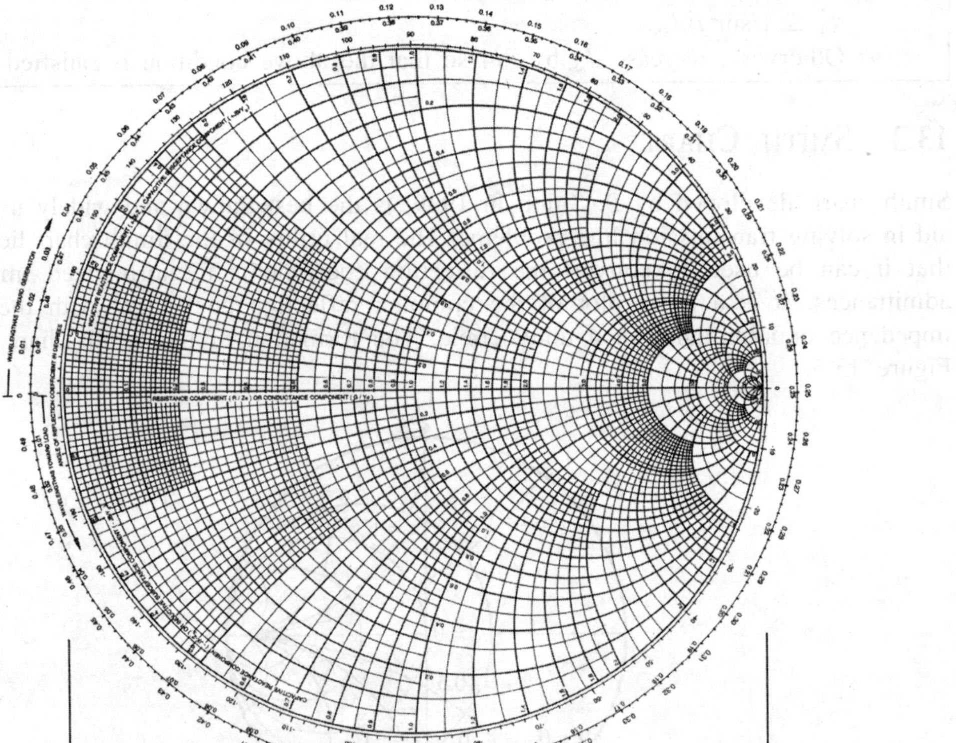

FIGURE 13.6 A widely used Smith chart.

Theory: The impedance **z** and reflection coefficient Γ at any arbitrary point over a lossless line are related through

$$\mathbf{z} = \frac{1+\Gamma}{1-\Gamma} \tag{13.23}$$

Both the impedance and reflection coefficient are complex having real and imaginary parts.

$$\mathbf{z} = \frac{1+\Gamma}{1-\Gamma} \quad \Rightarrow \quad r + jx = \frac{1+\Gamma_r + j\Gamma_i}{1-\Gamma_r - j\Gamma_i}$$

$$r = \frac{1-\Gamma_r^2 - \Gamma_i^2}{(1-\Gamma_r)^2 + \Gamma_i^2} \tag{13.24a}$$

and

$$x = \frac{2\Gamma_i}{(1-\Gamma_r)^2 + \Gamma_i^2} \tag{13.24b}$$

These two equations can be rearranged as

$$\left(\Gamma_r - \frac{r}{1+r}\right)^2 + \Gamma_i^2 = \left(\frac{1}{1+r}\right)^2 \tag{13.25a}$$

$$(\Gamma_r - 1)^2 + \left(\Gamma_i - \frac{1}{x}\right)^2 = \left(\frac{1}{x}\right)^2 \tag{13.25b}$$

We can observe that these equations represent circles on a complex reflection coefficient plane. The first one represents a family of constant resistance circles with radius $1/(1 + r)$ and centre at $r/(1 + r)$ along the real axis. The second one represents a family of constant reactance circles with radius $1/x$ and centre at $\Gamma_r = 1$, $\Gamma_i = 1/x$. Figure 13.7 illustrates the salient features of the Smith chart.

- The 'constant SWR circle' is a circle drawn on the chart with SWR as radius and '1' of the horizontal line as centre.
  - The impedances represented by various points over the SWR circle denote the impedances of the line within a length of $\lambda_g/2$.
  - Distance along the line is represented by angular distance around the chart, total circumference or 360° corresponding to the line length of $\lambda_g/2$.
  - The point at which the constant SWR circle intersects the horizontal straight line corresponds to the SWR of the load.
- Constant SWR ρ circles are concentric with their centre lying over the '1' of the chart. These circles intersect the zero reactance or susceptance horizontal line at points $z_0\rho$ and z_0/ρ.
- The horizontal radius to the right of the chart centre corresponds to V_{max}, I_{min}, z_{max} and ρ and the left of the chart centre corresponds to V_{min}, I_{max}, z_{min} and $1/\rho$. Also the location of V_{max} corresponds to $z_{max} = \rho$ on the line and that of the V_{min} corresponds to a point of $z_{min} = 1/\rho$.

- Radial lines represent loci of the constant line angle βz. In the chart wavelength scales corresponding to the line angle are included around the outside edge of the chart.
- For a lossy line not terminated in its characteristic impedance, the path of travel on the chart from the load to the generator is a decreasing logarithmic spiral.

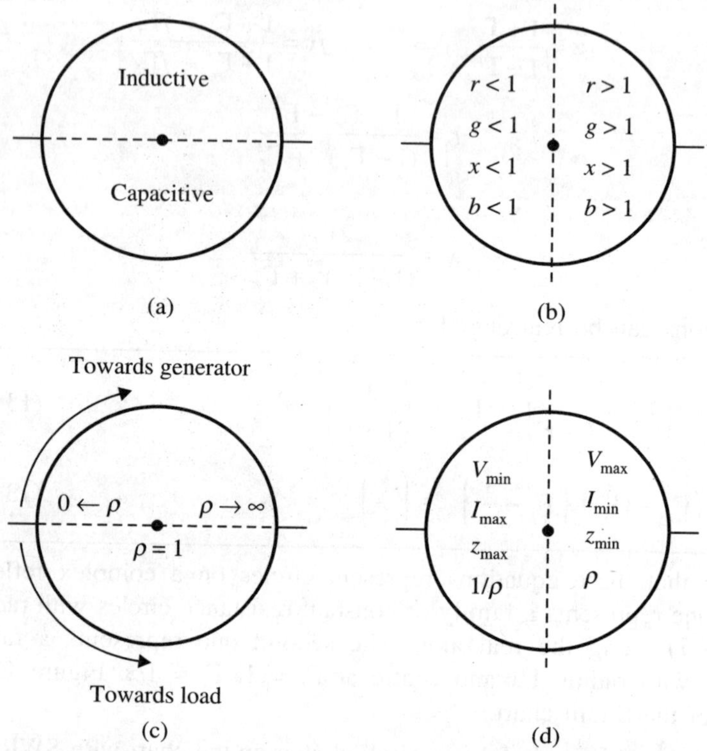

FIGURE 13.7 Illustrating salient features of Smith chart.

13.3.1 Single-stub Matching

The location and length of the stub to be placed over the line for matching purpose can be calculated using the Smith chart. The procedural steps are given hereunder.

- Plot the normalized impedance and draw the constant SWR circle on the Smith chart.
- Move a distance of $\lambda/4$ along the constant SWR circle to locate load admittance. Let it be P_1.
- On the SWR circle nearest to the load admittance point, locate a point, which represents admittance $1 \pm jb$. This is the point of the intersection of constant SWR circle and $r = 1$ circle. Let it be P_2.
- Read the distance between P_1 and P_2 using the scale provided at the circumference of the chart. This gives the distance in wavelengths where the stub has to be placed from the load.

- Starting from the point (∞, $j\infty$), find the distance of the point at which the susceptance is $\pm jb$. This gives the length of the short-circuited stub in wavelengths to be connected for matching.

13.3.2 Double-stub Matching

The single-stub matching is impractical when the stub is not being able to be placed physically in the ideal location. Particularly it is very difficult to place the stub at the exact required location in the case of coaxial lines.

In double-stub matching, the stubs are connected as near to the load as possible and the lengths of the stubs are adjusted to get proper matching. For the sake of convenience, let us call the stub that is nearest to the load as near-stub and the other one as far-stub. The spacing between the stubs is usually $\lambda/8$, $3\lambda/8$ or $5\lambda/8$ but most often it is $\lambda/4$. The spacing of $\lambda/2$ is avoided because it places the two stubs in parallel, resulting in the availability of single effective adjustment. The close spacing of the stubs is also avoided for the same reason.

Spacing circle: It is a circle obtained by rotating the constant conductance circle around the centre of the Smith chart. The amount of rotation depends upon the spacing of the stubs. For a spacing of either $\lambda/8$ or $5\lambda/8$, the amount of rotation is 90° anticlockwise and in the case of $3\lambda/8$, the rotation is 90° clockwise. If the spacing happens to be $\lambda/8$, then the amount of rotation is 180°.

To understand the significance of spacing circle, we have to view the line in between the stubs as a transformer, converting the admittances at the far-stub location into different admittance values at the near-stub location. The admittance of the line at the far-stub location is $1 \pm jb$ and therefore the constant conductance circle $g = 1$ describes these admittances. The admittances at the near-stub location then can be sketched over the Smith chart by moving the admittance points along the constant SWR circles by an angle corresponding to the length of the transformer. If we join all the admittance points at the location stub 2, it also results in the so-called 'spacing circle'.

Forbidden region: The double-stub matching technique does not work for certain values of load admittances. The region of the Smith chart that represents all these values of load admittances which are not amenable for the double-stub technique is called *forbidden region.* This region basically depends upon the spacing of the stubs.

When the first stub is right over the load, and for a stub spacing of $\lambda/8$, $3\lambda/8$ or $5\lambda/8$, or the forbidden region consists of the entire area encircled by constant conductance $g = 2$ circle. If it is $\lambda/4$, then the forbidden region is the entire area encircled by the constant conductance $g = 1$ circle.

Location of stubs: For a reflectionless operation of the line, the input admittance of the line looking towards the load at the far-stub location will be equal to the characteristic impedance. Or at the location of the far-stub, the real part of the admittance of the line is required to be one. The far-stub is used to cancel the imaginary part of the line admittance at its location, resulting in normalized admittance of one.

It is always desirable for the near-stub to be connected at or near the load. But it is not always possible to have such a connection because of the existence of forbidden regions.

Functioning: stub 1 transforms the input admittance of the line and load to the right of its location into an admittance which lies over the spacing circle. The line transformer further transforms this admittance into a value which lies over the unity conductance circle. The susceptance of the line at this point is tuned out using stub 2.

Procedural steps:

Step I: Locate the normalized admittance over the chart. Let it be *P*.

Step II: From *P*, move clockwise over the constant SWR circle, a distance corresponding to the distance of the first stub from the load. Let the point be *Q*. If the first stub is to be placed right over the load, then this step becomes redundant.

Step III: From *Q*, move over the constant *g* circle until the spacing circle is met. Let the point be *R*. In the case of two possibilities, i.e. clockwise and anticlockwise, either one can be selected. From the susceptance value of the point *R*, the length of the first stub can be calculated.

Step IV: From *R*, move over the constant SWR circle until $g = 1$ circle is met. Let it be *S*. From the susceptance value of the point *S*, the length of the second stub can be calculated. Note that the SWR circle in the present step is different from that of step II.

SOLVED PROBLEMS

1. A distortionless line of 60 Ω has and $\alpha = 20$ mNp/m a velocity of 60% of light velocity. Find the primary constants at 100 MHz.

Solution: Distortion line is characterized by

$$\frac{R}{L} = \frac{G}{C}$$

leading to

$$\mathbf{Z}_0 = \sqrt{\frac{R + j\omega L}{G + j\omega C}} = \sqrt{\frac{L}{C}}$$

$$\alpha = \sqrt{RG}, \quad \beta = \omega\sqrt{LC} \quad \text{and} \quad v = \frac{1}{\sqrt{LC}}$$

Now consider the following products:

$$v\mathbf{Z}_0 = \frac{1}{\sqrt{LC}}\sqrt{\frac{L}{C}} = \frac{1}{C}$$

$$\frac{\mathbf{Z}_0}{v} = \sqrt{LC}\sqrt{\frac{L}{C}} = L$$

$$\alpha\mathbf{Z}_0 = \sqrt{RG}\sqrt{\frac{L}{C}} = \sqrt{RG}\sqrt{\frac{R}{G}} = R$$

$$\frac{\alpha}{\mathbf{Z}_0} = \sqrt{RG}\sqrt{\frac{C}{L}} = \sqrt{RG}\sqrt{\frac{G}{R}} = G$$

From the above derivations and given values

$$C = \frac{1}{v\mathbf{Z}_0} = \frac{100}{60 \times 3 \times 10^8 \times 60} = 9.256 \times 10^{-11} \text{ F/m}$$

$$L = \frac{\mathbf{Z}_0}{v} = \frac{100 \times 60}{60 \times 3 \times 10^8} = 0.33256 \times 10^{-6} \text{ H/m}$$

$$R = \alpha\mathbf{Z}_0 = 20 \times 10^{-3} \times 60 = 1.2 \ \Omega/\text{m}$$

$$G = \frac{\alpha}{\mathbf{Z}_0} = \frac{20 \times 10^{-3}}{60} = 0.33 \times 10^{-3} \ \mho/\text{m}$$

2. For a lossless line of 75 Ω, find the location nearest to the load and the characteristic impedance of the QWT required to achieve matching for each of the following values of reflection coefficient over the load, $\mathbf{\Gamma}_l$:

(a) 1/9,

(b) $-j0.5$ and

(c) $j/3$.

Solution: In the solution procedure, the first step is to ascertain whether it is the minimum or maximum that occurs nearer to the load.

(a) In this case, the reflection coefficient over the load is pure real. Therefore, its angle is zero, i.e. $\theta_l = 0$. Therefore,

$$d_q = \frac{\lambda}{4\pi}(\theta_l + \pi) = \frac{\lambda}{4\pi}(0 + \pi) = \frac{\lambda}{4}$$

Hence, the minimum occurs at a distance of $\lambda/4$ from the load and the maximum right over the load. Therefore, the QWT can be inserted right at the load. Its characteristic impedance is computed now.

$$\mathbf{Z}_{0,q} = \mathbf{Z}_0\sqrt{\frac{(1+\Gamma_l)}{(1-\Gamma_l)}}$$

$$= 75\sqrt{\frac{\left[1+\left(\frac{1}{9}\right)\right]}{\left[1-\left(\frac{1}{9}\right)\right]}} = 75\sqrt{\frac{10}{8}} = 83.85\ \Omega$$

(b) In this case, the reflection coefficient over the load is negative and pure imaginary. Therefore, its angle $\theta_l = -\pi/2$. Therefore,

$$d_q = \frac{\lambda}{4\pi}\left[\left(\frac{-\pi}{2}\right) + \pi\right] = \frac{\lambda}{8}$$

Hence, the minimum occurs at a distance of $\lambda/8$ from the load and the nearest maximum will be at a distance of $\lambda/4$ still farther away. Therefore, the minimum is nearest to the load and the QWT is inserted at the minimum. Its characteristic impedance is computed now.

$$\mathbf{Z}_{0,q} = \mathbf{Z}_0\sqrt{\frac{(1-\Gamma_l)}{(1+\Gamma_l)}}$$

$$= 75\sqrt{\frac{[1-0.5]}{[1+0.5]}} = 75\sqrt{\frac{0.5}{1.5}} = 43.30\ \Omega$$

(c) In this case, the reflection coefficient over the load is pure imaginary. Therefore, its angle is $\theta_l = \pi/2$. Thus,

$$d_q = \frac{\lambda}{4\pi}\left[\left(\frac{\pi}{2}\right)+\pi\right] = \frac{3\lambda}{8}$$

Hence, minimum occurs at a distance of $3\lambda/8$ and it is more than $\lambda/4$. Therefore, it is the maximum that occurs near to the load and it is at a distance of $3\lambda/8 - \lambda/4 = \lambda/8$ from the load. Thus, the maximum is nearest to the load and the QWT is inserted at the maximum. Its characteristic impedance is computed now.

$$\mathbf{Z}_{0,q} = \mathbf{Z}_0\sqrt{\frac{(1+\Gamma_l)}{(1-\Gamma_l)}}$$

$$= 75\sqrt{\frac{\left[1+\left(\frac{1}{3}\right)\right]}{\left[1-\left(\frac{1}{3}\right)\right]}} = 75\sqrt{\frac{4}{2}} = 106.066\ \Omega$$

3. A lossline of characteristic impedance $\mathbf{Z}_0 = 60\ \Omega$ is to be matched to the loads of

(a) 30 Ω and

(b) $(12 - j24)\Omega$

by means of a short-circuited stub of characteristic impedance 60 Ω. Find the stub position closest to the load and its length so that the match is obtained.

Solution: We use the following formulae to find the required quantities:

$$b = \pm\frac{2\Gamma_l}{\sqrt{1-\Gamma_l^2}}$$

$$d_s = \frac{\lambda}{4\pi}\left(\theta_l \mp \frac{\pi}{2} \mp \tan^{-1}\frac{|b|}{2} - 2n\pi\right)$$

$$l_s = \begin{cases} -\dfrac{\lambda}{2\pi}\left(\tan^{-1}\dfrac{1}{b}\right) + \dfrac{\lambda}{2} & \text{for } b > 0 \\[2ex] \dfrac{\lambda}{2\pi}\left(\tan^{-1}\dfrac{1}{|b|}\right) & \text{for } b < 0 \end{cases}$$

(a) $\mathbf{Z}_l = 30\ \Omega$

For the given values

$$\mathbf{\Gamma}_l = \frac{\mathbf{Z}_l - \mathbf{Z}_0}{\mathbf{Z}_l + \mathbf{Z}_0} = \frac{30-60}{30+60} = -\frac{1}{3} = 0.33\angle\pi$$

$$b = \pm\frac{2\Gamma_l}{\sqrt{1-\Gamma_l^2}}$$

$$= \pm\frac{2\times 0.333}{\sqrt{1-0.333^2}} = \pm 0.706$$

$$d_s = \frac{\lambda}{4\pi}\left(\pi \mp \frac{\pi}{2} \mp \tan^{-1}\frac{0.706}{2} - 2n\pi\right)$$

$$= \lambda(0.025 \mp 0.125 \mp 0.027 - 0.5n)$$

$$= \begin{cases}\lambda(0.098 - 0.5n) & \text{for } b > 0\\ \lambda(0.402 - 0.5n) & \text{for } b < 0\end{cases}$$

In the above, we have selected the value of n such that the distance is within half wavelength. In the present case, appropriate value is zero. Therefore,

$$d_s = \begin{cases}\lambda 0.098 & \text{for } b > 0\\ \lambda 0.402 & \text{for } b < 0\end{cases}$$

The minimum of the above two will be 0.098λ which happens when $b > 0$. The length of the stub for such a case is

$$l_s = -\frac{\lambda}{2\pi}\left(\tan^{-1}\frac{1}{0.706}\right) + \frac{\lambda}{2}$$

$$= \frac{\lambda}{2\pi}(-0.956) + \frac{\lambda}{2} = 0.347\lambda$$

(b) $\mathbf{Z}_l = (12 - j24)\ \Omega$

$$\mathbf{\Gamma}_l = \frac{\mathbf{Z}_l - \mathbf{Z}_0}{\mathbf{Z}_l + \mathbf{Z}_0} = \frac{(12-j24)-60}{(12-j24)+60}$$

$$= \frac{(12-j24)-60}{(12-j24)+60} = 0.707\angle -2.35$$

$$b = \pm\frac{2\Gamma_l}{\sqrt{1-\Gamma_l^2}}$$

$$= \pm\frac{2\times 0.707}{\sqrt{1-0.707^2}} = \pm 1.999$$

$$d_s = \frac{\lambda}{4\pi}\left(-2.35 \mp \frac{\pi}{2} \mp \tan^{-1}\frac{1.999}{2} - 2n\pi\right)$$

$$= \lambda(-2.35 \mp 1.57 \mp 0.785 - 2n\pi)$$

$$= \begin{cases} \lambda(-0.374 - 0.5n) & \text{for } b > 0 \\ \lambda(3.97 \times 10^{-4} - 0.5n) & \text{for } b < 0 \end{cases}$$

In the above, we have selected the value of n such that the distance is within half wavelength. In the present case, the appropriate value is zero. Therefore,

$$d_s = \begin{cases} \lambda(-0.374 - 0.5n) & \text{for } b > 0 \\ \lambda(3.97 \times 10^{-4} - 0.5n) & \text{for } b < 0 \end{cases}$$

The minimum of the above two will be $3.97 \times 10^{-4}\ \lambda$ which happens when $b < 0$. The length of the stub for such a case is

$$l_s = \frac{\lambda}{2\pi}\left(\tan^{-1}\frac{1}{1.999}\right) = \frac{\lambda}{2\pi}(0.463)$$

$$= 0.0738\,\lambda$$

4. For each of the sets of the values given, find whether the double-stub matching technique is possible or not.

(a) $d_n = 0$, $d_{ss} = 3\lambda/8$ with $\mathbf{z}_l = 0.3 + j0.4$

(b) $d_n = \lambda/8$, $d_{ss} = 3\lambda/8$ with $\mathbf{z}_l = 0.5$

(c) $d_n = \lambda/4$, $d_{ss} = 5\lambda/8$ with $\mathbf{z}_l = 2.5 - j5.0$

Solution: It is not possible to achieve match with double stubs if the real part of the input admittance of the line section from the location of stub '1' terminated over the load is more than $1/\sin^2\beta d_{ss}$.

We know that the input impedance of a lossless line of length d_n is

$$\mathbf{Z}_{in} = \mathbf{Z}_0 \frac{\mathbf{Z}_l + j\mathbf{Z}_0 \tan \beta d_n}{\mathbf{Z}_0 + j\mathbf{Z}_l \tan \beta d_n}$$

$$\mathbf{z}_{in} = \frac{\mathbf{Z}_{in}}{\mathbf{Z}_0} = \frac{\mathbf{Z}_l + j\mathbf{Z}_0 \tan \beta d_n}{\mathbf{Z}_0 + j\mathbf{Z}_l \tan \beta d_n}$$

Its admittance then becomes

$$\mathbf{y}_{in} = \frac{1}{\mathbf{z}_{in}} = \frac{\mathbf{Z}_0 + j\mathbf{Z}_l \tan \beta d_n}{\mathbf{Z}_l + j\mathbf{Z}_0 \tan \beta d_n}$$

$$= \frac{\mathbf{Z}_0\left[1 + j\left(\dfrac{\mathbf{Z}_l}{\mathbf{Z}_0}\right)\tan \beta d_n\right]}{\mathbf{Z}_0\left[\left(\dfrac{\mathbf{Z}_l}{\mathbf{Z}_0}\right) + j \tan \beta d_n\right]}$$

$$= \frac{(1 + j\mathbf{z}_l \tan \beta d_n)}{(\mathbf{z}_l + j \tan \beta d_n)}$$

(a) Given that $d_n = 0$, $d_{ss} = 3\lambda/8$, with $\mathbf{z}_l = 0.3 + j0.4$. With $d_n = 0$, $\tan \beta d_n$ becomes zero. It gives

$$\mathbf{y}_{\text{in}} = \frac{(1 + j\mathbf{z}_l \tan \beta d_n)}{(\mathbf{z}_l + j \tan \beta d_n)}$$

$$= \frac{1}{\mathbf{z}_l} = \frac{1}{0.3 + j0.4} = 1.2 - j1.6$$

The real part of the admittance then is $\text{Re}[\mathbf{y}_{\text{in}}] = 1.2$. Next, we calculate the value of $1/\sin^2 \beta d_{ss}$.

$$\beta d_{ss} = \frac{2\pi}{\lambda} \times \frac{3\lambda}{8} = \frac{3\pi}{4} = 0.75\pi$$

It gives

$$\frac{1}{\sin^2 \beta d_{ss}} = \frac{1}{\sin^2 0.75\pi} = 2$$

From the obtained values, as $1.2 < 2$, i.e. $\text{Re}[\mathbf{y}_{\text{in}}] = 1.2 < 1/\sin^2 \beta d_{ss}$, the double- stub matching technique is possible.

(b) Given that $d_n = \lambda/8$, $d_{ss} = 3\lambda/8$ with $\mathbf{z}_l = 0.5$. With $d_n = \lambda/8$, then $\tan \beta d_n = \tan 2\pi/\lambda \times \lambda/8 = \tan \pi/4 = 1$. It gives

$$\mathbf{y}_{\text{in}} = \frac{(1 + j\mathbf{z}_l \tan \beta d_n)}{(\mathbf{z}_l + j \tan \beta d_n)}$$

$$= \frac{1 + j0.5}{0.5 + j} = 1.5 - j0.5$$

The real part of the admittance then is $\text{Re}[\mathbf{y}_{\text{in}}] = 1.5$. Next, we calculate the value of $1/\sin^2 \beta d_{ss}$.

$$\beta d_{ss} = \frac{2\pi}{\lambda} \times \frac{3\lambda}{8} = \frac{3\pi}{4} = 0.75\pi$$

It gives

$$\frac{1}{\sin^2 \beta d_{ss}} = \frac{1}{\sin^2 0.75\pi} = 2$$

From the obtained values, as $1.5 < 2$, i.e. $\text{Re}[\mathbf{y}_{\text{in}}] < 1/\sin^2 \beta d_{ss}$, the double-stub matching technique is possible.

(c) Given that $d_n = \lambda/4$, $d_{ss} = 5\lambda/8$, with $\mathbf{z}_l = 2.5 - j5.0$. With $d_n = \lambda/4$, then $\tan \beta d_n = (\tan 2\pi/\lambda) \times (\lambda/4) = \tan \pi/2 \rightarrow \infty$. It gives

$$\mathbf{y}_{\text{in}} = \frac{(1 + j\mathbf{z}_l \tan \beta d_n)}{(\mathbf{z}_l + j \tan \beta d_n)}$$

$$= \frac{\left[\left(\frac{1}{\tan \beta d_n}\right) 1 + j\mathbf{z}_l\right]}{\left[\left(\frac{\mathbf{z}_l}{\tan \beta d_n}\right) + j\right]}$$

With $\tan \beta d_n \to \infty$

$$\mathbf{y}_{\text{in}} = \frac{0 + j\mathbf{z}_l}{j} = \mathbf{z}_l = 2.5 - j5.0$$

The real part of the admittance then is $\text{Re}[\mathbf{y}_{\text{in}}] = 2.5$. Next, we calculate the value of $1/\sin^2 \beta d_{\text{ss}}$.

$$\beta d_{\text{ss}} = \frac{2\pi}{\lambda} \times \frac{5\lambda}{8} = \frac{5\pi}{4} = 1.25\pi$$

It gives

$$\frac{1}{\sin^2 \beta d_{\text{ss}}} = \frac{1}{\sin^2 1.25\pi} = 2$$

From the obtained values, as 2.5 > 2, i.e. $\text{Re}[\mathbf{y}_{\text{in}}] > 1/\sin^2 \beta d_{\text{ss}}$, the double-stub matching technique is not possible.

5. A loss line working at 100 MHz of characteristic impedance 50 Ω with $\varepsilon_r = 2.1$ and $\mu_r = 1$ is to be matched to a load of 100 Ω by means of a short-circuited stub. Find the stub position closest to the load and the length so that the match is obtained.

Solution: Propagation velocity is

$$v = \frac{c}{\sqrt{\varepsilon_r \mu_r}} = \frac{3 \times 10^8}{\sqrt{2.1}} = 2.07 \times 10^8 \text{ m/s}$$

The wavelength is

$$\lambda = \frac{2.07 \times 10^{10}}{100 \times 10^6} = 207 \text{ cm}$$

The distance of d_s from the load side is given by

$$d_s = \frac{\lambda}{2\pi} \tan^{-1} \sqrt{\frac{\mathbf{Z}_l}{\mathbf{Z}_0}}$$

$$= \frac{207}{2\pi} \tan^{-1} \sqrt{\frac{100}{50}} = 32.9 \times 0.615 = 20.25 \text{ cm}$$

The stub length l_s is given by

$$l_s = \frac{\lambda}{2\pi} \tan^{-1} \frac{\sqrt{\mathbf{Z}_l \mathbf{Z}_0}}{\mathbf{Z}_l - \mathbf{Z}_0}$$

$$= \frac{207}{2\pi} \tan^{-1} \frac{\sqrt{100 \times 50}}{100 - 50} = 32.9 \times 0.955 = 31.42 \text{ cm}$$

6. A loss line of characteristic impedance 50 Ω is to be matched to a load of (30 – j40)Ω by means of a short-circuited stub of characteristic impedance 50 Ω. Find the stub position closest to the load and the length so that the match is obtained.

Solution: Here the load is complex. We have to use the following formulae to find the required quantities:

$$b = \pm\frac{2\Gamma_l}{\sqrt{1-\Gamma_l^2}}$$

$$d_s = \frac{\lambda}{4\pi}\left(\theta_l \mp \frac{\pi}{2} - \tan^{-1}\frac{b}{2} - 2n\pi\right)$$

$$l_s = \begin{cases} \dfrac{\lambda}{2\pi}\left(\tan^{-1}\dfrac{-1}{b}\right) + \dfrac{\lambda}{2} & \text{for } b > 0 \\ \dfrac{\lambda}{2\pi}\left(\tan^{-1}\dfrac{-1}{b}\right) & \text{for } b < 0 \end{cases}$$

For the given values

$$\Gamma_l = \left|\frac{\mathbf{Z}_l - \mathbf{Z}_0}{\mathbf{Z}_l + \mathbf{Z}_0}\right| = \left|\frac{30 - j40 - 50}{30 - j40 + 50}\right|$$

$$= \left|\frac{-20 - j40}{80 - j40}\right| = \left|\frac{20 + j40}{80 - j40}\right| = 0.5$$

$$b = \pm\frac{2\Gamma_l}{\sqrt{1-\Gamma_l^2}}$$

$$= \pm\frac{2\times 0.5}{\sqrt{1-0.5^2}} = \pm\frac{1}{\sqrt{0.75}} = \pm 1.154$$

$$\theta_l = \angle\frac{\mathbf{Z}_l - \mathbf{Z}_0}{\mathbf{Z}_l + \mathbf{Z}_0}$$

$$= \angle\frac{-20 - j40}{80 - j40} = -90^\circ = -\frac{\pi}{2}\text{rad}$$

$$d_s = \frac{\lambda}{4\pi}\left(-\frac{\pi}{2} \mp \frac{\pi}{2} - \tan^{-1}\frac{\pm 1.154}{2} - 2n\pi\right)$$

$$= \begin{cases} \dfrac{\lambda}{4\pi}(-\pi - 0.523 - 2n\pi) & \text{for } b > 0 \\ \dfrac{\lambda}{4\pi}(0.523 - 2n\pi) & \text{for } b < 0 \end{cases}$$

$$= \begin{cases} \lambda(-0.10416 - 0.5n) & \text{for } b > 0 \\ \lambda(0.041 - 0.5n) & \text{for } b < 0 \end{cases}$$

$$= \begin{cases} \lambda(-0.10416 + 0.5) & \text{for } b > 0 \\ \lambda(0.041) & \text{for } b < 0 \end{cases}$$

$$= \begin{cases} \lambda(0.3958) & \text{for } b > 0 \\ \lambda(0.041) & \text{for } b < 0 \end{cases}$$

In the above, we have selected the value of n such that the distance is within half wavelength.

$$l_s = \begin{cases} \dfrac{\lambda}{2\pi}\left(\tan^{-1}\dfrac{-1}{1.154}\right) + \dfrac{\lambda}{2} & \text{for } b > 0 \\ \dfrac{\lambda}{2\pi}\left(\tan^{-1}\dfrac{1}{1.154}\right) & \text{for } b < 0 \end{cases}$$

$$= \begin{cases} \dfrac{\lambda}{2\pi}(-0.714) + \dfrac{\lambda}{2} & \text{for } b < 0 \\ \dfrac{\lambda}{2\pi}(0.714) & \text{for } b < 0 \end{cases}$$

$$= \begin{cases} -0.1136\lambda + 0.5\lambda & \text{for } b > 0 \\ 0.1136\lambda & \text{for } b < 0 \end{cases}$$

$$= \begin{cases} 0.386\lambda & \text{for } b > 0 \\ 0.1136\lambda & \text{for } b < 0 \end{cases}$$

7. A 50 Ω transmission line is terminated by an unknown impedance. The VSWR is 4 and the first minimum is formed at 2 cm from the load end. The frequency of operation is 1 GHz. Design a single-stub line matching for the given conditions.

Solution: The wavelength corresponding to 1 GHz frequency is

$$\lambda = \frac{c}{f} = \frac{3 \times 10^{10}}{1 \times 10^{9}} = 30 \text{ cm}$$

To find the distance and length of the stub, we need the magnitude of the reflection coefficient and angle of the reflection coefficient over the load. We obtain these two parameters from the given data in the following manner:

$$\Gamma_l = \frac{\rho - 1}{\rho + 1} = \frac{4 - 1}{4 + 1} = 0.6$$

$$\theta_l - 2\,\beta d_{\min} = -\pi \quad \Rightarrow \quad \theta_l = 2\,\beta d_{\min} - \pi$$

$$\theta_l = 2 \times \frac{2\pi}{\lambda} \times 2 - \pi = 2 \times \frac{2\pi \times 1 \times 10^{9}}{3 \times 10^{10}} \times 2 - \pi$$

$$= \frac{8\pi}{30} - \pi = -\frac{11\pi}{15} \text{ rad}$$

$$b = \pm\frac{2\Gamma_l}{\sqrt{1-\Gamma_l^2}}$$

$$= \pm\frac{2\times 0.6}{\sqrt{1-0.6^2}} = \pm\frac{1}{\sqrt{0.75}} = \pm 1.15$$

$$d_s = \frac{\lambda}{4\pi}\left(-\frac{11\pi}{15} \mp \frac{\pi}{2} - \tan^{-1}\frac{\pm 1.15}{2} - 2n\pi\right)$$

$$= \begin{cases} \dfrac{\lambda}{4\pi}(-1.23\pi - 0.522 - 2n\pi) & \text{for } b > 0 \\ \dfrac{\lambda}{4\pi}(-0.23\pi + 0.522 - 2n\pi) & \text{for } b < 0 \end{cases}$$

$$= \begin{cases} \dfrac{\lambda}{4\pi}(-4.386 - 2n\pi) & \text{for } b > 0 \\ \dfrac{\lambda}{4\pi}(-0.2 - 2n\pi) & \text{for } b < 0 \end{cases}$$

$$= \begin{cases} \lambda(-0.349 - 0.5n) & \text{for } b > 0 \\ \lambda(-0.016 - 0.5n) & \text{for } b < 0 \end{cases}$$

In the above, we have to select the value of n such that the distance is within half wavelength. Here the appropriate value is –1.

$$d_s = \begin{cases} 0.151\lambda & \text{for } b > 0 \\ 0.484\lambda & \text{for } b < 0 \end{cases}$$

$$= \begin{cases} 4.53 \text{ cm} & \text{for } b > 0 \\ 14.52 \text{ cm} & \text{for } b < 0 \end{cases}$$

Now we compute the length of the stub.

$$l_s = \begin{cases} \dfrac{\lambda}{2\pi}\left(\tan^{-1}\dfrac{-1}{1.15}\right) + \dfrac{\lambda}{2} & \text{for } b > 0 \\ \dfrac{\lambda}{2\pi}\left(\tan^{-1}\dfrac{1}{1.15}\right) & \text{for } b < 0 \end{cases}$$

$$= \begin{cases} \dfrac{\lambda}{2\pi}(-0.716) + \dfrac{\lambda}{2} & \text{for } b > 0 \\ \dfrac{\lambda}{2\pi}(0.716) & \text{for } b < 0 \end{cases}$$

$$= \begin{cases} -0.1139\lambda + 0.5\lambda & \text{for } b > 0 \\ 0.1139\lambda & \text{for } b < 0 \end{cases}$$

$$= \begin{cases} 0.3861\lambda & \text{for } b > 0 \\ 0.1139\lambda & \text{for } b < 0 \end{cases}$$

$$= \begin{cases} 9.12 \text{ cm} & \text{for } b > 0 \\ 3.39 \text{ cm} & \text{for } b < 0 \end{cases}$$

8. Derive the expressions for stub length and stub position valid only when the load is pure resistance.

Solution: The impedance of the line varies from point to point and in general it is a complex quantity having periodic variations along its length.

Reflected wave does not exist over a line if its termination is characteristic impedance. In other words, if the line sees normalized admittance of unity or in the case of lossless line normalized conductance of unity with normalized susceptance of zero, then reflections do not occur over the line. This fact can be extended to infer that the reflected wave can be eliminated over a portion of the lossless line whose normalized line conductance at its terminal point is unity.

In general, the characteristic impedance and terminating impedance are complex quantities. In the present derivation, we assume the line to be lossless and it results in a pure resistive characteristic impedance. Therefore, $\mathbf{Z}_0 = |\mathbf{Z}_0| = Z_0$. We also assume the terminating impedance as a pure resistive quantity. It also results in $\mathbf{Z}_l = |\mathbf{Z}_l| = Z_l$.

The line impedance over the line at a point P which is at a distance of l from the load end is

$$\mathbf{Z}_P = Z_0 \frac{Z_l + jZ_0 \tan \beta d_s}{Z_0 + jZ_l \tan \beta d_s}$$

$$\mathbf{z}_P = \frac{\mathbf{Z}_P}{Z_0} = \frac{Z_l + jZ_0 \tan \beta d_s}{Z_0 + jZ_l \tan \beta d_s}$$

$$= \frac{z_l + j \tan \beta d_s}{1 + jz_l \tan \beta d_s}$$

$$\mathbf{y}_P = \frac{1}{\mathbf{z}_P} = \frac{1 + jz_l \tan \beta d_s}{z_l + j \tan \beta d_s}$$

$$= \frac{y_l + j \tan \beta d_s}{1 + jy_l \tan \beta d_s}$$

$$= g' + jb'$$

We multiply the numerator and denominator with the complex conjugate of the denominator to rationalize the above expression for normalized admittance $\mathbf{y}_P$.

$$\mathbf{y}_P = \frac{y_l + j\tan\beta d_s}{1 + jy_l\tan\beta d_s}\,\frac{1 - jy_l\tan\beta d_s}{1 - jy_l\tan\beta d_s}$$

$$= \frac{y_l(1+\tan^2\beta d_s) + j(1-y_l^2)\tan\beta d_s}{(1+y_l^2\tan^2\beta d_s)}$$

By equating the real and imaginary parts, we obtain

$$g' = \frac{y_l(1+\tan^2\beta d_s)}{1+y_l^2\tan^2\beta d_s}$$

$$b' = \frac{(1-y_l^2)\tan\beta d_s}{1+y_l^2\tan^2\beta d_s}$$

If the normalized admittance $\mathbf{y}_P$ is made one indicating a match, then there would not be any reflected wave over the line left of the point P. The parameter $\mathbf{y}_P$ is made one, i.e. $\mathbf{y}_P = g' = 1$ by proper selection of l and nullifying b' by shunting the line at the point P with a short-circuited stub whose input admittance is pure susceptance which is the same as b' magnitude-wise but sign-wise opposite.

The distance l of the point P from the load end which makes $g' = 1$ can be found from

$$g' = \frac{y_l(1+\tan^2\beta d_s)}{1+y_l^2\tan^2\beta d_s} = 1$$

$$y_l(1+\tan^2\beta d_s) = 1 + y_l^2\tan^2\beta d_s$$

$$\tan^2\beta d_s(y_l - y_l^2) = 1 - y_l$$

$$\tan^2\beta d_s = \frac{1-y_l}{y_l(1-y_l)} = \frac{1}{y_l} = \frac{Z_l}{Z_0}$$

$$\beta d_s = \tan^{-1}\sqrt{\frac{Z_l}{Z_0}} \quad \text{or}$$

$$d_s = \frac{\lambda}{2\pi}\tan^{-1}\sqrt{\frac{Z_l}{Z_0}}$$

This is the expression for the distance of the point over the line where the stub is to be connected. The length of the short-circuited stub which requires to nullify susceptance b' of the line can be found from the fact that, for nullification, its input admittance must be equal to $-b'$.

Assuming that the stub has characteristic impedance equal to that of the main line, the input admittance of a short-circuited stub of length l' is pure susceptance equal to

$$\mathbf{Z}_{in}\Big|_{Z_l=0} = \mathbf{Z}_{sc} = Z_0\left.\frac{Z_l + jZ_0\tan\beta l_s}{Z_0 + jZ_l\tan\beta l_s}\right|_{Z_l=0} = jZ_0\tan\beta l_s$$

$$\mathbf{Y}_{sc} = \frac{1}{\mathbf{Z}_{sc}} = -jY_0 \cot \beta l_s \text{ and}$$

$$\mathbf{y}_{sc} = -j \cot \beta l_s$$

For nullification, $\mathbf{y}_{sc} = -jb'$

$$\cot \beta l_s = \frac{(1 - y_l^2) \tan \beta d_s}{1 + y_l^2 \tan^2 \beta d_s}$$

Substituting $\tan^2 \beta d_s = Z_l/Z_0 = z_l$ in the above relation, we get

$$\cot \beta l_s = \frac{(1 - y_l^2)\sqrt{z_l}}{1 + y_l^2 z_l} = \frac{(1 - y_l^2)\sqrt{z_l}}{1 + y_l}$$

$$= \frac{(1 - y_l)(1 + y_l)\sqrt{z_l}}{(1 + y_l)} = (1 - y_l)\sqrt{z_l}$$

$$= \left(1 - \frac{1}{z_l}\right)\sqrt{z_l} = \frac{(z_l - 1)\sqrt{z_l}}{z_l}$$

$$= \frac{(z_l - 1)}{\sqrt{z_l}} = \frac{\left(\frac{Z_l}{Z_0} - 1\right)}{\sqrt{\frac{Z_l}{Z_0}}}$$

$$= \frac{Z_l - Z_0}{\sqrt{Z_l Z_0}}$$

It results in

$$\tan \beta l_s = \frac{\sqrt{Z_l Z_0}}{Z_l - Z_0}$$

Solving the above equation for l_s, we obtain

$$l_s = \frac{\lambda}{2\pi} \tan^{-1} \frac{\sqrt{Z_l Z_0}}{Z_l - Z_0}$$

If the stub used has a characteristic impedance Z_{0s} which is different from that of the line, then the calculations are to be proceeded in a slightly different manner. For matching, the input susceptance of the stub must be equal to the susceptance of the line at the point of stub attachment magnitude-wise but sign-wise they have to differ. The input susceptance of a short-circuited stub is $-Y_{0s} \cot \beta l_s$. If b' is the normalized susceptance, then the susceptance of the line at the stub connection point becomes $Y_0 b'$. Therefore, for matching,

$$Y_{0s} \cot \beta l_s = Y_0 b'$$

$$Y_{0s} \cot \beta l_s = Y_0 b' = Y_0 \frac{(1 - y_l^2) \tan \beta d_s}{1 + y_l^2 \tan^2 \beta d_s}$$

$$\cot \beta l_s = \frac{Y_0}{Y_{0s}} \frac{(1 - y_l^2) \tan \beta d_s}{(1 + y_l^2 \tan^2 \beta d_s)}$$

Substituting $\tan^2 \beta d_s = Z_l/Z_0 = z_l$ in the above relation and then solving for d_s, we get

$$d_s = \frac{\lambda}{2\pi} \tan^{-1} \frac{Z_0}{Z_{0s}} \frac{\sqrt{Z_l Z_0}}{Z_l - Z_0}$$

9. A transmission line of characteristic impedance 5 Ω is terminated over a load of impedance $(100 + j50)\,\Omega$. Using the Smith chart, determine

(a) VSWR ρ,

(b) reflection coefficient,

(c) distance of voltage minimum from the load,

(d) line impedance at a distance of 0.15λ from the load,

(e) line admittance at a distance of 0.15λ from the load end

(f) the location of the nearest point to the load where the real part of the line admittance is equal to the line characteristic admittance.

Solution: The normalized impedance is

$$z_l = \frac{100 + j50}{50} = 2 + j1$$

(a) Locate the normalized impedance over the Smith chart and with '1' centre and from '1' to the point representing the normalized impedance, draw the constant SWR circle. The circle cuts the real or horizontal axis at two points. The value against the point right of the centre is SWR. From the Smith chart operation (see Figure 13.8), it is found 2 to be 2.32.

(b) From the horizontal scale provided over the bottom of the chart, the reflection coefficient can be computed. It is 0.48.

(c) The distance between the load point to the point of intersection of real axis and SWR circle on the left side of the centre can be read using the scale provided over the periphery of the chart. The load point location is 0.212 and the voltage minimum point is 0.50. Therefore, the distance of the first voltage minimum from the load is 0.50 – 0.212 = 0.288 times wavelength.

(d) The required line impedance can be found by moving a distance of 0.15λ over the SWR circle in clockwise direction. The point represents normalized impedance of $z_d = 0.82 - j0.7$. The absolute impedance then is $Z_d = 50 \times (0.82 - j0.7) = 41 - j35$.

(e) Locate the point over the SWR circle diametrically opposite to the point located in the previous step. It represents the required admittance. Its

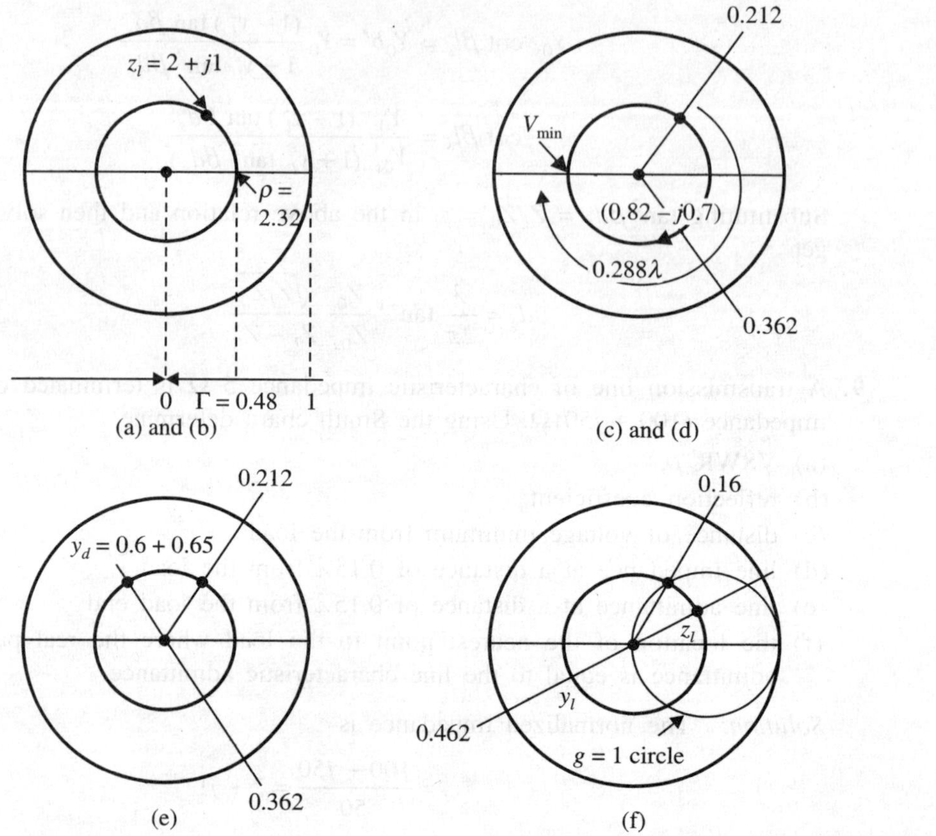

FIGURE 13.8 Illustrating Smith chart operations pertaining to problem.

normalized value from the chart is $y_d = 0.6 + j6.5$. The absolute admittance then is $Y_d = (0.6 + j6.5)/50 = 0.012 + j0.13$.

(f) The distance of the nearest point to the load where the real part of the line admittance is equal to the line characteristic admittance can be found by locating the load admittance over the SWR circle and then finding the distance in clockwise between this point and the point of intersection of the SWR circle with $g = 1$ circle. From the chart, the load admittance point is 0.462 and the intersection point is 0.16. The distance between them then can be found to be 0.198 times wavelength.

10. Using the Smith chart, determine VSWR ρ, the location of the first V_{max} and first V_{min} from the load over the line connected to a normalized impedance of $(2 + j2)\,\Omega$ and functioning and operating at the wavelength of 6 cm.

Solution: The solution procedure involves several steps which are mentioned here.

- Take a Smith chart and locate the given normalized impedance over it (see Figure 13.9). Let that point be *P*.

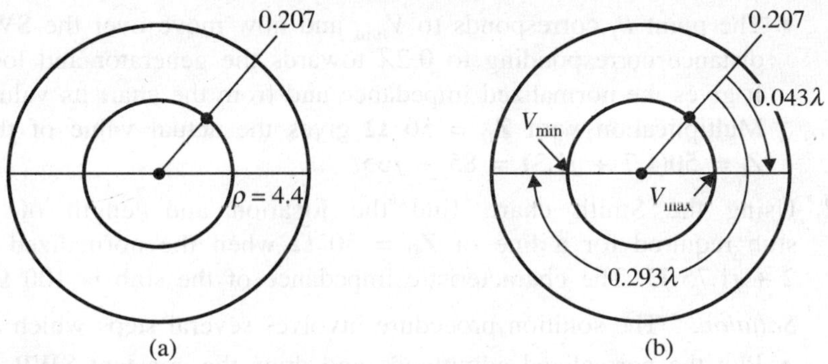

FIGURE 13.9 Illustrating Smith chart operations pertaining to Problem 10.

- With centre of the chart as centre and centre to point P as radius, draw the constant SWR circle.
- The constant SWR circle cuts the horizontal line of the chart at two points, one on the left of the centre P_l and the other on the right of the centre P_r. The point on the right of the centre gives the SWR over the line. From the chart it is 4.4.
- From P, read the angular distance in clockwise direction over the SWR circle to P_l and P_r and convert them into linear distances. The first one gives the distance of the first V_{min} and the second one gives the first V_{max} from the load. From the chart, they are 0.293λ and 0.043λ, respectively.

11. Using the Smith chart, find the terminated impedance of a line of $\mathbf{Z}_0 = 50\ \Omega$, SWR = 3 and when the load is shorted, the shift in minima is 0.2λ towards the generator.

Solution: The solution procedure involves several steps which are mentioned here.

- Take a Smith chart and with centre of the chart as centre and centre to 3 over the horizontal axis as radius, draw the constant SWR circle (see Figure 13.10).
- The constant SWR circle cuts the horizontal line of the chart at two points, one on the left of the centre P_l and the other on the right of the centre P_r.

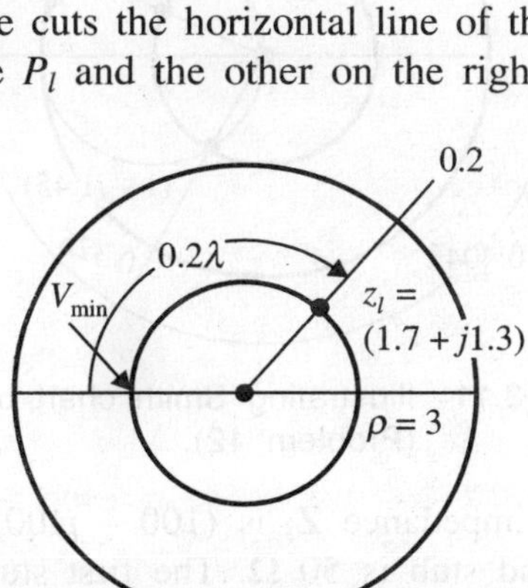

FIGURE 13.10 Illustrating Smith chart operations pertaining to Problem 11.

- The point P_l corresponds to V_{min} and now move over the SWR circle from P_l a distance corresponding to 0.2λ towards the generator and locate the load point. It gives the normalized impedance and from the chart its value is $z_l = 1.7 + j1.3$. Multiplication with $\mathbf{Z}_0 = 50\ \Omega$ gives the actual value of the impedance. It is $Z_l = 50(1.7 + j1.3) = 85 + j65$.

12. Using the Smith chart, find the location and length of the short-circuited stub required for a line of $\mathbf{Z}_0 = 50\ \Omega$ when the normalized load admittance is $2 + j1.75$ ℧. The characteristic impedance of the stub is 100 Ω.

Solution: The solution procedure involves several steps which are mentioned here.

- Plot the normalized admittance and draw the constant SWR circle on the Smith chart (see Figure 13.11).
- On the SWR circle nearest to the load admittance point, locate a point, in clockwise direction, which represents admittance $1 \pm jb$. This is the point of intersection of the constant SWR circle and circle $g = 1$. For the given data, from the Smith chart it is $(1 - j1.45)$.
- Read the distance between the points identified in the previous steps using the scale provided at the circumference of the chart. This gives the distance in wavelengths where the stub has to be placed from the load. From the chart it is $0.325 - 0.208 = 0.117$ times wavelength.
- Starting from the point $(\infty, j\infty)$, find the distance of the point at which the susceptance is $j1.45$. This gives the length of the short-circuited stub in wavelengths to be connected for matching. From the chart, it is 0.404λ.

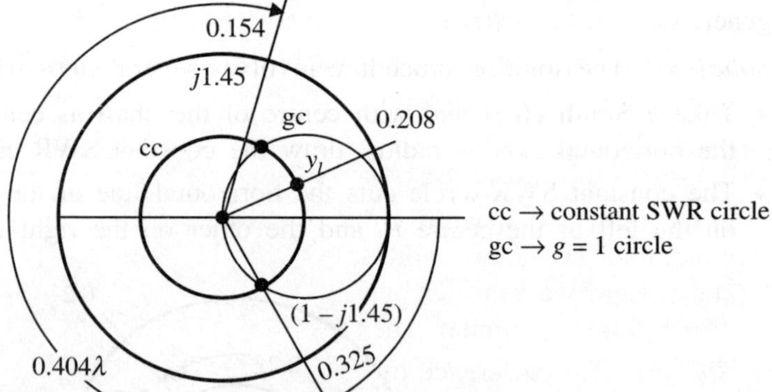

FIGURE 13.11 Illustrating Smith chart operations for single-stub matching (Problem 12).

13. The terminating impedance $\mathbf{Z}_l$ is $(100 - j100)\,\Omega$ and the characteristic impedance $\mathbf{Z}_0$ of the line and stub is 50 Ω. The first stub is placed at $0.05\,\lambda$ away from the load. The spacing between the two stubs is $(3/8)\lambda$. Determine the length of the short-circuited stubs when the match is achieved.

Solution:

Procedural steps:

Step I: Locate the normalized admittance over the chart. From the Smith chart, the normalized admittance is (0.24 + *j*0.25) (see Figure 13.12).

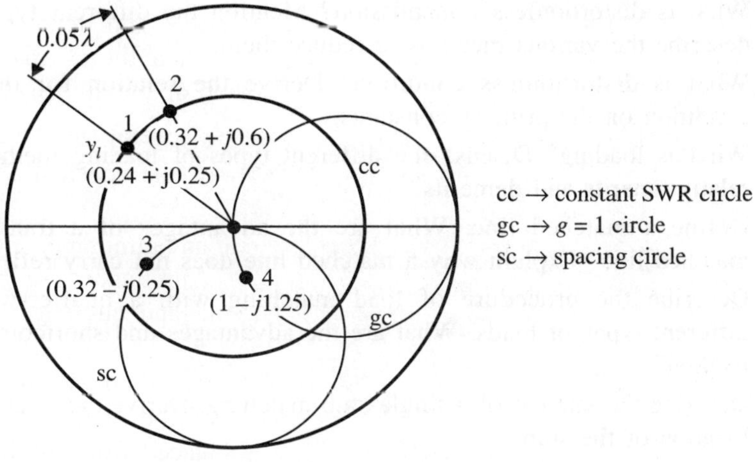

FIGURE 13.12 Illustrating Smith chart operations fo double-stub matching (Problem 13).

Step II: From the point located in the first step, move clockwise over the constant SWR circle, a distance corresponding to the distance of the first stub from the load. It is in the given problem. The admittance of the point reached, from the Smith chart, is (0.32 + *j*0.6).

Step III: Draw the spacing circle of $(3/8)\lambda$ by rotating the constant-conductance unity circle ($g = 1$) through a phase angle of $2\beta d = 2\beta(3/8)\lambda + (3/2)\pi$ towards the load.

Step IV: From the point located in the previous step, move over the constant g circle until the spacing circle is met. We come across two points meeting this condition. These two points in general give two different sets of lengths of the stubs. Here we consider one point, i.e. (0.32 – *j*0.25) and find the solution lengths. The solution is similar when the other point is selected.

Step V: The difference in the susceptance values of previous two points has to be nullified by the first stub. Here it is –*j*0.25 – *j*0.65 = –*j*0.85. The stub susceptance has to be *j*0.85. From the chart, the corresponding length of the first stub can be calculated. It is 0.14λ.

Step VI: From the point located in the previous step, i.e. (0.32 – *j*0.25), move over the constant SWR circle and in clockwise direction until its corresponding point over $g = 1$ circle is met. From the susceptance value of this point, which is *j*1.25, the length of the second stub can be calculated. From the Smith chart, 0.395λ. Note that the SWR circle in the present step is different from that of step II.

EXERCISE SECTION

Essay Type Questions

1. What is line distortion? Discuss the different types of distortions and their remedies.
2. What is distortionless transmission? Mention the different types of distortions and describe the various methods to reduce them.
3. What is distortionless condition? Derive the relation for the distortionless line condition on the primary constants.
4. What is loading? Discuss the different types of loading methods mentioning their relative merits and demerits.
5. Define a matched line. What are the advantages of a transmission line over a matched line? Explain why a matched line does not carry reflected wave.
6. Describe the procedure of load matching with a quarter-wave transformer for different types of loads. What are the advantages and shortcomings involved in this method?
7. Describe the method of a single-stub matching. Derive the relation for the length and location of the stub.
8. What is Smith chart? Describe its salient features.
9. Derive the relations connecting the complex reflection coefficient with normalized impedance for the curves of the Smith chart.
10. Describe the procedures of single-stub and double-stub matching techniques using the Smith chart.

Numerical Problems

1. A distortionless line of 60 Ω has $\alpha = 25$ mNp/m and a velocity of 70% of light velocity. Find the primary constants at 200 MHz.
2. For a lossless line of 150 Ω, find the location nearest to the load and the characteristic impedance of the QWT required to achieve matching for each of the following values of the reflection coefficient over the load Γ_l:
   (a) 1/5
   (b) $-j0.5$
   (c) $j/5$
3. A loss line working at 180 MHz of characteristic impedance 50 Ω with ε_r and $\mu_r = 1$ is to be matched to a load of 150 Ω by means of a short-circuited stub. Find the stub position closest to the load and the length so that the match is obtained.
4. A lossless line of characteristic impedance equal to 60 Ω is connected to a load consisting of a series combination of $R = 25$ Ω, $L = 1.5$ μH and $C = 150$ pF. Find SWR and the distance of the first minimum from the load for the following frequencies:

(a) $\omega = 1.2 \times 10^8$
(b) $\omega = 1.5 \times 10^8$ and
(c) $\omega = 0.6 \times 10^8$

5. A loss line working at 150 MHz of characteristic impedance 75 Ω with $\varepsilon_r = 2$ and $\mu_r = 1$ is to be matched to a load of 100 Ω by means of a short-circuited stub. Find the stub position closest to the load and the length so that the match is obtained.
6. A loss line of characteristic impedance $\mathbf{Z}_0 = 75\ \Omega$ is to be matched to the loads of
   (a) 25 Ω and
   (b) $(10 - j20)\ \Omega$
   by means of a short-circuited stub of characteristic impedance 60 Ω. Find the stub position closest to the load and its length so that the match is obtained.
7. For each of the sets of the values given, find whether the double-stub matching technique is possible or not.
   (a) $d_n = 0$, $d_{ss} = 5\lambda/8$ with $\mathbf{z}_l = 0.3 + j0.5$
   (b) $d_n = \lambda/16$, $d_{ss} = 3\lambda/8$ with $\mathbf{z}_l = 0.55$
   (c) $d_n = \lambda/4$, $d_{ss} = 7\lambda/8$ with $\mathbf{z}_l = 3 - j6.0$

Fill in the Blanks

1. The dependence of attenuation on frequency causes ________.
2. Different phase delays of different components cause ________.
3. To avoid delay distortion, the condition to be satisfied is ________.
4. Distortionless condition is ________.
5. Loading is the addition of ________ to achieve ________ condition.
6. Load matching refers to the termination of the line with ________.
7. The characteristic impedance of a quarter-wave transformer is ________.
8. In a single-stub matching, the length of the stub is ________.
9. In a single-stub matching, the location of the stub is ________.
10. Double-stub matching is not possible when the load falls in ________ region.
11. Over the Smith chart, the full circles represent ________.
12. Over the Smith chart, the arcs of the circles represent ________.
13. The equation representing the circles of the chart is ________.
14. The equation representing the arcs of the chart is ________.
15. Over the Smith chart, the upper half is ________ and the lower half is ________.
16. The Smith chart uses only ________ and describes the line for ________.
17. The centre of constant SWR circle is always the ________ of the chart.
18. V_{max}, I_{min} and ρ correspond to ________ half of the chart and I_{max}, V_{min} and $1/\rho$ correspond to ________ half of the chart.

Answers

1. frequency distorion
2. delay distortion
3. $\beta \propto \omega$
4. $RC = GL$
5. inductive coils, distortionless condition
6. characteristic impedance
7. $\mathbf{Z}_0' = \sqrt{\mathbf{Z}_0 \mathbf{Z}_l}$
8. $l = \dfrac{\lambda}{2\pi} \tan^{-1} \sqrt{\dfrac{\mathbf{Z}_l}{\mathbf{Z}_0}}$
9. $l' = \dfrac{\lambda}{2\pi} \tan^{-1} \dfrac{\sqrt{\mathbf{Z}_l \mathbf{Z}_0}}{\mathbf{Z}_l - \mathbf{Z}_0}$
10. forbidden region
11. normalized resistance/conductance
12. normalized reactance/susceptance
13. $\left(\Gamma_r - \dfrac{r}{1+r}\right)^2 + \Gamma_i^2 = \left(\dfrac{1}{1+r}\right)^2$
14. $(\Gamma_r - 1)^2 + \left(\Gamma_i - \dfrac{1}{x}\right)^2 = \left(\dfrac{1}{x}\right)^2$
15. inductive, capacitive
16. normalized impedance/admittance, half wavelength
17. centre
18. right, left

Match the Following

Set I

1. Load matching — (a) $RC = GL$
2. Distortionless condition — (b) $Z_l = Z_0$
3. Spacing circle — (c) double-stub matching
4. Quarter-wave transformer — (d) normalized conductance of unity
5. Single-stub matching — (e) voltage min or max

Set II

1. Smith chart — (a) reactance >1
2. Upper half — (b) reactance <1
3. Lower half — (c) capacitive reactance
4. Left half — (d) inductive reactance
5. Right half — (e) impedance

Set III

1. Admittance — (a) upper half
2. Capacitive susceptance — (b) Smith chart
3. Inductive susceptance — (c) lower half

4. Left half	(d) conductance >1
5. Right half	(e) conductance <1

Set IV

1. Circles	(a) towards load
2. Arcs	(b) towards generator
3. Line impedance	(c) constant SWR circle
4. Clockwise	(d) reactance/susceptance
5. Anticlockwise	(e) resistance/conductance

Set V

1. Circles	(a) $(\Gamma_r - 1)^2 + \left(\Gamma_i - \frac{1}{x}\right)^2 = \left(\frac{1}{x}\right)^2$
2. Arcs	(b) $\left(\Gamma_r - \frac{r}{1+r}\right)^2 + \Gamma_i^2 = \left(\frac{1}{1+r}\right)^2$
3. Centres	(c) $\Gamma_r = \frac{r}{(1+r)\Gamma_i} = \frac{1}{x}$ and $\Gamma_r = 1\Gamma_i = \frac{1}{x}$
4. Radii	(d) line of length $\frac{\lambda}{2}$
5. Entire chart	(e) $\frac{1}{(1+r)}, \frac{1}{x}$

Answers

Set I:	1. (b)	2. (a)	3. (c)	4. (e)	5. (d)
Set II:	1. (e)	2. (d)	3. (c)	4. (b)	5. (a)
Set III:	1. (b)	2. (a)	3. (c)	4. (e)	5. (d)
Set IV:	1. (e)	2. (d)	3. (c)	4. (b)	5. (a)
Set V:	1. (b)	2. (a)	3. (c)	4. (e)	5. (d)

Multiple Choice Questions

1. Distortionless line condition is

(a) $\frac{R}{L} = \frac{G}{C}$ (b) $\frac{R}{L} > \frac{G}{C}$

(c) $\frac{R}{L} < \frac{G}{C}$ (d) none of these

2. Loading of the line refers to the connection of
   (a) inductive coils (b) capacitive boxes
   (c) both (a) and (b) (d) none of these
3. The Smith chart
   (a) is a polar plot (b) represents complex reflection coefficient
   (c) is inscribed in a unity circle (d) all of these
4. The complete circles in the Smith chart represent
   (a) normalized resistance (b) normalized conductance
   (c) both (a) and (b) (d) none of these
5. The arcs in the Smith chart represent
   (a) normalized reactance (b) normalized susceptance
   (c) both (a) and (b) (d) none of these
6. The circles and arcs over the Smith chart are
   (a) orthogonal (b) opposite to each other
   (c) at 45° (d) none of these
7. The upper half of the Smith chart represents
   (a) inductive reactance (b) inductive susceptance
   (c) both (a) and (b) (d) none of these
8. The lower half of the Smith chart represents
   (a) capacitive reactance (b) capacitive susceptance
   (c) both (a) and (b) (d) none of these
9. The radius of the constant SWR circle is
   (a) voltage SWR (b) current SWR
   (c) both (a) and (b) (d) none of these
10. The centre of the constant SWR circle is
   (a) '1' of the horizontal line (b) centre of the chart
   (c) both (a) and (b) (d) none of these
11. The Smith chart represents the line for a length of
   (a) $\frac{\lambda_g}{2}$ (b) $\frac{\lambda}{2}$
   (c) λ (d) none of these
12. In the left half of the Smith chart, the resistance and reactance values are
   (a) less than 1 (b) more than 1
   (c) both (a) and (b) (d) none of these
13. In the right half of the Smith chart, the resistance and reactance values are
   (a) less than 1 (b) more than 1
   (c) both (a) and (b) (d) none of these

14. The leftmost point of the Smith chart represents
    (a) (0, 0) (b) (∞, ∞)
    (c) either (a) or (b) (d) none of these
15. The rightmost point of the Smith chart represents
    (a) (0, 0) (b) (∞, ∞)
    (c) either (a) or (b) (d) none of these
16. The topmost point of the Smith chart represents
    (a) (1, 1) (b) (–1, –1)
    (c) either (a) or (b) (d) none of these
17. The bottommost point of the Smith chart represents
    (a) (1, 1) (b) (–1, –1)
    (c) either (a) or (b) (d) none of these
18. The Smith chart is always used with
    (a) normalized impedances (b) normalized admittances
    (c) both (a) and (b) (d) none of these
19. The Smith chart is useful to analyse
    (a) lossless lines (b) lossy lines
    (c) both (a) and (b) (d) none of these
20. The horizontal line left of the centre represents
    (a) V_{max}, I_{min} (b) V_{min}, I_{max}
    (c) both (a) and (b) (d) none of these
21. The horizontal line right of the centre represents
    (a) V_{max}, I_{min} (b) V_{min}, I_{max}
    (c) both (a) and (b) (d) none of these
22. Towards load over the line corresponds to
    (a) clockwise rotation over the Smith chart
    (b) anticlockwise rotation over the Smith chart
    (c) both (a) and (b)
    (d) none of these
23. Towards source over the line corresponds to
    (a) clockwise rotation over the Smith chart
    (b) anticlockwise rotation over the Smith chart
    (c) both (a) and (b)
    (d) none of these
24. The point over constant SWR circle diametrically opposite to the load impedance point represents
    (a) load admittance (b) load impedance
    (c) either (a) or (b) (d) none of these

25. The point over constant SWR circle diametrically opposite to the load admittance point is
    (a) load admittance (b) load impedance
    (c) either (a) or (b) (d) none of these
26. Travel of length $\lambda_g/2$ over the line corresponds to rotation over the chart by
    (a) 180° (b) 360°
    (c) 90° (d) none of these
27. If the load is pure reactance, then the load point over the Smith chart stays
    (a) at the periphery (b) over the horizontal line
    (c) in the lower half (d) in the upper half
28. If the load is pure resistance, then the load point over the Smith chart stays
    (a) at the periphery (b) over the horizontal line
    (c) in the lower half (d) in the upper half
29. The centre of the arcs of the Smith chart is
    (a) $\left(\frac{r}{1+r}, 0\right)$ (b) $\left(1, \frac{1}{x}\right)$
    (c) $\left(\frac{r}{1+r}, 1\right)$ (d) $\left(1, \frac{r}{1+r}\right)$
30. The centre of the circles of the Smith chart is
    (a) $\left(\frac{r}{1+r}, 0\right)$ (b) $\left(1, \frac{1}{x}\right)$
    (c) $\left(\frac{r}{1+r}, 1\right)$ (d) $\left(1, \frac{r}{1+r}\right)$
31. If the load impedance is pure resistance equal to the characteristic impedance of the line, then the load point over the Smith chart will be
    (a) at the centre (b) at the periphery
    (c) in the upper half (d) in the lower half

Answers

1. (a)	2. (a)	3. (d)	4. (c)	5. (c)
6. (a)	7. (c)	8. (c)	9. (c)	10. (c)
11. (a)	12. (a)	13. (b)	14. (a)	15. (b)
16. (a)	17. (b)	18. (c)	19. (c)	20. (b)
21. (d)	22. (b)	23. (a)	24. (a)	25. (a)
26. (b)	27. (a)	28. (b)	29. (b)	30. (a)
31. (a)				

Appendix

Table A.1 Names and units of the electric and magnetic quantities

Symbol	Name	Unit	Abbreviation
$\mathbf{v}$	Velocity	metre/second	m/s
$\mathbf{F}$	Force	newton	N
Q	Charge	coulomb	C
$\mathbf{r}$, $\mathbf{R}$	Distance	metre	m
ε_0, ε	Permittivity	farad/metre	F/m
$\mathbf{E}$	Stearic field intensity	volt/metre	V/m
ρ	Volume charge density	coulomb/metre^3	C/m^3
v	Volumc	metre^3	m^3
λ	Linear charge density	coulomb/metre	C/m
σ	Surface charge density	coulomb/metre^2	C/m^2
ψ	Electric flux	coulomb	C
$\mathbf{D}$	Electric flux density	coulomb/metre^2	C/m^2
$\mathbf{a}$	Area	metre^2	m^2
W	Work, energy	joule	J
L	Length	metre	m
V	Potential	volt	V
$\mathbf{P}$	Dipole moment	coulomb-metre	C·m
$\mathbf{I}$	Current	ampere	A
$\mathbf{J}$	Current density	ampere/metre^2	A/m^2
μ_e, μ_h	Mobility	metre^2/volt-second	m^2/V·s
e	Electronic charge	coulomb	C
σ	Conductivity	Siemens/metre	S/m
R	Resistance	ohm	Ω
$\mathbf{P}$	Polarization	coulomb/metre^2	C/m^2
C	Capacitance	farad	F

(Contd...)

Table A.1 Names and units of the electric and magnetic quantities (*Contd...*)

Symbol	Name	Unit	Abbreviation
R	Sheet resistance	ohm per square	Ω/□
H	Magnetic field intensity	ampere/metre	A/m
K	Surface current density	ampere/metre	A/m
B	Magnetic flux density	tesla (or weber/metre^2)	T(or Wb/m^2)
μ_0, μ	Permeability	henry/metre	H/m
Φ	Magnetic flux	weber	Wb
V_m	Magnetic scalar potential	ampere	A
A	Vector magnetic potential	weber/metre	Wb/m
T	Torque	newton-metre	N·m
m	Magnetic moment	ampere-metre^2	A·m^2
M	Magnetization	ampere-metre	A/m
R	Reluctance	ampere-turn/weber	A·t/Wb
L	Inductance	henry	H
M	Mutual inductance	henry	H
ω	Radian frequency	radian/second	rad/s
c	Velocity of light	metre/second	m/s
λ	Wavelength	metre	m
η	Intrinsic impedance	ohm	Ω
k	Wave number	metre^{-1}	m^{-1}
α	Attenuation constant	neper/metre	Np/m
β	Phase constant	radian metre	rad/m
f	Frequency	hertz	Hz
P	Poynting vector	watt/metre^2	W/m^2
P	Power	watt	Ω
δ	Skin depth	metre	m
Γ	Reflection coefficient		
ρ	Standing-wave ratio		
γ	Propagation constant	meter^{-1}	m$^-$1
G	Conductance	siemen	S
Z	Impedance	ohm	Ω
Y	Admittance	siemen	S
Q	Quality factor		

Table A.2 Dielectric constants ε_γ and $\varepsilon''/\varepsilon'$ of different materials

Material	ε_γ	$\varepsilon''/\varepsilon'$
Air	1.0005	
Alcohol, ethyl	25	0.1
Aluminium oxide	8.8	0.0006
Amber	2.7	0.002
Bakelite	4.74	0.022
Barium titanate	1200	0.013
Carbon dioxide	1.001	
Ferrite (NiZn)	12.4	0.00025
Germanium	16	
Glass	4–7	0.002
Ice	4.2	0.05
Mini	5.4	0.0006
Neoprene	6.6	0.011
Nylon	3.5	0.02
Paper	3	0.008
Plexiglas	3.45	0.03
Polyethylene	2.26	0.0002
Polypropylene	2.25	0.0003
Polystyrene	2.56	0.00005
Porcelain (dry process)	6	0.014
Pyranol	4.4	0.0005
Pyrex glass	4	0.0006
Quartz (fused)	3.8	0.00075
Rubber	2.5–3	0.002
Silica or SiO_2 (fused)	3.8	0.00075
Silicon	11.8	
Snow	3.3	0.5
Sodium chloride	5.9	0.0001
Soil (dry)	2.8	0.05
Steatite	5.8	0.003
Styrofoam	1.03	0.0001
Teflon	2.1	0.0003
Titanium dioxide	100	0.0015
Water (distilled)	80	0.04
Water (sea)		4
Water (dehydrated)	1	0
Wood (dry)	1.5–4	0.01

Table A.3 Conductivities (σ) of different materials

Material	σ, S/m	Material	σ, S/m
Silver	6.17×10^7	Nichrome	0.1×10^7
Copper	5.80×10^7	Graphite	7×10^4
Gold	4.10×10^7	Silicon	2300
Aluminium	3.82×10^7	Ferrite (typical)	100
Tungsten	1.82×10^7	Water (sea)	5
Zinc	1.67×10^7	Limestone	10^{-2}
Brass	1.5×10^7	Clay	5×10^{-3}
Nickel	1.45×10^7	Water (fresh)	10^{-3}
Iron	1.03×10^7	Water (distilled)	10^{-4}
Phosphor bronze	1×10^7	Soil (sandy)	10^{-5}
Solder	0.7×10^7	Granite	10^{-6}
Carbon steel	0.6×10^7	Marble	10^{-8}
German silver	0.3×10^7	Bakelite	10^{-9}
Manganin	0.227×10^7	Porcelain (dry process)	10^{-10}
Constantan	0.226×10^7	Diamond	2×10^{-11}
Germanium	0.22×10^7	Polystyrene	10^{-16}
Stainless steel	0.11×10^7	Quartz	10^{-17}

Table A.4 Relative permeabilities (μ_r) of different materials

Material	μ_r
Bismuth	0.9999986
Paraffin	0.99999942
Wood	0.9999995
Silver	0.99999981
Aluminium	1.00000065
Beryllium	1.00000079
Nickel chloride	1.00004
Manganese sulphate	1.0001
Nickel	50
Cast iron	60
Cobalt	60
Powdered iron	100
Machine steel	300
Ferrite (typical)	1000
Permalloy 45	2500
Transformer iron	3000
Silicon iron	3500
Iron (pure)	4000
Mumetal	20,000
Sendust	30,000
Supermalloy	1,00,000

Table A.5 Physical constants of different materials

Quantity	Value
Electron charge	$e = (1.60217733 \pm 0.00000046) \times 10^{-19}$ C
Electron mass	$m = (9.1093897 \pm 0.0000054) \times 10^{-31}$ kg
Permittivity of free space	$\varepsilon_0 = 8.854\,187817 \times 10^{-12}$ F/m
Permeability of free space	$\mu_0 = 4\pi \times 10^{-7}$ H/m
Velocity of light	$c = 2.99792458 \times 10^8$ m/s

Series Expansion

$$e^x = 1 + \frac{x}{1!} + \frac{x^2}{2!} + \frac{x^3}{3!} + \cdots$$

$$\cos\theta = 1 - \frac{\theta^2}{2!} + \frac{\theta^4}{4!} + \cdots$$

$$\sin\theta = \theta - \frac{\theta^3}{3!} + \frac{\theta^5}{5!} + \cdots$$

$$\log(1 + x) = x - \frac{x^2}{2} + \frac{x^3}{3} + \cdots \quad \text{if } |x| < 1$$

Differentiation

$$x^n \to nx^{n-1}$$

$$e^x \to e^x$$

$$a^x \to a^x \log a$$

$$\sin x \to \cos x$$

$$\cos x \to -\sin x$$

$$\tan x \to \sec^2 x$$

$$\cot x \to -\text{cosec}^2 x$$

$$\sec x \to \sec x \tan x$$

$$\cot^{-1} x \to \frac{-1}{(1 + x^2)}$$

$$\text{cosec}\, x \to -\text{cosec}\, x \cot x$$

$$\log x \to \frac{1}{x}$$

$$\sin^{-1} x \to \frac{1}{\sqrt{1 - x^2}}$$

$$\cos^{-1} x \to \frac{-1}{\sqrt{1 - x^2}}$$

$$\tan^{-1} x \to \frac{1}{(1 + x^2)}$$

Integration

$$\int e^x\,dx = e^x + c$$

$$\int a^x\,dx = \left(\frac{a^x}{\log a}\right) + c$$

$$\int x^n\,dx = \left(\frac{x^{n+1}}{n+1}\right) + c \qquad n \neq -1$$

$$\int \left(\frac{1}{x}\right) dx = \log x + c$$

$$\int \sin x\,dx = -\cos x + c$$

$$\int \cos x\,dx = \sin x + c$$

$$\int \tan x\,dx = -\log \cos x + c$$

$$\int \cot x\,dx = \log \sin x + c$$

$$\int \sec x\,dx = \log(\sec x + \tan x) + c$$

$$\int \operatorname{cosec} x\,dx = \log(\operatorname{cosec} x - \cot x) + c$$

$$\int \sec^2 x\,dx = \tan x + c$$

$$\int \operatorname{cosec}^2 x\,dx = -\cot x + c$$

$$\int \sec x \tan x\,dx = \sec x + c$$

$$\int \operatorname{cosec} x \cot x\,dx = -\operatorname{cosec} x + c$$

Indefinite Integrals

$$\int x \sin ax\,dx = \frac{1}{a^2}(\sin ax - ax \cos ax)$$

$$\int x \cos ax\,dx = \frac{1}{a^2}(\cos ax + ax \sin ax)$$

$$\left.\begin{aligned} \int \sin ax \sin bx\,dx &= \frac{\sin(a-b)x}{2(a-b)} - \frac{\sin(a+b)x}{2(a+b)} \\ \int \cos ax \cos bx\,dx &= \frac{\sin(a-b)x}{2(a-b)} + \frac{\sin(a+b)x}{2(a+b)} \\ \int \sin ax \cos bx\,dx &= -\left[\frac{\cos(a-b)x}{2(a-b)} + \frac{\cos(a+b)x}{2(a+b)}\right] \end{aligned}\right\} a^2 \neq b^2$$

$$\int xe^{ax}\,dx = \frac{e^{ax}}{a^2}(ax-1)$$

$$\int e^{ax}\sin bx\,dx = \frac{e^{ax}}{a^2+b^2}(a\sin bx - b\cos bx)$$

$$\int e^{ax}\cos bx\,dx = \frac{e^{ax}}{a^2+b^2}(a\cos bx + b\sin bx)$$

$$\int \frac{dx}{x^2+a^2} = \frac{1}{a}\tan^{-1}\frac{x}{a} + c$$

$$\int \frac{dx}{x^2-a^2} = \frac{1}{2a}\log\left|\frac{x-a}{x+a}\right| + c$$

$$\int \frac{x}{x^2+a^2}\,dx = \frac{1}{2}\ln(x^2+a^2)$$

$$\int \frac{1}{\sqrt{(x^2 \pm a^2)}}\,dx = \ln\left[x + \sqrt{(x^2 \pm a^2)}\right]$$

$$\int \frac{x}{\sqrt{(x^2+a^2)}}\,dx = \sqrt{(x^2+a^2)}$$

$$\int \frac{1}{(x^2+a^2)^{3/2}}\,dx = \frac{\frac{x}{a^2}}{\sqrt{(x^2+a^2)}}$$

$$\int \frac{x}{(x^2+a^2)^{3/2}}\,dx = -\frac{1}{\sqrt{(x^2+a^2)}}$$

$$\int \frac{x^2}{(x^2+a^2)^{3/2}}\,dx = \ln\left(\frac{\sqrt{(x^2+a^2)}}{a} + \frac{x}{a}\right) - \frac{x}{\sqrt{(x^2+a^2)}}$$

$$\int \frac{dx}{\sqrt{1-x^2}} = \sin^{-1} x + c = -\cos^{-1} x + c$$

$$\int \frac{dx}{1+x^2} = \tan^{-1} x + c = -\cot^{-1} x + c$$

$$\int \frac{dx}{\sqrt{x^2 \pm a^2}} = \log\left(x + \sqrt{x^2 \pm a^2}\right) + c$$

$$\int \frac{dx}{\sqrt{a^2 - x^2}} = \sin^{-1}\frac{x}{a} + c$$

Hyperbolic Functions

$$\sinh x = \frac{e^x - e^{-x}}{2}$$

$$\cosh x = \frac{e^x + e^{-x}}{2}$$

$$\tanh x = \frac{e^x - e^{-x}}{e^x + e^{-x}}$$

$$\sinh(x \pm y) = \sinh x \cosh y \pm \cosh x \sinh y$$

$$\cosh(x \pm y) = \cosh x \cosh y \pm \sinh x \sinh y$$

$$\tanh(x \pm y) = \frac{\tanh x \pm \tanh y}{1 \pm \tanh x \tanh y}$$

$$\sinh jx = j \sin x$$

$$\cosh jx = \cos x$$

$$\tanh jx = j \tan x$$

Logarithmic Units

- Two voltages, or currents, differ by one neper when one of them is 'e' times as large as the other.

$$N \text{ nepers} = \ln\left|\frac{V_1}{V_2}\right| = \ln\left|\frac{I_1}{I_2}\right|$$

The ratio of input to output power can be expressed as $P_1/P_2 = e^{2N}$.

- The bel is defined as the logarithm of a power ratio.

$$\text{Number of bels} = \log\frac{P_1}{P_2}$$

The decibel (dB) is a dimensionless number that expresses the ratio of two power levels.

As it is more convenient to handle with a smaller unit, decibel abbreviated 'dB' which is one tenth of a bel has been defined.

$$\text{dB} = 10 \log\frac{P_1}{P_2}$$

In terms of voltages

$$\text{dB} = 20 \log \left| \frac{V_1}{V_2} \right|$$

The voltage definition of dB has no meaning at all unless the two voltages under consideration appear across equal impedances.

- Nepers and dBs are related through

$$e^{2N} = 10^{\text{dB}/10}$$

resulting in

$$1 \text{ neper} = 8.686 \text{ dB}$$

Important Equations

Coulomb's law:

$$\mathbf{F} = \frac{1}{4\pi\varepsilon}\frac{Qq}{R^2}\hat{\mathbf{R}}$$

Gauss law:

$$\oint_s \mathbf{D}\cdot d\mathbf{a} = Q_{\text{enc}}$$

Electric field intensity–volume charge density

$$\mathbf{E}(\mathbf{r}) = \frac{1}{4\pi\varepsilon}\int \frac{\rho(\mathbf{r}')d\tau'}{R^2}\hat{\mathbf{R}}$$

$$\nabla\cdot\mathbf{E} = \frac{\rho}{\varepsilon};\quad \nabla\times\mathbf{E} = \mathbf{0}$$

Electric field intensity–scalar potential:

$$\mathbf{E} = -\nabla V;\quad V = -\int_{\infty}^{P}\mathbf{E}\cdot d\mathbf{l}$$

Scalar potential–volume charge density

$$V(\mathbf{r}) = \frac{1}{4\pi\varepsilon}\int \frac{\rho(r')d\tau'}{R}$$

$$\nabla^2 V = -\frac{\rho}{\varepsilon}$$

Potential energy

Volume charge distribution $\quad W = \frac{1}{2}\int \rho V d\tau$

General field expression $\quad W = \frac{\varepsilon_0}{2}\int_{\text{all space}} E^2 d\tau$

Dielectric systems $$W = \frac{1}{2} \int_{\text{all space}} \mathbf{D} \cdot \mathbf{E} d\tau$$

Polarization:

Surface charge density $$\sigma_b = \mathbf{P} \cdot \hat{\mathbf{n}}$$

Volume charge density $$\rho_b = -\nabla \cdot \mathbf{P}$$

Relative permittivity $$\varepsilon_r = 1 + \chi_e$$

Lorentz's force law:

$$\mathbf{F} = Q\mathbf{E} + Q(\mathbf{v} \times \mathbf{B})$$

Ampere's circuital law:

$$\oint \mathbf{H} \cdot d\mathbf{l} = I_{\text{enc}}$$

Magnetic field intensity–volume current density:

$$\mathbf{H}(\mathbf{r}) = \frac{1}{4\pi} \int \frac{\mathbf{J}(\mathbf{r}') \times \hat{\mathbf{R}}}{R^2} d\tau'$$

$$\nabla \cdot \mathbf{H} = 0; \ \nabla \times \mathbf{H} = \mathbf{J}_{\text{enc}}$$

Magnetic field intensity–vector potential:

$$\mathbf{A}(\mathbf{r}) = \frac{\mu_0}{4\pi} \int \frac{\mathbf{H}(\mathbf{r}') \times \hat{\mathbf{R}}}{R^2} d\tau'$$

$$\nabla \cdot \mathbf{A} = 0; \ \nabla \times \mathbf{A} = \mathbf{B}$$

Vector potential–volume current density:

$$\mathbf{A}(\mathbf{r}) = \frac{\mu_0}{4\pi} \int \frac{\mathbf{J}(\mathbf{r}') d\tau'}{R}$$

$$\nabla^2 \mathbf{A} = -\mu_0 \mathbf{J}$$

Magnetic energy:

$$W = \frac{\mu_0}{2} \int_{\text{all space}} H^2 d\tau$$

Magnetization:

Surface current density $$\mathbf{K}_b = \mathbf{M} \times \hat{\mathbf{n}}$$

Volume current density $$\mathbf{J}_b = \nabla \times \mathbf{M}$$

Relative permeability $$\mu_r = 1 + \chi_m$$

Inductance:

Self-inductance $$L = \frac{N\Phi}{I}$$

Mutual inductance $$M_{12} = \frac{N_2 \Phi_{12}}{I_1}$$

Maxwell's equations

$$\nabla \times \mathbf{H} = \mathbf{J} + \frac{\partial \mathbf{D}}{\partial t} \qquad \oint \mathbf{H} \cdot d\mathbf{l} = \int \left(\mathbf{J} + \frac{\partial \mathbf{D}}{\partial t} \right) \cdot d\mathbf{a}$$

$$\nabla \times \mathbf{E} = -\frac{\partial \mathbf{B}}{\partial t} \qquad \oint \mathbf{E} \cdot d\mathbf{l} = -\int \frac{\partial \mathbf{B}}{\partial t} \cdot d\mathbf{a}$$

$$\nabla \cdot \mathbf{D} = \rho \qquad \oint \mathbf{D} \cdot d\mathbf{a} = Q_{\text{enc}}$$

$$\nabla \cdot \mathbf{B} = 0 \qquad \oint \mathbf{B} \cdot d\mathbf{a} = 0$$

Continuity equation:

$$\nabla \cdot \mathbf{J} = -\frac{\partial \rho}{\partial t} \qquad \oint_s \mathbf{J} \cdot d\mathbf{a} = -\frac{d}{dt} \int_v \rho \, d\tau$$

Retarded potentials–sources

$$\mathbf{A}(\mathbf{r}, t) = \frac{\mu}{4\pi} \int \frac{\mathbf{J}\left(\mathbf{r}', t - \frac{R}{c} \right)}{R} d\tau'$$

$$V(\mathbf{r}, t) = \frac{1}{4\pi\varepsilon} \int \frac{\rho\left(\mathbf{r}', t - \frac{R}{c} \right)}{R} d\tau'$$

Retarded potentials–fields

$$\mathbf{E} = -\nabla V - \frac{\partial \mathbf{A}}{\partial t}$$

$$\nabla \times \mathbf{A} = \mathbf{B}$$

Poynting vector

$$\mathbf{P} = \mathbf{E} \times \mathbf{H} \qquad \mathbf{P} = \frac{1}{2}(\mathbf{E} \times \mathbf{H}^*)$$

Boundary conditions

$$\hat{\mathbf{n}} \times (\mathbf{E}_1 - \mathbf{E}_2) = 0$$

$$\hat{\mathbf{n}} \times (\mathbf{H}_1 - \mathbf{H}_2) = 0$$

$$\hat{\mathbf{n}} \cdot (\varepsilon_1 \mathbf{E}_1 - \varepsilon_2 \mathbf{E}_2) = 0$$

$$\hat{\mathbf{n}} \cdot (\mu_1 \mathbf{H}_1 - \mu_1 \mathbf{H}_2) = 0$$

Wave equations

$$\nabla^2 \mathbf{E} = \gamma^2 \mathbf{E}$$

$$\nabla^2 \mathbf{H} = \gamma^2 \mathbf{H}$$

Propagation constant

$$\boldsymbol{\gamma} = \sqrt{j\omega\mu(\sigma + j\omega\varepsilon)}$$

Transmission line equations

$$-\frac{\partial v}{dx} = Ri(x,t) + L\frac{\partial i}{dt} \qquad -\frac{\partial i}{\partial x} = Gv(x,t) + C\frac{\partial v}{\partial t}$$

Wave equations

$$\frac{d^2\mathbf{V}}{dx^2} = \gamma^2\mathbf{V} \qquad \frac{d^2\mathbf{I}}{dx^2} = \gamma^2\mathbf{I}$$

Secondary constants

$$\boldsymbol{\gamma} = \sqrt{(R + j\omega L)(G + j\omega C)}$$

$$\mathbf{Z}_0 = \sqrt{\frac{R + j\omega L}{G + j\omega C}}$$

Input impedance

$$\mathbf{Z}_{\text{in}} = \mathbf{Z}_0 \frac{\mathbf{Z}_l + \mathbf{Z}_0 \tanh \boldsymbol{\gamma} l}{\mathbf{Z}_0 + \mathbf{Z}_l \tanh \boldsymbol{\gamma} l}$$

Reflection coefficient

$$\boldsymbol{\Gamma} \triangleq \frac{\mathbf{V}_{\text{ref}}}{\mathbf{V}_{\text{inc}}}$$

$$\boldsymbol{\Gamma}_l = \frac{\mathbf{Z}_l - \mathbf{Z}_0}{\mathbf{Z}_l + \mathbf{Z}_0}$$

$$\boldsymbol{\Gamma}_s = \frac{\mathbf{Z}_s - \mathbf{Z}_0}{\mathbf{Z}_s + \mathbf{Z}_0}$$

$$\boldsymbol{\Gamma} = \boldsymbol{\Gamma}_l e^{-2\gamma d} = \boldsymbol{\Gamma}_s e^{2\gamma x}$$

Standing wave ratio

$$\rho \triangleq \frac{|\mathbf{V}_{\max}|}{|\mathbf{V}_{\min}|}$$

$$\rho = \frac{1+\Gamma}{1-\Gamma}; \qquad \Gamma = \frac{\rho - 1}{\rho + 1}$$

Distortion-less transmission

$$\frac{R}{L} = \frac{G}{C}$$

Matched termination

$$\mathbf{Z}_l = \mathbf{Z}_0$$

Smith chart

$$\left(\Gamma_r - \frac{r}{1+r}\right)^2 + \Gamma_i^2 = \left(\frac{1}{1+r}\right)^2$$

$$(\Gamma_r - 1)^2 + \left(\Gamma_i - \frac{1}{x}\right)^2 = \left(\frac{1}{x}\right)^2$$

Index